Springer Proceedings in Mathematics & Statistics

Volume 90

For further volumes:
http://www.springer.com/series/10533

Springer Proceedings in Mathematics & Statistics

This book series features volumes composed of select contributions from workshops and conferences in all areas of current research in mathematics and statistics, including OR and optimization. In addition to an overall evaluation of the interest, scientific quality, and timeliness of each proposal at the hands of the publisher, individual contributions are all refereed to the high quality standards of leading journals in the field. Thus, this series provides the research community with well-edited, authoritative reports on developments in the most exciting areas of mathematical and statistical research today.

Bourama Toni
Editor

New Frontiers of Multidisciplinary Research in STEAM-H (Science, Technology, Engineering, Agriculture, Mathematics, and Health)

Springer

Editor
Bourama Toni
Department of Mathematics
and Computer Sciences
Virginia State University
Petersburg, VA, USA

ISSN 2194-1009 ISSN 2194-1017 (electronic)
ISBN 978-3-319-07754-3 ISBN 978-3-319-07755-0 (eBook)
DOI 10.1007/978-3-319-07755-0
Springer Cham Heidelberg New York Dordrecht London

Library of Congress Control Number: 2014947518

Mathematics Subject Classification (2010): 00A06, 01-06, 00A71, 00B10, 62-02, 62-07

Printed on acid-free paper

Springer is part of Springer Science+Business Media (www.springer.com)

To
My wife Emi Takada,
and our wonderful children:
Emma Akira and Andy Shingo

To
Assita Toni and Safia Anita Toni

To
My mother Madani Dembélé,
and my late father Adama Toni:
pour une trop longue absence . . .

Preface

This multidisciplinary book brings together leading researchers in the STEAM-H disciplines (Science, Technology, Engineering, Agriculture, Mathematics and Health) to present their own work in the perspective to advance their specific fields, and in a way to generate a genuine interdisciplinary interaction transcending disciplinary boundaries. All chapters therein were carefully edited, peer-reviewed; they are reasonably self-contained and pedagogically exposed for a multidisciplinary readership.

Contributions are invited only, and reflect the most recent advances delivered in a high standard, self-contained. The goals are:

1. To foster student interest in science, technology, engineering, agriculture, mathematics, and health.
2. To enhance multidisciplinary understanding between the disciplines, including through a participative seminar series by showing how some new advances in a particular discipline can be of interest to the other discipline, or how different disciplines contribute to a better understanding of a relevant issue at the interface of mathematics and the sciences.
3. To promote the spirit of inquiry so characteristic of mathematics for the advances of the natural, physical, and behavioral sciences by featuring leading experts and outstanding presenters.
4. To encourage diversity in the attendees and readers' background and expertise, while at the same time structurally fostering genuine interdisciplinary interactions and networking.

Current disciplinary boundaries do not encourage effective interactions between scientists; researchers from different fields usually occupy different buildings on university campuses, publish in journals specific to their field, and attend different scientific meetings. Existing scientific meetings usually fall into either small gatherings specializing on specific questions, targeting specific and small group of scientists already aware of each other's work and potentially collaborating, or large meetings covering a wide field and targeting a diverse group of scientists but usually not allowing specific interactions to develop due to their large size and a crowded program. Traditional departmental seminars are becoming so technical as to be largely inaccessible to anyone who did not coauthor the research being presented. Here contributors focus on how to make their work intelligible, accessible to a diverse audience, which in the process enforces mastery of their own field of expertise.

This volume strongly advocates multidisciplinarity with the goal to generate new interdisciplinary approaches, instruments, and models including new knowledge, transcending scientific boundaries to adopt a more holistic approach. For instance, it should be acknowledged, following Nobel laureate and president of the UK's Royal Society of Chemistry, Professor Sir Harry Kroto, "that the traditional chemistry, physics, biology departmentalised university infrastructures—which are now clearly out-of-date and a serious hindrance to progress—must be replaced by new ones which actively foster the synergy inherent in multidisciplinarity." The National Institutes of Health and the Howard Hughes Medical Institute have strongly recommended that undergraduate biology education should incorporate mathematics, physics, chemistry, computer science, and engineering until "interdisciplinary thinking and work become second nature." Young physicists and chemists are encouraged to think about the opportunities waiting for them at the interface with the life sciences. Mathematics is playing an ever more important role in the physical and life sciences, engineering, and technology, blurring the boundaries between scientific disciplines.

This book will be a reference of choice for established interdisciplinary scientists and mathematicians, and a source of inspiration for a broad spectrum of researchers and research students, graduate and postdoctoral fellows; the shared emphasis of these carefully selected and refereed contributed chapters is on important methods, research directions, and applications of analysis including within and beyond mathematics. As such the volume promotes mathematical sciences, physical and life sciences, engineering, and technology education, as well as interdisciplinary, industrial, and academic genuine cooperation.

Towards such goals the following chapters are featured in the current volume.

Chapter "Controlling Chaos in the Heart: Some Mathematics Behind Terminating Cardiac Arrhythmia" by John W. Cain describes two vastly different methods for controlling cardiac arrhythmia and how those methods can be modeled mathematically. The traditional method, point stimulation, involves the delivery of spatially localized electrical shocks through the tip of an electrode, and is the basis for medical devices such as the implantable cardioverter defibrillator (ICD). A newer approach, known as far-field pacing (FFP), involves application of a pulsed electric field across the entire heart. FFP exploits tissue heterogeneity, such as interfaces between regions of healthy cells and dead (electrically non-conducting) ones, as a means of creating "virtual electrodes."

Chapter " Working Memory and Transfer: Theoretical and Practical Considerations" by Susanne M. Jaeggi and Martin Buschkuehl provides evidence for the efficacy of several working memory interventions developed in their laboratories and reviews the emerging literature from other groups. It discusses data that demonstrate transfer to non-trained tasks throughout the lifespan, that is, in young adults, in old adults, in typically developing children, as well as children with Attention-Deficit Hyperactivity Disorder (ADHD). It also presents the neural correlates that underlie improvements observed with working memory training. The authors argue that, even though transfer effects can be elusive, and some of the effects seemingly not easy to replicate, instead of taking inconsistencies as a proof for a lack of efficacy, researchers need to develop innovative approaches to move the cognitive training literature beyond the simple question of whether or not training is effective, and to address questions of underlying mechanisms, individual differences, and training features and parameters that might mediate and moderate the efficacy of training.

Chapter "Partial Functional Differential Equations, Reduction of Complexity and Applications" by Khalil Ezzinbi aims at reducing the complexity of partial functional differential equations, assuming that the undelayed part is not necessarily densely defined and satisfies the Hille-Yosida condition. The delayed part is continuous. The author proves the dynamic of solutions are obtained through an ordinary differential equation that is well-posed in a finite dimensional space. He then shows the existence of almost automorphic solutions for partial functional differential equations. For illustration, he provides an application to the Lotka-Volterra model with diffusion and delay.

Chapter "Characterizations of Convex Quadrics in Terms of Midsurfaces and Shadow-Boundaries" by Valeriu Soltan discusses the middle points of any family of parallel chords of a real quadric surface in the Euclidean space R^n known to belong to a hyperplane, property holding as well for the shadow-boundaries of that surface. The author reviews the existing results and adds some new ones which characterize convex quadrics among convex hypersurfaces in R^n, possibly unbounded, in terms of plane quadric sections, hyperplanarity of their midsurfaces and shadow-boundaries.

Chapter "Classifying Normal, Nevus, and Primary Melanoma Skin Samples Using Penalized Ordinal Regression" by Kellie J. Archer, Jiayi Hou, and André A.A. Williams looks into translational research that is developing multigenic classifiers using data from high-throughput genomic experiments. While often the class to be predicted is nominal, sometimes it may be inherently ordinal. For example, tissue samples may be collected with the goal of classifying them as normal < pre-malignant < malignant. In this case, molecular features monotonically associated with the ordinal response may be important to disease development. While one can apply nominal response classification methods to ordinal response data, in so doing some information is lost that may improve the predictive performance of the classifier. The authors developed an R package, glmpathcr, capable of fitting a penalized continuation ratio model when the outcome to be predicted is ordinal. And they demonstrate application of their method by predicting progression to melanoma using microarray gene expression data.

Chapter "Structure–Activity Relationship Analysis of 7-Deazaadenosines as Anticancer Agents" by Josue A. Nava-Bello, Ewa Wasilewski, Angelica M. Bello, and Alejandro A. Nava-Ocampo considers the lengthy and costly complex process to develop a successful therapeutic. In order to accelerate this process, molecular modeling has become a key component of drug design. Methods used in computational chemistry vary from *ab initio* quantum chemistry methods to semi-empirical calculations and molecular mechanics. A study of the anticancer activity of a series of 7-aryl- and 7-hetaryl-7-deazaadenosines showed that nucleosides with 5-member heterocycles at the position 7 were more potent *in vitro* cytostatic agents against hematological and solid tumor cell lines than molecules with 6-member heterocycles. The authors present a quantitative structure–activity relationship (QSAR) analysis of these chemical moieties in order to have a better understanding of their structural properties and identify their molecular descriptors explaining their biological activities. They found that 5-member cyclic structures have three energy molecular descriptors that were negatively correlated to their biological activity, in particular, compounds with higher energies had higher biological potency represented by lower IC_{50} values. CLogP, a parameter of lipophilicity, was also found to be positively correlated to their biological activity, i.e., compounds with lower CLogP values had higher biological potency represented by lower concentrations inhibiting the growth of cancer cells by 50 %. Qualitatively, 5-member-ring heterocycles of 7-deazaadenosine had lower steric hindrance, i.e., were structurally smaller, than their 6-member counterparts. They made the case that, such a context, a QSAR analysis could be extraordinarily helpful in studying the mode of action of molecules with potential pharmacological or toxicological relevance.

Chapter “More than an African American Facilitator and a Prayer: Integrating Culture and Community into HIV Prevention Programs for African American Girls” by Faye Z. Belgrave, Jasmine Abrams, Sarah Javier, and Morgan Maxwell focuses on the need for prevention and intervention programs to address health disparity within a culturally sensitive and developmentally appropriate framework, in the case of sexually active African American adolescent females at a heightened risk for contracting sexually transmitted infections including HIV/AIDS. Research has shown that culturally integrated interventions can be effective at reducing HIV risk The goals of this chapter are to: (1) define culture, cultural competency, and cultural integration; (2) discuss community integration in HIV prevention programs; and (3) discuss ways in which culture can be attended to and integrated in prevention and intervention efforts. The chapter addresses each goal in order, beginning with an overview of relevant concepts.

Chapter “Dynamics of Niche Construction in Models ‘Consumers-Renewable Resource’ and ‘Prey-Predators-Renewable Resource’” by Faina S. Berezovskaya and Georgiy P. Karev deals with the question of “how much over-consumption a renewable resource can tolerate” using mathematical models, where a consumer population compete for the common resource, can contribute to resource restoration, and is subject to attacks of predators. The bifurcation analysis of the systems shows that well-adapted predators can keep the system in a stable equilibrium even for “strong” prey over-consumption, when the initial system of resource-consumer goes extinct. It means that predators may extend the domain of the total system coexistence.

Chapter "Recent Advances in Approaches to the Study of Gene Locus Control Regions" by Benjamin D. Ortiz contributes to the decades long investigation into the regulation of gene transcription in vertebrates, with the locus control region (LCR) emerging as perhaps the most powerful *cis*-acting regulatory DNA element that one can envision. An LCR element is unique in that it supports both specific spatiotemporal regulation of transcription during development and a poorly understood "insulation capacity" that prevents genomic interference with the gene regulatory program it would impose upon a linked transgene. As such, it represents a complete, compact, and portable package of the DNA sequence information required to establish an independently and predictably regulated gene locus in native chromatin of a whole animal. Both *in vivo* and cell culture models have contributed significantly to building the field of LCRs. Nevertheless, the gold standard experimental approach to LCR study is transgenic mice, which has been dominant in the progress made in the field over the past 25 years. However, recent technological advances are resulting in a re-emergence of cell culture-based approaches to LCR study, portending a coming era of more rapid progress in this significant but understudied field. The investigation of the unique and powerful gene regulatory activities supported by LCR elements offers unparalleled opportunities to gain insight into *cis*-mediated transcriptional regulation at the single gene locus level. Furthermore, such insights are critical to advancing the safety and efficacy of gene therapy.

Chapter "Dynamical Roles of Jacobian Feedback Loops and Qualitative Modeling" by Bourama Toni presents a mathematical methodology for the qualitative modeling of differential systems using the feedback loops encoded in the Jacobian matrix, and described by the products of the Jacobian entries under cyclic permutations of the indices. The technique is easy to implement and could quickly demarcate both parameter and phase spaces into exciting regions (limit cycle, multiple equilibria, chaotic behavior), non-exciting ones (single stable fixed points), hard-instance regions (ergodic behavior). As such it could be useful in surveying dynamical responses of models simulating physico-chemical, biological, biochemical, economical systems and game theory. It efficiently asserts the possibility of multistationarity, periodicity, self-sustained oscillations, chaotic behavior using strictly the qualitative relations and assumptions of the systems, to achieve primarily qualitative understanding rather than quantitative numerical prediction. To illustrate the author includes a complete loop analysis of the celebrated Lorenz and Rossler systems predicting their global dynamics.

Chapter "Forecasting of Time Series Data Using Multiple Break Points and Mixture Distribution" by Rajan Lamichhane, Norou Diawara, and Cynthia M. Jones deals with special classes of stochastic processes with time series of sparse data. Studies in such cases focus on the analysis, construction, and prediction in parametric models. Here, the authors assume several nonlinear time series with additive noise components, and the model fitting is proposed in two stages. The first stage identifies the density using all the clusters information, without specifying any prior knowledge of the underlying distribution function of the time series. In the second stage, they partition the time series into consecutive non-overlapping intervals of quasi stationary increments where the coefficients shift from one stable regression relationship to a different one using a breakpoints detection algorithm. These breakpoints are estimated by minimizing the likelihood from the residuals. The authors approach time series prediction through the mixture distribution of combined error components. Parameter estimation of mixture distribution is done by using the EM algorithm. The method is then applied to a simulated data.

Chapter "Direct Differentiation of Human Pluripotent Stem Cells into Advanced Spermatogenic Cells: In Search of an *In Vitro* System to Model Male Factor Infertility" by Charles A. Easley, Calvin R. Simerly, and Gerald Schatten focuses on Assisted Reproductive Technology (ART) which has gained worldwide acceptance, and on Intracytoplasmic Sperm Injection (ICSI) which has aided couples with severe male factor infertility to achieve pregnancies. While ICSI has circumvented some defects in *in vitro* fertilization (IVF), numerous patients still fail to achieve pregnancies. Even with patients with known causes for male factor infertility (Klinefelter Syndrome, Sertoli Cell Only Syndrome, DAZ family deletions, etc.), root causes are still being investigated, although there is no *in vitro* model for human spermatogenesis to examine intracellular root causes. Differentiation of stem cells into spermatogenic lineages *in vitro* provides a unique window into the biological mechanisms responsible for driving pluripotent stem cells into essential progeny—haploid spermatids and viable sperm—as well as provides an innovative approach for determining novel root causes for male infertility. Our recent work outlined a novel approach for differentiating human embryonic stem cells (hESCs) and induced pluripotent stem cells (hiPSCs) into advanced spermatogenic lineages including haploid spermatids with correct parent-of-origin genomic imprints on two loci. The work provides herein a foundation for building a true *in vitro* model for human spermatogenesis with which to model, diagnose, and potentially treat male factor infertility.

Chapter "Stepanov-Like Pseudo-Almost Periodic Functions in Lebesgue Spaces with Variable Exponents $L^{p(x)}$" by Toka Diagana and Mohamed Zitane introduces and studies a new class of functions called Stepanov-like pseudo-almost periodic spaces with variable exponents, which generalizes in a natural way the space of Stepanov-like pseudo-almost periodic spaces. Basic properties of these new spaces are established. The existence of pseudo-almost periodic solutions to some first-order differential equations with $S^{p,q(x)}$-pseudo-almost periodic coefficients will also be studied.

Chapter "Group Circle Systems on Conics" by Raymond R. Fletcher studies a *group circle system*, a collection of points and circles in the Euclidean plane determined by the elements of an abelian group mapped injectively to the plane, where no five points in the range set are cocyclic. Here the attention is confined to group circle systems all of whose points (or vertices) lie on a noncircular conic.

Chapter "Remanufacturing Processes, Planning and Control" by Jianzhi Li and Zhenhua Wu provides a summary of critical issues in remanufacturing process and its planning and control. The chapter starts with an introduction of the special characteristics and the associated problems in remanufacturing. Typical remanufacturing processes such as cleaning, testing, and disassembly are then discussed in detail. The chapter also provides a discussion of process sequencing for product disassembly to minimize cost and energy consumption. Due to the stochastic nature in the material arrival process, production planning represents another main challenge for remanufacturers. Based on a case study of a business in Austin TX, a simulation model with a prioritized stochastic batch arrival mechanism, considering factors that affect the total profit, is also discussed. The chapter also presents a genetic algorithm (GA) to optimize the production planning and control policies for dedicated remanufacturing.

The concluding chapter "Viscous Interfacial Motion: Analysis and Computation" by Jin Wang considers the interfacial flows between two viscous incompressible fluids. After formulating the mathematical framework, the author first presents analytical solutions to the linearized problem, discusses some results from linear asymptotic analysis, and then describes a numerical method for computing the nonlinear motion which ensures a high accuracy on and near the moving interface. Simulation results on viscous Stokes waves are presented to demonstrate the advantages of this method. In addition, as an example of nonlinear asymptotic study, the authors conduct a perturbation series analysis for Stokes waves with small viscosity, the results of which provide an analytical justification to the numerical observation.

The book as a whole will certainly enhance the overall objective of the series (seminars and previous volumes), that is, to foster student interest and enthusiasm in the STEAM-H disciplines (Science, Technology, Engineering, Agriculture, Mathematics and Health), stimulate graduate and undergraduate research, and generate collaboration among researchers on a genuine interdisciplinary basis.

Virginia State University is in an area that is socially, economically, intellectually very dynamic, and home to some of the most important research centers in the USA, including NASA Langley Research Center, manufacturing companies (Rolls-Royce, Canon, Chromalloy, Sandvik, Siemens, Sulzer Metco, NN Shipbuilding, Aerojet) and their academic consortium (CCAM), University of Virginia, Virginia Tech, the Virginia Logistics Research Center (CCAL), Virginia Nanotechnology Center, Aerospace Corporation, C3I Research and Development Center, Defense Advanced Research Projects Agency, Naval Surface Warfare Center, National Accelerator Facility, and the Homeland Security Institute. The series, the seminars and the written thematic continuation published by Springer a world-renowned publisher, is now well established and is expected to become a national and international reference in interdisciplinary STEAM-H education and research.

Petersburg, VA, USA Bourama Toni

Acknowledgments

We would like to express our sincere appreciation of the President of Virginia State University Dr. Keith Miller's encouragement, and support of the seminar series and its publications in line with his vision of "Building a Better World."

We gratefully acknowledge the following supports: the Office of the Provost, Dr. W. Weldon Hill, Mr. Daniel Roberts; the Office of the Dean, School of Engineering, Science and Technology, Professor Keith M. Williamson, Professor Dawit Haile, Professor Larry Brown, Ms. Victoria Perkins, Mrs. Bonnie Grant; the Department of Education HBCU MS Program and Professor Pamela Leigh-Mack; the NSF/HBCU-UP and Professor Ali Ansari.

We would like to thank very much Melissa Watts, Taurus Richardson, whose tireless efforts have contributed to the smooth organization, presentations, recording, and attendance.

We sincerely appreciate the promotional support by the VSU Radio Station WVST 91.3 and Station Manager Ms. Jennifer Williamson. Her professionalism and keen interest have contributed greatly to well establish the STEAM-H seminar series.

Our many thanks to the VSU webmaster team, Quentin Bolton and Shawna Brown, for contributing an excellent seminar website to effectively expose the event to the community at large.

Special thanks are extended to all the contributors, the faculty and student participants, in particular to Professor Oliver Hill, Professor Emeritus Walter Elias, Professor Gerald Burton, Dr. Reginald Hopkins, Dr. Cheryl Talley, Mr. Leroy Lane, Dr. Yahya Njai, Dr. Giti Javidi, Dr. Zhifu Xie, Mr. Andrew Wynn, Dr. Victor Vilchiz, Dr. Chandan Samantaray, Dr. Kostadin Damevski, Dr. Brian Sayre, Dr. Laban Rutto, Dr. Eshan Sheybani, Ms. Eleanor Poarch-Wall, Dr. Robert Wieman, Dr. Sarah Witiak, Mr. Daniel Fritz, Mr. Calvin Smith, Mr. Leonard Pearlman, all great supporters and/or frequent presenters at the seminar since its inception. Their genuine efforts to engage VSU faculty in interdisciplinary research and to encourage the students into the STEAM-H disciplines are very much appreciated.

We would like to express our sincere thanks to all the anonymous referees for their professionalism. They all made the seminar and its published thematic continuation a reality for the greater benefice of the community of Science, Technology, Engineering, Agriculture, Mathematics, and Health.

Petersburg, VA, USA Bourama Toni

Contents

Contributors

Jasmine Abrams Department of Psychology, Virginia Commonwealth University, Richmond, VA, USA

Kellie J. Archer Department of Biostatistics, Virginia Commonwealth University, Richmond, VA, USA

Faye Z. Belgrave Department of Psychology, Virginia Commonwealth University, Richmond, VA, USA

Angelica M. Bello Center for Molecular Design and Preformulations, Toronto General Research Institute, University Health Network, Toronto, ON, Canada

Faina S. Berezovskaya Department of Mathematics, Howard University, Washington, DC, USA

Martin Buschkuehl School of Education, University of California, Irvine, CA, USA

John W. Cain Department of Mathematics and Computer Science, University of Richmond, Richmond, VA, USA

Toka Diagana Department of Mathematics, Howard University, Washington, DC, USA

Norou Diawara Department of Mathematics and Statistics, Old Dominion University, Norfolk, VA, USA

Charles A. Easley IV Laboratory of Translational Cell Biology, Department of Cell Biology, Emory University, Atlanta, GA, USA

Khalil Ezzinbi Faculty of Sciences Semlalia, Department of Mathematics, Cadi Ayyad University, Marrakesh, Morocco

Raymond R. Fletcher Department of Mathematics and Computer Science, Virginia State University, Petersburg, VA, USA

Jiayi Hou Department of Biostatistics, Virginia Commonwealth University, Richmond, VA, USA

Susanne M. Jaeggi School of Education, University of California, Irvine, CA, USA

Sarah Javier Department of Psychology, Virginia Commonwealth University, Richmond, VA, USA

Cynthia M. Jones Department of Mathematics and Statistics, Old Dominion University, Norfolk, VA, USA

Georgiy P. Karev National Center for Biotechnology Information, Bethesda, MD, USA

Rajan Lamichhane Department of Mathematics, Texas A&M University-Kingsville, Kingsville, TX, USA

Jianzhi Li Manufacturing Engineering Department, The University of Texas-Pan American, Edinburg, TX, USA

Morgan Maxwell Department of Psychology, Virginia Commonwealth University, Richmond, VA, USA

Josue A. Nava-Bello Center for Molecular Design and Preformulations, Toronto General Research Institute, University Health Network, Toronto, ON, Canada

Alejandro A. Nava-Ocampo Faculty of Medicine, Department of Pharmacology and Toxicology, University of Toronto, Toronto, ON, Canada

Benjamin D. Ortiz Department of Biological Sciences, Hunter College and Graduate Center, City University of New York, New York, NY, USA

Gerald Schatten Department of OB/GYN and Reproductive Sciences, University of Pittsburgh, Pittsburgh, PA, USA

Calvin R. Simerly Department of OB/GYN and Reproductive Sciences, University of Pittsburgh, Pittsburgh, PA, USA

Valeriu Soltan Department of Mathematical Sciences, George Mason University, Fairfax, VA, USA

Bourama Toni Department of Mathematics and Computer Science, Virginia State University, Petersburg, VA, USA

Jin Wang Department of Mathematics and Statistics, Old Dominion University, Norfolk, VA, USA

Ewa Wasilewski Leslie Dan Faculty of Pharmacy, University of Toronto, Toronto, ON, Canada

André A.A. Williams Division of Biostatistics and Bioinformatics, National Jewish Health, Denver, CO, USA

Zhenhua Wu Department of Engineering, Virginia State University, Petersburg, VA, USA

Mohamed Zitane Department of Mathematics, Laboratory of An. Math and NCG, Faculty of Science, Ibn Tofaïl University, Kenitra, Morocco

Controlling Chaos in the Heart: Some Mathematics Behind Terminating Cardiac Arrhythmias

John W. Cain

Abstract Precisely coordinated rhythmic contraction of heart muscle tissue is essential for the effective pumping of blood, and abnormal cardiac rhythms (arrhythmias) can be fatal. Patients with certain types of arrhythmias receive surgically-implanted devices which are designed to intervene when severe abnormalities are detected.

Here, we shall describe two vastly different methods for controlling cardiac rhythm and how those methods can be modeled mathematically. The traditional method, point stimulation, involves the delivery of spatially localized electrical shocks through the tip of an electrode, and is the basis for medical devices such as the implantable cardioverter defibrillator (ICD). A newer approach, known as far-field pacing (FFP), involves application of a pulsed electric field across the entire heart. FFP exploits tissue heterogeneity, such as interfaces between regions of healthy cells and dead (electrically non-conducting) ones, as a means of creating "virtual electrodes." Importantly, studies suggest that FFP can successfully terminate arrhythmias such as fibrillation using far less energy than point stimulation, potentially sparing patients from the excruciating pain associated with traditional ICD intervention.

J.W. Cain (✉)
Department of Mathematics and Computer Science, University of Richmond, 28 Westhampton Way, Richmond, Virginia 23173, USA

Department of Mathematics, Harvard University, One Oxford Street, Cambridge, Massachusetts 02138, USA
e-mail: jcain2@richmond.edu

B. Toni (ed.), *New Frontiers of Multidisciplinary Research in STEAM-H (Science, Technology, Engineering, Agriculture, Mathematics, and Health)*,
Springer Proceedings in Mathematics & Statistics 90, DOI 10.1007/978-3-319-07755-0_1,

1 Introduction

Closed-loop feedback can be an incredibly useful and powerful tool for guiding dynamical systems towards desirable equilibrium states that might otherwise be unstable. Ott, Grebogi, and Yorke (OGY) [18] designed a feedback control technique which has been used successfully to suppress both periodic and aperiodic responses in physical systems. Subsequent experimental and theoretical analyses [20, 21, 26, 27] have advanced more robust methods for controlling bifurcations and chaos. Adaptations of the OGY method have been used by numerous authors [3–5, 9, 10, 12] as a means of preventing bifurcations that lead to cardiac arrhythmias.

Our primary interest in feedback control lies in its ability to terminate oscillatory or chaotic behavior in cardiac rhythm—behavior that can be catastrophic if left unchecked. The principal idea underlying OGY control and its variants is that, by applying small perturbations to an accessible system parameter μ, it is sometimes possible to force the dynamical variables $\boldsymbol{x}$ to converge to an equilibrium $\boldsymbol{x}^*$ that would be unstable in the absence of control. Here are three examples to illustrate what we have in mind:

- The upward vertical position of a pendulum is an example of an unstable equilibrium. In order to stabilize that equilibrium using feedback control, we might apply a sequence of small "kicks", pushing the pendulum clockwise whenever it attempts to fall counterclockwise and vice-versa. If control is successful, both the angular velocity of the pendulum and the magnitude of the kicks should tend to zero, leaving the pendulum precariously balanced along the upward vertical. Stabilizing an inverted pendulum is actually a classic feedback control problem; see, for example, [15].
- Suppose that quasi-static variation of the parameter μ causes a supercritical Hopf bifurcation in which $\boldsymbol{x}^*$ loses stability, leading to "undesirable" oscillations. It may be possible to stabilize $\boldsymbol{x}^*$ via tiny perturbations to μ, ultimately terminating the oscillatory behavior. See also [1] for an alternative approach[1] toward controlling Hopf bifurcations.
- A discrete-time, purely academic example: The discrete logistic mapping

$$x_{n+1} = \mu^* x_n (1 - x_n), \tag{1}$$

  where $\mu^* \in [0, 4]$ and $x_0 \in (0, 1)$, is among the best-known examples of system with chaotic solutions. The fixed point $x^* = 1 - (\mu^*)^{-1}$ is asymptotically stable for $1 < \mu^* < 3$. If μ^* is increased quasi-statically, a cascade of period-doubling

[1]Some readers may be familiar with general nonlinear control systems $x' = f(x, u)$ where x is a state variable, u is a control, and f depends smoothly on both arguments. Let us emphasize: in this article, we shall not consider continuous controls $u(x)$, and feedback is always applied by perturbing a *parameter*.

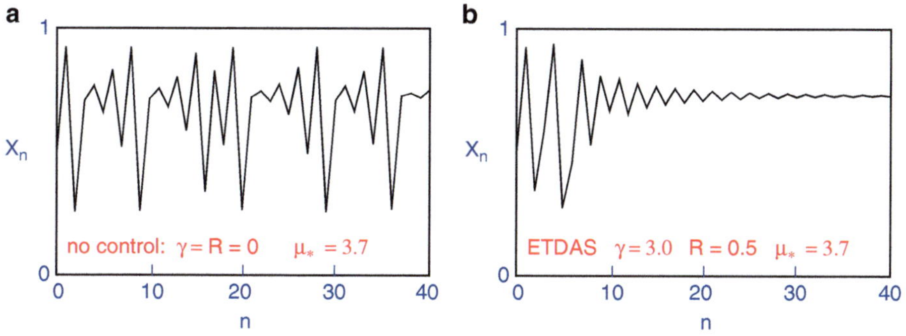

Fig. 1 (**a**) Chaotic behavior of iterates of the mapping (1) for the particular choice of $\mu^* = 3.7$. (**b**) Termination of chaos via ETDAS feedback control

bifurcations occurs, the first at $\mu^* = 3$ and the second at $\mu^* = 1 + \sqrt{6}$. The sequence of bifurcation values of μ^* has an accumulation point at approximately 3.5699, beyond which chaotic solutions can exist. For $\mu^* \in [3, 4]$, OGY-type feedback control can be used to stabilize x^* via small perturbations to μ^*, even if the "baseline" μ^* happens to lie in the chaos regime. Figure 1 illustrates the use of a specific feedback control algorithm (see next section) as a means of achieving the desired results.

Throughout this survey article, there are two different aspects of cardiac arrhythmia control that we shall consider, one involving the timing of the electrical "shocks" applied by a medical device and the other involving the actual experimental setup for applying those shocks. Regarding the former, in Sect. 2 we shall describe a specific feedback control algorithm known as *extended time-delay autosynchronization (ETDAS)* [26] that will serve as our basis for timing the shocks. Regarding the latter, Sect. 1.1 provides preliminary descriptions of two vastly different experimental setups for delivering the shocks: point stimulation and far-field pacing. Results of numerical simulations of both setups are reported in Sect. 4. In order to understand (a) how ETDAS works and (b) the important distinctions between point stimulation and far-field pacing, it will be helpful to model tissue samples of different dimensions: "zero-dimensional" single-cell samples (Sect. 3), "one-dimensional" fibers of cells joined end-to-end (Sect. 4.1), and "two-dimensional" thin sheets of cardiac tissue (Sect. 4.2).

We focus on a particular rhythm known as APD *alternans*, a beat-to-beat alternation of action potential duration (APD) (see Fig. 2). APD alternans is believed to serve as a substrate through which cardiac rhythm can degrade into potentially deadly rhythms such as ventricular fibrillation [2, 6, 14, 19, 22, 25]. In the language of nonlinear dynamics, high-amplitude alternans can induce breakup of spiral waves of propagating action potentials, which can degrade into turbulent patterns. Turbulent electrical wave patterns cause the cardiac muscle tissue to quiver erratically, preventing coordinated contraction of the tissue and impairing or preventing the heart's ability to pump blood.

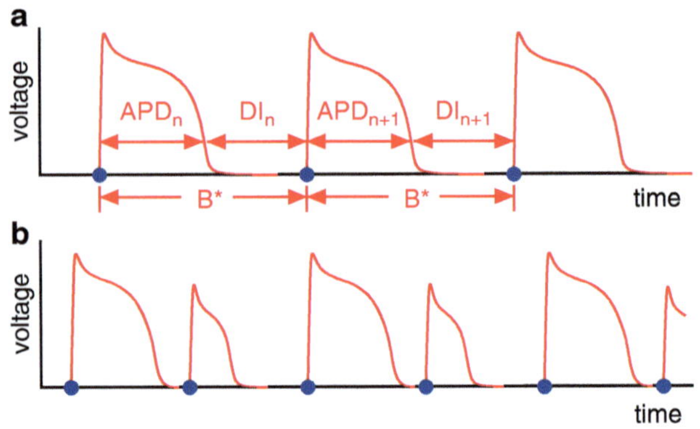

Fig. 2 Schematic action potentials in a periodically stimulated cardiac cell. Brief stimuli (indicated by *dots* on the time axis) are applied with period B^*, resulting in a sequence of action potentials (prolonged elevation of voltage across the cell membrane). (**a**) If B^* is large, the result is a sequence if identical action potentials. (**b**) If B^* is short, APD alternans can occur

The onset of APD alternans is typically associated with an instability that occurs when the heart rate becomes critically fast. More exactly, if a patch of cells is paced (stimulated periodically) with period B^*, then one of several [approximate] steady-state responses may occur. If B^* is large, then the sequence of APD values typically converges[2] to some limit A*. This steady-state APD value is a function of B^*, a feature of cardiac tissue known as APD *restitution*. If heart rate becomes faster (i.e., if B^* decreases), then A* may lose stability via a period-doubling bifurcation [17] or a border-collision bifurcation [23], resulting in the period-2 response of alternans. Further decreasing B^* can lead to pattern known as 2:1 conduction block: every 2 stimuli elicit only one action potential—the cells ignore every other stimulus because the rapid pacing does not given them sufficient time to recover their excitability.

1.1 Point-Stimulation vs. Far-Field Pacing

Most previous attempts to control alternans with OGY-type methods involve the use of an electrode that applies stimuli in a localized region of tissue, a technique that we will refer to as *point stimulation*. The intent of the electrode is to "reset" the heart's native electrical activity when an abnormal rhythm is detected. In circumstances

---

[2]Throughout this article, we neglect small beat-to-beat variations in APD due to background noise. This simplifying assumption gives us license to adopt deterministic models as opposed to stochastic ones, facilitating mathematical analysis and computer simulations.

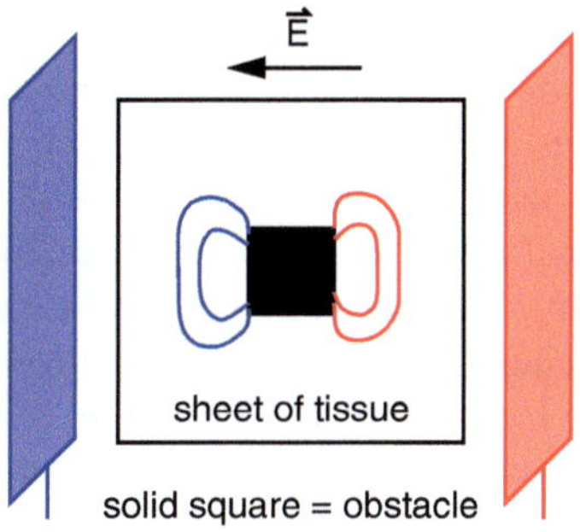

Fig. 3 Schematic diagram of FFP. An electric field $\boldsymbol{E}$ is applied across a square sheet of tissue containing a smaller (again square) non-conducting obstacle. The field induces depolarization on one side of the obstacle and hyperpolarization on the other side, turning it into a "virtual" electrode

when pacing with period B^* induces alternans, control can be applied by having the electrode perturb B^* on a beat-to-beat basis. The perturbations are typically chosen proportional to the difference between the two preceding APD values, and experiments (see, for example, Hall and Gauthier [10]) have shown that point stimulation can terminate alternans in tiny, "zero-dimensional" patches of cardiac cells. In Sect. 3, we explain how routine linear stability analysis of a single-cell alternans model can be used to approximate the ranges of parameter values under which ETDAS control is expected to succeed.

In spatially extended tissue, using point stimulation with feedback control applied to the pacing period B^* appears to be far less effective. For example, Echebarria and Karma [5] performed numerical simulations in which each cell in a "one-dimensional" fiber (consisting of cells joined end-to-end) experienced alternans. They found that when a special case of ETDAS is implemented via point stimulation at some specific spatial location along the fiber, alternans could only be suppressed in the cells within some small distance from the [simulated] electrode. Later experiments appeared to confirm the predictions in [5], calling into question whether point-stimulation could ever be effective enough to achieve whole-heart control. In Sect. 4.1, we will show results indicating that the findings in [5] are still observed even when the [full, unrestricted] ETDAS method is used.

A recent study of Fenton et al. [8] offers an alternative to point stimulation—an alternative that appears to be quite successful in terminating arrhythmias in the whole heart. *Far-field pacing (FFP)* is a technique in which a pulsed electric field is applied across the entire heart, using two plate electrodes, a cathode and an anode (see Fig. 3). The idea behind FFP is that any anatomical obstacles (e.g., regions of dead, non-conducting tissue) in the heart can be turned into "virtual" electrodes. The field depolarizes cells on one side of an obstacle and hyperpolarizes cells on another side and, if the field is sufficiently strong, the depolarized cells will fire a propagating action potential. In [8], the authors note that FFP successfully terminated atrial fibrillation in 69 of 74 occurrences in 8 experimental preparations on canine hearts. The impressive success rate is only one noteworthy aspect of their study. Importantly, their electric field strengths involve energies that are near the "pain threshold", far weaker than the high-energy, pain-inducing anti-fibrillation pacing associated with the implantable defibrillators that some patients receive. Regarding the timing of the electric field pulses, the authors of [8] used *overdrive* pacing to combat the atrial fibrillation. That is, they issued a very rapid train of electric field pulses at a frequency designed to overdrive the dominant frequency of the spatiotemporal chaos associated with fibrillation. In doing so, they "reset" the electrical behavior of the tissue, the idea being that the heart's native pacemaker cells would subsequently take over and resume a normal rhythm. Because overdrive pacing could cause tissue fatigue not to mention reduced battery life for an implantable device, one might ask whether there are equally-successful alternative ways to administer trains of FFP pulses. In Sect. 4.2, we briefly describe some numerical experiments in which ETDAS, as opposed to overdrive pacing, is used to automate the timing of those pulses.

2 ETDAS for Discrete Systems

The ETDAS method was originally introduced in [26]. Although ETDAS can be used to stabilize unstable equilibria of differential equations, here we will use it to stabilize unstable fixed points of discrete-time systems (as we shall be regarding individual heartbeats as discrete-time events).

2.1 *One-Dimensional Mappings*

Consider a one-dimensional mapping $x_{n+1} = f_\mu(x_n)$ where f_μ is a continuously differentiable function of the variable x_n and the parameter μ. Assume that the mapping has an isolated fixed point x^*, the value of which may depend on μ. Further suppose that a period-doubling bifurcation occurs at some critical value of μ, at which point x^* loses stability and a stable 2-cycle is born. ETDAS attempts to terminate the alternation and stabilize x^* by perturbing μ to $\mu + \epsilon_n$, with the perturbations ϵ_n chosen according to the rule

$$\epsilon_n = \gamma\,(x_n - x_{n-1}) + R\epsilon_n. \tag{2}$$

The parameter γ is sometimes referred to as the *feedback gain*. The effect of the parameter R is more subtle: it adjusts the perturbations according to the history of all previous iterates [26] and, in addition to offering added robustness, the added flexibility it offers relative to the original OGY schemes is valuable in reducing sensitivity to noise [3]. It is instructive to inspect Eq. (2) in the special case that $R = 0$ and $\gamma \neq 0$. In that case, ETDAS merely modifies the parameter μ by an amount proportional to the difference between the two previous iterates—the magnitude of the perturbations is based upon the amplitude of the alternation.

The beauty of ETDAS and most other techniques adapted from the original OGY formalism is that they do not require advance knowledge of the value of x^*, the

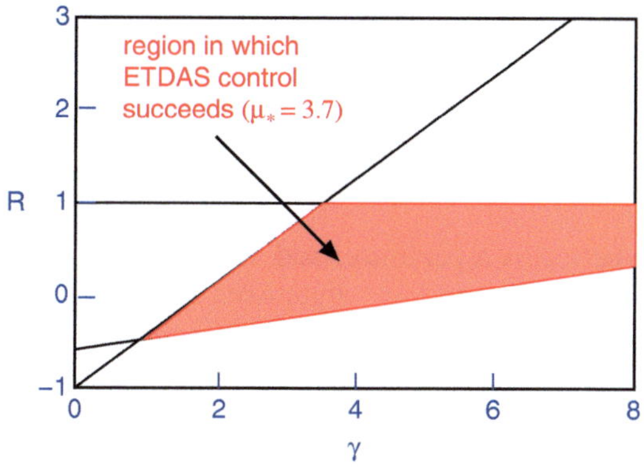

Fig. 4 Domain in R-γ parameter space in which linear stability analysis of (3) predicts that ETDAS may stabilize the unstable fixed point x^* if $\mu^* = 3.7$

unstable fixed point being targeted for stabilization, and they do not "move" the fixed point. As an academic example, consider the discrete logistic mapping (1) with $\mu^* \in (3, 4)$. In that parameter regime, the fixed point $x^* = 1 - (\mu^*)^{-1}$ is unstable. If we apply ETDAS control by modifying μ^* according to (2), we obtain

$$\begin{aligned} x_{n+1} &= (\mu^* + \epsilon_n)x_n(1 - x_n) \\ \epsilon_{n+1} &= \gamma\left[(\mu^* + \epsilon_n)x_n(1 - x_n) - x_n\right] + R\epsilon_n. \end{aligned} \tag{3}$$

The application of ETDAS increases the dimension of the original mapping by 1. The new system has fixed point $(x^*, 0)$ which, if γ and R are chosen suitably, can be stabilized even if $\mu^* > 3$. Linearizing (3) about this fixed point, we may predict the region of γ-R parameter space in which ETDAS may successfully stabilize the fixed point. If J denotes the Jacobian matrix associated with the right-hand side of (3) evaluated at $(x^*, 0)$, requiring that each eigenvalue of J have modulus less than 1 guarantees local asymptotic stability of the fixed point. The following Lemma applies:

Lemma 2.1. *If J is a 2×2 matrix, then its eigenvalues have modulus less than 1 if and only if (i)* $\det(J) < 1$*; (ii)* $\mathrm{tr}(J) - \det(J) < 1$*; and (iii)* $\mathrm{tr}(J) + \det(J) > -1$.

Applying those criteria to our academic example (3) with $\mu^* = 3.7$, we obtain inequalities on R and γ that yield the region sketched in Fig. 4. Note that the [uncontrolled] discrete logistic mapping (1) has chaotic solutions when $\mu^* = 3.7$. By picking an (R, γ) pair as suggested by Fig. 4, we may actually control the chaos for stabilize x^* as illustrated in Fig. 1.

2.2 *Higher-Dimensional Mappings*

Using ETDAS for prevention of bifurcations and chaos in higher-dimensional mappings is equally straightforward, and comes with the relatively minor expense of raising the dimension of the underlying system by 1. In the next section, we will demonstrate the success of feedback control in a two-dimensional system of the form

$$\begin{aligned} x_{n+1} &= f_\mu(x_n, y_n) \\ y_{n+1} &= g_\mu(x_n, y_n), \end{aligned} \tag{4}$$

where μ is a parameter and the functions f and g are continuously differentiable with respect to their arguments as well as μ. Assuming that the [uncontrolled] system (4) has an isolated fixed point (x^*, y^*) that loses stability as μ is varied, ETDAS aims to prevent oscillation of the iterates x_n by perturbing μ to $\mu + \epsilon_n$, with ϵ_n chosen as in (2) above. The resulting system has three dynamical variables, (x_n, y_n, ϵ_n) and a fixed point $(x^*, y^*, 0)$. Linearizing about that fixed point leads to a 3×3 Jacobian matrix J, and our goal is to choose R and γ in such a way that all eigenvalues of J move into the open unit disc in the complex plane. For reference, we state a Lemma that can help in our exploration of parameter space:

Lemma 2.2. *If J is a 3×3 matrix, then its eigenvalues satisfy the characteristic equation*

$$p(\lambda) \equiv \lambda^3 + a_1\lambda^2 + a_2\lambda + a_3 = 0,$$

where $a_1 = -\text{tr}(J)$, $a_3 = -\det(J)$, and

$$a_2 = \frac{1}{2}\left[(\text{tr}(J))^2 - \text{tr}(J^2)\right].$$

The eigenvalues lies in the open unit disc of the complex plane if and only if (i) $p(1) > 0$; (ii) $p(-1) < 0$; (iii) $3+a_1-a_2-3a_3 > 0$; and (iv) $1+a_1a_3-a_2-a_3^2 > 0$.

3 Restitution and ETDAS in Zero Dimensions

Often, analysis of cardiac rhythm is distilled to the problem of examining sequences of time intervals that represent when a cell can be regarded as "excited" (i.e., transmembrane voltage v elevated above some threshold v_{thr}) or "recovered" ($v < v_{\text{thr}}$). Given a threshold reference voltage v_{thr} slightly above the cell's resting potential, action potential duration APD can be regarded as the amount of time that $v > v_{\text{thr}}$ in a given beat. As indicated in Fig. 2, we will let A_n denote[3] the duration of the action potential following the nth electrical stimulus. The *diastolic interval (DI)* essentially measures the amount of recovery time that the cell is allowed between the end of an action potential and the application of the next stimulus. We let D_n

[3]Henceforth, we shall use one-letter abbreviations when mathematical notation is required, preferring A_n and D_n to APD_n and DI_n.

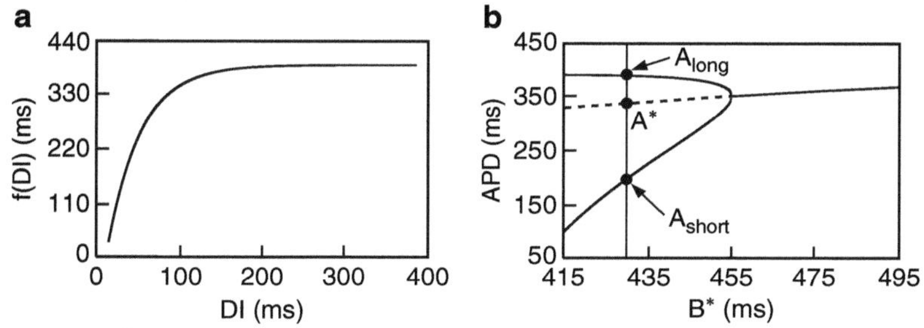

Fig. 5 (**a**) Example of a restitution curve from Eq. (5). (**b**) Bifurcation to alternans in Eq. (6) using the restitution function from Eq. (5)

denote the DI that follows the nth action potential. When pacing is periodic, we will let B^* denote the pacing period, sometimes referred to as the *basic cycle length (BCL)*. Notice that $\mathrm{A}_n + \mathrm{D}_n = B^*$ for each n.

There is a vast literature dedicated to models of the cardiac action potential, most of which are based upon the formalism originally posed by Hodgkin and Huxley [11] in their model of the nerve action potential in giant squid axon. The differential equations in those sorts of models track the voltage v as well as the various currents associated with passage of sodium, potassium, and calcium ions across the cell membranes. That level of detail is often not necessary for the purposes of analyzing and/or controlling rhythm, and bifurcations leading to alternans are more easily understood by extracting mappings that relate A_{n+1} to A_n. *Restitution* of APD can be defined as the tendency for steady-state APD, A^*, to decrease with faster pacing (shorter B^*). It can also be thought of as the tendency for APD to increase if the preceding DI is increased: allowing a cell more time to rest prolongs the duration of its next action potential. Mathematically, this relationship can be written [17] in the form $\mathrm{A}_{n+1} = f(\mathrm{D}_n)$, and the graph of f is called a *restitution curve*. An example of a restitution curve

$$\mathrm{A}_{n+1} = f(\mathrm{D}_n) = 392 - 525\exp(-\mathrm{D}_n/40), \tag{5}$$

that was fit to bullfrog restitution data [10] is graphed in Fig. 5a (all quantities measured in milliseconds). Notice that lengthening DI yields diminished return on the investment of additional recovery time.

The restitution relationship $\mathrm{A}_{n+1} = f(\mathrm{D}_n)$ can be written as a one-dimensional mapping

$$\mathrm{A}_{n+1} = f(\mathrm{D}_n) = f(\mathrm{B}^* - \mathrm{A}_n). \tag{6}$$

For the restitution function given in Eq. (5), a period-doubling bifurcation to alternans occurs when $B^* = 455$, as shown in Fig. 5b.

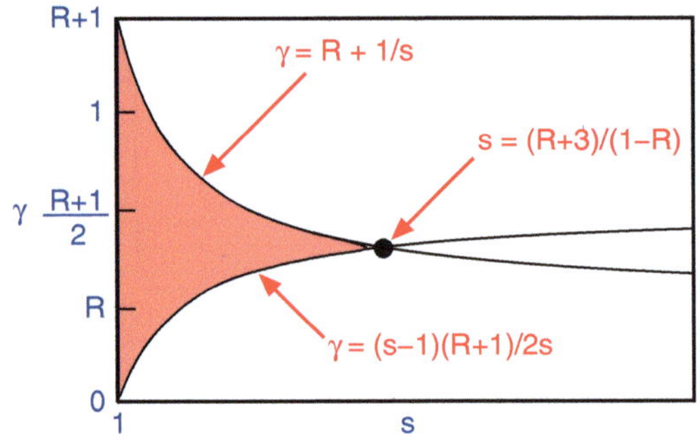

Fig. 6 Domain in which ETDAS is projected to succeed for the system (7), (8). The parameter R is assumed to be less than 1 as explained in the text

3.1 ETDAS in a Restitution Mapping

Applied to the restitution mapping (6), ETDAS modifies the pacing period B^* on a beat-to-beat basis according to

$$\mathrm{A}_{n+1} = f(\mathrm{B}^* + \epsilon_n - \mathrm{A}_n), \tag{7}$$

and the perturbations ϵ_n are updated recursively by

$$\epsilon_{n+1} = \gamma\left[f(\mathrm{B}^* + \epsilon_n - \mathrm{A}_n) - \mathrm{A}_n\right] + R\epsilon_n. \tag{8}$$

If B^* is large enough that alternans does not occur, control is turned off by setting $\gamma = R = 0$ and the sequence of APD values converges to a stable steady-state A^*. Routine linear stability analysis of the uncontrolled map predicts that this will occur if $|f'(\mathrm{B}^* - \mathrm{A}^*)| < 1$; i.e., if the slope of the restitution function f has magnitude smaller than 1. If the slope exceeds 1, alternans may occur, with APD alternating between two numbers $\mathrm{A}_{\mathrm{long}}$ and $\mathrm{A}_{\mathrm{short}}$ that lie on either side of A^*. Linearizing Eqs. (7) and (8) about $(\mathrm{A}^*, 0)$ and applying Lemma 2.1, it is possible to predict the ranges of γ and R for which ETDAS is predicted to stop alternans. In fact, one may show that this domain is defined by the inequalities $0 \leq R < 1$ and

$$\left(\frac{R+1}{2}\right)\left(1 - \frac{1}{s}\right) < \gamma < R + \frac{1}{s}, \tag{9}$$

where $s = f'(\mathrm{B}^* - \mathrm{A}^*)$ measures the slope of the restitution function. The details of that calculation appear in [3], and the results are summarized in Fig. 6. Notice that as the slope s becomes steeper, the region in parameter space in which control is projected to succeed shrinks.

3.2 ETDAS with Memory

The restitution mapping (6) is sometimes a useful predictive tool when studying cardiac rhythm, but using a one-dimensional mapping as a caricature for a high-dimensional system like the heart is bound to have limitations. Indeed, it is well-known [13, 28, 29] that APD is influenced not only by the preceding DI, but also by the recent pacing history of the tissue. In other words, A_{n+1} should really be regarded as a function of $\mathrm{A}_n, \mathrm{A}_{n-1}, \ldots, \mathrm{A}_{n-k+1}$ for some $k > 1$, a phenomenon known as *short-term memory*. Schaeffer et al. [24] use asymptotic and perturbation methods to derive a $k = 2$-dimensional restitution mapping model from a differential equation model of the action potential. Their two-dimensional model can exhibit a surprisingly rich variety of dynamical behavior, capturing far more of the physiologically relevant phenomena than one might expect from such a low-dimensional mapping. With the Schaeffer restitution model as a basis, ETDAS is implemented just as easily as in the previous subsection. Again, ETDAS raises the dimension of the mapping by 1, this time resulting in a 3×3 system. Linear stability analysis coupled with Lemma 2.2 leads to a system of four inequalities which indicate how R and γ should be chosen for control to succeed.

Example. The Fox-Bodenschatz-Gilmour memory model for cardiac restitution is given by

$$\begin{aligned} \mathrm{A}_{n+1} &= (1 - \alpha M_{n+1}) G(\mathrm{B}^* - \mathrm{A}_n) \\ M_{n+1} &= [1 - (1 - M_n) e^{-\mathrm{A}_n/\tau}] e^{-(\mathrm{B}^* - \mathrm{A}_n)/\tau}, \end{aligned} \tag{10}$$

where

$$G(x) = A + \frac{E}{1 + e^{-(x-C)/D}}, \tag{11}$$

the memory variable M_n and the parameter α are dimensionless, and the parameters A, C, D, and E have units of time (milliseconds, in our case). We have written the equations in the form that appears in [29] and, in order to have a large window of alternans, we adapt their parameters as follows: $\alpha = 0.2$, $A = 88$, $C = 280$, $D = 28$, $E = 250$, and $\tau = 1{,}000$. Alternans occurs if B^* between approximately 340 and 615, as illustrated in the "bubble" bifurcation diagram of Fig. 7. In order to control alternans, we apply ETDAS replacing B^* with $\mathrm{B}^* + \epsilon_n$ as in the previous subsection. The results of ETDAS with $R = 0.3$, $\gamma = 0.4$ (selected under the guidance of our aforementioned linear stability analysis) are also shown in Fig. 7. Observe that alternans is completely suppressed throughout the window in which it had previously occurred.

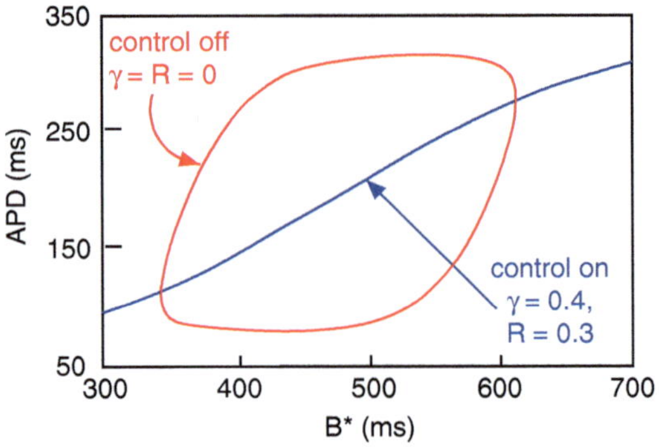

Fig. 7 Bifurcation diagram for the memory model (10) both with and without control. ETDAS ($R = 0.3$, $\gamma = 0.4$) completely suppresses alternans throughout the window $340 < B^* < 615$

4 Control in Spatially Extended Systems

As we explained in the Introduction, studies indicate that point stimulation control of alternans in one-dimensional fibers of cardiac cells succeeds only within some limited distance from the pacing electrode. In their numerical simulations, Echebarria and Karma [5] applied OGY control via point stimulation to one end of a fiber that initially exhibited alternans at every point along the fiber. For their particular choice of feedback gain and cell membrane model, they found that it was only possible to suppress alternans in cells whose distance from the stimulus electrode was on the order of 0.5–1.0 cm. In their simulations, they use a restrictive special case of ETDAS in which the parameter R is set to 0, and their simple model of the cell membrane is unable to exhibit short-term memory. Here, we recreate their numerical experiments, but using the full ETDAS method with two different models of the action potential—one without memory [16] and one with memory [7].

4.1 ETDAS Via Point Stimulation in One-Dimensional Fibers

Following Echebarria and Karma [5], we use a standard cable model of action potential propagation in one spatial dimension. Assuming that one end of the fiber is subjected to point stimulation via a pacing electrode, we will let $x \geq 0$ denote the distance along the fiber to the pacing site, measured in centimeters. The cable equation

$$\frac{\partial v}{\partial t} = D\frac{\partial^2 v}{\partial x^2} - \frac{I_{ion} + I_{stim}}{C_m} \tag{12}$$

models the transmembrane voltage $v = v(x,t)$ in each cell along the fiber. The membrane capacitance C_m is a constant that can be measured experimentally, and the diffusion coefficient D incorporates the cell surface-to-volume ratio and the intracellular resistivity. We test two different formulations of the ionic currents I_{ion}, one obtained by solving the differential equations of a memoryless two-current model [16] and another obtained from an adaptation of the Fenton-Karma model [7] appearing in the appendix of [30]. The stimulus current I_{stim} is chosen to be a periodic (period B^*) train of square-wave impulses, each of duration 1.0 ms. Stimuli are applied via a simulated electrode to cells in the 1-mm wide region $0 \leq x \leq 0.1$, and the amplitude of I_{stim} is chosen sufficiently strong to elicit a propagating action potential during each stimulus, provided that B^* is not too small.

Although we have not yet performed a comprehensive search of parameter space, we discuss our preliminary findings regarding the ability of ETDAS to suppress alternans in a fiber. Because the results were independent of whether the model exhibited memory (as in [7]) or did not (as in [16]), we report the results from the latter. Guided by the bifurcation diagrams in [16], we selected model parameters that would induce alternans in a pacing period of $B^* = 300$ ms. Then, we initiated alternans in the fiber modeled by Eq. (12) by using a stimulus current I_{stim} with period 300 ms. If the fiber is sufficiently long (we used a fiber of length 10 cm), the fiber exhibits spatially *discordant alternans* after several beats: at some distance from the pacing site $x = 0$, the cells abruptly transition from long-short APD alternation to short-long APD alternation. This behavior is illustrated in Fig. 8, which shows the last two seconds out of 40 beats of discordant alternans without ETDAS control ($\gamma = R = 0$). The left panel shows a space-time plot of wave fronts (approximately straight lines) and wave backs (curves with noticeable oscillations) of those last few action potentials. APD can be measured from the vertical gap between a wave front and the subsequent wave back, and clearly APD oscillates as a function of distance x from the pacing site. The right panel of Fig. 8 shows a graph of $(APD - A^*)$ as a function of x during the last two beats of this discordant alternans pattern. Note the presence of several "nodes" marking transition points between regions of long-short and short-long APD alternans.

In order to simulate the use of ETDAS control, we repeated the simulations that were used to generate Fig. 8 except that after 20 beats, ETDAS was implemented using $\gamma = 0.6$ and $R = 0.3$. The results are shown in Fig. 9, which parallels the previous figure. Within a distance of approximately 1 cm of the pacing site ($x = 0$), ETDAS does manage to suppress alternans. The pattern of discordant alternans resumes for $x > 1$, with fewer nodes than in the absence of control. Our results appear to confirm the limited applicability of point stimulation described in [5], and we are not optimistic that a more comprehensive exploration of $\gamma - R$ parameter space would do anything to change this.

4.2 *Point Stimulation Versus FFP in Two-Dimensional Sheets*

The simulations in Sect. 4.1 indicate that the primary finding of Echebarria and Karma [5] holds even when tissue memory is taken into account and a more robust feedback control method (ETDAS) is applied. Namely, point stimulation does a poor job of achieving whole-heart control of discordant alternans. Given

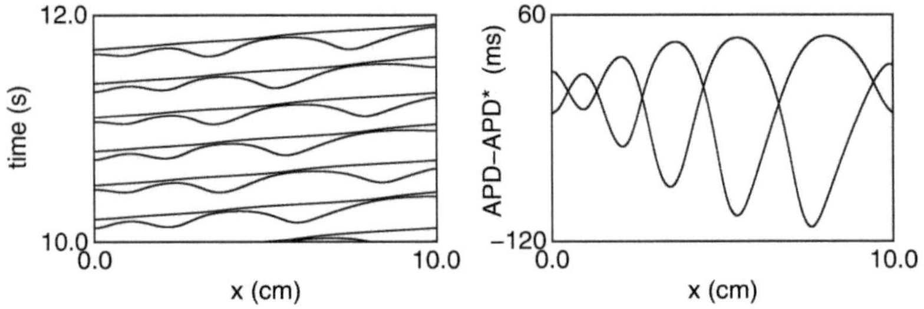

Fig. 8 *Left panel*: Space-time plot of wave fronts (*nearly straight lines*) and wave backs (*curves with noticeable oscillations*) of action potentials during spatially discordant alternans without ETDAS control ($\gamma = R = 0$). The last two seconds out of 40 beats with $B^* = 300$ ms are shown, out to a distance of 10 cm from the stimulus site. *Right panel*: Deviation of APD from A^* during the last two beats

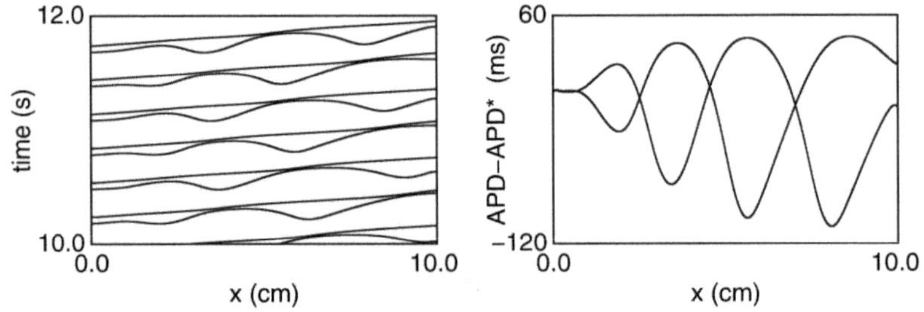

Fig. 9 Same as Fig. 8 except that after 20 beats without control ($\gamma = R = 0$), ETDAS is applied during beats 21–40 using $\gamma = 0.6$ and $R = 0.3$. Although APD alternans is terminated close to the stimulus site $x = 0$, discordant alternans persists elsewhere

the previously-reported [8] effectiveness of FFP as an alternative, we turn our attention to the problem of automating the timing of the electric field pulses. Modeling the implementation of FFP requires more care: the effects of pulsed electric field stimulation near a non-conducting obstacle are best understood if the extracellular and intracellular potentials v_i and v_e are tracked separately. (Note: The transmembrane voltage v that we considered previously is defined as $v_i - v_e$.) To model propagation of action potentials in a 2-D sheet of tissue with a non-conducting obstacle in the center, we begin with the bidomain model which appears in [8]:

$$
\begin{aligned}
C_m \frac{\partial(v_i - v_e)}{\partial t} &= \nabla D_i \nabla (v_i - v_e) - I_{ion} \\
C_m \frac{\partial(v_i - v_e)}{\partial t} &= -\nabla D_e \nabla (v_i - v_e) + I_{ion}.
\end{aligned} \tag{13}
$$

As before, C_m denotes cell membrane capacitance. The intracellular and extracellular spaces in the tissue are each equipped with their own diffusion tensors D_i and

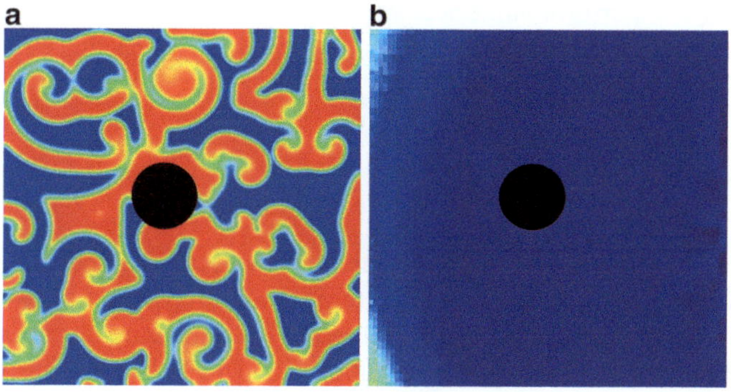

Fig. 10 Regions of high (*red*) and low (*blue*) transmembrane potential $v = v_i - v_e$ in simulated square, 2-D sheets of tissue with circular, non-conducting obstacles (*dark circles*) near the middle of the sheet. (**a**) Spatiotemporal chaos reminiscent of fibrillation. (**b**) Resetting the sheet in panel (**a**) by using FFP with stimuli timed according to ETDAS

D_e. No-flux boundary conditions are used around the outer boundary of a square domain, and non-conducting obstacles can be simulated by setting the diffusion coefficients to 0 in some region near the center of the sheet.

Previous implementations of FFP used overdrive pacing, a rapid train of electric field impulses, to reset the electrical activity when an abnormal rhythm was detected. Here, we attempted an alternative to overdrive pacing in which ETDAS is used to time the field pulses. In order to determine an "effective" pacing period B^* amidst spatiotemporal chaos, the dominant pacing frequency can be extracted by use of Fourier transform methods and a power spectral density plot. Setting B^* to be the period corresponding to the dominant frequency, we apply ETDAS with B^* as the reference pacing period.[4] The perturbations ϵ_n are chosen based upon the difference between the two previous APD values in the cells used to approximate B^*.

Figure 10 illustrates the use of FFP to reset spatiotemporally chaotic electrical activity in a sheet of tissue. In panel (a), disorganized electrical activity occurs in a sheet containing an anatomical obstacle that is non-conducting (dark circle near middle of sheet). After measuring the effective B^* as explained in the previous paragraph, a train of 20 FFP pulses was applied, timed according to the ETDAS protocol. Panel (b) shows the results: complete resetting of the electrical activity. It is unclear whether this encouraging result is robust, or whether it was a consequence of luck—many more simulations would be required to determine which is the case. Supposing optimistically that ETDAS can achieve the resetting shown in Fig. 10b for a broad range of γ and R values, there are important follow-up questions. Can ETDAS reset the tissue using the same number of stimuli that overdrive pacing would require? If so, the [lower-frequency] ETDAS protocol ought to avoid tissue fatigue associated with the [higher-frequency] overdrive pacing protocol. Does either protocol perform better if the dominant period B^* is reduced? In any case,

[4] Admittedly, this method of approximating B^* would require measurement of a long time series of APD and DI, wasting precious time in the event of a life-threatening rhythm.

it is worthwhile to perform numerical simulations of various pacing protocols (not only ETDAS) that might provide an alternative to overdrive pacing.

5 Discussion

We have surveyed some of the mathematical aspects of feedback control of abnormal cardiac rhythms, using APD alternans and fibrillation as test cases. We described two very different experimental setups for delivering electrical stimuli to terminate abnormal rhythms: point stimulation and FFP. The former is the basis for traditional implantable pacemaker devices such as the implantable cardioverter defibrillator, while the latter may someday offer a new breed of devices which reduce discomfort to the patient and have improved battery life. In our simulations of both point stimulation and FFP, we used ETDAS feedback control as way to time the application of the electrical stimuli. The relative simplicity of ETDAS makes it amenable to mathematical analysis, and the size of the control domain (Fig. 6) compared to those of its predecessors suggests that ETDAS is quite robust. It remains to be seen whether these sorts of feedback control algorithms might provide a "smart" alternative to overdrive FFP pacing.

Acknowledgements The author gratefully acknowledges the financial support of a University of Richmond Faculty Research Council Fellowship.

References

1. Abed, E. H., & Fu, J.-H. (1986). Local feedback stabilization and bifurcation control, I. Hopf bifurcation. *Systems & Control Letters, 7*, 11–17.
2. Adam, D. R., Smith, J. M., Akselrod, S., Nyberg, S., Powell, A. O., & Cohen, R. J. (1984). Fluctuations in T-wave morphology and susceptibility to ventricular fibrillation. *Journal of Electrocardiology, 17*, 209–218.
3. Berger, C. M., Cain, J. W., Socolar, J. E. S., & Gauthier, D. J. (2007). Control of electrical alternans in simulations of paced myocardium using extended time-delay autosynchronization. *Physical Review E, 76*, 041917.
4. Christini, D. J., Riccio, M. L., Culianu, C. A., Fox, J. J., Karma, A., & Gilmour, R. F. Jr. (2006). Control of electrical alternans in canine cardiac purkinje fibers. *Physical Review Letters, 96*, 104101.
5. Echebarria, B., & Karma, A. (2002). Spatiotemporal control of cardiac alternans. *Chaos, 12*, 923–930.
6. Fenton, F. H., Cherry, E. M., Hastings, H. M., & Evans, S. J. (2002). Multiple mechanisms of spiral wave breakup in a model of cardiac electrical activity. *Chaos, 12*, 852–892.
7. Fenton, F. H., & Karma, A. (1998). Vortex dynamics in three-dimensional continuous myocardium with fiber rotation: filament instability and fibrillation. *Chaos, 8*, 20–47.
8. Fenton, F. H., Luther, S., Cherry, E. M., Otani, N. F., Krinsky, V., Pumir, A., et al. (2009). Termination of atrial fibrillation using pulsed low-energy far-field stimulation. *Circulation, 120*, 467–476.

9. Garfinkel, A., Spano, M. L., Ditto, W. L., & Weiss, J. N. (1992). Controlling cardiac chaos. *Science, 257*, 1230–1235.
10. Hall, G. M., & Gauthier, D. J. (2002). Experimental control of cardiac muscle alternans. *Physical Review Letters, 88*, 198102.
11. Hodgkin, A. L., & Huxley, A. F. (1952). A quantitative description of membrane current and its application to conduction and excitation in nerve. *The Journal of Physiology, 117*, 500–544.
12. Jordan, P. N., & Christini, D. J. (2004). Adaptive diastolic interval control of cardiac action potential duration alternans. *Journal of Cardiovascular Electrophysiology, 15*, 1177–1185.
13. Kalb, S. S., Dobrovolny, H. M., Tolkacheva, E. G., Idriss, S. F., Krassowska, W., & Gauthier, D. J. (2004). The restitution portrait: a new method for investigating rate-dependent restitution. *Journal of Cardiovascular Electrophysiology, 15*, 698–709.
14. Karma, A. (1993). Spiral breakup in model equations of action potential propagation in cardiac tissue. *Physical Review Letters, 71*, 1103–1106.
15. Landry, M., Campbell, S. A., Morris, K., & Aguilar, C. O. (2005). Dynamics of an inverted pendulum with delayed feedback control. *SIAM Journal on Applied Dynamical Systems, 4*, 333–351.
16. Mitchell, C. C., & Schaeffer, D. G. (2003) A two-current model for the dynamics of cardiac membrane, *Bulletin of Mathematical Biology, 65*, 767–793.
17. Nolasco, J. B., & Dahlen, R. W. (1968). A graphic method for the study of alternation in cardiac action potentials. *Journal of Applied Physiology, 25*, 191–196.
18. Ott, E., Grebogi, C., & Yorke, J. A. (1990). Controlling chaos. *Physical Review Letters, 64*, 1196–1199.
19. Pastore, J. M., Girouard, S. D., Laurita, K. R., Akar, F. G., & Rosenbaum, S. (1999). Mechanism linking t-wave alternans to the genesis of cardiac fibrillation. *Circulation, 99*, 1385–1394.
20. Pyragas, K. (1992). Continuous control of chaos by self-controlling feedback. *Physics Letters A, 170*, 421–428.
21. Pyragas, K. (1995). Control of chaos via extended delay feedback. *Physics Letters A, 206*, 323–330.
22. Rosenbaum, D. S., Jackson, L. E., Smith, J. M., Garan, H., Ruskin, J. N., & Cohen, R. J. (1994). Electrical alternans and vulnerability to ventricular arrhythmias. *The New England Journal of Medicine, 330*, 235–241.
23. Schaeffer, D. G., Berger, C., Gauthier, D. J., Dobrovolny, H., Krassowska, W., & Zhao, X. (2007). Period-doubling bifurcation to alternans in paced cardiac tissue: crossover from smooth to border-collision characteristics. *Physical Review Letters, 99*, 058101.
24. Schaeffer, D. G., Cain, J. W., Gauthier, D. J., Kalb, S. S., Oliver, R. A., Tolkacheva, E. G., et al. (2007). An ionically based mapping model with memory for cardiac restitution. *Bulletin of Mathematical Biology, 69*, 459–482.
25. Smith, J. M., Clancy, E. A., Valeri, R., Ruskin, J. N., & Cohen, R. J. (1988). Electrical alternans and cardiac electrical instability. *Circulation, 77*, 110–121.
26. Socolar, J. E. S., Sukow, D. W., & Gauthier, D. J. (1994). Stabilizing unstable periodic orbits in fast dynamical systems. *Physical Review E, 50*, 3245–3248.
27. Socolar, J. E. S., & Gauthier, D. J. (1998). Analysis and comparison of multiple-delay schemes for controlling unstable fixed points of discrete maps. *Physical Review E, 57*, 6589–6595.
28. Tolkacheva, E. G., Romeo, M. M., & Gauthier, D. J. (2004). Control of cardiac alternans in a mapping model with memory. *Physica D, 194*, 385–391.
29. Tolkacheva, E. G., Romeo, M. M., Guerraty, M., & Gauthier, D. J. (2004). Condition for alternans and its control in two-dimensional mapping model of paced cardiac tissue. *Physical Review E, 69*, 031904.
30. Tolkacheva, E. G., Schaeffer, D. G., Gauthier, D. J., & Krassowska, W. (2003). Condition for alternans and stability of the 1:1 response pattern in a "Memory" model of paced cardiac dynamics. *Physical Review E, 67*, 031904.

Working Memory Training and Transfer: Theoretical and Practical Considerations

Susanne M. Jaeggi and Martin Buschkuehl

Abstract The study of transfer and brain plasticity is currently one of the hot topics in cognitive science. Transfer refers to performance improvements in tasks that were not part of an intervention. In this chapter, we will provide evidence for the efficacy of several working memory (WM) interventions developed in our laboratories and review the emerging literature from other groups. We will discuss data that demonstrate transfer to non-trained tasks throughout the lifespan, that is, in young adults, in older adults, in typically developing children, as well as children with Attention-Deficit Hyperactivity Disorder (ADHD). We will also briefly discuss the neural correlates that underlie improvements as a function of WM training. In addition to describing successful instances of transfer, we will also point out that transfer effects can be elusive, and that some of the effects do not seem to be easily replicated. We argue that instead of taking inconsistencies as a proof for a lack of efficacy, researchers need to develop innovative approaches to move the cognitive training literature beyond the simple question of whether or not training is effective, and to address questions of underlying mechanisms, individual differences, and training features and parameters that might mediate and moderate the efficacy of training.

S.M. Jaeggi (✉) • M. Buschkuehl
School of Education, University of California, 3452 Education, Irvine, CA 92697-5500, USA
e-mail: smjaeggi@uci.edu

M. Buschkuehl
MIND Research Institute, 111 Academy, Suite 100, Irvine, CA 92617, USA
e-mail: mbuschkuehl@mindresearch.net

B. Toni (ed.), *New Frontiers of Multidisciplinary Research in STEAM-H (Science, Technology, Engineering, Agriculture, Mathematics, and Health)*,
Springer Proceedings in Mathematics & Statistics 90, DOI 10.1007/978-3-319-07755-0_2,

1 Introduction

The study of transfer and brain plasticity is currently one of the hot topics in cognitive science but it has been an issue in educational research for many years. Transfer generally refers to the application of learning and skills in other contexts and new situations, and more specifically, in the context of cognitive interventions, transfer refers to performance improvements in tasks that were not part of the intervention. While some have argued that there is no evidence for transfer as a function of cognitive training, we and others have pointed out that certain kinds of interventions can be, indeed, effective, but that there are important boundary conditions that have to be taken into account when evaluating training success. In this chapter, we will provide evidence for the efficacy of several working memory (WM) interventions developed in our laboratories and review the emerging literature from other groups. We will discuss data that demonstrate transfer to non-trained tasks throughout the lifespan, that is, in young adults, in older adults, in typically developing children, as well as children with Attention-Deficit Hyperactivity Disorder (ADHD). We will also briefly discuss the neural correlates that underlie improvements observed as a function of WM training. In addition to describing successful instances of transfer, we will also point out that transfer effects can be elusive, and that some of the effects do not seem to be easily replicated. We argue that instead of taking inconsistencies as a proof for a lack of efficacy, researchers need to develop innovative approaches to move the cognitive training literature beyond the simple question of whether or not training is effective, and to address questions of underlying mechanisms, individual differences, and training features and parameters that might mediate and moderate the efficacy of training.

2 On Transfer

There are numerous commercial training interventions claiming to make us smarter. Some of the available interventions are increasingly used in the classroom environment with the hope of improving the users' cognitive ability and scholastic achievement. The common assumption and hope of those interventions are that the skills and knowledge acquired by playing such games and tasks will generalize and become applicable in new situations and domains, a process that is called 'transfer'. Transfer is an essential concept in the domain of education and learning because the main goal of education is to teach new generations of students to master professional and life demands, and not just to solve a specific math problem or know how to conjugate French verbs. But what do we mean by 'transfer'? Consider the following analogy—a driver might become proficient in backing up the car she is used to drive with in her narrow driveway. Now, if she borrows her neighbor's car, which is bigger, the driver may be able to apply her driving skills to the new situation. Although her parking may not be as fast or precise as it is in her own vehicle, she will probably still be successful. However, things would become very difficult if she would be asked to back up a trailer truck—although there are clearly similarities between a car and a trailer truck (steering wheel, break, gas, etc.), the trailer truck has other features which will make the transition difficult (e.g. the fact that there is a trailer that makes backing up decidedly harder for most people). Another example is if you would start to exercise on a regular basis by going running. As a consequence, you would not only improve in running, but as you are strengthening your cardiovascular system and leg muscles, you would also improve other functions that rely on a healthy cardiovascular system and stronger leg muscles, such as climbing stairs, biking, or swimming (e.g. [151]).

Although the existence of transfer in the physical domain is hardly surprising to anyone, demonstrating transfer in the cognitive domain has been difficult, and for over 100 years, arguments have been made about whether transfer exists or not [40, 113, 126, 128, 161]. Nonetheless, as Perkins and Salomon [117] have pointed out, any learning involves transfer in at least a trivial sense: there is no such thing as learning if there is no demonstration of the learning outcome in a different context, even if the context is very similar. The main question is thus how to distinguish between trivial transfer and transfer in which there is a meaningful generalization effect. Usually, researchers conceptually divide transfer into categories of “near” and “far” [128, 142, 170]. Near transfer refers to an effect of the trained task on a non-trained task that is closely related to it; far transfer refers to an effect of the trained task on a non-trained task that is quite different, perhaps sharing very few features (applying the concept to the car driving analogy, near transfer would refer to parking your car vs. your neighbor’s car, and far transfer would refer to parking your car vs. parking a trailer truck). Unfortunately, there is neither a formal definition nor an operational method to measure the distance of transfer, although there are some attempts to do so [8]. Nevertheless, it may be most useful to understand near and far transfer effects as two points on a continuum and to use the distinction as a descriptive means to get at an intervention’s impact [173].

Why should transfer occur in the first place? We and others have argued that transfer depends on the degree of process-overlap between the training task and the outcome measures—the more similar processes there are between the tasks, the higher the chances for transfer, which also relates to the argument of near and far transfer above [34, 71, 100]. Those overlaps can occur neurally in the form of shared brain areas or networks between training and outcome measures. Cognitively, such overlaps can occur in the form of common processing demands (e.g. attentional control), similar strategies that can be applied in the training and the transfer tasks (e.g. chunking), or an acquired “mindset” during training which facilitates transfer (e.g. increased self-confidence). Of course, those overlaps are not easily disentangled, and transfer can occur due through either one, or through a combination of different mechanisms [128].

3 Brain Training and Transfer: The Case of Working Memory

In this chapter, we will not focus on education in the broad sense as a means to investigate transfer, but rather, on a relatively narrow set of interventions that aim to improve certain specific cognitive skills over a relatively short timeframe. Such interventions are often referred to as 'brain training' tools, and there is a growing demand and market for such products (cf. [139]). Unfortunately, the scientific evidence demonstrating the efficacy of commercial interventions is rather sparse in that the effects (if they are assessed at all) rarely go beyond tasks that were specifically trained (cf. [63, 104] for recent meta-analyses). Nevertheless, there is accumulating evidence that certain cognitive interventions may indeed be effective. For example, there are a number of studies that demonstrated improvements in non-trained cognitive tasks after some form of WM, executive function, or attention training (see e.g. [20, 41, 70] for recent reviews). Not surprisingly, performance improvements are most often observed in tasks that are quite closely related to the trained task. For example, interventions designed to target WM skills typically result in improvements in non-trained measures of WM, i.e. they show near transfer effects ([98, 114], e.g. [66, 96, 127]). Nonetheless, there is also work demonstrating evidence for far transfer, for example, there is an accumulating number of studies reporting improvements in measures of fluid intelligence (Gf) after training on WM and related skills (see e.g. [77] for a recent overview). The concept of Gf has been introduced by Cattell [25] in that he described Gf as the ability to reason and to solve novel, abstract problems without relying on previously acquired skills or knowledge. Gf is contrasted with crystallized intelligence (Gc), i.e. the ability to use skills, knowledge, and experience. It has been argued that Gf facilitates learning in a general sense. Indeed, there is a lot of empirical evidence showing that Gf is the most reliable predictor for achievement, that is, individual differences in Gf predict successful performance in educational and professional settings (e.g. [38, 53]). As such, developing means to improve Gf is of particular relevance. Of course, improvements in Gf tasks can be easily obtained by practicing the Gf test themselves; however, such effects are highly specific and the tests lose their predictive value for other tasks [158], and furthermore, such improvements would not be considered as transfer but practice effects.

Researchers have used a wide variety of measures to assess improvements in Gf as a function of cognitive training, but most commonly, they have used visuospatial matrix reasoning measures such as Raven's progressive matrices [121]. The reason for this interest in matrices tests is because they are seen as being the most representative of Spearman's g [143], that is, a global measure of cognitive ability [55]. Nonetheless, other measures have been used in WM training studies as well, for example more verbal measures, such as analogies or inferences [76]. Furthermore, WM training research has begun to use multiple measures to assess Gf as a composite measure in order to reduce task-specific variance (e.g. [32, 76, 123, 131, 146]). Although the number of studies finding improvements in measures of Gf is still relatively small, a pattern seems to emerge revealing larger effect sizes in the visuospatial domain as compared to the verbal domain [22, 32, 77, 146]. This dissociation might have emerged as a function of the tests used (along with their psychometric properties), by the participants' familiarity with the material [77], or more generally by the fact that one domain might be more malleable to change than another (e.g. [86]). At the present stage of research, however, further work is needed to clarify this issue.

Apart from improvements in Gf, researchers have also observed transfer to basic attentional skills and visual processes [56, 57], language-related skills [29, 46, 99, 111], arithmetic and numeracy skills [94, 172], measures of academic achievement [63, 138], or even to self-regulatory behavior such as ADHD symptoms or drinking behavior in alcoholics and delay discounting in stimulant addicts [10, 14, 66, 93]. Thus, at the current stage of research, it seems like interventions that target skills related to WM and attentional control can be effective tools to improve higher cognitive skills—why might that be?

WM is the cognitive mechanism that supports active maintenance of task-relevant information during the performance of a cognitive task [6]. WM underlies the performance of virtually all complex cognitive activities [136]. Imagine yourself mentally calculating the 18 % tip for your dinner, participating in a conversation with your parents while simultaneously texting to your friend, or reading a complex paragraph in your History textbook. All of these tasks rely on deliberate WM processes in that they require multiple processing steps and temporary storage of intermediate results, going back and forth between different tasks, as well as resisting distracting information. People differ in how much information they can hold in WM, and how well they can maintain that information in the face of distraction [43, 84]. These individual differences predict how well individuals perform in school-relevant tasks such as mathematics and reading comprehension (e.g. [37, 49, 116]). WM capacity is also crucial for our ability to acquire new knowledge and skills [118]. Research has shown that WM is a better predictor of scholastic achievement than intelligence, especially in young children [2]. Deficits in WM are considered a primary source of cognitive impairment in numerous special-needs populations ranging from ADHD to mathematics disability [105]. WM also has significant effects on classroom behavior. For example, teachers are more likely to rate children with poor WM capacity as disruptive and inattentive [3, 48].

In sum, WM is a fundamental cognitive system that is highly relevant for success in and out of schools. Given the relevance of WM to daily life and educational settings, it is not surprising that many cognitive interventions target WM skills with the ultimate goal to obtain transfer in relevant areas such as scholastic achievement. Referring back to the analogy in the physical domain described earlier, we can characterize WM as taking the place of the cardiovascular system that underlies performance of many different activities. That is, we see WM as an underlying entity that determines the performance of a multitude of tasks, and thus, strengthening WM skills should lead to performance improvements in tasks that rely on the functioning of the WM system [85].

This idea of strengthening underlying processes to improve general performance is not new at all. Indeed, over 120 years ago, William James proposed that improving attention could have high practical importance by stating that[1]

> "An education which would improve attention would be the education par excellence" (James, 1890, The Principles of Psychology, Vol. 1, p. 424).

James explained the importance of strengthening attentional skills by referring to the crucial role of attentional control for human performance:

> "(. . .) the faculty of voluntarily bringing back a wandering attention, over and over again, is the very root of judgment, character, and will. (. . .)" (p. 424).

He also pointed out some of the major practical difficulties that accompany the design of interventions, and he suggested that the most promising approach would be to somehow capture a person's interest and motivation, one of the critical features of programs that aim to keep participants training for longer than just one or two sessions [74].

Klingberg and colleagues were among the first to use a WM training based on the premises outlined above [92, 93; but see 1 for a very early and pioneering example]. They developed an intervention that consisted of a battery of computerized tasks targeting mainly WM processes. Those tasks were embedded into an interesting videogame environment to make the intervention engaging and motivating. Another critical feature was that their tasks were adaptive. That is, the tasks became incrementally harder as the participants improved, and rewards were provided based on performance. The authors targeted children with ADHD as training population because WM deficits are often among the core symptoms in ADHD [169]. The authors' rationale was that training WM should reduce ADHD symptoms, and in addition, yield transfer to other tasks that rely on WM. Indeed, both studies demonstrated that a 5-week intervention resulted in reduced ADHD symptoms, as well as in transfer effects to non-trained variants of the trained tasks, in a measure of executive control (the Stroop task), and, finally, to Ravens' matrix reasoning, a common proxy for Gf. Since then, this intervention has been further developed and used by other researchers, and it is currently marketed under the name 'Cogmed' (most recently distributed by Pearson). Unfortunately, although near transfer effects on non-trained measures of WM are consistently observed using this particular intervention, the far transfer effects do not seem to be easily replicated, neither by the Klingberg group [13, 160], nor by other groups using the same program [18, 64, 65]. Nonetheless, there is a recent study by an independent group that replicated the improvement on parent-rated ADHD symptoms [10], the improvements on Gf [124], and another recent study even reported improvements in scholastic achievement measures in an applied school setting [63]. Taken together,

[1]Note that by referring to James' quote it is not our intention to equate WM with attention without a proper and detailed discussion of the matter. Instead we want to emphasize the idea entailed in the quote that training underlying processes is likely to affect not only the trained process but all other cognitive functions that rely at least in part on those functions.

it seems that the training regimen developed by Klingberg and colleagues does have important benefits, even though those benefits are most consistently expressed as near transfer effects (see also [104]). From an applied point of view, however, near transfer effects can be very useful given the importance of WM for scholastic achievement.

In terms of ameliorating ADHD symptoms, one of the major goals of the intervention program, a recent review classified Cogmed as a 'Possibly Efficacious Treatment for youth with ADHD' using established evidence-based treatment criteria as proposed by the Society for Clinical Child and Adolescent Psychology [28]. Nonetheless, despite the promising effects, the intervention remains controversial as documented in a recent special issue on the topic that appeared in the Journal of Applied Research in Memory and Cognition (Volume 1; see e.g. [140]). Although we agree that the evidence for Cogmed's efficacy to date is mixed, we have argued that it is probably too early for a final verdict, especially since there are over 60 ongoing studies using the program [76]. Furthermore, the few studies that have been published have populations that are rather diverse and hardly comparable across studies (e.g. ranging from typically developing children to stroke patients), and often, the outcome measures were not comparable either. Thus, we have emphasized that it is likely that those factors along with individual differences could account for the mixed effects observed to date [137]. On another level, the intervention has also been criticized as 'kitchen sink approach' as it contains many different tasks, which might or might not contribute to transfer, and as such, it is not possible to get at the underlying mechanisms of transfer. Such a criticism also applies to other interventions that rely on a diverse battery of tasks ([131], e.g. [90]). This critique is certainly appropriate from an experimental standpoint, however, such an approach has merits from a practical point of view since it is certainly more interesting for participants to train on a diverse battery of task rather than repeating the same task over and over again [69, 76]. Thus, if researchers are interested in generalized improvements and do not necessarily care *why* the improvements occur, it might be better to rely on the combination of multiple components, hoping that one or more of them will be successful.

On the other hand, other research groups have taken a different approach by relying on a more narrow set of tasks, for example on so-called WM capacity tasks such as reading span tasks [16, 22, 29, 99, 164]. Those interventions are usually adaptive as well in that they adjust to the participants' performance. All of those interventions consistently observed transfer within the trained WM domain. In addition, some report far transfer effects, e.g. to reading-related processes [29, 99], or to measures of intelligence [16, 166]. However, improvements in fluid intelligence are not consistently observed [29, 99], and it has been argued that such span-type interventions might be restricted in their generalizing ability.

Although we have used span-type interventions in some of our studies as well [22, 99], more often we have been taking yet another approach by relying on an n-back task as the intervention vehicle. In this task, participants are required to process a continuous stream of stimuli (e.g. letters, shapes, or locations; presented in 3 s intervals) and decide for each stimulus whether or not it matches the one that was presented *n* items previously. For example, if participants are asked to do a 2-back task, the following letter stream contains two targets: L-K-P-**K**-F-R-K-**R**-R (i.e., the second K and the second R; highlighted in bold for illustration purposes). This task has been widely used in the neuroimaging literature to investigate the underlying mechanisms of WM load (cf. [112] for a meta-analysis). For our interventions, we have made the task adaptive in that its difficulty (i.e. load) varies from one block of trials to another by changing the level of *n* [83]. That is, adjustments are made continuously based on the trainee's performance: As performance improves, the level of *n* will be increased in the next block (for example, from a 2-back to a 3-back); as it worsens, the level of *n* will be decreased in the next block (for example, from a 2-back to a 1-back). As such, the task always remains demanding and tailored to individual performance (cf. [71] for a first description of the intervention). We have argued that some of the task's features are highly relevant for the concept of training, that is, this task involves *multiple* WM processes, such as storage, but also interference resolution, attentional control, as well as sustained attention [83, 152]. We found the n-back task to be promising as a training vehicle because n-back performance reliably predicts Gf and measures of executive control [55, 72, 73, 87], and furthermore, it is highly predictive for academic achievement and teacher-reported behavioral problems such as impulsivity and hyperactivity (e.g. [5]).

To date, we have used our adaptive n-back intervention in multiple studies with young adults, and we observed improvements in various matrix reasoning tasks that are strongly related to Gf [71, 73, 77]. We have also shown that the longer participants train on the task, the more improvements they show in Gf, that is, we have demonstrated a dose–response effect of training. Recently, we have replicated this finding in a sample of healthy older adults [146]. In addition, several other research groups have successfully replicated transfer to measures of Gf using the n-back task as training vehicle [32, 78, 120, 125, 134, 147, 155], and others have observed performance improvements in other domains as well, such as executive control and WM capacity [4, 96, 98, 127].

More recently, we adapted an adult version of the n-back task for children and created a video game-like context by incorporating features garnered from the video-game literature such as points, high scores, and appealing graphics and themes [50, 101, 119, 144]. We found that this video-game-like intervention led to improvements in non-trained measures of WM, but also in measures of sustained attention and inhibition in typically developing children and children with ADHD [75]. In addition, we also found transfer to matrix reasoning tasks, but critically, only in children who showed considerable gains in the training task [74]. Similar patterns have also been observed by Zhao and colleagues who trained typically developing children on a WM updating task related to n-back, that is, they demonstrated improvements in Gf, which correlated with the improvement in the training task [175]. Finally, another group has demonstrated a relationship between training performance and outcome in the domain of language skills in young adults [111], in sum, training *quality* seems to be an important feature to determine training success.

As for the interventions discussed further above, this work is not without controversy either (see e.g. [141]). For example, there are a few studies from other research groups that fail to show transfer to Gf after n-back training. Notably, there are two studies that fail to find improvements in Gf, however, they observe transfer in other non-trained measures, and the failure to find group differences in Gf could be attributed to either the selection of the control task, or the fact that the intervention time was too short ([96, 127]; cf. [77] for further discussion). But there are also studies that did not observe transfer in *any* of their outcome measures [30, 123, 159]. Those studies are difficult to interpret since there are of course many reasons that could give rise to null-effects, such as sample size, population differences, the selection of outcome measures, measurement issues (e.g. lack of reliability in the outcome measures), lack of training quality (i.e. lack of training improvement), individual differences, motivational or other issues (cf. [77] for further discussion). Nonetheless, such null-effects can be informative to further investigate important boundary conditions and to get at the underlying mechanisms of training and transfer; issues that we will address in the next sections.

4 Why is There Transfer? In Search of Underlying Mechanisms

Despite the accumulating evidence that there are generalizing effects after WM training, we only have a very vague idea *why* transfer effects occur. Thus, to date we can only speculate about the possible underlying mechanisms of training and transfer. As Chein and Morrison pointed out [29], there are many reasons why participants could perform better after training, such as changes in strategies, improved executive control, speed of processing, pre-existing individual differences and motivational factors, or simply familiarity with the stimuli and improved test-taking skills. What makes it even more difficult for research is the fact that transfer could also occur by a combination of those factors. Nonetheless, the question is whether we can derive some general principles from the existing literature that might shed some light on the underlying mechanisms. In the following, we will separately describe potential cognitive and neural mechanisms, although we acknowledge that the two domains are certainly intertwined.

5 Cognitive Mechanisms

We and others have argued that in order for transfer to occur, one important mechanism might be that training and transfer tasks need to share a common processing basis [71, 82]. What could be the common cognitive mechanism that could drive transfer effects from n-back training to such diverse tasks such as executive control and reasoning? One prominent feature of the n-back task is that participants have to resolve interference in that they frequently encounter so-called "lure" trials. For example, a stimulus that appeared three or one items back during a 2-back task is considered a lure. Going back to our earlier example of a 2-back task, L-K-P-**K**-F-R-*K*-**R**-*R*, the third K as well as the third R are lure trials (indicated in italics for illustration purposes). That is, those are items that were presented three and/or one positions back in the sequence rather than the required two back, and thus, they promote a sense of familiarity that participants have to suppress.

Due to the restricted set of unique stimuli that we have been using in our task versions (6 or 8), lure trials are a frequent occurrence, and thus, the participants are required to resolve interference and resist distraction while doing the task. It is conceivable that the participants' ability to resolve interference is strengthened by our form of training, which in turn, might be responsible for the transfer to other domains that require interference resolution, such as matrix reasoning. Indeed, in matrix reasoning tasks, such as Ravens, participants have to discriminate between target patterns and patterns that are quite similar but are missing one or two important components of the correct solution. In addition to pattern discrimination, the Ravens task also requires participants to discriminate between current rules and rules that are no longer relevant. Thus, as in the n-back task, performing well in Ravens requires resisting distraction and interference resolution [36, 168]. This hypothesis is further strengthened by previous findings showing that individual differences in Gf are predicted by lure interference in n-back tasks [54, 87]. More direct evidence for such a model comes from recent intervention work showing that training on an n-back task with a controlled (and high) number of lure trials predicted performance on a reading task which explicitly required interference resolution (i.e. decoding garden path sentences), a result which was not present in a group that trained on an n-back task without lures [111]. Another issue that might drive the generalization effect is the ability for sustained attention and response inhibition, both of which are presumably involved in successful n-back performance. Consistent with this notion, we have repeatedly found robust training-related improvements in n-back lure trials, in addition to sustained attention and response inhibition tasks such as the continuous performance test (CPT) in both typically developing children as well as children with ADHD [75].

Of course, there might be other underlying cognitive mechanisms that drive transfer from n-back to higher cognitive functions. The two potential mechanisms might not be the only mechanisms driving transfer, but they are the ones that we are currently exploring in our ongoing research. Additional research from other groups will hopefully contribute to shed more light on the underlying cognitive mechanisms of transfer and ultimately provide models that can be used to further refine the existing interventions.

6 Neural Mechanisms

Another path to investigate underlying mechanisms of WM training is provided by the field of neuroscience using various methods. It is currently assumed that chances for transfer increase if the training task activates identical or at least comparable brain areas as the transfer task, which is a similar assumption as the one discussed in the previous section concerning cognitive mechanisms. And indeed, it has been demonstrated that transfer occurs if the training and the transfer task engage overlapping brain regions, but not if the training and transfer task engage different brain regions [34].

In terms of quantitative brain activation changes as a result of WM training, there are currently different hypotheses concerning the direction of the effects (cf. [21]). For example, it is conceivable that the same brain areas are active before and after training, but as a result of the intervention, there is *less* activation in these areas after training which suggests an increase in neural efficiency. Another possible outcome is that the same brain areas are active as well, but now there is *more* activation after training, suggesting that the brain cells are now working harder. A third possibility is a combination of these two potential effects, i.e., a simultaneous increase and decrease of activations, which could vary by brain region. Such an outcome could reflect practice-related changes in cortical representations in task-related areas resulting in an activation increase in those areas, whereas activation decreases in other brain areas that serve more general processes such as attentional control could reflect more automatic and more efficient processing. Another and last potential outcome is that as a result of training, old and new brain areas are active, suggesting that the training induced new ways to deal with a task, for example by developing new task-related strategies. An excellent discussion of these hypothesized effects can be found in Kelly et al. [88].

To date, there are only a handful of published studies that examined activation changes as a result of n-back training [23, 60, 132, 133, 135]. These studies seem to provide converging evidence in two ways. First, n-back training leads to activation changes mainly in prefrontal and also parietal brain regions (especially right Brodmann areas 40, 6 and 9); regions that are assumed as being part of the WM network as well as in reasoning and attentional control. Second, activations seem to increase in the beginning of the training, i.e., when the training task is still fairly new to participants (for example, in the first 1–2 weeks of training), but they decrease with prolonged training (for example in the final 3–4 weeks of training). This pattern suggests that at the beginning of training, the brain has to work harder to cope with the task demands, but with increased time on task, the neural processing becomes more efficient [23].

Besides investigating activation changes with fMRI methods, researchers have also been investigating other neural correlates of cognitive training, such as changes in functional connectivity (e.g. [96]), volume changes in gray matter (e.g. [154]), changes in fiber tracts via diffusion tensor imaging (DTI) (e.g. [153]), changes in dopaminergic functions (e.g. [103]), or even the effects of certain genotypes or polymorphisms on training outcome (e.g. [12, 19]).

Despite the emerging literature, there are still relatively few studies available that use neuroscience methods to get at the underlying mechanisms of WM training. Additionally, the training and transfer tasks used in these studies vary considerably, and therefore, the current result patterns are still rather inconclusive (cf. [21, 68, 80] for recent reviews). Further research is clearly needed to elucidate the neural correlates of training, but we see neuroimaging as an invaluable approach to deepen our understanding of the underlying mechanisms of WM training and transfer.

7 How Can We Make Training Effective? Issues for Future Research

What are the critical conditions that have to be met in order for training to be successful? Apart from further investigating the underlying mechanisms of training and transfer, there are other factors that will deserve the attention of future research. We have already discussed the importance of targeting WM and related processes as underlying mechanism for complex cognition, and further, we have outlined the beneficial effects of targeting multiple processes during training as it will increase the chances for process and neural network overlap, and as such, the chances for transfer. But there are also other factors that might be important for transfer to occur, and we will outline a few of those factors in the following.

7.1 Minimizing Strategy Use and Maximizing Variability

As outlined above, our intervention approach can be described as 'process-specific', that is, rather than improving a strategy or practicing a specific task to perfection, it has been our aim to improve the underlying processing system, and in particular, WM skills [100]. That is, we have argued that in order to obtain transfer, the intervention should minimize the development of explicit strategies and skills that are specific to the task in question. Indeed, it has been shown that strategy training usually only leads to very narrow transfer ([44, 108, 109], but see [24, 145]).

A related principle is that there should be variability during training so that individuals may develop more flexible ways to approach the task in contrast to developing strategies that are only applicable to one training task [52, 130]. For example, participants could be exposed to different tasks in different contexts in order to maximize transfer. This principle may account for some of the success of intervention studies that rely on batteries of tasks as training interventions (such as [13, 93, 102, 131, 160]), however, as discussed above, this kitchen-sink approach is not ideal from an experimental point of view.

Another approach to induce task variability within the same task is to incorporate various difficulty levels [71, 73, 77], as well as varying material and contexts within the same training task [74]. This can be achieved by implementing an adaptive training method that adjusts the training difficulty to the performance of each subject (see [156] for a pioneering study). This principle adds to the motivational features of the task by keeping it constantly challenging across the entire intervention period. The balance of task engagement and challenge of this principle may be important for training success. That is, the goal of our adaptive procedure has been to make sure that the task is not too easy for participants in order to avoid repeated practice and automaticity, which would trigger the development of specific strategies, and further, it has been our aim to prevent participants from becoming bored with the task. But on the other hand, we make sure that the task is not becoming too difficult either, in order to prevent that participants are overwhelmed and become discouraged and lose motivation to train. Indeed, studies that have not used adaptive training programs failed to show transfer (cf. [92, 93]—control groups, [33, 97]).

7.2 *Distribution of Training*

Another open question concerns the optimal scheduling and duration of training. We and others have shown that there is a dose–response effect of training in which larger transfer effects occur with longer training time [9, 35, 71, 146]. Interventions that last about a month are most frequently used, although there are shorter WM interventions that have proved to be beneficial as well (e.g. [16, 99, 176]). Overall, the optimal duration and spacing of a successful intervention is still largely unknown, and to our knowledge, there are no studies to date that have investigated the role of spacing and frequency of WM training, although the role of spacing has been extensively investigated in the domain of skill acquisition and learning (see e.g. [27, 27], for reviews).

7.3 *Motivation*

It makes intuitive sense that motivation should play an important role in any kind of training. For example, let us consider we wanted to get in better cardiovascular shape. In order to substantially improve our fitness level, it is not enough to just walk, we actually have to run. We believe that the same principle applies to cognitive training as well. Thus, if you do not want to engage and get better, you will likely not improve as much as someone who puts a lot of effort into training. As we discussed before, both training quantity and quality are important. Only children who improve in the training task demonstrate transfer [74], and transfer increases with increasing training time [9, 35, 71, 146]. Therefore, intrinsic motivation and persistence is necessary in order to achieve a high quality of training over a long period of time. One of the challenges for intervention developers is to design the tasks so that the participants remain interested and motivated to stay engaged for more than one session. However, the development of motivational features is not an easy endeavor, as we want to avoid participants to be motivated only by extrinsic factors, which ultimately, is detrimental for performance [39, 77].

In a similar vein, self-efficacy beliefs and beliefs in the malleability of intelligence seem important for training success [77]. Previous research found that individuals who believe that intelligence is fixed are more likely to disengage and withdraw from tasks that are perceived as being too challenging. In contrast, individuals with a malleable mindset about intelligence are more likely to pull through challenging tasks (e.g. [15], see also [42]).

7.4 *The Role of Age and Individual Differences*

Transfer effects following WM training have been observed over a wide population range, from typically developing preschoolers (e.g. [13, 160]), school-aged children (e.g. [74, 99, 175]), to young adults (e.g. [29, 71, 73, 77, 147]), and older adults (e.g. [16, 22, 34, 97, 131, 146, 176]). Further, there is evidence that WM training is also effective in special-needs populations with pre-existing WM deficits, such as ADHD (e.g. [65, 93]), learning disabilities (e.g. [64, 124, 164]), Cochlear Implant users [95], and Schizophrenia [102]. The question is whether there is a particular population for whom the training might work best (cf. [20, 74]). For example, it seems harder to demonstrate transfer in older than young adults, and furthermore, there are even differences within old age in that the effect sizes decrease as a function of age [17, 34, 97, 131, 176]. Age-related limitations in plasticity might be a restricting factor for training and transfer. Consequently, it might be that transfer is more likely in younger adults and children [47], and also for those participants who are still highly functioning (cf. [165]). However, the successful training studies with special needs populations and children with WM deficits have thus far suggested otherwise, and further, our own and other groups' research has shown that it is usually those individuals who start off with the lowest scores who profit the most, presumably because they have more room to improve [71, 74, 146, 163, 176]. Those findings are of particular interest when it comes to the application of this line of work in older adults. What we have to consider is that even though there is reduced plasticity in old age, it does not mean that the brain and cognition are not malleable after a certain age. As has been shown in various studies now, there is ample evidence that it is still possible to improve cognition in old age, and that those improvements are maintained over several years, and a lot of this evidence comes from the ACTIVE study ([171], e.g. [81]). Nonetheless, the differential training and transfer effects that are observed across age groups can serve as a model to study developmental trajectories and further inform the design of targeted interventions that can be modified to reach specific age groups [68].

In addition to age and pre-existing ability, there might be other factors that drive training success, such as personality, need for cognition, and beliefs in the malleability of intelligence (e.g. [77, 150]). To conclude, the issue of individual differences and training has been largely overlooked until very recently [137].

8 How Broad Is the Transfer?

Unfortunately, to date, there is minimal evidence that WM training extends beyond laboratory tasks to direct measures of scholastic achievement or real-world outcomes. Nonetheless, there are notable exceptions in important clinical domains related to executive control, such as symptom reductions in ADHD, alcohol abuse, or psychosocial functioning in schizophrenia (e.g. [10, 66, 92, 93, 102, 124]).

Furthermore, there is evidence for improved reading skills in typically developing children and adults [29, 46, 99, 111], and there are studies that have demonstrated improvements in scholastic achievement after training on WM or attentional skills involving executive control [63, 89, 138, 172], or, finally, there are reports demonstrating improvements in daily living activities in older adults [122, 171]. Nonetheless, such reports are still rare, and future research will have to further determine the real-life applications of brain training. Translating WM training from the laboratory into the real world, for example by bringing it into the classroom and by assessing its efficacy with measures of scholastic achievement will be a challenging undertaking: The application in classroom settings will come with unique problems and will certainly have an impact on training quality and fidelity, which are among the key issues for transfer to occur [77]. But as others have shown, this endeavor is not impossible [63].

9 How Long Do the Effects Last?

Unfortunately, we still do not know whether transfer effects last beyond the training period, and if so, for how long. Only a handful of studies have tested the long-term effects of training by re-testing the experimental and control groups several months or even years after training completion [16, 22, 74, 93, 122, 164, 171, 176]. The few studies that have looked into this issue provide encouraging evidence that some of the effects are long lasting. More difficult to interpret are a set of studies that found transfer effects only at follow-up several months after training completion while there were no effects at post-test right at the end of the intervention [64, 164]. The mechanisms of such effects have been described as "sleeper effects", although it is not clear how they arise and further research in needed to elucidate this issue. It has been argued that transfer might be maintained or even increased by cascading effects, for example by improved self-efficacy beliefs, which are then applicable in various situations [59, 91]. On the other hand, if we assume that WM training processes are comparable to processes that occur with cardiovascular training, the longevity of the effects will probably be limited: If you stop running on a regular basis, your fitness levels will dissipate quicker than one might hope. Therefore, a potential approach to maximize long-term retention is to include booster sessions after training completion (e.g. [7, 11, 26, 58, 167]). However, future research will have to determine the frequency and duration of such booster session in order to maximize retention effects.

10 Could There Be Negative Effects of Training?

Given the many positive effects that might accompany WM training, there is a legitimate question of whether there are any downsides of training WM skills as well. One could argue that WM training might take away precious time from other important activities, such as physical exercise, socializing with your friends, or practicing your musical instrument. And of course, it will be certainly more efficient to just sit down and study your multiplication table or study for the LSAT instead of training WM, and we are by no means suggesting that n-back training should *replace* any of those approaches. But on the other hand, if WM training is used *in addition* to those activities, it might be indeed a valuable endeavor to facilitate learning, and in our various approaches, our daily training time has been limited to as little as 10 min a day [74]; hardly a significant time investment even if it would turn out not to be working for an individual, which is of course always a possibility. Another downside might be the cost of some of the marketed products, that is, training WM might come with a significant financial investment. Nonetheless, there are many free or very affordable alternatives available either online or as applications.[2]

Other potential downsides have been suggested as well, for example, whether improving cognitive control could potentially reduce performance of other functions that require *less* cognitive control, such as creativity or early language development [68]. As discussed above, there are developmental periods that might have to be taken into consideration as which might influence training efficacy, however, at the current stage of research, very little is known about those issues, and so far, we are not aware of any detrimental effects of WM training on cognition.

11 Conclusion

In sum, current research indicates that there is good reason to conclude that training WM skills can be beneficial, not only to improve WM skills themselves, but also to improve skills that rely on the integrity of WM functions, such as attentional control, language-related abilities, Gf, or scholastic achievement. Nonetheless, there are many open questions when it comes to the underlying mechanisms of transfer, as well as the extent of transfer and the longevity of the effects. Furthermore, there have been some concerns regarding the effect sizes and replicability of far transfer effects. Thus, one of the foremost goals of future research should be to shed light on those issues by systematically exploring the underlying mechanisms and determining the variables that make an intervention most effective, as well as disentangling the mediators (*why* participants benefit) and moderators (*who* might benefit) of training and transfer. Furthermore, it is still an open question to what extent WM training affects measures of academic achievement and daily life.

It is important to note that we are not suggesting that there is anything "magical" about WM training, that is, it requires hard work and engagement from both, participants and researchers in order to be effective. To reiterate our analogy

[2]Note that we are neither supporting nor endorsing any of the marketed products. The software that we developed in our laboratory and that is described in published articles is freely available for research purposes.

from physical training, it is certainly not enough to leisurely stroll to improve cardiovascular fitness, but rather, you have to run and challenge yourself. We think that the same is true in the domain of WM and general cognitive function.

Finally, we would like to emphasize that we do not want to imply that WM training is the only approach to improve cognition. WM training has been serving as our model to explore near and far transfer effects, as well as to determine the relationship between WM and higher cognitive function. That said, there are certainly other approaches that are just as valuable. While education seems to be by far the most effective approach to improve cognitive ability (cf. [69]), there are other interventions that might serve as supplement to boost and/or maintain cognitive function, either separately, or in combination with other approaches (cf. also [41]). Examples for such approaches are cognitive enrichment and stimulation [45, 61, 106, 110, 115, 148, 162], musical training [107, 129], physical exercise [31], mediation [79, 157], social interaction [174], but also nutrition [51], or pharmacological interventions [149]. To conclude, WM training is one of many approaches that could be used to improve cognitive function. There is certainly no one-size-fits-it-all approach in the domain of cognitive improvement, and future research will hopefully shed more light on what interventions work best for which individual, and under which particular circumstances.

References

1. Aiken, C. (1895). Methods of mind-training: concentrated attention and memory. New York: *American Book Company.*
2. Alloway, T. P., & Alloway, R. G. (2010). Investigating the predictive roles of working memory and IQ in academic attainment. *Journal of Experimental Child Psychology, 106*(1), 20–29.
3. Alloway, T. P., et al. (2009). The cognitive and behavioral characteristics of children with low working memory. *Child Development, 80*(2), 606–621.
4. Anguera, J. A., et al. (2012). The effects of working memory resource depletion and training on sensorimotor adaptation. *Behavioural Brain Research, 228*(1), 107–115.
5. Aronen, E. T., et al. (2005). Working memory, psychiatric symptoms, and academic performance at school. *Neurobiology of Learning and Memory, 83*(1), 33–42.
6. Baddeley, A. (1992). Working memory. *Science, 255*(5044), 556–559.
7. Ball, K., et al. (2002). Effects of cognitive training interventions with older adults: A randomized controlled trial. *JAMA, 288*(18), 2271–2281.
8. Barnett, S. M., & Ceci, S. J. (2002). When and where do we apply what we learn? A taxonomy for far transfer. *Psychological Bulletin, 128*(4), 612–637.
9. Basak, C., et al. (2008). Can training in a real-time strategy video game attenuate cognitive decline in older adults? *Psychology and Aging, 23*(4), 765–777.
10. Beck, S. J., et al. (2010). A controlled trial of working memory training for children and adolescents with ADHD. *Journal of Clinical Child and Adolescent Psychology, 39*(6), 825–836.
11. Bell, M., Bryson, G., & Wexler, B. E. (2003). Cognitive remediation of working memory deficits: Durability of training effects in severely impaired and less severely impaired schizophrenia. *Acta Psychiatrica Scandinavica, 108*(2), 101–109.
12. Bellander, M., et al. (2011). Preliminary evidence that allelic variation in the LMX1A gene influences training-related working memory improvement. *Neuropsychologia, 49*(7), 1938–1942.

13. Bergman Nutley, S., et al. (2011). Gains in fluid intelligence after training non-verbal reasoning in 4-year-old children: A controlled, randomized study. *Developmental Science, 14*(3), 591–601.
14. Bickel, W. K., et al. (2011). Remember the future: Working memory training decreases delay discounting among stimulant addicts. *Biological Psychiatry, 69*(3), 260–265.
15. Blackwell, L. S., Trzesniewski, K. H., & Dweck, C. S. (2007). Implicit theories of intelligence predict achievement across an adolescent transition: A longitudinal study and an intervention. *Child Development, 78*(1), 246–263.
16. Borella, E., et al. (2010). Working memory training in older adults evidence of transfer and maintenance effects. *Psychology and Aging, 25*(4), 767–778.
17. Borella, E., et al. (2013). Working memory training in old age: An examination of transfer and maintenance effects. *Archives of Clinical Neuropsychology, 28*(4), 331–347.
18. Brehmer, Y., Westerberg, H., & Backman, L. (2012). Working-memory training in younger and older adults: Training gains, transfer, and maintenance. *Frontiers in Human Neuroscience, 6*, 63.
19. Brehmer, Y., et al. (2009). Working memory plasticity modulated by dopamine transporter genotype. *Neuroscience Letters, 467*(2), 117–120.
20. Bryck, R. L., & Fisher, P. A. (2012). Training the brain: Practical applications of neural plasticity from the intersection of cognitive neuroscience, developmental psychology, and prevention science. *American Psychologist, 67*(2), 87–100.
21. Buschkuehl, M., Jaeggi, S.M., & Jonides, J. (2012) Neuronal effects following working memory training. *Developmental Cognitive Neuroscience, 2*(Supplement 1), S167–S179.
22. Buschkuehl, M., et al. (2008). Impact of working memory training on memory performance in old–old adults. *Psychology and Aging, 23*(4), 743–753.
23. Buschkuehl, M., et al. (2014). Neural effects of short-term training on working memory. *Cognitive, Affective, & Behavioral Neuroscience, 14*(1), 147–160.
24. Carretti, B., Borella, E., & De Beni, R. (2007). Does strategic memory training improve the working memory performance of younger and older adults? *Experimental Psychology, 54*(4), 311–320.
25. Cattell, R. B. (1963). Theory of fluid and crystallized intelligence: A critical experiment. *Journal of Educational Psychology, 54*(1), 1–22.
26. Cepeda, N. J., et al. (2006). Distributed practice in verbal recall tasks: A review and quantitative synthesis. *Psychological Bulletin, 132*(3), 354–380.
27. Cepeda, N. J., et al. (2008). Spacing effects in learning: A temporal ridgeline of optimal retention. *Psychological Science, 19*(11), 1095–1102.
28. Chacko, A., et al. (2013). Cogmed working memory training for youth with ADHD: A closer examination of efficacy utilizing evidence-based criteria. *Journal of Clinical Child and Adolescent Psychology, 42*(6), 769–783.
29. Chein, J. M., & Morrison, A. B. (2010). Expanding the mind's workspace: Training and transfer effects with a complex working memory span task. *Psychonomic Bulletin & Review, 17*(2), 193–199.
30. Chooi, W. T., & Thompson, L. A. (2012). Working memory training does not improve intelligence in healthy young adults. *Intelligence, 40*(6), 531–542.
31. Colcombe, S., & Kramer, A. F. (2003). Fitness effects on the cognitive function of older adults: A meta-analytic study. *Psychological Science, 14*(2), 125–130.
32. Colom, R., et al. (2013). Adaptive n-back training does not improve fluid intelligence at the construct level: Gains on individual tests suggest training may enhance visuospatial processing. *Intelligence, 41*(5), 712–727.
33. Craik, F. I., et al. (2007). Cognitive rehabilitation in the elderly: Effects on memory. *Journal of the International Neuropsychological Society, 13*(1), 132–142.
34. Dahlin, E., et al. (2008). Transfer of learning after updating training mediated by the striatum. *Science, 320*(5882), 1510–1512.
35. Dahlin, E., et al. (2009). Training of the executive component of working memory: Subcortical areas mediate transfer effects. *Restorative Neurology and Neuroscience, 27*(5), 405–419.

36. Darowski, E. S., et al. (2008). Age-related differences in cognition: The role of distraction control. *Neuropsychology, 22*(5), 638–644.
37. de Jonge, P., & de Jong, P. F. (1996). Working memory, intelligence and reading ability in children. *Personality and Individual Differences, 21*(6), 1007–1020.
38. Deary, I. J., et al. (2007). Intelligence and educational achievement. *Intelligence, 35*(1), 13–21.
39. Deci, E. L., Koestner, R., & Ryan, R. M. (1999). A meta-analytic review of experiments examining the effects of extrinsic rewards on intrinsic motivation. *Psychological Bulletin, 125*(6), 627–668.
40. Detterman, D. K. (1993). The case for prosecution: Transfer as an epiphenomenon. In D. K. Detterman & R. J. Sternberg (Eds.), *Transfer on trial: Intelligence, cognition, and instruction* (pp. 1–24). Norwood, NJ: Ablex Publishing Corporation.
41. Diamond, A., & Lee, K. (2011). Interventions shown to aid executive function development in children 4 to 12 years old. *Science, 333*(6045), 959–964.
42. Duckworth, A. L., et al. (2007). Grit: Perseverance and passion for long-term goals. *Journal of Personality and Social Psychology, 92*(6), 1087–1101.
43. Engle, R. W., Kane, M. J., & Tuholski, S. W. (1999). Individual differences in working memory capacity and what they tell us about controlled attention, general fluid intelligence, and functions of the prefrontal cortex. In A. Miyake & P. Shah (Eds.), *Models of working memory: Mechanisms of active maintenance and executive control* (pp. 102–134). Cambridge: Cambridge University Press.
44. Ericsson, K. A., & Delaney, P. F. (1998). Working memory and expert performance. In R. Logie & K. J. Gilhooly (Eds.), *Working memory and thinking* (pp. 93–114). Hillsdale, NJ: Erlbaum.
45. Feuerstein, R. (1980). *Instrumental enrichment: An intervention program for cognitive modifiability*. Baltimore, MD: University Park Press.
46. García-Madruga, J. A., et al. (2013). Reading comprehension and working memory's executive processes: An intervention study in primary school students. *Reading Research Quarterly, 48*(2), 155–174.
47. Garlick, D. (2002). Understanding the nature of the general factor of intelligence: The role of individual differences in neural plasticity as an explanatory mechanism. *Psychological Review, 109*(1), 116–136.
48. Gathercole, S. E., Lamont, E., & Packiam Alloway, T. (2006). Working memory in the classroom. In S. Pickering (Ed.), *Working memory and education* (pp. 219–240). Oxford, UK: Elsevier Press.
49. Gathercole, S. E., et al. (2006). Working memory in children with reading disabilities. *Journal of Experimental Child Psychology, 93*(3), 265–281.
50. Gee, J. P. (2007). *What video games have to teach us about learning and literacy: Revised and updated edition*. New York, NY: Palgrave Macmillan.
51. Gomez-Pinilla, F. (2008). Brain foods: The effects of nutrients on brain function. *Nature Reviews Neuroscience, 9*(7), 568–578.
52. Gopher, D. (2007). Emphasis change as a training protocol for high-demand tasks. In A. F. Kramer, D. A. Wiegmann, & A. Kirlik (Eds.), *Attention: From theory to practice* (pp. 209–224). New York, NY: Oxford University Press.
53. Gottfredson, L. S. (1997). Why g matters: The complexity of everyday life. *Intelligence, 24*(1), 79–132.
54. Gray, J. R., Chabris, C. F., & Braver, T. S. (2003). Neural mechanisms of general fluid intelligence. *Nature Neuroscience, 6*(3), 316–322.
55. Gray, J. R., & Thompson, P. M. (2004). Neurobiology of intelligence: Science and ethics. *Nature Reviews Neuroscience, 5*(6), 471–482.
56. Green, C. S., & Bavelier, D. (2003). Action video game modifies visual selective attention. *Nature, 423*(6939), 534–537.
57. Green, C. S., & Bavelier, D. (2007). Action-video-game experience alters the spatial resolution of vision. *Psychological Science, 18*(1), 88–94.

58. Haskell, W. L., et al. (2007). Physical activity and public health – updated recommendation for adults from the American college of sports medicine and the American heart association. *Circulation, 116*(9), 1081–1093.
59. Hayslip, B. (1989). Fluid ability training with aged people: A past with a future? *Educational Gerontology, 15*, 573–595.
60. Hempel, A., et al. (2004). Plasticity of cortical activation related to working memory during training. *American Journal of Psychiatry, 161*(4), 745–747.
61. Herrnstein, R. J., et al. (1986). Teaching thinking skills. *American Psychologist, 41*(11), 1279–1289.
62. Hindin, S. B., & Zelinski, E. M. (2012). Extended practice and aerobic exercise interventions benefit untrained cognitive outcomes in older adults: A meta-analysis. *Journal of the American Geriatrics Society, 60*(1), 136–141.
63. Holmes, J., & Gathercole, S.E. (2013) Taking working memory training from the laboratory into schools. *Educational Psychology*.
64. Holmes, J., Gathercole, S. E., & Dunning, D. L. (2009). Adaptive training leads to sustained enhancement of poor working memory in children. *Developmental Science, 12*(4), F9–F15.
65. Holmes, J., et al. (2010). Working memory deficits can be overcome: Impacts of training and medication on working memory in children with ADHD. *Applied Cognitive Psychology, 24*(6), 827–836.
66. Houben, K., Wiers, R. W., & Jansen, A. (2011). Getting a grip on drinking behavior: Training working memory to reduce alcohol abuse. *Psychological Science, 22*(7), 968–975.
67. Hsu, N. S., & Jaeggi, S. M. (2014). The emergence of cognitive control abilities in childhood. *Current Topics in Behavioral Neurosciences, 16*, 149–166.
68. Hsu, N. S., Novick, J. M., & Jaeggi, S. M. (2014). The development and malleability of executive control abilities. *Frontiers in Behavioral Neuroscience, 8*(221).
69. Hunt, E., & Jaeggi, S. M. (2013). Challenges for research on intelligence. *Journal of Intelligence, 1*(1), 36–54.
70. Hussey, E. K., & Novick, J. M. (2012). The benefits of executive control training and the implications for language processing. *Frontiers in Psychology, 3*, 158.
71. Jaeggi, S. M., et al. (2008). Improving fluid intelligence with training on working memory. *Proceedings of the National Academy of Sciences of the United States of America, 105*(19), 6829–6833.
72. Jaeggi, S. M., et al. (2010). The concurrent validity of the N-back task as a working memory measure. *Memory, 18*(4), 394–412.
73. Jaeggi, S. M., et al. (2010). The relationship between n-back performance and matrix reasoning – implications for training and transfer. *Intelligence, 38*(6), 625–635.
74. Jaeggi, S. M., et al. (2011). Short- and long-term benefits of cognitive training. *Proceedings of the National Academy of Sciences of the United States of America, 108*(25), 10081–10086.
75. Jaeggi, S.M., et al. (2011). *Working memory training in typically developing children and children with attention deficit hyperactivity disorder: Evidence for plasticity in executive control processes*. Eighteenth Annual Cognitive Neuroscience Society Meeting, San Francisco, CA.
76. Jaeggi, S. M., et al. (2012). Cogmed and working memory training – current challenges and the search for underlying mechanisms. *Journal of Applied Research in Memory and Cognition, 1*, 211–213.
77. Jaeggi, S. M., et al. (2014). The role of individual differences in cognitive training and transfer. *Memory and Cognition, 42*(3), 464–480.
78. Jausovec, N., & Jausovec, K. (2012). Working memory training: Improving intelligence – changing brain activity. *Brain and Cognition, 79*(2), 96–106.
79. Jha, A. P., Krompinger, J., & Baime, M. J. (2007). Mindfulness training modifies subsystems of attention. *Cognitive, Affective, & Behavioral Neuroscience, 7*(2), 109–119.
80. Jolles, D. D., & Crone, E. A. (2012). Training the developing brain: A neurocognitive perspective. *Frontiers in Human Neuroscience, 6*, 76.

81. Jones, R. N., et al. (2013). The ACTIVE cognitive training interventions and trajectories of performance among older adults. *Journal of Aging and Health, 25*(8 Suppl), 186S–208S.
82. Jonides, J. (2004). How does practice makes perfect? *Nature Neuroscience, 7*(1), 10–11.
83. Jonides, J., et al. (1997). Verbal working memory load affects regional brain activation as measured by PET. *Journal of Cognitive Neuroscience, 9*(4), 462–475.
84. Jonides, J., et al. (2008). The mind and brain of short-term memory. *Annual Review of Psychology, 59*, 193–224.
85. Jonides, J., et al. (2012). Building better brains. *Scientific American Mind, 23*(4), 59–63.
86. Kan, K. J., et al. (2013). On the nature and nurture of intelligence and specific cognitive abilities: The more heritable, the more culture dependent. *Psychological Science, 24*(12), 2420–2428.
87. Kane, M. J., et al. (2007). Working memory, attention control, and the N-back task: A question of construct validity. *Journal of Experimental Psychology. Learning, Memory, and Cognition, 33*(3), 615–622.
88. Kelly, C., Foxe, J. J., & Garavan, H. (2006). Patterns of normal human brain plasticity after practice and their implications for neurorehabilitation. *Archives of Physical Medicine and Rehabilitation, 87*(12 Suppl 2), S20–S29.
89. Kerns, A. K., Eso, K., & Thomson, J. (1999). Investigation of a direct intervention for improving attention in young children with ADHD. *Developmental Neuropsychology, 16*, 273–295.
90. Kesler, S., et al. (2013). Cognitive training for improving executive function in chemotherapy-treated breast cancer survivors. *Clinical Breast Cancer, 13*(4), 299–306.
91. Klauer, K. J., & Phye, G. D. (2008). Inductive reasoning: A training approach. *Review of Educational Research, 78*(1), 85–123.
92. Klingberg, T., Forssberg, H., & Westerberg, H. (2002). Training of working memory in children with ADHD. *Journal of Clinical and Experimental Neuropsychology, 24*(6), 781–791.
93. Klingberg, T., et al. (2005). Computerized training of working memory in children with ADHD – a randomized, controlled trial. *Journal of the American Academy of Child and Adolescent Psychiatry, 44*(2), 177–186.
94. Kroesbergen, E. H., van't Noordende, J. E., & Kolkman, M. E. (2014). Training working memory in kindergarten children: Effects on working memory and early numeracy. *Child Neuropsychology, 20*(1), 23–37.
95. Kronenberger, W. G., et al. (2011). Working memory training for children with cochlear implants: A pilot study. *Journal of Speech, Language, and Hearing Research, 54*(4), 1182–1196.
96. Kundu, B., et al. (2013). Strengthened effective connectivity underlies transfer of working memory training to tests of short-term memory and attention. *Journal of Neuroscience, 33*(20), 8705–8715.
97. Li, S. C., et al. (2008). Working memory plasticity in old age: Practice gain, transfer, and maintenance. *Psychology and Aging, 23*(4), 731–742.
98. Lilienthal, L., et al. (2013). Dual n-back training increases the capacity of the focus of attention. *Psychonomic Bulletin & Review, 20*(1), 135–141.
99. Loosli, S. V., et al. (2012). Working memory training improves reading processes in typically developing children. *Child Neuropsychology, 18*(1), 62–78.
100. Lustig, C., et al. (2009). Aging, training, and the brain: A review and future directions. *Neuropsychology Review, 19*(4), 504–522.
101. Malone, T. W., & Lepper, M. R. (1987). Making learning fun: A taxonomy of intrinsic motivations for learning. In R. E. Snow & M. J. Farr (Eds.), *Aptitude, learning and instruction: III. Conative and affective process analyses* (pp. 223–253). Hilsdale, NJ: Erlbaum.
102. McGurk, S. R., et al. (2007). A meta-analysis of cognitive remediation in schizophrenia. *American Journal of Psychiatry, 164*(12), 1791–1802.
103. McNab, F., et al. (2009). Changes in cortical dopamine D1 receptor binding associated with cognitive training. *Science, 323*(5915), 800–802.

104. Melby-Lervåg, M., & Hulme, C. (2013). Is working memory training effective? A meta-analytic review. *Developmental Psychology, 49*(2), 270–291.
105. Minear, M., & Shah, P. (2006). Sources of working memory deficits in children and possibilities for remediation. In S. Pickering (Ed.), *Working memory and education* (pp. 274–307). Oxford, UK: Elsevier Press.
106. Mitchell, M. B., et al. (2012). Cognitively stimulating activities: Effects on cognition across four studies with up to 21 years of longitudinal data. *Journal of Aging Research, 2012*, 461592.
107. Moreno, S., et al. (2011). Short-term music training enhances verbal intelligence and executive function. *Psychological Science, 22*(11), 1425–1433.
108. Neely, A. S., & Backman, L. (1993). Maintenance of gains following multifactorial and unifactorial memory training in late adulthood. *Educational Gerontology, 19*(2), 105–117.
109. Neely, A. S., & Backman, L. (1995). Effects of multifactorial memory training in old-age – generalizability across tasks and individuals. *The Journals of Gerontology. Series B, Psychological Sciences and Social Sciences, 50*(3), P134–P140.
110. Noice, H., & Noice, T. (2009). An arts intervention for older adults living in subsidized retirement homes. *Aging, Neuropsychology, and Cognition, 16*(1), 56–79.
111. Novick, J. M., et al. (2013). Clearing the garden-path: Improving sentence processing through cognitive control training. *Language and Cognitive Processes, 28*, 1–44.
112. Owen, A. M., et al. (2005). N-back working memory paradigm: A meta-analysis of normative functional neuroimaging studies. *Human Brain Mapping, 25*(1), 46–59.
113. Owen, A. M., et al. (2010). Putting brain training to the test. *Nature, 465*(7299), 775–778.
114. Owens, M., Koster, E. H., & Derakshan, N. (2013). Improving attention control in dysphoria through cognitive training: Transfer effects on working memory capacity and filtering efficiency. *Psychophysiology, 50*(3), 297–307.
115. Park, D. C., et al. (2014). The impact of sustained engagement on cognitive function in older adults: The synapse project. *Psychological Science, 25*(1), 103–112.
116. Passolunghi, M. C., & Siegel, L. S. (2001). Short-term memory, working memory, and inhibitory control in children with difficulties in arithmetic problem solving. *Journal of Experimental Child Psychology, 80*, 44–57.
117. Perkins, D. N., & Salomon, G. (1994). Transfer of learning. In T. Husen & T. N. Postelwhite (Eds.), *International handbook of educational research* (pp. 6452–6457). Oxford: Pergamon Press.
118. Pickering, S. (Ed.). (2006). *Working memory and education*. Oxford, UK: Elsevier Press.
119. Prensky, M. (2001). *Digital game-based learning*. New York: McGraw-Hill.
120. Qiu, F., et al. (2009). *Study on improving fluid intelligence through cognitive training system based on Gabor stimulus*. ICISE 2009.
121. Raven, J. C. (1990). *Advanced progressive matrices. Sets I, II*. Oxford: Oxford University Press.
122. Rebok, G. W., et al., (2014). Ten-year effects of the advanced cognitive training for independent and vital elderly cognitive training trial on cognition and everyday functioning in older adults. *Journal of the American Geriatrics Society.*
123. Redick, T. S., et al. (2013). No evidence of intelligence improvement after working memory training: A randomized, placebo-controlled study. *Journal of Experimental Psychology: General, 142*(2), 359–379.
124. Roughan, L., & Hadwin, J. A. (2011). The impact of working memory training in young people with social, emotional and behavioural difficulties. *Learning and Individual Differences, 21*, 759–764.
125. Rudebeck, S. R., et al. (2012). A potential spatial working memory training task to improve both episodic memory and fluid intelligence. *PLoS One, 7*(11), e50431.
126. Saczynski, J. S., Willis, S. L., & Schaie, K. W. (2002). Strategy use in reasoning training with older adults. *Aging, Neuropsychology, and Cognition, 9*(1), 48–60.
127. Salminen, T., Strobach, T., & Schubert, T. (2012). On the impacts of working memory training on executive functioning. *Frontiers in Human Neuroscience, 6*, 166.

128. Salomon, G., & Perkins, D. N. (1989). Rocky roads to transfer: Rethinking mechanisms of a neglected phenomenon. *Educational Psychologist, 24*(2), 113–142.
129. Schellenberg, E. G. (2004). Music lessons enhance IQ. *Psychological Science, 15*(8), 511–514.
130. Schmidt, R. A., & Bjork, R. A. (1992). New conceptualizations of practice: Common principles in three paradigms suggest new concepts for training. *Psychological Science, 3*(4), 207–217.
131. Schmiedek, F., Lövdén, M., & Lindenberger, U. (2012). Hundred days of cognitive training enhance broad cognitive abilities in adulthood: Findings from the COGITO study. *Frontiers in Aging Neuroscience, 2*(27).
132. Schneiders, J. A., et al. (2011). Separating intra-modal and across-modal training effects in visual working memory: An fMRI investigation. *Cerebral Cortex, 21*(11), 2555–2564.
133. Schneiders, J. A., et al. (2012). The impact of auditory working memory training on the fronto-parietal working memory network. *Frontiers in Human Neuroscience, 6*, 173.
134. Schweizer, S., Hampshire, A., & Dalgleish, T. (2011). Extending brain-training to the affective domain: Increasing cognitive and affective executive control through emotional working memory training. *PLoS One, 6*(9), e24372.
135. Schweizer, S., et al. (2013). Training the emotional brain: Improving affective control through emotional working memory training. *Journal of Neuroscience, 33*(12), 5301–5311.
136. Shah, P., & Miyake, A. (1999). Models of working memory: An introduction. In A. Miyake & P. Shah (Eds.), *Models of working memory: Mechanism of active maintenance and executive control* (pp. 1–26). New York: Cambridge University Press.
137. Shah, P., et al. (2012). Cognitive training for ADHD: The importance of individual differences. *Journal of Applied Research in Memory and Cognition, 1*, 204–205.
138. Shalev, L., Tsal, Y., & Mevorach, C. (2007). Computerized progressive attentional training (CPAT) program: Effective direct intervention for children with ADHD. *Child Neuropsychology, 13*(4), 382–388.
139. SharpBrains (2013). *Digital Brain Health Market Report. Executive summary: Infographic on the Digital Brain Health Market 2012–2020 [cited 2013 September 16]*. Available from http://sharpbrains.com/executive-summary
140. Shipstead, Z., Hicks, K. L., & Engle, R. W. (2012). Cogmed working memory training: Does the evidence support the claims? *Journal of Applied Research in Memory and Cognition, 1*, 185–193.
141. Shipstead, Z., Redick, T. S., & Engle, R. W. (2012). Is working memory training effective? *Psychological Bulletin, 138*(4), 628–654.
142. Singley, M. K., & Anderson, J. R. (1989). *The transfer of cognitive skill*. Cambridge, MA: Harvard University Press.
143. Snow, R. E., Kyllonen, P. C., & Marshalek, B. (1984). The topography of ability and learning correlations. In R. J. Sternberg (Ed.), *Advances in the psychology of human intelligence* (pp. 47–103). Hillsdale, NJ: Lawrence Erlbaum Associates.
144. Squire, K. (2003). Video games in education. *International Journal of Intelligent Simulations and Gaming, 2*, 1–16.
145. St Clair-Thompson, H., et al. (2010). Improving children's working memory and classroom performance. *Educational Psychology, 30*(2), 203–219.
146. Stepankova, H., et al. (2014). Dose–response relationship of working memory training and improvements in fluid intelligence: A randomized controlled study in old adults. *Developmental Psychology, 50*(4), 1049–1059.
147. Stephenson, C. L., & Halpern, D. F. (2013). Improved matrix reasoning is limited to training on tasks with a visuospatial component. *Intelligence, 41*, 341–357.
148. Stine-Morrow, E. A., et al. (2008). The effects of an engaged lifestyle on cognitive vitality: A field experiment. *Psychology and Aging, 23*(4), 778–786.
149. Stough, C., et al. (2011). Improving general intelligence with a nutrient-based pharmacological intervention. *Intelligence, 39*, 100–107.

150. Studer-Luethi, B., et al. (2012). Influence of neurotisicm and conscientiousness on working memory training outcome. *Personality and Individual Differences, 53*(1), 44–49.
151. Suter, E., Marti, B., & Gutzwiller, F. (1994). Jogging or walking – comparison of health effects. *Annals of Epidemiology, 4*(5), 375–381.
152. Szmalec, A., et al. (2011). Control of interference during working memory updating. *Journal of Experimental Psychology. Human Perception and Performance, 37*(1), 137–151.
153. Takeuchi, H., et al. (2010). Training of working memory impacts structural connectivity. *Journal of Neuroscience, 30*(9), 3297–3303.
154. Takeuchi, H., et al. (2011). Working memory training using mental calculation impacts regional gray matter of the frontal and parietal regions. *PLoS One, 6*(8), e23175.
155. Takeuchi, H., et al. (2013). Effects of working memory training on functional connectivity and cerebral blood flow during rest. *Cortex, 49*(8), 2106–2125.
156. Tallal, P., et al. (1996). Language comprehension in language-learning impaired children improved with acoustically modified speech. *Science, 271*(5245), 81–84.
157. Tang, Y. Y., et al. (2007). Short-term meditation training improves attention and self-regulation. *Proceedings of the National Academy of Sciences of the United States of America, 104*(43), 17152–17156.
158. te Nijenhuis, J., van Vianen, A. E. M., & van der Flier, H. (2007). Score gains on g-loaded tests: No g. *Intelligence, 35*, 283–300.
159. Thompson, T. W., et al. (2013). Failure of working memory training to enhance cognition or intelligence. *PLoS One, 8*(5), e63614.
160. Thorell, L. B., et al. (2009). Training and transfer effects of executive functions in preschool children. *Developmental Science, 12*(1), 106–113.
161. Thorndike, E. L., & Woodworth, R. S. (1901). The influence of improvement in one mental function upon the efficiency of other functions. *Psychological Review, 8*, 247–261.
162. Tranter, L.J., & Koutstaal, W. (2007). Age and flexible thinking: An experimental demonstration of the beneficial effects of increased cognitively stimulating activity on fluid intelligence in healthy older adults. *Neuropsychology, Development, and Cognition. Section B, Aging, Neuropsychology and Cognition*, 1–24.
163. Twamley, E. W., Burton, C. Z., & Vella, L. (2011). Compensatory cognitive training for psychosis: Who benefits? Who stays in treatment? *Schizophrenia Bulletin, 37*(Suppl 2), S55–S62.
164. Van der Molen, M. J., et al. (2010). Effectiveness of a computerised working memory training in adolescents with mild to borderline intellectual disabilities. *Journal of Intellectual Disability Research, 54*(4), 433–447.
165. Verhaeghen, P., & Marcoen, A. (1996). On the mechanisms of plasticity in young and older adults after instruction in the method of loci: Evidence for an amplification model. *Psychology and Aging, 11*(1), 164–178.
166. von Bastian, C. C., & Oberauer, K. (2013). Distinct transfer effects of training different facets of working memory capacity. *Journal of Memory and Language, 69*, 36–58.
167. Whisman, M. A. (1990). The efficacy of booster maintenance sessions in behavior therapy: Review and methodological critique. *Clinical Psychology Review, 10*(2), 155–170.
168. Wiley, J., et al. (2011). New rule use drives the relation between working memory capacity and Raven's advanced progressive matrices. *Journal of Experimental Psychology. Learning, Memory, and Cognition, 37*(1), 256–263.
169. Willcutt, E. G., et al. (2005). Validity of the executive function theory of attention-deficit/hyperactivity disorder: A meta-analytic review. *Biological Psychiatry, 57*(11), 1336–1346.
170. Willis, S. L. (2001). Methodological issues in behavioral intervention research with the elderly. In J. E. Birren & K. W. Schaie (Eds.), *Handbook of the psychology of aging* (pp. 78–108). San Diego: Academic Press.
171. Willis, S. L., et al. (2006). Long-term effects of cognitive training on everyday functional outcomes in older adults. *JAMA, 296*(23), 2805–2814.

172. Witt, M. (2011). School based working memory training: Preliminary finding of improvement in children's mathematical performance. *Advances in Cognitive Psychology, 7*, 7–15.
173. Woltz, D. J., Gardner, M. K., & Gyll, S. P. (2000). The role of attention processes in near transfer of cognitive skills. *Learning and Individual Differences, 12*, 209–251.
174. Ybarra, O., et al. (2008). Mental exercising through simple socializing: Social interaction promotes general cognitive functioning. *Personality and Social Psychology Bulletin, 34*(2), 248–259.
175. Zhao, X., et al. (2011). Effect of updating training on fluid intelligence in children. *Chinese Science Bulletin, 56*(21), 2202–2205.
176. Zinke, K., et al. (2014). Working memory training and transfer in older adults: Effects of age, baseline performance, and training gains. *Developmental Psychology, 50*(1), 304–315.

Partial Functional Differential Equations: Reduction of Complexity and Applications

Khalil Ezzinbi

Abstract The aim of this work is to reduce the complexity of partial functional differential equations. We suppose that the undelayed part is not necessarily densely defined and satisfies the Hille-Yosida condition. The delayed part is continuous. We prove the dynamic of solutions are obtained through an ordinary differential equations that is well-posed in a finite dimensional space. The powerty of this results is used to show the existence of almost automorphic solutions for partial functional differential equations. For illustration, we provide an application to the Lotka-Volterra model with diffusion and delay.

1 Introduction

Partial functional differential equations are an important area of research in applied mathematics, since many phenomenas in physical and biological systems are modeled using the history of the system. Then a system using delay is well-posed in infinite dimensional spaces and many classical results in differential equations well-posed in finite dimensional spaces cannot be applied. The aim of this chapter is to reduce the complexity of partial functional differential equations. We prove the existence of an ordinary differential equation that is well-posed in finite dimensional spaces and give all the fundamental properties on the qualitative analysis for the whole partial functional differential equations. Recall that the theory of partial functional differential equations was initiated in [23], for more details we refer to the book [25].

K. Ezzinbi (✉)
Faculty of Sciences Semlalia, Department of Mathematics, Cadi Ayyad University,
B.P. 2390 Marrakesh, Morocco
e-mail: ezzinbi@uca.ma

B. Toni (ed.), *New Frontiers of Multidisciplinary Research in STEAM-H (Science, Technology, Engineering, Agriculture, Mathematics, and Health)*,
Springer Proceedings in Mathematics & Statistics 90, DOI 10.1007/978-3-319-07755-0_3,

Here we are concerned with the following partial functional differential equation with finite delay

$$\begin{cases} \dfrac{d}{dt}u(t) = Au(t) + L(u_t) + f(t) \text{ for } t \geq 0, \\ u_0 = \varphi \in C := C([-r,0], X), \end{cases} \tag{1}$$

where A is a linear operator on a Banach space X not necessarily densely defined and satisfies the Hille-Yosida condition: there exist $M \geq 0$, $\omega \in \mathbb{R}$ such that $(\omega, +\infty) \subset \rho(A)$, and

$$|R(\lambda, A)^n| \leq \frac{M}{(\lambda - \omega)^n} \text{ for } n \in \mathbb{N} \text{ and } \lambda > \omega,$$

where $\rho(A)$ is the resolvent set of A and $R(\lambda, A) = (\lambda - A)^{-1}$, C is the space of continuous functions from $[-r, 0]$ to the observable X endowed with the uniform norm topology. L is a bounded linear operator from C into X and f is an almost automorphic function from $\mathbb{R}$ to X, the history function $u_t \in C$ is defined by

$$u_t(\theta) = u(t+\theta) \text{ for } \theta \in [-r, 0].$$

As an example of Eq. (1), we propose the following model arising in many problems in population dynamics and physical systems

$$\begin{cases} \dfrac{\partial}{\partial t}v(t,x) = \dfrac{\partial^2}{\partial x^2}v(t,x) + \displaystyle\int_{-r}^{0} G(\theta)v(t+\theta,x)d\theta + h(t,x) \text{ for } t \geq 0 \text{ and } x \in [0,\pi], \\ u(t,x) = 0 \text{ for } x = 0, \pi \text{ and } t \geq 0, \\ u(\theta,x) = \varphi(\theta,x) \text{ for } \theta \in [-r,0] \text{ and } x \in [0,\pi], \end{cases}$$

We use the reduction of complexity to prove that the existence of almost automorphic solution of Eq. (1) is equivalent to the existence of a bounded solution on $\mathbb{R}^+$. To achieve this goal, we use the variation of constants formula obtained in [2] and we develop new fundamental results about the spectral decomposition of solutions.

Almost automorphic functions are more general than almost periodic functions and they were introduced by S. Bochner [6], for more details about this topics we refer to the recent book [18] where the author give an important overview about the theory of almost automorphic functions and their applications to differential equations. The existence of almost automorphic solutions for differential equations in infinite dimensional space has been studied by several authors. For example in [19], the author studied the existence of almost automorphic solutions for the following semilinear abstract differential equation

$$\frac{d}{dt}x(t) = \mathcal{C}x(t) + \theta(t) \text{ for } t \geq 0, \tag{2}$$

where $\mathcal{C}$ generates an exponentially stable semigroup on a Banach space Y and θ is an almost automorphic function from $\mathbb{R}$ to Y. The author proved that the only bounded mild solution of Eq. (2) on $\mathbb{R}$ is almost automorphic.

Recently in [10], the authors studied the existence of almost automorphic solutions for the following partial functional differential equations with infinite delay

$$\begin{cases} \frac{dx}{dt}(t) = \mathcal{D}x(t) + \mathcal{L}(t)x_t + \mathcal{K}(t) \text{ for } t \geq 0, \\ x_0 = \varphi \in \mathcal{B}, \end{cases} \tag{3}$$

where $\mathcal{D}$ is the generator of a strongly continuous semigroup of linear operators on a Banach space E which is equivalent by Hille-Yosida's theorem that $\mathcal{D}$ satisfies the Hille-Yosida condition and $\overline{D(\mathcal{D})} = E$. The phase space $\mathcal{B}$ is a linear space of functions mapping $(-\infty, 0]$ into E satisfying some axioms introduced by Hale and Kato [10], for all $t \geq 0$, $\mathcal{L}(t)$ is a bounded linear operator form $\mathcal{B}$ to E and periodic in t. For every $t \geq 0$, the history function $x_t \in \mathcal{B}$ is defined by

$$x_t(\theta) = x(t + \theta) \text{ for } \theta \leq 0.$$

The function $\mathcal{K}$ is an almost automorphic function from $\mathbb{R}$ to E. The authors proved that the existence of a bounded mild solution on $\mathbb{R}^+$ of Eq. (3) is equivalent to the existence of an almost automorphic solution.

2 Variation of Constants Formula for Partial Functional Differential Equations with Finite Delay

Throughout this chapter, we suppose that

($\mathbf{H_0}$) A satisfies the Hille-Yosida condition.

We consider the following definition and results which are taken from [1].

Definition 2.1 ([1]). We say that a continuous function u from $[-r, \infty)$ into X is an integral solution of Eq. (1), if the following conditions hold

(i) $\int_0^t u(s)ds \in D(A)$ for $t \geq 0$,

(ii) $u(t) = \varphi(0) + A\int_0^t u(s)ds + \int_0^t [L(u_s) + f(s)]\, ds$ for $t \geq 0$,

(iii) $u_0 = \varphi$.

If $\overline{D(A)} = X$, the integral solutions coincide with the known mild solutions. We can see that if u is an integral solution of Eq. (1), then $u(t) \in \overline{D(A)}$ for all $t \geq 0$, in particular $\varphi(0) \in \overline{D(A)}$. Let us introduce the part A_0 of the operator A in $\overline{D(A)}$ defined by

$$\begin{cases} D(A_0) = \left\{x \in D(A) : Ax \in \overline{D(A)}\right\} \\ A_0 x = Ax \text{ for } x \in D(A_0). \end{cases}$$

Lemma 2.2 ([4, Lemma 3.3.12, pp. 140]). *A_0 generates a strongly continuous semigroup $(T_0(t))_{t\geq 0}$ on $\overline{D(A)}$.*

For the existence of the integral solutions, one has the following result.

Theorem 2.3 ([1]). *Assume that* $(\mathbf{H_0})$ *holds, then for all $\varphi \in C$ such that $\varphi(0) \in \overline{D(A)}$, Eq. (1) has a unique integral solution u on $[-r, +\infty)$. Moreover u is given by*

$$u(t) = T_0(t)\varphi(0) + \lim_{\lambda \to +\infty} \int_0^t T_0(t-s)B_\lambda[L(u_s) + f(s)]ds \text{ for } t \geq 0,$$

where $B_\lambda = \lambda R(\lambda, A)$ for $\lambda > \omega$.

In the sequel of this work, we call integral solutions as solutions

Let C_0 be the phase space of Eq. (1):

$$C_0 = \left\{\varphi \in C : \varphi(0) \in \overline{D(A)}\right\}.$$

For each $t \geq 0$, we define the linear operator $\mathcal{U}(t)$ on C_0 by

$$\mathcal{U}(t)\varphi = v_t(.,\varphi),$$

where $v(.,\varphi)$ is the solution of the following linear equation

$$\begin{cases} \dfrac{d}{dt}v(t) = Ay(t) + L(y_t) \text{ fort } \geq 0, \\ v_0 = \varphi \in C, \end{cases}$$

Proposition 2.1 ([1]). *The family $(\mathcal{U}(t))_{t\geq 0}$ is a strongly continuous semigroup of linear operators on C_0:*

(*i*) *for all $t \geq 0$, $\mathcal{U}(t)$ is a bounded linear operator on C_0,*
(*ii*) *$\mathcal{U}(0) = I$,*
(*iii*) *$\mathcal{U}(t+s) = \mathcal{U}(t)\mathcal{U}(s$, for all $t, s \geq 0$,*
(*iv*) *for all $\varphi \in C_0$, $\mathcal{U}(t)\varphi$ is a continuous function of $t \geq 0$ with values in C_0. Moreover,*
(*v*) *$(\mathcal{U}(t))_{t\geq 0}$ satisfies, for $t \geq 0$ and $\theta \in [-r, 0]$, the following translation property*

$$(\mathcal{U}(t)\varphi)(\theta) = \begin{cases} (\mathcal{U}(t+\theta)\varphi)(0) & \text{if} t+\theta \geq 0 \\ \varphi(t+\theta) & \text{if} t+\theta \leq 0. \end{cases}$$

Theorem 2.5 ([2, Theorem 3]). *Let $\mathcal{A}_u$ be defined on C_0 by*

$$\begin{cases} D(\mathcal{A}_u) = \Big\{\varphi \in C^1([-r,0];X) : \varphi(0) \in D(A), \varphi'(0) \in \overline{D(A)} \text{ and } \varphi'(0) \\ \qquad = A\varphi(0) + L(\varphi)\Big\} \\ \mathcal{A}_u\varphi = \varphi' \text{ for } \varphi \in D(\mathcal{A}_u). \end{cases}$$

Then $\mathcal{A}_u$ is the infinitesimal generator of the semigroup $(\mathcal{U}(t))_{t\geq 0}$ on C_0.

In order to give a variation of constants formula, we need to recall some notations and results which are taken from [2]. Let $\langle X_0\rangle$ be the space defined by

$$\langle X_0\rangle = \{X_0 c : c \in X\},$$

where the function $X_0 c$ is defined by

$$(X_0 c)(\theta) = \begin{cases} 0 \text{ if } \theta \in [-r,0), \\ c \text{ if } \theta = 0. \end{cases}$$

The space $C_0 \oplus \langle X_0\rangle$ is equipped with the norm $\|\phi + X_0 c\| = |\phi|_C + |c|$ for $(\phi, c) \in C_0 \times X$, is a Banach space and consider the extension $\tilde{\mathcal{A}}_{\mathcal{U}}$ of the operator $\mathcal{A}_u$ defined on $C_0 \oplus \langle X_0\rangle$ by

$$\begin{cases} D\left(\tilde{\mathcal{A}}_{\mathcal{U}}\right) = \Big\{\varphi \in C^1([-r,0];X) : \varphi(0) \in D(A) \text{ and } \varphi'(0) \in \overline{D(A)}\Big\}, \\ \tilde{\mathcal{A}}_{\mathcal{U}}\varphi \quad = \varphi' + X_0\left(A\varphi(0) + L\varphi - \varphi'(0)\right). \end{cases}$$

Lemma 2.6 ([2, Theorem 13 and Lemma 15]). *Assume that* $(\mathbf{H_0})$. *Then $\tilde{\mathcal{A}}_{\mathcal{U}}$ satisfies the Hille-Yosida condition on $C_0 \oplus \langle X_0\rangle$: there exists $\tilde{M} \geq 0$, $\tilde{\omega} \in \mathbb{R}$ such that $(\tilde{\omega}, +\infty) \subset \rho(\tilde{\mathcal{A}}_{\mathcal{U}})$, and*

$$\left|R(\lambda, \tilde{\mathcal{A}}_{\mathcal{U}})^n\right| \leq \frac{\tilde{M}}{(\lambda - \tilde{\omega})^n} \text{ for } n \in \mathbb{N},\ \lambda > \tilde{\omega}$$

where $R(\lambda, \tilde{\mathcal{A}}_{\mathcal{U}}) = (\lambda - \tilde{\mathcal{A}}_{\mathcal{U}})^{-1}$. Moreover, the part of $\tilde{\mathcal{A}}_{\mathcal{U}}$ on $\overline{D\left(\tilde{\mathcal{A}}_{\mathcal{U}}\right)} = C_0$ is exactly the operator $\mathcal{A}_u$.

Theorem 2.7 ([2, Theorem 16]). *Assume that* $(\mathbf{H_0})$ *holds. Then for all* $\varphi \in C_0$*, the solution u of Eq. (1) is given by the following variation of constants formula*

$$u_t = \mathcal{U}(t)\varphi + \lim_{\lambda \to +\infty} \int_0^t \mathcal{U}(t-s)\tilde{B}_\lambda (X_0 f(s))\,ds \text{ for } t \geq 0,$$

where $\tilde{B}_\lambda = \lambda R(\lambda, \tilde{A}_{\mathcal{U}})$ *for* $\lambda > \tilde{\omega}$.

3 Reduction of Complexity for Functional Differential Equations with Finite Delay

In the following, we assume that:

$(\mathbf{H_1})$ The operator $T_0(t)$ is compact on $\overline{D(A)}$ for every $t > 0$.

Theorem 3.1. *Assume that* $(\mathbf{H_0})$ *and* $(\mathbf{H_1})$ *hold, then* $\mathcal{U}(t)$ *is compact for* $t > r$.

As a consequence from the compactness property and [8, Theorem 5.3.7, pp. 333], we have the following spectral decomposition result.

Corollary 3.2 ([2]). C_0 *is decomposed as follows:*

$$C_0 = S \oplus V,$$

where S *is* $\mathcal{U}$*-invariant and there are positive constants* α *and* N *such that*

$$|\mathcal{U}(t)\varphi|_C \leq Ne^{-\alpha t}|\varphi|_C \quad for\ each\ t \geq 0\ and\ \varphi \in S. \tag{4}$$

V *is a finite dimensional space and the restriction of* $\mathcal{U}$ *to* V *becomes a group.*

In the sequel, $\mathcal{U}^s(t)$ and $\mathcal{U}^v(t)$ denote the restriction of $\mathcal{U}(t)$ respectively on S and V which correspond to the above decomposition.

Let $d = \dim V$ with a basis vectors $\Phi = \{\phi_1, \ldots, \phi_d\}$. Then, there exist d-elements $\{\psi_1, \ldots, \psi_d\}$ in C_0^* such that

$$\begin{cases} \langle \psi_i, \phi_j \rangle = \delta_{ij}, \\ \langle \psi_i, \phi \rangle = 0 \text{ for all } \phi \in S \text{and} i \in \{1, \ldots, d\}, \end{cases} \tag{5}$$

where $\langle .,. \rangle$ denotes the duality pairing between C_0^* and C_0 and

$$\delta_{ij} = \begin{cases} 1 \text{ if } i = j \\ 0 \text{ if } i \neq j \end{cases}$$

Let $\Psi = col\{\psi_1, \ldots, \psi_d\}$, $\langle \Psi, \Phi \rangle$ is a $d \times d$-matrix, where the (i, j)-component is $\langle \psi_i, \phi_j \rangle$. Denote by Π^s and Π^v the projections respectively on S and V. For each $\varphi \in C_0$, we have

$$\Pi^v \varphi = \Phi \langle \Psi, \varphi \rangle .$$

In fact, for $\varphi \in C_0$, we have $\varphi = \Pi^s \varphi + \Pi^v \varphi$ with $\Pi^v \varphi = \sum_{i=1}^{d} \alpha_i \phi_i$ and $\alpha_i \in \mathbb{R}$. By (5), we conclude that

$$\alpha_i = \langle \psi_i, \varphi \rangle .$$

Hence

$$\begin{aligned} \Pi^v \varphi &= \sum_{i=1}^{d} \langle \psi_i, \varphi \rangle \phi_i \\ &= \Phi \langle \Psi, \varphi \rangle . \end{aligned}$$

Since $(\mathcal{U}^v(t))_{t \geq 0}$ is a group on V, then there exists a $d \times d$-matrix G such that

$$\mathcal{U}^v(t)\, \Phi = \Phi e^{tG} \text{for} t \in \mathbb{R}.$$

Moreover, $\sigma(G) = \{\lambda \in \sigma(\mathcal{A}_u) : \operatorname{Re}(\lambda) \geq 0\}$.

For $n, n_0 \in N$ such that $n \geq n_0 \geq \tilde{\omega}$ and $i \in \{1, \ldots, d\}$, we define the linear mapping $x^*_{i,n}$ by

$$x^*_{i,n}(a) = \langle \psi_i, \tilde{B}_n X_0 a \rangle \text{for } a \in X.$$

Since $|\tilde{B}_n| \leq \frac{n}{n-\tilde{\omega}} \tilde{M}$, for any $n \geq n_0$, then $x^*_{i,n}$ is a bounded linear operator from X to $\mathbb{R}$ with

$$|x^*_{i,n}| \leq \frac{n}{n - n_0} \tilde{M}\, |\psi_i| \text{ for any } n \geq n_0.$$

Define the d-column vector $x^*_n = col\left(x^*_{1,n}, \ldots, x^*_{d,n}\right)$, then

$$\langle x^*_n, a \rangle = \langle \Psi, \tilde{B}_n X_0 a \rangle \text{for } a \in X,$$

with

$$\langle x^*_n, a \rangle_i = \langle \psi_i, \tilde{B}_n X_0 a \rangle \text{for } i = 1, \ldots, d \text{and } a \in X.$$

Consequently,

$$\sup_{n\geq n_0} |x_n^*| < \infty,$$

which implies that $\left(x_n^*\right)_{n\geq n_0}$ is a bounded sequence in $\mathcal{L}(X,\mathbb{R}^d)$. We have the following important result of this work.

Theorem 3.3. *There exists $x^* \in \mathcal{L}(X,\mathbb{R}^d)$, such that $\left(x_n^*\right)_{n\geq n_0}$ converges weakly to x^* in the sense that*

$$\langle x_n^*, x\rangle \to \langle x^*, x\rangle \text{ as } n\to\infty \text{ for all } x\in X.$$

For the proof, we need the following fundamental Theorem in functional analysis.

Theorem 3.4 ([24, pp. 776] (Banach-Alaoglu-Bourbaki)). *Let Y be any separable Banach space and $\left(z_n^*\right)_{n\in\mathbb{N}}$ any bounded sequence in Y^*. Then there exists a subsequence $\left(z_{n_k}^*\right)_{k\in\mathbb{N}}$ of $\left(z_n^*\right)_{n\in\mathbb{N}}$ which converges weakly in Y^*in the sense that there exists $z^*\in Y^*$ such that*

$$\left\langle z_{n_k}^*, x\right\rangle \to \langle z^*, x\rangle \text{ as } n\to\infty \text{ for all } x\in Y.$$

Proof. Let Z_0 be any closed separable subspace of X. Since $\left(x_n^*\right)_{n\geq n_0}$ is a bounded sequence, then by Theorem 3.4 we get that the sequence $\left(x_n^*\right)_{n\geq n_0}$ has a subsequence $\left(x_{n_k}^*\right)_{k\in\mathbb{N}}$ which converges weakly to some $x_{Z_0}^*$ in Z_0. We claim that all the sequence $\left(x_n^*\right)_{n\geq n_0}$ converges weakly to $x_{Z_0}^*$ in Z_0. In fact, we proceed by contradiction and suppose that there exists a subsequence $\left(x_{n_p}^*\right)_{p\in\mathbb{N}}$ of $\left(x_n^*\right)_{n\geq n_0}$ which converges weakly to some $\tilde{x}_{Z_0}^*$ with $\tilde{x}_{Z_0}^* \neq x_{Z_0}^*$. Let $u_t(.,\sigma,\varphi,f)$ denote the solution of Eq. (1). Then

$$\Pi^\nu u_t(.,\sigma,0,f) = \lim_{n\to+\infty}\int_\sigma^t \mathcal{U}^\nu(t-\xi)\,\Pi^\nu\left(\tilde{B}_n X_0 f(\xi)\right)d\xi,$$

and

$$\Pi^\nu\left(\tilde{B}_n X_0 f(\xi)\right) = \Phi\left\langle\Psi, \tilde{B}_n X_0 f(\xi)\right\rangle = \Phi\left\langle x_n^*, f(\xi)\right\rangle.$$

It follows that

$$\begin{aligned}\Pi^\nu u_t(.,\sigma,0,f) &= \lim_{n\to+\infty}\Phi\int_\sigma^t e^{(t-\xi)G}\left\langle\Psi, \tilde{B}_n X_0 f(\xi)\right\rangle d\xi,\\ &= \lim_{n\to+\infty}\Phi\int_\sigma^t e^{(t-\xi)G}\left\langle x_n^*, f(\xi)\right\rangle d\xi.\end{aligned}$$

For any $a \in Z_0$, set $f(.) = a$, then

$$\lim_{k\to+\infty}\int_{\sigma}^{t} e^{(t-\xi)G}\left\langle x_{n_k}^*, a\right\rangle d\xi = \lim_{p\to+\infty}\int_{\sigma}^{t} e^{(t-\xi)G}\left\langle x_{n_p}^*, a\right\rangle d\xi \text{ for } a \in Z_0,$$

which implies that

$$\int_{\sigma}^{t} e^{(t-\xi)G}\left\langle x_{Z_0}^*, a\right\rangle d\xi = \int_{\sigma}^{t} e^{(t-\xi)G}\left\langle \tilde{x}_{Z_0}^*, a\right\rangle d\xi \text{ for } a \in Z_0,$$

consequently $x_{Z_0}^* \equiv \tilde{x}_{Z_0}^*$, which gives a contradiction. We conclude that the whole sequence $\left(x_n^*\right)_{n\geq n_0}$ converges weakly to $x_{Z_0}^*$ in Z_0. Let Z_1 be another closed separable subspace of X, by using the same argument as above, we get that $\left(x_n^*\right)_{n\geq n_0}$ converges weakly to $x_{Z_1}^*$ in Z_1. Since $Z_0 \cap Z_1$ is a closed separable subspace of X, we get that $x_{Z_1}^* \equiv x_{Z_0}^*$ in $Z_0 \cap Z_1$. For any $x \in X$, we define x^* by

$$\langle x^*, x\rangle = \langle x_Z^*, x\rangle,$$

where Z is any closed separable subspace of X such that $x \in Z$. Then x^* is well defined on X and x^* is a bounded linear from X to $\mathbb{R}^d$ such that

$$|x^*| \leq \sup_{n\geq n_0} |x_n^*| < \infty,$$

and $\left(x_n^*\right)_{n\geq n_0}$ converges weakly to x^* in X.

As a consequence, we conclude that

Corollary 3.5. *For any continuous function $h : \mathbb{R} \to X$, we have*

$$\lim_{n\to+\infty}\int_{\sigma}^{t} \mathcal{U}^{\nu}(t-\xi)\,\Pi^{\nu}\left(\tilde{B}_n X_0 h(\xi)\right) d\xi = \Phi\int_{\sigma}^{t} e^{(t-\xi)G}\left\langle x^*, h(\xi)\right\rangle d\xi \text{ for all } t,\sigma \in \mathbb{R}.$$

Theorem 3.6. *Assume that* $(\mathbf{H_0})$ *and* $(\mathbf{H_1})$ *hold. Let u be a solution of Eq. (1) on $\mathbb{R}$. Then $z(t) = \langle \Psi, u_t\rangle$ is a solution of the ordinary differential equation*

$$\frac{d}{dt}z(t) = Gz(t) + \langle x^*, f(t)\rangle \text{ for } t \in \mathbb{R}. \tag{6}$$

Conversely, if f is a bounded function on $\mathbb{R}$ and z is a solution of Eq. (6) on $\mathbb{R}$, then the function u given by

$$u(t) = \left[\Phi z(t) + \lim_{n\to+\infty}\int_{-\infty}^{t} \mathcal{U}^{s}(t-\xi)\,\Pi^{s}\left(\tilde{B}_n X_0 f(\xi)\right) d\xi\right](0) \text{ for } t \in \mathbb{R},$$

is a solution of Eq. (1) on $\mathbb{R}$.

Let u be a solution of Eq. (1) on $\mathbb{R}$. Then

$$u_t = \Pi^s u_t + \Pi^v u_t \quad \text{for all } t \in \mathbb{R},$$

and

$$\Pi^v u_t = \mathcal{U}^v (t-\sigma)\, \Pi^v u_\sigma + \lim_{n\to+\infty} \int_\sigma^t \mathcal{U}^v (t-\xi)\, \Pi^v \left(\tilde{B}_n X_0 f(\xi)\right) d\xi \text{ for } t, \sigma \in \mathbb{R}.$$

Since $\Pi^v u_t = \Phi \langle \Psi, u_t \rangle$ and by Corollary 3.5, we get that

$$\begin{aligned}\Phi \langle \Psi, u_t \rangle &= \mathcal{U}^v (t-\sigma)\, \Phi \langle \Psi, u_\sigma \rangle + \Phi \int_\sigma^t e^{(t-\xi)G} \langle x^*, f(\xi) \rangle\, d\xi \text{ for } t, \sigma \in \mathbb{R}, \\ &= \Phi e^{(t-\sigma)G} \langle \Psi, u_\sigma \rangle + \Phi \int_\sigma^t e^{(t-\xi)G} \langle x^*, f(\xi) \rangle\, d\xi \text{ for } t, \sigma \in \mathbb{R}.\end{aligned}$$

Let $z(t) = \langle \Psi, u_t \rangle$. Then

$$z(t) = e^{(t-\sigma)G} z(\sigma) + \int_\sigma^t e^{(t-\xi)G} \langle x^*, f(\xi) \rangle\, d\xi \text{ for } t, \sigma \in \mathbb{R}.$$

Consequently, z is a solution of the ordinary differential equation (6) on $\mathbb{R}$. Conversely, assume that f is bounded on $\mathbb{R}$, then $\int_{-\infty}^t \mathcal{U}^s (t-\xi)\, \Pi^s \left(\tilde{B}_n X_0 f(\xi)\right) d\xi$ is well defined on $\mathbb{R}$. Let z be a solution of (6) on $\mathbb{R}$ and v be defined by

$$v(t) = \Phi z(t) + \lim_{n\to+\infty} \int_{-\infty}^t \mathcal{U}^s (t-\xi)\, \Pi^s \left(\tilde{B}_n X_0 f(\xi)\right) d\xi \text{ for } t \in \mathbb{R}.$$

Since

$$z(t) = e^{(t-\sigma)G} z(\sigma) + \int_\sigma^t e^{(t-\xi)G} \langle x^*, f(\xi) \rangle\, d\xi \text{ for } t, \sigma \in \mathbb{R},$$

Using Corollary 3.5, the function v_1 given by

$$v_1(t) = \Phi z(t) \text{ for } t \in \mathbb{R},$$

satisfies

$$v_1(t) = \mathcal{U}^v (t-\sigma)\, v_1(\sigma) + \lim_{n\to+\infty} \int_\sigma^t \mathcal{U}^v (t-\xi)\, \Pi^v \left(\tilde{B}_n X_0 f(\xi)\right) d\xi \quad \text{for } t, \sigma \in \mathbb{R}.$$

Moreover, the function v_2 given by

$$v_2(t) = \lim_{n\to+\infty} \int_{-\infty}^{t} \mathcal{U}^s(t-\xi)\,\Pi^s\left(\tilde{B}_n X_0 f(\xi)\right) d\xi \text{ for } t \in \mathbb{R},$$

satisfies

$$v_2(t) = \mathcal{U}^s(t-\sigma)\,v_2(\sigma) + \lim_{n\to+\infty} \int_{\sigma}^{t} \mathcal{U}^s(t-\xi)\,\Pi^s\left(\tilde{B}_n X_0 f(\xi)\right) d\xi \text{ for all } t \geq \sigma.$$

Then, for all $t \geq \sigma$ with $t, \sigma \in \mathbb{R}$, one has

$$\begin{aligned}
\mathcal{U}(t-\sigma)\,v(\sigma) &= \mathcal{U}^v(t-\sigma)\,v_1(\sigma) + \mathcal{U}^s(t-\sigma)\,v_2(\sigma),\\
&= v_1(t) - \lim_{n\to+\infty}\int_{\sigma}^{t} \mathcal{U}^v(t-\xi)\,\Pi^v\left(\tilde{B}_n X_0 f(\xi)\right) d\xi + v_2(t) -\\
&\qquad \lim_{n\to+\infty}\int_{\sigma}^{t} \mathcal{U}^s(t-\xi)\,\Pi^s\left(\tilde{B}_n X_0 f(\xi)\right) d\xi,\\
&= v(t) - \lim_{n\to+\infty}\int_{\sigma}^{t} \mathcal{U}(t-\xi)\left(\tilde{B}_n X_0 f(\xi)\right) d\xi.
\end{aligned}$$

Therefore

$$v(t) = \mathcal{U}(t-\sigma)\,v(\sigma) + \lim_{n\to+\infty} \int_{\sigma}^{t} \mathcal{U}(t-\xi)\left(\tilde{B}_n X_0 f(\xi)\right) \xi \text{ for } t \geq \sigma.$$

By Theorem 2.7, we obtain that the function u defined by $u(t) = v(t)(0)$ is a solution of Eq. (1) on $\mathbb{R}$.

4 Partial Functional Differential Equations with Infinite Delay and Variation of Constants Formula

The first part of this work is to establish a new variation of constants formula for the following partial functional differential equation with infinite delay

$$\begin{cases} \frac{d}{dt}x(t) = Ax(t) + Lx_t + f(t) \text{ for } t \geq 0, \\ x_0 = \phi \in \mathcal{B}, \end{cases} \tag{7}$$

where $A : D(A) \to X$ is a nondensely defined linear operator on a complex Banach space $(X, |.|)$, $\mathcal{B}$ is a normed linear space of functions mapping $(-\infty, 0]$ into X and satisfying some fundamental Axioms, x_t is an element of $\mathcal{B}$ defined by

$$x_t(\theta) = x(t + \theta) \text{ for } \theta \in (-\infty, 0],$$

L is a bounded linear operator from $\mathcal{B}$ into X, and f is a continuous X-valued function on $\mathbb{R}^+$. We assume that A is a Hille-Yosida operator.

Variation of constants formulas for partial functional differential equations plays an important role to study qualitative analysis for this kind of equations. We refer to [25] in the case of finite delay and to [13] in the case of infinite delay. Recently, in [2] it has been established a new variation of constants formula for neutral partial functional differential equations. This formula has been used to get some behavior results of solutions. In this work we will use the same method and techniques used in [2] to establish the same formula for partial functional differential equations with infinite delay whose linear part is nondensely defined. The variation of constants formula will be used in order to study the existence of almost periodic solutions when f is almost periodic. We establish the equivalence between the existence of an almost periodic solution for Eq. (7) on $\mathbb{R}$ and the existence of a bounded solution on $\mathbb{R}^+$. In this direction, Hino et al. [14, 16] have established a new variation of constants formula for Eq. (7) where A is densely defined and generates a strongly continuous semigroup on X. Let $x_t(\sigma, \varphi)$ be the mild solution of Eq. (7). Then they proved that $x_t(\sigma, \varphi)$ is represented by this formula

$$x_t(\sigma, \varphi) = U(t - \sigma)\varphi + \lim_{n \to \infty} \int_\sigma^t U(t - s)\Gamma^n f(s)\, ds \text{ for } t \geq \sigma, \tag{8}$$

where $\Gamma^n f(s)$ is defined by

$$(\Gamma^n f(s))(\theta) = \begin{cases} (n\theta + 1) f(s) & \text{for } -\frac{1}{n} \leq \theta \leq 0, \\ 0 \text{ for } \theta \leq -\frac{1}{n}, \end{cases}$$

and $(U(t))_{t \geq 0}$ is the solution semigroup of Eq. (7) with $f = 0$. Recall that if $\mathcal{B}$ satisfies the Hale and Kato's Axioms then $\Gamma^n f(s) \in \mathcal{B}$. The authors used formula (8) in order to establish the existence of almost periodic solution of Eq. (7). This work presents an extension of the works [14] and [16]. We will show that the density of the domain $D(A)$ is not needed here to get a new variation of constants formula, when $\mathcal{B}$ is a uniform fading memory space we establish a spectral decomposition of $\mathcal{B}$ which allows us to study the existence of bounded solutions.

The problem of finding periodic and almost periodic solutions of differential equations has been studied by several authors we refer to [5, 9, 15–17, 21] and the references therein. Consider the case of ordinary differential equations of the form

$$\frac{d}{dt} x(t) = Bx(t) + g(t) \text{ for } t \in \mathbb{R}, \tag{9}$$

where B is a $n \times n$ matrix, and g is a continuous function from $\mathbb{R}$ to X. Bohr and Neugebauer established in [9], that if f is almost periodic, then the existence of a bounded solution on $\mathbb{R}^+$ of Eq. (9) is equivalent to the existence of an almost periodic solution of Eq. (9). Moreover every bounded solution on the whole line $\mathbb{R}$ is almost periodic, for more details we refer to [9, Th.5.8]. For functional differential equations in finite dimensional space with finite delay r, the existence of a bounded solution is equivalent to the existence of periodic or almost periodic solutions, since the solution semigroup becomes compact whenever $t > r$. See [11]. The last condition becomes false for more general partial functional differential equations. In the case of infinite dimensional state space and finite delay, Travis and Webb [23] have shown that the compactness of the solution semigroup for $t > r$ remains true when A generates a compact semigroup, this property could be used in order to prove the existence of periodic or almost periodic solutions. When the delay is infinite, the compactness is not enough to deal with the existence of periodic or almost periodic solutions, we have to make more assumptions on the abstract phase space, like "uniform fading memory space", recently in [21], the authors studied the existence of a periodic solution of Eq. (7), where A generates a compact C_0-semigroup on X and f is periodic, using Hale and Chow fixed point Theorem, the authors proved that the Poincare map has at least one fixed point which gives a periodic solution.

In this work, we employ an axiomatic definition of the phase space $\mathcal{B}$ which has been introduced at first by Hale and Kato [12]. In the following, we assume that $\mathcal{B}$ is a normed space of functions mapping $]-\infty, 0]$ into X satisfying the following fundamental axioms:

(A): There exist a positive constant N, a locally bounded functions $M(\cdot)$ on $[0, +\infty)$ and a continuous function $K(\cdot)$ on $[0, +\infty[$, such that if $x :]-\infty, a] \to X$ is continuous on $[\sigma, a]$ with $x_\sigma \in \mathcal{B}$, for some $\sigma < a$, then for all $t \in [\sigma, a]$,

- (i) $x_t \in \mathcal{B}$,
- (ii) $t \mapsto x_t$ is continuous with respect to the norm of $\mathcal{B}$ on $[\sigma, a]$,
- (iii) $N|x(t)| \leq |x_t| \leq K(t-\sigma) \sup_{\sigma \leq s \leq t} |x(s)| + M(t-\sigma)|x_\sigma|$.

(B) : $\mathcal{B}$ is a Banach space.

We assume that

(D_1) : if $(\phi_n)_{n\geq 0}$ is a sequence in $\mathcal{B}$ such that $\phi_n \to 0$ in $\mathcal{B}$ as $n \to +\infty$, then for all $\theta \leq 0$, $(\phi_n(\theta))_{n\geq 0}$ converges to 0 in X.
Let $C(]-\infty, 0], X)$ be the space of continuous functions from $]-\infty, 0]$ into X. We make the following assumptions:

(D_2) : $\mathcal{B} \subset C(]-\infty, 0], X)$,

(D_3): there exists $\lambda_0 \in \mathbb{R}$ such that, for all $\lambda \in \mathbb{C}$ with $\operatorname{Re}\lambda > \lambda_0$ and $x \in X$, we have that $e^{\lambda \cdot}x \in \mathcal{B}$ and

$$K_0 := \sup_{\substack{\mathrm{Re}\,\lambda > \lambda_0,\, x \in X \\ x \neq 0}} \frac{\left|e^{\lambda \cdot} x\right|}{|x|} < \infty,$$

where $\left(e^{\lambda \cdot} x\right)(\theta) = e^{\lambda \theta} x$ for $\theta \in]-\infty, 0]$ and $x \in X$.

The following results are taken from [3].

Definition 4.1 ([3]). A function $u : \mathbb{R} \to X$ is called an integral solution of Eq. (7) on $\mathbb{R}^+$ if the following conditions hold

(i) u is continuous on $\mathbb{R}^+$,
(ii) $u_0 = \phi$,
(iii) $\int_0^t u(s)\,ds \in D(A)$ for $t \geq 0$,
(iv) $u(t) = \phi(0) + A \int_0^t u(s)\,ds + \int_0^t L u_s ds + \int_0^t f(s)\,ds$ for $t \geq 0$.

If the operator A is densely defined, then the integral solution coincides with the mild solution given in [14].

Theorem 4.2 ([3, pp. 336]). *Assume that $\mathcal{B}$ satisfies* **(A)** *and* **(B)**. *Then for all $\phi \in \mathcal{B}$ such that $\phi(0) \in \overline{D(A)}$, Eq. (7) has a unique integral solution $u(., \phi, L, f)$ on $\mathbb{R}^+$ given by*

$$u(t) = \begin{cases} T_0(t)\phi(0) + \lim_{\lambda \to +\infty} \int_0^t T_0(t-s)\lambda R(\lambda, A)\left[Lu_s + f(s)\right] ds \text{ for } t \geq 0, \\ \phi(t) \text{ for } \quad t \leq 0. \end{cases}$$

A continuous function u on $\mathbb{R}$ is said to be an integral solution of Eq. (7) on $\mathbb{R}$ if $u_s \in \mathcal{B}$ for $s \in \mathbb{R}$ and

$$u(t) = T_0(t-\sigma)u(\sigma) + \lim_{\lambda \to +\infty} \int_\sigma^t T_0(t-s)\lambda R(\lambda, A)[Lu_s + f(s)]\ ds \text{ for any } t \geq \sigma.$$

Let $\mathcal{B}_A := \left\{\phi \in \mathcal{B} : \phi(0) \in \overline{D(A)}\right\}$ be the phase space corresponding to Eq. (7). Define $U(t)$ for $t \geq 0$ by

$$U(t)\phi = u_t(\cdot, \phi, L) \text{ for } \phi \in \mathcal{B}_A,$$

where $u(\cdot, \phi, L)$ is the integral solution of Eq. (7) with $f = 0$.

Proposition 4.3 ([3, Proposition 2]). $(U(t))_{t \geq 0}$ *is a strongly continuous semigroup on $\mathcal{B}_A$, that's*

(i) $U(0) = Id$,
(ii) $U(t+s) = U(t)U(s)$ *for* $t, s \geq 0$,

(iii) *for all* $\phi \in \mathcal{B}_A$, $t \mapsto U(t)\phi$ *is continuous.*
Moreover $(U(t))_{t\,0}$ *satisfies the translation property*

$$(U(t)\phi)(\theta) = \begin{cases} U(t+\theta)\phi(0) \text{ for } t+\theta \geq 0 \\ \\ \phi(t+\theta) \text{ for } t+\theta \leq 0. \end{cases}$$

In order to establish a new variation of constant formula, we follow the same approach used in [2]. Before we need to recall the following results.

Lemma 4.4 ([3, Proposition 5]). *Let* $\mathcal{B}$ *satisfy Axioms* **(A)**, **(B)**, **(D**$_1$**)** *and* **(D**$_2$**)**. *Then the infinitesimal generator* A_U *of* $(U(t))_{t\geq 0}$ *is given by:*

$$\begin{cases} D(A_U) = \left\{ \begin{matrix} \phi \in C^1(]-\infty,0],X) \cap \mathcal{B}_A : \phi' \in \mathcal{B}_A, \phi(0) \in D(A) \, and \\ \phi'(0) = A\phi(0) + L(\phi) \end{matrix} \right\}, \\ A_U\phi \quad = \phi'. \end{cases}$$

By Axiom (**D**$_3$), we define for each complex number λ such that $\mathcal{R}e(\lambda) > \lambda_0$, the linear operator $\Delta(\lambda) : D(A) \to X$ by

$$\Delta(\lambda) = \lambda \mathrm{I} - A - L\left(e^{\lambda \cdot}\mathrm{I}\right).$$

Consider the space $\mathfrak{X} := \mathcal{B}_A \oplus \langle X_0 \rangle$, where $\langle X_0 \rangle = \{X_0 x : x \in X\}$ and $X_0 x$ is a function defined by

$$(X_0 x)(\theta) = \begin{cases} 0 \text{ if } \theta \in]-\infty, 0[, \\ x \text{ if } \theta = 0. \end{cases}$$

Then $\mathfrak{X}$ endowed with the norm $\|\phi + X_0 x\| = \|\phi\| + |x|$ is a Banach space.

Theorem 4.5. *Assume that* $\mathcal{B}$ *satisfies Axioms* **(A)**, **(B)**, **(D**$_1$**)**, **(D**$_2$**)** *and* **(D**$_3$**)**. *Then the extension* $\widetilde{A_U}$ *of the operator* A_U *defined on* $\mathfrak{X}$ *by*

$$\begin{cases} D\left(\widetilde{A_U}\right) = \{\phi \in \mathcal{B}_A : \phi' \in \mathcal{B}_A, and \ \phi(0) \in D(A)\}, \\ \widetilde{A_U}\phi \quad = \phi' + X_0(A\phi(0) + L\phi - \phi'(0)), \end{cases}$$

is a Hille-Yosida operator on $\mathfrak{X}$.

For the proof we need the following fundamental lemma.

Lemma 4.6. *There exist* $\omega_1 > \lambda_0$ *and* $M_1 \in \mathbb{R}$ *such that for* $\lambda > \omega_1$ *we have*

(i) $\Delta(\lambda)$ *is invertible and* $\left|\Delta(\lambda)^{-1}\right| \leq \frac{M_0}{\lambda - \omega_1}$.
(ii) $D\left(\widetilde{A_U}\right) = D(A_U) \oplus \left\langle e^{\lambda \cdot} \right\rangle$, *where*

$$\left\langle e^{\lambda \cdot} \right\rangle = \left\{ e^{\lambda \cdot} x : x \in D(A) \right\}.$$

(iii) $\lambda \in \rho\left(\widetilde{A_U}\right)$, *and for* $n \in \mathbb{N}^*$, $(\phi, x) \in \mathcal{B}_A \times X$, *one has*

$$R\left(\lambda, \widetilde{A_U}\right)^n (\phi + X_0 x) = R(\lambda, A_U)^n \phi + R(\lambda, A_U)^{n-1}\left(e^{\lambda \cdot}\Delta(\lambda)^{-1} x\right).$$

Proof of the Lemma. a) For $\lambda > \overline{\omega} := \max\{0, \omega_0, \lambda_0\}$, one has

$$\Delta(\lambda) = \lambda \mathrm{I} - A - L\left(e^{\lambda \cdot}\mathrm{I}\right) = (\lambda \mathrm{I} - A)\left(\mathrm{I} - R(\lambda, A) L\left(e^{\lambda \cdot}\mathrm{I}\right)\right),$$

and

$$\left|R(\lambda, A) L\left(e^{\lambda \cdot}x\right)\right| \le \frac{M_0 |L|}{\lambda - \omega_0}\left|e^{\lambda \cdot}x\right| \le \frac{M_0 K_0 |L|}{\lambda - \omega_0}|x| \text{ for } x \in X.$$

Consequently

$$\left|R(\lambda, A) L\left(e^{\lambda \cdot}\mathrm{I}\right)\right| \le \frac{\overline{M}}{\lambda - \omega_0} < 1 \quad \text{for all } \lambda > \omega_1 := \overline{\omega} + \overline{M},$$

where $\overline{M} := M_0 K_0 |L|$. We conclude that the operator $\left(\mathrm{I} - R(\lambda, A) L\left(e^{\lambda \cdot}\mathrm{I}\right)\right)$ is invertible, and

$$\left|\left(\mathrm{I} - R(\lambda, A) L\left(e^{\lambda \cdot}\mathrm{I}\right)\right)^{-1}\right| \le \frac{1}{1 - \left|R(\lambda, A) L\left(e^{\lambda \cdot}\mathrm{I}\right)\right|} \le \frac{\lambda - \omega_0}{\lambda - \omega_0 - \overline{M}}.$$

Consequently, $\Delta(\lambda)$ is invertible for $\lambda > \omega_1$ and

$$|\Delta(\lambda)^{-1}| \le \frac{M_0}{\lambda - \bar{\omega}}.$$

b) Let $\lambda > \omega_1$ and $\left(e^{\lambda \cdot}x\right) \in D(A_U) \cap \left\langle e^{\lambda \cdot}\right\rangle$. Then $\lambda x = Ax + L\left(e^{\lambda \cdot}x\right)$, that is

$$\Delta(\lambda)x = 0.$$

Since $\Delta(\lambda)$ is invertible for $\lambda > \omega_1$, we conclude that $D(A_U) \cap \left\langle e^{\lambda \cdot}\right\rangle = \{0\}$. On the other hand, let $\tilde{\psi} \in D\left(\widetilde{A_U}\right)$ and ψ given by

$$\psi = \tilde{\psi} + e^{\lambda \cdot}\Delta(\lambda)^{-1}\left(A\tilde{\psi}(0) + L\tilde{\psi} - \tilde{\psi}'(0)\right).$$

Then

$$\begin{aligned}
A\psi(0) + L\psi &= A\tilde{\psi}(0) + L\tilde{\psi} + A\Delta(\lambda)^{-1}\left(A\tilde{\psi}(0) + L\tilde{\psi} - \tilde{\psi}'(0)\right) \\
&\quad + L\left(e^{\lambda\cdot}\Delta(\lambda)^{-1}\left(A\tilde{\psi}(0) + L\tilde{\psi} - \tilde{\psi}'(0)\right)\right) \\
&= A\tilde{\psi}(0) + L\tilde{\psi} - \Delta(\lambda)\Delta(\lambda)^{-1}\left(A\tilde{\psi}(0) + L\tilde{\psi} - \tilde{\psi}'(0)\right) \\
&\quad + \lambda\Delta(\lambda)^{-1}\left(A\tilde{\psi}(0) + L\tilde{\psi} - \tilde{\psi}'(0)\right) \\
&= \tilde{\psi}'(0) + \lambda\Delta(\lambda)^{-1}\left(A\tilde{\psi}(0) + L\tilde{\psi} - \tilde{\psi}'(0)\right) \\
&= \psi'(0).
\end{aligned}$$

Hence $\psi \in D(A_U)$, which implies that $D\left(\widetilde{A_U}\right) = D(A_U) \oplus \left\langle e^{\lambda\cdot}\right\rangle$.

c) Let $\lambda > \omega_1$ and $\tilde{\psi} \in \mathfrak{X}$. Then $\tilde{\psi} = \psi + X_0 x$ for some $\psi \in \mathcal{B}_A$ and $x \in X$. We seek for $\tilde{\phi} = \phi + e^{\lambda\cdot}a \in D\left(\widetilde{A_U}\right)$ such that $\left(\lambda I - \widetilde{A_U}\right)\tilde{\phi} = \tilde{\psi}$, where $\phi \in D(A_U)$ and $a \in D(A)$. We have $\left(\lambda I - \widetilde{A_U}\right)\left(\phi + e^{\lambda\cdot}a\right) = \psi + X_0 x$, which is equivalent to find $(a,\phi) \in D(A) \times D(A_U)$ such that

$$\begin{cases} (\lambda I - A_U)\phi = \psi, \\ \Delta(\lambda)\, a \qquad = x. \end{cases}$$

For ω_1 large enough, it follows that, $\left(\lambda I - \widetilde{A_U}\right)^{-1}$ exists for $\lambda > \omega_1$, and

$$\left(\lambda I - \widetilde{A_U}\right)^{-1}(\psi + X_0 x) = (\lambda I - A_U)^{-1}\psi + e^{\lambda\cdot}\Delta(\lambda)^{-1}x.$$

Consequently, for $n \in \mathbb{N}^*$, we have

$$R\left(\lambda, \widetilde{A_U}\right)^n(\psi + X_0 x) = R(\lambda, A_U)^n\psi + R(\lambda, A_U)^{n-1}\left(e^{\lambda\cdot}\Delta(\lambda)^{-1}x\right). \qquad \square$$

Proof of Theorem 4.5. Since A_U is the generator of the semigroup $(U(t)_{t\geq 0})$ on $\mathcal{B}_A$, by Hille and Yosida's Theorem [20] there exists a positive constant $\tilde{M}$ such that

$$\sup_{n\in\mathbb{N},\, \lambda>\omega_1} |(\lambda - \omega_1)^n R(\lambda, A_U)^n| \leq \tilde{M}.$$

By Lemma 4.6, there exist ω_1 and $M_1 > 0$ such that

$$\sup_{n\in\mathbb{N},\, \lambda>\omega_1} \left|(\lambda - \omega_1)^n R\left(\lambda, \widetilde{A_U}\right)^n\right| \leq M_1. \qquad \square$$

Lemma 4.7. *The part of $\widetilde{A_U}$ in $\overline{D\left(\widetilde{A_U}\right)}$ is the operator A_U.*

Proof. From Lemma 4.4, the operator A_u generates a strongly continuous semigroup on $\mathcal{B}_A$, by Hille and Yosida's Theorem $\overline{D(A_U)} = \mathcal{B}_A$. Since, $D(A_U) \subset D\left(\widetilde{A_U}\right) \subset \mathcal{B}_A$, then

$$\overline{D(A_U)} = \overline{D\left(\widetilde{A_U}\right)} = \mathcal{B}_A.$$

Let C be the part of $\widetilde{A_U}$ in $\overline{D\left(\widetilde{A_U}\right)}$, which is defined by

$$\begin{cases} D(C) = \left\{\phi \in D\left(\widetilde{A_U}\right) : \widetilde{A_U}\phi \in \mathcal{B}_A\right\}, \\ C\phi \quad = \widetilde{A_U}\phi. \end{cases}$$

Then $D(A_U) \subseteq D(C)$ and $A_U\phi = C\phi$ for all $\phi \in D(A_U)$.

Conversely, let $\phi \in D(C)$. Then

$$\begin{cases} \phi \in C^1(]-\infty, 0], X) \cap \mathcal{B}_A,\ \phi' \in \mathcal{B}_A, \phi(0) \in D(A) \\ \qquad \phi' + X_0(A\phi(0) + L\phi - \phi'(0)) \in \mathcal{B}_A. \end{cases}$$

By assumption $(\mathbf{D}_2)$, it follows that

$$\begin{cases} \phi \in D\left(\widetilde{A_U}\right) \text{ and } \phi'(0) = A\phi(0) + L\phi \\ C\phi = \phi'. \end{cases}$$

From which we conclude that $C = A_U$. □

Consider the following evolution equation

$$\begin{cases} \dfrac{d}{dt}\xi(t) = \widetilde{A_U}\xi(t) + X_0 f(t) \text{ for } t \geq 0 \\ \xi(0) \quad = \tilde{\phi} \in \mathfrak{X}. \end{cases} \tag{10}$$

Definition 4.8. A continuous function $\xi : [0, +\infty[\to \mathcal{B}_A$ is called an integral solution of Eq. (10) if

(i) $\displaystyle\int_0^t \xi(s)\,ds \in D\left(\widetilde{A_U}\right)$ for $t \geq 0$,

(ii) $\displaystyle\xi(t) = \tilde{\phi} + \widetilde{A_U}\int_0^t \xi(s)\,ds + \int_0^t X_0 f(s)\,ds$ for $t \geq 0$.

Theorem 4.9. *Assume that* $(\mathbf{D}_1)$, $(\mathbf{D}_2)$ *and* $(\mathbf{D}_3)$ *hold. If u is an integral solution of Eq. (7), then the function given by $\xi(t) = u_t$, $t \geq 0$, is an integral solution of Eq. (10) for $\tilde{\phi} = \phi$. Conversely, if ξ is an integral solution of Eq. (10) with $\tilde{\phi} = \phi$, then the function u defined by*

$$u(t) = \begin{cases} \xi(t)(0) \ if\ t \geq 0 \\ \phi(t) \quad\ if\ t \leq 0 \end{cases}$$

is an integral solution of Eq. (7).

Proof. Let $\phi \in \mathcal{B}_A$ and u be the integral solution of Eq. (7). Define $\xi : [0, \infty) \to \mathcal{B}_A$ by

$$\xi(t) = u_t \text{ for } t \geq 0.$$

To compute the integral in $\mathcal{B}$ in term of the integral in X, we need the following lemma.

Lemma 4.10 ([3]). *Assume that* $(\mathbf{D}_1)$ *holds, and* $F : [0,a] \to \mathcal{B}$ *is continuous, then*

$$\left(\int_0^a F(s)\,ds\right)(\theta) = \int_0^a F(s)(\theta)\,ds \textit{ for all } \theta \leq 0.$$

By Lemma 4.10, we have

$$\begin{aligned}\frac{d}{d\theta}\left(\int_0^t u_s ds\right)(\theta) &= \frac{d}{d\theta}\left(\int_0^t u(s+\theta)\,ds\right)\\ &= \frac{d}{d\theta}\left(\int_\theta^{t+\theta} u(s)\,ds\right)\\ &= u_t(\theta) - \phi(\theta).\end{aligned}$$

Then

$$\widetilde{A_U}\left(\int_0^t \xi(s)\,ds\right) = u_t - \phi + X_0\left(A\int_0^t u(s)\,ds + L\left(\int_0^t u_s ds\right) - u(t) - \phi(0)\right).$$

Since u is an integral solution of Eq. (7), it follows that

$$u(t) = \phi(0) + A\int_0^t u(s)\,ds + L\left(\int_0^t u_s ds\right) + \int_0^t f(s)\,ds,$$

which implies that

$$\xi(t) = \phi + \widetilde{A_U}\int_0^t \xi(s)\,ds + X_0\int_0^t f(s)\,ds \quad \text{for } t \geq 0.$$

Consequently ξ is an integral solution of Eq. (10). Conversely, let ξ be an integral solution of Eq. (10) for $\tilde{\phi} = \phi$. Then ξ satisfies the following translation property

$$\xi(t)(\theta) = \begin{cases} \xi(t+\theta)(0) \text{ if } t+\theta \geq 0,\\ \phi(t+\theta) \quad \text{if } t+\theta \leq 0,\end{cases}$$

In fact, for $t+\theta \geq 0$,

$$\xi(t)(\theta) = (U(t)\phi)(\theta) + \lim_{\lambda\to+\infty}\int_0^t \left(U(t-s)\lambda R\left(\lambda,\widetilde{A_U}\right)X_0 f(s)\right)(\theta)\,ds.$$

Then

$$\xi(t)(\theta) = (U(t+\theta)\phi)(0) + \lim_{\lambda\to+\infty}\int_0^{t+\theta} \left(U(t+\theta-s)\lambda R\left(\lambda,\widetilde{A_U}\right)X_0 f(s)\right) \times (0)\,ds + \lim_{\lambda\to+\infty}\int_{t+\theta}^{t} \left(U(t-s)\lambda R\left(\lambda,\widetilde{A_U}\right)X_0 f(s)\right)(\theta)\,ds.$$

Since

$$\begin{aligned}\lim_{\lambda\to+\infty}\int_{t+\theta}^{t} &\left(U(t-s)\lambda R\left(\lambda,\widetilde{A_U}\right)X_0 f(s)\right)(\theta)\,ds \\ &= \lim_{\lambda\to+\infty}\int_{t+\theta}^{t}\left(\lambda R\left(\lambda,\widetilde{A_U}\right)X_0 f(s)\right)(t-s+\theta)\,ds \\ &= \lim_{\lambda\to+\infty}\int_{t+\theta}^{t} e^{\lambda(t-s+\theta)}\lambda\Delta(\lambda)^{-1} f(s)\,ds \\ &= 0.\end{aligned}$$

which gives that

$$\begin{aligned}\xi(t)(\theta) &= (U(t+\theta)\phi)(0) + \lim_{\lambda\to+\infty}\int_0^{t+\theta}\left(U(t+\theta-s)\lambda R\left(\lambda,\widetilde{A_U}\right)X_0 f(s)\right) \\ &\quad\times(0)\,ds \\ &= \xi(t+\theta)(0).\end{aligned}$$

If we consider the function

$$u(t) = \begin{cases} \xi(t)(0) & \text{if } t > 0, \\ \phi(t) & \text{if } t \le 0. \end{cases}$$

Then $\xi(t) = u_t$ for all t 0 and

$$u_t = \phi + \widetilde{A_U}\left(\int_0^t u_s ds\right) + \int_0^t X_0 f(s)\,ds \text{ for } t \ge 0.$$

Which implies that u is an integral solution of Eq. (7). □

Theorem 4.11. *Assume that* $(\mathbf{D}_1)$, $(\mathbf{D}_2)$ *and* $(\mathbf{D}_3)$ *hold. Then the integral solution* x *of Eq. (7) is given by the following variation of constants formula*

$$x_t = U(t)\phi + \lim_{n\to+\infty}\int_0^t U(t-s)\widetilde{B_n}(X_0 f(s))\,ds \; for \; t \ge 0, \tag{11}$$

where $\widetilde{B_n} = n\left(n-\widetilde{A_u}\right)^{-1}$.

Proof. This theorem is a consequence from Theorem 4.9 and the following lemma.

Lemma 4.12 ([22]). *Let C be a Hille-Yosida operator on a Banach space Y and $\alpha : \mathbb{R}^+ \to Y$ be a continuous function. Consider the following problem*

$$\begin{cases} \frac{d}{dt}x(t) = Cx(t) + \alpha(t) \text{ for } t \geq 0, \\ x(0) = x_0 \in Y. \end{cases}$$

If $x_0 \in \overline{D(C)}$, then there exists a unique continuous function x such that

(i) $\int_0^t x(s)\,ds \in D(C)$ *for* $t \geq 0$

(ii) $x(t) = x_0 + C\int_0^t x(s)\,ds + \int_0^t \alpha(s)\,ds$ *for* $t \geq 0$.

Moreover, x is given by

$$x(t) = S_0(t)x_0 + \lim_{\lambda\to+\infty}\int_0^t S_0(t-s)C_\lambda \alpha(s)\,ds \text{ for } t \geq 0,$$

where $C_\lambda := \lambda(\lambda I - C)^{-1}$ and $(S_0(t))_{t\geq 0}$ is the semigroup generated by the part of C in $\overline{D(C)}$.

5 Reduction of Complexity of Partial Functional Differential Equations in Fading Memory Spaces

Let C_{00} be the space of X-valued continuous function on $]-\infty, 0]$ with compact support.

(C) : If a uniformly bounded sequence $(\varphi_n)_{n\in\mathbb{N}}$ in C_{00} converges to a function φ compactly on $]-\infty, 0]$, then φ is in $\mathcal{B}$ and $|\varphi_n - \varphi| \to 0$ as $n \to \infty$.

Let $(S_0(t))_{t\geq 0}$ be the strongly continuous semigroup defined on the subspace

$$\mathcal{B}_0 := \{\phi \in \mathcal{B} : \phi(0) = 0\}$$

by

$$(S_0(t)\phi)(\theta) = \begin{cases} \phi(t+\theta) & \text{if } t+\theta \leq 0, \\ 0 & \text{if } t+\theta \geq 0. \end{cases}$$

Definition 5.1. Assume that the space $\mathcal{B}$ satisfies Axioms **(B)** and **(C)**. $\mathcal{B}$ is said to be a fading memory space if for all $\phi \in \mathcal{B}_0$,

$$S_0(t)\phi \underset{t\to\infty}{\longrightarrow} 0 \text{ in } \mathcal{B}_0.$$

Moreover, $\mathcal{B}$ is said to be a uniform fading memory space if

$$|S_0(t)| \underset{t\to\infty}{\longrightarrow} 0, \text{ with respect to the operator norm.}$$

Lemma 5.2 ([13, pp 190]). *The following statements hold:*

(*i*) *If $\mathcal{B}$ is a fading memory space, then the functions $K(\cdot)$ and $M(\cdot)$ in axiom* (**A**) *can be chosen to be constants.*
(*ii*) *If $\mathcal{B}$ is a uniform fading memory space, then we can choose the function $K(\cdot)$ constant and the function $M(\cdot)$ such that $M(t) \to 0$ as $t \to \infty$.*

Proposition 5.3 ([13]). *If the phase space $\mathcal{B}$ is a fading memory space, then the space $BC(]-\infty,0],X)$ of bounded continuous X-valued functions on $]-\infty,0]$ endowed with the uniform norm topology is continuously embedding in $\mathcal{B}$. In particular $\mathcal{B}$ satisfies* ($\mathbf{D}_3$)*, for $\lambda_0 > 0$.*

In this section, we assume that

($\mathbf{H}_2$) $\mathcal{B}$ is a uniform fading memory space.

Let V be a bounded subset of a Banach space Y, the Kuratowski measure of noncompactness $\alpha(V)$ of V is given by

$$\alpha(v) = \inf\left\{ \begin{array}{c} d > 0 \text{ such that there exists a finite number of sets } V_1,\ldots,V_n \text{ with} \\ diam(V_i) \le d \text{ such that } V \subseteq \bigcup_{i=1}^{n} V_i \end{array} \right\},$$

and for a bounded linear operator F on Y, we define $|F|_\alpha$ by $|F|_\alpha = \inf\{k > 0 : \alpha(F(V)) \le k\alpha(V), \text{forallboundedset} V \text{of} Y\}$. For a strongly continuous semigroup $(S(t))_{t\ge 0}$, we define the essential growth bound $\omega_{ess}(S)$ by

$$\omega_{ess}(S) = \lim_{t\to\infty} \frac{1}{t}\log|S(t)|_\alpha.$$

Theorem 5.4 ([5]). *Assume that $\mathcal{B}$ satisfies Axioms* (**A**), (**B**), ($\mathbf{D}_1$) *and assumptions* ($\mathbf{H}_0$), ($\mathbf{H}_1$), ($\mathbf{H}_2$) *hold. Then*

$$\omega_{ess}(U) < 0.$$

From [8, Corollary IV.2.11], it follows that

$$\sigma_u(A_U) := \{\lambda \in \sigma(A_U) : \mathrm{Re}(\lambda) \ge 0\}$$

is a finite subset and $\mathcal{B}_A$ is decomposed as follows:

$$\mathcal{B}_A = \mathcal{S} \oplus \mathcal{V},$$

where $\mathcal{S}$, $\mathcal{V}$ are two closed subspaces of $\mathcal{B}_A$ which are invariant by $(U(t))_{t\geq 0}$. Let $U^{\mathcal{S}}(t)$ be the restriction of $U(t)$ on $\mathcal{S}$, then there exist positive constants N and μ such that

$$\left|U^{\mathcal{S}}(t)\phi\right| \leq Ne^{-\mu t}|\phi| \text{ for all } \phi \in \mathcal{S},$$

$\mathcal{V}$ is a finite dimensional space and the restriction $U^{\mathcal{V}}(t)$ of $U(t)$ on $\mathcal{V}$ becomes a group. Let $\Pi^{\mathcal{S}}$ and $\Pi^{\mathcal{V}}$ denote the projections on $\mathcal{S}$ and $\mathcal{V}$ respectively. Let $d = \dim \mathcal{V}$ and take a basis $\{\phi_1, \ldots, \phi_d\}$ in $\mathcal{V}$. Then there exist d-elements $\{\psi_1, \ldots, \psi_d\}$ in the dual space $\mathcal{B}_A^*$ of $\mathcal{B}_A$, such that $\langle \psi_i, \phi_j \rangle = \delta_{ij}$, where

$$\delta_{ij} = \begin{cases} 1 \text{ if } i = j \\ 0 \text{ if } i \neq j, \end{cases}$$

and $\psi_i = 0$ on $\mathcal{S}$, where $\langle \cdot, \cdot \rangle$ denotes the canonical pairing between the dual space and the original space. Denote by $\Phi := (\phi_1, \ldots, \phi_d)$ and Ψ is the transpose of $(\psi_1, \ldots, \psi_d)$, in particular one has

$$\Psi\Phi = \mathbb{I}_{\mathbb{R}^d},$$

where $\mathbb{I}_{\mathbb{R}^d}$ is the identity $d \times d$ matrix. For each $\phi \in \mathcal{B}_A$, $\Pi^{\mathcal{V}}\phi$ is computed by:

$$\begin{aligned} \Pi^{\mathcal{V}}\phi &= \Phi \langle \Psi, \phi \rangle \\ &= \sum_{i=1}^{d} \langle \psi_i, \phi \rangle \phi_i. \end{aligned}$$

Let $\zeta(t) := (\zeta_1(t), \ldots, \zeta_d(t))$ be the component of $\Pi^{\mathcal{V}} x_t$ in the basis vector Φ, then

$$\Pi^{\mathcal{V}} x_t = \Phi\zeta(t), \text{and } \zeta(t) = \langle \Psi, x_t \rangle.$$

Since $\left(U^{\mathcal{V}}(t)\right)_{t\geq 0}$ is a group on a finite dimensional space $\mathcal{V}$, then there exists a $d \times d$ matrix G such that

$$U^{\mathcal{V}}(t)\phi = \Phi e^{Gt} \langle \Psi, \phi \rangle \text{ for all } t \in \mathbb{R} \text{ and } \phi \in \mathcal{V},$$

which means that

$$U^{\mathcal{V}}(t)\Phi = \Phi e^{Gt} \text{ for all } t \in \mathbb{R}.$$

For $n > \omega_1$ and $i \in \{1, \ldots, d\}$, we define the functional x_n^{*i} by

$$\langle x_n^{*i}, x \rangle = \langle \psi_i, \widetilde{B_n}(X_0 x) \rangle \text{ for all } x \in X.$$

Then x_n^{*i} is a bounded linear operator on X with $\left|x_n^{*i}\right| \leq K_0 M_1 |\psi_i|$. Define the d-column vector x_n^* as an element of $\mathcal{L}(X, \mathbb{R}^d)$ (the space of bounded linear operator from X into $\mathbb{R}^d$) given by the transpose of $\left(x_n^{*1}, \ldots, x_n^{*d}\right)$. Then, for all $n \geq 1$, $x \in X$

$$\langle x_n^*, x\rangle = \left\langle \Psi, \widetilde{B_n}\left(X_0 x\right)\right\rangle \text{and} \sup_{n\geq\omega_1} |x_n^*| \leq K_0 M_1 \sup_{i=1,\ldots,d} |\psi_i| < \infty.$$

Theorem 5.5. *The sequence $\left(x_n^*\right)_{n\geq 0}$ converges weakly in $\mathcal{L}(X, \mathbb{R}^d)$, in the sense that*

$$\langle x_n^*, x\rangle \underset{n\to\infty}{\longrightarrow} \langle x^*, x\rangle \textit{ for all } x \in X.$$

Let Y_0 be any separable closed subspace of X. By Theorem 3.4, the restriction $\left(x_n^{Y_0^*}\right)_{n\geq 0}$ of $\left(x_n^*\right)_{n\geq 0}$ in Y_0 has a subsequence $\left(x_{n_k}^{Y_0^*}\right)_{k\geq 0}$ such that

$$\lim_{k\to\infty} \left\langle x_{n_k}^{Y_0^*}, y\right\rangle = \left\langle x^{Y_0^*}, y\right\rangle \text{for all } y \in Y_0,$$

where $x^{Y_0^*} \in Y_0^*$. We claim that the whole sequence $\left(x_n^{Y^*}\right)_{n\geq 0}$ converges weakly in Y_0^* to $x^{Y_0^*}$. We proceed by contradiction and assume that there exists a subsequence $\left(x_{m_k}^{Y^*}\right)_{k\geq 0}$ of $\left(x_n^{Y^*}\right)_{n\geq 0}$ such that $x_{m_k}^{Y_0^*} \underset{k\to\infty}{\longrightarrow} x_1^{Y_0^*}$ weakly in Y_0, with $x^{Y_0^*} \neq x_1^{Y_0^*}$. To conclude we need the following lemma.

Lemma 5.6. *For any continuous function $h : \mathbb{R}^+ \to X$ one has:*

$$\lim_{n\to\infty} \int_0^t U^\nu(t-s)\,\Pi^\nu\left(\widetilde{B_n}\left(X_0 h(s)\right)\right) ds = \Phi \lim_{n\to\infty} \int_0^t e^{(t-s)G} \langle x_n^*, h(s)\rangle\, ds.$$

Proof of the Lemma. In fact, we have

$$\begin{aligned}
&\lim_{n\to\infty} \int_0^t U^\nu(t-s)\,\Pi^\nu\left(\widetilde{B_n}\left(X_0 h(s)\right)\right) ds \\
&\quad = \lim_{n\to\infty} \int_0^t \left(U^\nu(t-s)\,\Phi\right)\left\langle \Psi, \widetilde{B_n}\left(X_0 h(s)\right)\right\rangle ds, \\
&\quad = \lim_{n\to\infty} \int_0^t \Phi e^{(t-s)G} \langle x_n^*, h(s)\rangle\, ds, \\
&\quad = \Phi \lim_{n\to\infty} \int_0^t e^{(t-s)G} \langle x_n^*, h(s)\rangle\, ds.
\end{aligned}$$

Let $h(\cdot) = y$ for any $y \in Y_0$. Then

$$\int_0^t e^{(t-s)G}\left\langle x^{Y_0^*}, y\right\rangle ds = \int_0^t e^{(t-s)G}\left\langle x_1^{Y_0^*}, y\right\rangle ds \text{ for any } y \in Y_0.$$

This is true if and only if $\left\langle x^{Y_0^*}, y\right\rangle = \left\langle x_1^{Y_0^*}, y\right\rangle$, for all $y \in Y_0$, which gives a contradiction. Consequently the whole sequence $\left(x_n^{Y_0^*}\right)_{n\geq 0}$ converges weakly in $\mathcal{L}(Y_0, \mathbb{R}^d)$ to $x^{Y_0^*}$.

Let Y_1 be another separable closed space of X. Then the restriction $\left(x_n^{Y_1^*}\right)_{n\geq 0}$ of $(x_n^*)_{n\geq 0}$ in Y_1 converges weakly to some $x^{Y_1^*} \in Y_1^*$, and we get that $x^{Y_0^*} = x^{Y_1^*}$ in $Y_0 \cap Y_1$. Since $(x_n^*)_{n\, 0}$ converges weakly in $Y_0 \cap Y_1$, and by the uniqueness of the limit we obtain that $x^{Y_0^*} = x^{Y_1^*}$ in $Y_0 \cap Y_1$. Let x^* be the operator defined by

$$\langle x^*, x\rangle = \left\langle x^{Y^*}, x\right\rangle,$$

for any separable closed space Y of X such that $x \in Y$. Then x^* is well defined and belongs to $\mathcal{L}(X, \mathbb{R}^d)$. Moreover

$$\langle x_n^*, x\rangle \underset{n\to\infty}{\longrightarrow} \langle x^*, x\rangle \text{ for all } x \in X. \qquad \square$$

Consequently, we get the following.

Corollary 5.7. *For any continuous function $h : [0, a] \to X$:*

$$\lim_{n\to\infty} \int_0^t U^{\mathcal{V}}(t-s)\,\Pi^{\mathcal{V}}\left(\widetilde{B_n}(X_0 h(s))\right) ds = \Phi \int_0^t e^{(t-s)G}\left\langle x^*, h(s)\right\rangle ds \textit{ for all } t \in [0, a].$$

Theorem 5.8. *Assume that* **(A)**, **(B)**, **(D$_1$)**, **(D$_2$)**, **(H$_0$)**, **(H$_1$)** *and* **(H$_2$)** *hold. Let u be an integral solution of Eq. (7) on $\mathbb{R}$. Then $\zeta(t) = \langle \Psi, u_t\rangle$, $t \in \mathbb{R}$ is a solution of the following ordinary differential equation*

$$\dot{\zeta}(t) = G\zeta(t) + \langle x^*, f(t)\rangle \text{ for } t \in \mathbb{R}. \tag{12}$$

Conversely, if f is bounded and ζ is a solution of Eq. (12), then the function

$$\left(\Phi\zeta(t) + \lim_{n\to+\infty}\int_{-\infty}^t U^{\mathcal{S}}(t-s)\,\Pi^{\mathcal{S}}\left(\widetilde{B_n}(X_0 f(s))\right) ds\right)(0) \tag{13}$$

is an integral solution of Eq. (7) on $\mathbb{R}$.

Proof. Using the variation of constants formula (11), we obtain that for $t \geq \sigma$

$$\begin{aligned}
\langle \Psi, x_t \rangle &= \langle \Psi, U(t-\sigma)x_\sigma \rangle + \left\langle \Psi, \lim_{n\to+\infty} \int_\sigma^t U(t-s)\left(\widetilde{B_n}(X_0 f(s))\right) ds \right\rangle, \\
&= e^{(t-\sigma)G} \langle \Psi, x_\sigma \rangle + \lim_{n\to\infty} \int_\sigma^t e^{(t-s)G} \left\langle \Psi, \left(\widetilde{B_n}(X_0 f(s))\right)\right\rangle ds, \\
&= e^{(t-\sigma)G} \langle \Psi, x_\sigma \rangle + \lim_{n\to\infty} \int_\sigma^t e^{(t-s)G} \langle x_n^*, f(s) \rangle ds, \\
&= e^{(t-\sigma)G} \langle \Psi, x_\sigma \rangle + \int_\sigma^t e^{(t-s)G} \langle x^*, f(s) \rangle ds,
\end{aligned}$$

which means that $\zeta(t) = \langle \Psi, x_t \rangle$, $t \in \mathbb{R}$ is a solution of the ordinary differential equation (12). Conversely, if we assume that f is bounded on $\mathbb{R}$, then formula (13) is well defined, since the restriction of the solution semigroup on $\mathcal{S}$ is exponentially stable. Let y be defined by:

$$y(t) := \lim_{\lambda\to+\infty} \int_{-\infty}^t U^{\mathcal{S}}(t-s)\Pi^{\mathcal{S}}\left(\widetilde{B_n}(X_0 f(s))\right) ds \text{ for } t \in \mathbb{R}.$$

Then for $t \geq \sigma$,

$$\begin{aligned}
U^{\mathcal{S}}(t-\sigma)y(\sigma) &+ \lim_{n\to+\infty} \int_\sigma^t U^{\mathcal{S}}(t-s)\Pi^{\mathcal{S}}\left(\widetilde{B_n}(X_0 f(s))\right) ds \\
&= \lim_{n\to+\infty} \left(\int_{-\infty}^\sigma U^{\mathcal{S}}(t-s)\Pi^{\mathcal{S}}\left(\widetilde{B_n}(X_0 f(s))\right) ds + \right. \\
&\qquad \left. \int_\sigma^t U^{\mathcal{S}}(t-s)\Pi^{\mathcal{S}}\left(\widetilde{B_n}(X_0 f(s))\right) ds \right) \\
&= y(t).
\end{aligned} \tag{14}$$

Moreover the solution ζ of Eq. (12) is given by

$$\zeta(t) = e^{(t-\sigma)G}\zeta(\sigma) + \int_\sigma^t e^{(t-s)G} \langle x^*, f(s) \rangle ds \text{ for } t \geq \sigma.$$

Corollary 5.7, gives that

$$\Phi\zeta(t) = \Phi e^{(t-\sigma)G}\zeta(\sigma) + \lim_{n\to\infty} \int_\sigma^t U^{\mathcal{V}}(t-s)\Pi^{\mathcal{V}}\left(\widetilde{B_n}(X_0 f(s))\right) ds \text{ for } t \geq \sigma,$$

and

$$\Phi\zeta(t) = U^{\mathcal{V}}(t-\sigma)\Phi\zeta(\sigma) + \lim_{n\to\infty} n \int_\sigma^t U^{\mathcal{V}}(t-s)\Pi^{\mathcal{V}}\left(\widetilde{B_n}(X_0 f(s))\right) ds \text{ for } t \geq \sigma. \tag{15}$$

Set $\xi(t) = \Phi\zeta(t) + y(t)$ on $\mathbb{R}$, by (14) and (15), we obtain that

$$\begin{aligned}\xi(t) &= U(t-\sigma)(\Phi\zeta(\sigma) + y(\sigma)) \\ &\quad + \lim_{n\to\infty} n\int_\sigma^t U(t-s)\left[\Pi^{\mathcal{V}} + \Pi^{\mathcal{S}}\right]\left(\widetilde{B_n}(X_0 f(s))\right) ds \text{ for } t \geq \sigma. \\ &= U(t-\sigma)\xi(\sigma) + \lim_{n\to\infty}\int_\sigma^t U(t-s)\left(\widetilde{B_n}(X_0 f(s))\right) ds \text{ for } t \geq \sigma.\end{aligned}$$

From Theorem 4.9, we conclude that the function

$$\left(\Phi\zeta(t) + \lim_{n\to+\infty}\int_{-\infty}^t U^{\mathcal{S}}(t-s)\,\Pi^{\mathcal{S}}\left(\widetilde{B_n}(X_0 f(s))\right) ds\right)(0)$$

is an integral solution of Eq. (7).

6 Almost Automorphic Solutions for Eq. (1)

We recall some properties about almost automorphic functions. Let $\mathcal{BC}(\mathbb{R}, X)$ be the space of all bounded continuous functions from $\mathbb{R}$ to X, provided with the uniform norm topology. Let $h \in \mathcal{BC}(\mathbb{R}, X)$ and $\tau \in \mathbb{R}$, we define the function h_τ by

$$h_\tau(s) = h(\tau + s) \text{ for all } s \in \mathbb{R}.$$

Definition 6.1 ([9, Definition 1.1.1, pp.1]). A bounded continuous function $h : \mathbb{R} \to X$ is said to be almost periodic if

$$\{h_\tau : \tau \in \mathbb{R}\} \text{ is relatively compact in } \mathcal{BC}(\mathbb{R}, X).$$

Definition 6.2 (Bochner, [18, Theorem 5.8, pp. 86]). A continuous function $h : \mathbb{R} \to X$ is said to be almost automorphic if for every sequence of real numbers $(s'_n)_n$ there exists a subsequence $(s_n)_n$ such that

$$\lim_{n\to\infty} h(t + s_n) = k(t) \text{ exists for all } t \text{ in } \mathbb{R}$$

and

$$\lim_{n\to\infty} k(t - s_n) = h(t) \text{ for all } t \text{ in } \mathbb{R}.$$

Remark. If the convergence in the both limits is uniform, then h is almost periodic. The concept of almost automorphy is much larger than almost periodicity. By the pointwise convergence, the function k is just measurable and not necessarily continuous.

Definition 6.3 (Bochner, [18, Theorem 5.8, pp. 86]). A continuous function $h : \mathbb{R} \to X$ is said to be compact almost automorphic if for every sequence of real numbers $(s'_n)_n$, there exists a subsequence $(s_n)_n$ such that

$$\lim_{m\to\infty}\lim_{n\to\infty} h(t + s_n - s_m) = h(t) \text{ exists uniformly on any compact set in } \mathbb{R}.$$

Theorem 6.4 ([18]). *If we equip $AA(X)$, the space of almost automorphic X-valued functions with the sup norm, then $AA(X)$ turns out to be a Banach space.*

Consider the following ordinary differential equation

$$\frac{d}{dt}x(t) = Gx(t) + e(t) \text{ for } t \in \mathbb{R} \tag{16}$$

where G is a constant $n \times n$-matrix and $e : \mathbb{R} \to \mathbb{R}^n$ is a continuous function.

Theorem 6.5 ([18, Theorem 5.8, pp. 86]). *Assume that e is an almost automorphic function. Then the following are equivalent:*

i) existence of a bounded solution on $\mathbb{R}^+$ of Eq. (16),
ii) existence of an almost automorphic solution of Eq. (16).
Moreover every bounded solution of Eq. (16) on the whole line is almost automorphic.

In the following, we assume that:

($\mathbf{H_3}$) f is an almost automorphic function.

Consider now the following equation in the whole line $\mathbb{R}$

$$\frac{d}{dt}u(t) = Au(t) + L(u_t) + f(t) \text{ for } t \in \mathbb{R}. \tag{17}$$

Theorem 6.6. *Assume that $(\mathbf{H_0}), (\mathbf{H_1})$ and $(\mathbf{H_3})$ hold. If there exists $\varphi \in C$ such that Eq. (1) has a bounded solution on $\mathbb{R}^+$. Then Eq. (17) has an almost automorphic integral solution.*

Proof. Let u be a bounded solution of Eq. (1) on $\mathbb{R}^+$. Then by Theorem 3.6, the function $z(t) = \langle \Psi, u_t \rangle$ for $t \geq 0$, is a solution of the ordinary differential equation (6) and z is bounded on $\mathbb{R}^+$. Moreover, the function

$$\varrho(t) = \langle x^*, f(t) \rangle \text{ for } t \in \mathbb{R},$$

is almost automorphic from $\mathbb{R}$ to $\mathbb{R}^d$. By Theorem 6.5, we get that the reduced system (6) has an almost automorphic solution $\tilde{z}$. Consequently $\Phi\tilde{z}(.)$ is an almost automorphic function on $\mathbb{R}$. By Theorem 3.6, the function $u(t) = v(t)(0)$, where

$$v(t) = \Phi\tilde{z}(t) + \lim_{n\to+\infty}\int_{-\infty}^{t} \mathcal{U}^s(t-\xi)\,\Pi^s\left(\tilde{B}_n X_0 f(\xi)\right) d\xi \text{ for } t \in \mathbb{R},$$

is a solution of Eq. (17) on $\mathbb{R}$. We claim that v is almost automorphic. In fact, consider the function y by

$$y(t) = \lim_{n\to+\infty} \int_{-\infty}^{t} \mathcal{U}^s(t-\xi)\,\Pi^s\left(\tilde{B}_n X_0 f(\xi)\right) d\xi \text{ for } t \in \mathbb{R},$$

Since f is almost automorphic, then for any sequence of real numbers $\left(\alpha'_p\right)_{p\geq 0}$ there exists a subsequence $\left(\alpha_p\right)_{p\geq 0}$ of $\left(\alpha'_p\right)_{p\geq 0}$ such that

$$\lim_{p\to\infty} f(t+\alpha_p) = h(t) \text{ for all } t \in \mathbb{R}$$

and

$$\lim_{p\to\infty} h(t-\alpha_p) = f(t) \text{ for all } t \in \mathbb{R}.$$

Now

$$y(t+\alpha_p) = \lim_{n\to+\infty} \int_{-\infty}^{t+\alpha_p} \mathcal{U}^s\left(t+\alpha_p-\xi\right)\Pi^s\left(\tilde{B}_n X_0 f(\xi)\right) d\xi \text{ for } t \in \mathbb{R},$$

which gives that

$$y(t+\alpha_p) = \lim_{n\to+\infty} \int_{-\infty}^{t} \mathcal{U}^s(t-\xi)\,\Pi^s\left(\tilde{B}_n X_0 f(\xi+\alpha_p)\right) d\xi \text{ for } t \in \mathbb{R},$$

By the Lebesgue's dominated convergence theorem, we get that

$$y(t+\alpha_p) \to w(t) \text{ as } p \to \infty,$$

where w is given by

$$w(t) = \lim_{n\to+\infty} \int_{-\infty}^{t} \mathcal{U}^s(t-\xi)\,\Pi^s\left(\tilde{B}_n X_0 h(\xi)\right) d\xi \text{ for } t \in \mathbb{R}.$$

Using the same argument as above, we prove that

$$w(t-\alpha_p) \to \lim_{n\to+\infty} \int_{-\infty}^{t} \mathcal{U}^s(t-\xi)\,\Pi^s\left(\tilde{B}_n X_0 f(\xi)\right) d\xi \text{ as } p \to \infty,$$

which implies that y is almost automorphic. Consequently, v is an almost automorphic integral solution of Eq. (17).

7 Lotka-Volterra Equation

In order to apply the previous results, we consider the model of Lotka-Volterra with diffusion which is taken from [23] and [25]

$$\begin{cases} \dfrac{\partial}{\partial t}v(t,x) = \dfrac{\partial^2}{\partial x^2}v(t,x) + \displaystyle\int_{-r}^{0} G(\theta)v(t+\theta,x)d\theta + h(t,x) \text{ for } t \geq 0 \text{ and } x \in [0,\pi], \\ u(t,x) = 0 \text{ for } x = 0,\pi \text{ and } t \geq 0, \\ u(\theta,x) = \varphi_0(\theta,x) \text{ for } \theta \in [-r,0] \text{ and } x \in [0,\pi], \end{cases} \tag{18}$$

where $G : [-r,0] \to \mathbb{R}$, $\varphi_0 : [-r,0] \times [0,\pi] \to \mathbb{R}$ and $h : \mathbb{R} \times [0,\pi] \to \mathbb{R}$ are continuous functions.

Let $X = C([0,\pi];\mathbb{R})$ be the space of continuous functions from $[0,\pi]$ to $\mathbb{R}$ endowed with the uniform norm topology. Define the operator $A : D(A) \subset X \to X$ by

$$\begin{cases} D(A) = \left\{y \in C^2([0,\pi];\mathbb{R}) : y(0) = y(\pi) = 0\right\}, \\ Ay = y''. \end{cases}$$

Lemma 7.1 ([7, Proposition 14.6 , pp. 319–320]).

$$(0,+\infty) \subset \rho(A) \textit{ and} \left|(\lambda - A)^{-1}\right| \leq \frac{1}{\lambda} \text{ for} \lambda > 0.$$

Moreover,

$$\overline{D(A)} = \{y \in X : y(0) = y(\pi) = 0\}.$$

This Lemma implies that condition $(\mathbf{H}_0)$ is satisfied.

We introduce $L : C := C([-r,0], X) \to X$ by

$$L(\phi)(x) = \int_{-r}^{0} G(\theta)\phi(\theta)(x)d\theta \text{ for } x \in [0,\pi] \text{ and } \phi \in C.$$

$f : \mathbb{R} \longrightarrow X$ is defined by

$$f(t)(x) = h(t,x) \text{ for } t \in \mathbb{R} \text{ and } x \in [0,\pi].$$

The initial data $\varphi \in C$ is provided by

$$\varphi(\theta)(x) = \varphi_0(\theta,x) \text{ for } (\theta,x) \in [-r,0]\text{x}[0,\pi]$$

L is a bounded linear operator from C to X and by form continuity of h, we get that f is a continuous function from $\mathbb{R}$ to X. Equation (18) takes the following abstract form

$$\begin{cases} \frac{d}{dt}u(t) = Au(t) + L(u_t) + f(t) \text{ for } t \geq 0, \\ u_0 = \varphi \in C. \end{cases} \tag{19}$$

Let A_0 be the part of A in $\overline{D(A)}$. Then, A_0 is given by

$$\begin{cases} D(A_0) = \left\{ y \in C^2([0,\pi];\mathbb{R}) : y(0) = y(\pi) = y''(0) = y''(\pi) = 0 \right\}, \\ A_0 y = Ay \text{ for } y \in D(A_0). \end{cases}$$

It is well known from [8, Example 1.4.34 , pp. 123], that A_0 generates a strongly continuous compact semigroup $(T_0(t))_{t\geq 0}$ on $\overline{D(A)}$ and

$$|T_0(t)| \leq e^{-t} \text{for } t \geq 0,$$

Let $\varphi_0 \in C([-r,0] \times [0,\pi];\mathbb{R})$ be such that

$$\varphi_0(0,0) = \varphi_0(0,\pi) = 0.$$

Then by Theorem 2.3, we deduce that Eq. (19) has a unique integral solution on $[-r,+\infty)$.

In order to study the existence of an almost automorphic solution of the following equation

$$\frac{d}{dt}u(t) = Au(t) + L(u_t) + f(t) \text{ for } t \in \mathbb{R}. \tag{20}$$

We suppose that

(H$_4$) h is almost automorphic in t uniformly for $x \in [0,\pi]$, which means that there exists a measurable function $g : \mathbb{R} \times [0,\pi] \to \mathbb{R}$ such that

$$\lim_{n\to\infty} h(t+s_n,x) = g(t,x) \text{ exists for all } t \text{ in } \mathbb{R} \text{ uniformly in } x \in [0,\pi]$$

and

$$\lim_{n\to\infty} g(t-s_n,x) = h(t,x) \text{ for all } t \text{ in } \mathbb{R} \text{ uniformly in } x \in [0,\pi]$$

Moreover, we suppose that:

(H$_5$) there exists a constant $\beta \in (0,1)$ such that

$$\int_{-r}^{0} |G(\theta)|\, d\theta \leq (1-\beta).$$

Proposition 7.2. *Assume that* $(\mathbf{H_4})$ *and* $(\mathbf{H_5})$ *hold. Then there exists* $\varphi \in C$ *such that Eq. (19) has a bounded solution on* $\mathbb{R}^+$*. Consequently Eq. (20) has an almost automorphic solution.*

Proof. The first goal is to prove that Eq. (19) has a bounded solution on $\mathbb{R}^+$. Let $\rho = \left(1 + \frac{|f|_\infty}{\beta}\right)$, where $|f|_\infty = \sup_{s\in\mathbb{R}} |f(s)|$. Consider $\varphi \in C_0$ such that $|\varphi|_C < \rho$. We claim that

$$|u(t)| \leq \rho \text{ for all } t \geq 0. \tag{21}$$

We proceed by contradiction. Let t_0 be the first time such that (21) is not true. Then

$$t_0 = \inf\{t > 0 : |u(t)| > \rho\}.$$

By continuity of u, one has

$$|u(t_0)| = \rho,$$

and there exists a positive constant $\varepsilon > 0$ such that

$$|u(t)| > \rho \text{ for } t \in (t_0, t_0 + \varepsilon).$$

We have,

$$u(t_0) = T_0(t_0)\varphi(0) + \lim_{\lambda\to+\infty} \int_0^{t_0} T_0(t_0 - s)B_\lambda[L(u_s) + f(s)]ds$$

which implies that

$$|u(t_0)| \leq e^{-t_0}\rho + \int_0^{t_0} e^{-(t_0-s)} \left[\int_{-r}^0 |G(\theta)|\,|u(s+\theta)|\,d\theta + |f|_\infty\right] ds.$$

Since $|u(t)| \leq \rho$ for $t \leq t_0$. Then

$$|u(t)| \leq \rho \text{ for } t \in [-r, t_0].$$

Therefore

$$|u(t_0)| \leq e^{-t_0}\rho + \left(1 - e^{-t_0}\right)\left[\int_{-r}^0 |G(\theta)|\,d\theta\rho + |f|_\infty\right].$$

Condition $(\mathbf{H_5})$ implies that

$$|u(t_0)| \leq e^{-t_0}\rho + \left(1 - e^{-t_0}\right)[(1-\beta)\rho + |f|_\infty],$$

and

$$|u(t_0)| \leq e^{-t_0}\rho + \left(1 - e^{-t_0}\right)\rho + \left(1 - e^{-t_0}\right)[-\beta\rho + |f|_\infty].$$

Consequently, we obtain that

$$|u(t_0)| \leq \rho - \left(1 - e^{-t_0}\right)\beta < \rho,$$

by continuity of u, there exists a positive ε_0 such that

$$|u(t)| < \rho \text{ for } t \in (t_0, t_0 + \varepsilon_0),$$

which gives a contradiction and we deduce that Eq. (19) has a bounded integral solution u on $\mathbb{R}^+$, and by Theorem 6.6, we get that Eq. (20) has an almost automorphic solution.

Acknowledgements The author would like to thank Professor Ovide Arino from whom he has learnt a lot about the theory of partial functional differential equations and its application. This work is dedicated to his memory.

References

1. Adimy, M., & Ezzinbi, K. (1999). Existence and linearized stability for partial neutral functional differential equations. *Differential Equations and Dynamical Systems, 7*, 371–417,
2. Adimy, M., Ezzinbi, K., & Laklach, M. (2001). Spectral decomposition for partial neutral functional differential equations. *Canadian Applied Mathematics Quarterly, 9*(1), 1–34. Spring.
3. Adimy, M., Bouzahir, H., & Ezzinbi, K. (2002). Local existence and stability for some partial functional differential equations with infinite delay. *Nonlinear Analysis, Theory, Methods and Application, 48*, 323–348.
4. Arendt, W., Batty, C. J. K., Hieber, M., & Neubrander, F. (2001). *Vector valued laplace transforms and cauchy problems*. Monographs in Mathematics, vol. 96. Basel: Birkhäuser.
5. Benkhalti, R., Bouzahir, H., & Ezzinbi, K. (2001). Existence of periodic solutions for some partial functional differential equations with infinite delay. *Journal of Mathematical Analysis and Applications, 256*, 257–280.
6. Bochner, S. (1964). Continuous mappings of almost automorphic and almost automorphic functions. *Proceedings of the National Academy of Sciences of the USA, 52*, 907–910.
7. Da Prato, G., & Sinestrari, E. (1987). Differential operators with nondense domains. *Annali Scuola Normale Superiore di Pisa, 14*(2), 285–344.
8. Engel, K. J., & Nagel, R. (1986) *One-parameter semigroups of positive operators*. Lecture Notes in Mathematics, vol. 1184. Berlin/New York: Springer-Verlag.
9. Fink, A. (1974). *Almost periodic differential equations*. Lectures Notes, vol. 377. New York: Springer-Verlag.
10. Hino, Y., & Murakami, S. (2003). Almost automorphic for abstract functional differential equations. *Journal of Mathematical Analysis and Applications, 286*, 741–752.
11. Hale, J. K. (1977). *Theory of functional differential equations*. New York: Springer-Verlag.

12. Hale, J., & Kato, J. (1978). Phase spaces for retarded equations with unbounded delay. *Funkcia Ekvac, 21*, 11–41.
13. Hino, Y., Murakami, S., & Naito, T. (1991). *Functional differential equations with infinite delay*. Lectures Notes in Mathematics, vol. 1473. Berlin-New York: Springer.
14. Hino, Y., Murakami, S., Naito, T., & Minh, N. V. (2002). A variation of constants formula for abstract functional differential equations in the phase spaces. *Journal of Differential Equations, 179*, 336–355.
15. Massera, J. L. (1950). The existence of periodic solutions of systems of differential equations. *Duke Mathematical Journal, 17*, 457–475.
16. Murakami, S., Naito, T., & Minh, N. V. (2004). Massera theorem for almost periodic solutions of functional differential equations. *Journal of the Mathematical Society of Japan, 56*(1), 247–268.
17. Naito, T., Van Minh, N., & Son Shin, J. (2001). New spectral criteria for almost periodic solutions of evolution equations. *Studia Mathematica, 142*, 97–111.
18. N'Guérékata, G. M. (2001). *Almost automorphic and almost automorphic functions in abstract spaces*. Amesterdam: Kluwer.
19. N'Guérékata, G. M. (2001). Almost auotmorphy, almost periodicity and stability of motions in Banach spaces. *Forum Maths, 13*, 581–588.
20. Pazy, A. (1983). *Semigroups of linear operators and applications to partial differential equations*. Applied Math. Sciences, vol. 44. New York: Springer-Verlag.
21. Shin, J. S., & Naito, T. (1999). Semi-Fredholm operators and periodic solutions for linear functional differential equations in Banach spaces. *Journal of Differential Equations, 153*, 407–441.
22. Thieme, H. R. (1990). Semiflows generated by Lipschitz perturbations of non-densely defined operators. *Differential and Integral Equations, 3*(6), 1035–1066.
23. Travis, C. C., & Webb, G. F. (1974). Existence and stability for partial functional differential equations. *Transactions American Mathematical Society, 200*, 395–418.
24. Zeidler, E. (1993). *Nonlinear functional analysis and it's applications, tome I, fixed point theorem*. New York: Springer-Verlag.
25. Wu, J. (1996). *Theory and applications of partial functional differential equations*. New York: Springer-Verlag.

Characterizations of Convex Quadrics in Terms of Plane Quadric Sections, Midsurfaces, and Shadow-Boundaries

Valeriu Soltan

Abstract It is well known that the middle points of any family of parallel chords of a real quadric surface Q in the Euclidean space R^n belong to a hyperplane, and that a similar property holds for the shadow-boundaries of Q. In this article we review the existing results and add some new ones which characterize convex quadrics among convex hypersurfaces in R^n, possibly unbounded, in terms of plane quadric sections, hyperplanarity of their midsurfaces and shadow-boundaries.

Mathematics Subject Classification (2000) Primary: 52A20

1 Introduction

In a standard way, a *quadric* (or a *second degree surface*) in the Euclidean space R^n, $n \geq 2$, is the locus of points $x = (\xi_1, \ldots, \xi_n)$ which satisfy a quadratic equation

$$F(x) \equiv \sum_{i,k=1}^{n} a_{ik}\xi_i\xi_k + 2\sum_{i=1}^{n} b_i\xi_i + c = 0, \tag{1}$$

where not all a_{ik} are zero. We say that a quadric $Q \subset R^n$ is a *hypersurface* provided its complement $R^n \setminus Q$ contains at least two components. The latter happens if and only if either $F(x)$ is a complete square describing a hyperplane or both open sets $\{x \in R^n : F(x) < 0\}$ and $\{x \in R^n : F(x) > 0\}$ are nonempty.

V. Soltan (✉)
George Mason University, Fairfax, VA 22030, USA
e-mail: vsoltan@gmu.edu

B. Toni (ed.), *New Frontiers of Multidisciplinary Research in STEAM-H (Science, Technology, Engineering, Agriculture, Mathematics, and Health)*,
Springer Proceedings in Mathematics & Statistics 90, DOI 10.1007/978-3-319-07755-0_4,

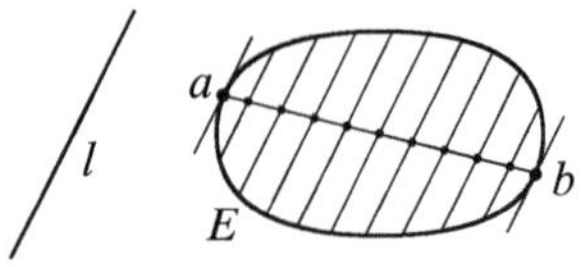

Fig. 1 Midsurface of a convex body

In what follows, an r-dimensional *plane* in R^n is a translate of an r-dimensional subspace, a *hyperplane* is a plane of dimension $n-1$. Given a pair of distinct points $a, c \in \mathrm{R}^n$, the *line* $\langle a, c\rangle$ through a, c and the *segment* $[a, c]$ with endpoints a, c are defined, respectively, by

$$\langle a, c\rangle = \{(1-\lambda)a + \lambda c : \lambda \in \mathrm{R}\}, \quad [a, c] = \{(1-\lambda)a + \lambda c : 0 \le \lambda \le 1\}.$$

It is easy to see that, given a quadric Q and a line l in R^n, either l lies within Q or $Q \cap l$ contains at most two points. If the line l meets Q at precisely two points, a and c, then the segment $[a, c]$ is called a *chord* of Q. It is a matter of common knowledge that the middle points of any family of parallel chords of a real quadric $Q \subset \mathrm{R}^n$ (called the *midsurface* of Q) belong to a hyperplane. (For reader's convenience, we provide the proof of this statement in Theorem 1 below.)

Similarly, we say that a line l *supports* the quadric hypersurface $Q \subset \mathrm{R}^n$, described by (1), provided $\emptyset \ne Q \cap l \ne l$ and l lies in one of the closed sets $\{x \in \mathrm{R}^n : F(x) \le 0\}$ and $\{x \in \mathrm{R}^n : F(x) \ge 0\}$. (This definition can be modified to accommodate the case when Q is any real quadric, assuming that l lies in the smallest plane containing Q.) The set of all points (possibly, empty) at which the quadric hypersurface Q is supported by translates of a given line l is called the *shadow-boundary* of Q with respect to l and is denoted $S_l(Q)$. It is known that each shadow-boundary of Q lies in a hyperplane (see Theorem 2).

The concepts of midsurface and shadow-boundary can be easily illustrated. For example, the middle points of all chords of an ellipse E, which are parallel to a given line l, fulfill the line segment $[a, b]$, while the shadow-boundary $S_l(E)$ consists of a and b (Fig. 1).

One might ask whether the hyperplanarity of midsurfaces or shadow-boundaries characterizes quadric hypersurfaces within a certain family $\mathscr{S}$ of hypersurfaces in R^n. There is a variety of results addressing this question for the cases when $\mathscr{S}$ is (i) the family of sufficiently regular hypersurfaces, or (ii) the family of bounded convex hypersurfaces.

The purpose of this article is to survey the existing results and to provide new statements which characterize convex quadrics within the family of all convex, possibly unbounded, hypersurfaces in R^n. The article may be considered as a sequel to the paper [42] (previously published in this series) which describes convex quadrics and their characteristic properties in terms of plane quadric sections. The main content of this article is divided into the following sections.

2. Properties of quadric surfaces
3. Convex quadrics and their plane sections

4. Convex hypersurfaces with hyperplanar midsurfaces
5. Convex hypersurfaces with hyperplanar shadow-boundaries
6. Orthogonal projections of convex quadrics

This paper is based on a talk given at the interdisciplinary Seminar on Mathematical Sciences and Applications of Virginia State University.

2 Properties of Quadric Surfaces

This section contains some results about geometric properties of quadric surfaces in R^n, which are further used throughout the text. Although these properties are often viewed as commonly known, their proofs are hardly accessible in mathematical literature, or are given in a more restricted setting. For reader's convenience, we provide proofs of these results.

Theorem 1. *The middle points of all chords of a real quadric surface $Q \subset R^n$ which are parallel to a given chord $[a, c]$ of Q belong to a hyperplane.*

Proof. Assume that Q is given by (1). The line l through a and c can be expressed as

$$l = \{z + tv : t \in R\}, \quad \text{with } v \neq o, \tag{2}$$

where z is the middle point of $[a, c]$ and $v = c - a$. Equivalently, $x = (\xi_1, \dots, \xi_n)$ belongs to l if and only if

$$\xi_i = \phi_i + t v_i, \quad t \in R, \quad \text{forall } i = 1, \dots, n, \tag{3}$$

where $z = (\phi_1, \dots, \phi_n)$ and $v = (v_1, \dots, v_n)$. To determine the values of t for which $x \in Q \cap l$, we substitute $\xi_1, \dots, \xi_n$ from (3) into (1) and arrange the powers of t. The result is a quadratic equation in t,

$$A(v)\, t^2 + 2B(v, z)\, t + C(z) = 0, \tag{4}$$

where

$$A(v) = \sum_{i,k=1}^{n} a_{ik} v_i v_k, \quad B(v, z) = \tfrac{1}{2} \sum_{i=1}^{n} \tfrac{\partial F(z)}{\partial \xi_i}\, v_i, \quad C(z) = F(z). \tag{5}$$

Then a and c correspond to opposite non-zero solutions, t_0 and $-t_0$, of (4), which is possible if and only if $A(v)\, C(z) < 0$ and $B(v, z) = 0$. The equality

$$\sum_{i=1}^{n} \Big(\sum_{k=1}^{n} a_{ik} \phi_k + b_i \Big) v_i \equiv B(v, z) = 0,$$

re-written in the form

$$\sum_{k=1}^{n}\Big(\sum_{i=1}^{n} a_{ik}v_i\Big)\phi_k + \sum_{i=1}^{n} b_i v_i = 0 \tag{6}$$

and interpreted as an equation in $\phi_1, \dots, \phi_n$, (6) describes a hyperplane, H. Indeed, at least one of the scalars

$$c_k = \sum_{i=1}^{n} a_{ik}v_i, \quad k = 1, \dots, n,$$

is distinct from zero, since otherwise

$$A(v) = c_1 v_1 + \dots + c_n v_n = 0,$$

which is impossible because of $A(v) \neq 0$.

If $[a', c']$ is a chord of Q parallel to $[a, c]$, then v is a nonzero multiple of $c' - a'$, implying that the line l' through a' and c' is given by

$$l' = \{z' + tv : t \in \mathrm{R}\},$$

where $z' = (\phi'_1, \dots, \phi'_n)$ is the middle point of $[a', c']$. Repeating the argument above, we obtain that $\phi'_1, \dots, \phi'_n$ satisfy (6), which gives $z' \in H$. Hence the midsurface of Q corresponding to $[a, c]$ lies in H.

Theorem 2. *Each shadow-boundary of a quadric hypersurface $Q \subset \mathrm{R}^n$ lies in a hyperplane.*

Proof. Let l be a line in R^n supporting Q. As in the proof of Theorem 1, we assume that Q is given by (1), and (2) describes l, where $z \in Q \cap l$. Equivalently, the coordinates of any point $x \in l$ are given in (3), and the values of t for which $x \in Q \cap l$ are the solutions of the quadratic equation (4). Without loss of generality, we suppose that l lies in the set $\{x \in \mathrm{R}^n : F(x) \geq 0\}$. Then the quadratic polynomial

$$f(t) = A(v)\,t^2 + 2B(v, z)\,t + C(z)$$

is non-negative over R and is not identically zero. Hence $t = 0$ is a unique solution of (4), due to the inclusion $z \in Q \cap l$. The latter is possible if and only if $A(v) \neq 0$ and $B(v, z) = C(z) = 0$, where $A(v)$, $B(v, z)$, and $C(z)$ are given by (5). In particular, the set $Q \cap l$ is a singleton.

Similarly to the proof of Theorem 1, the Eq. (6) describes a hyperplane, H, containing z. If l' is a translate of l supporting Q, then l' can be expressed as

$$l' = \{z' + tv : t \in \mathrm{R}\}, \quad \text{where} \quad z' \in Q \cap l'.$$

Repeating the argument above, we obtain the inclusion $z' \in H$. Hence the shadow-boundary $S_l(Q)$ lies in H.

Theorem 3. *Let E_1 and E_2 be non-degenerate quadric curves in* R^3*, which lie, respectively, in distinct planes L_1 and L_2 of* R^3 *such that $E_1 \cap E_2$ consists of a pair of points. For any point $v \in \mathrm{R}^3 \setminus (L_1 \cup L_2)$, there is a unique quadric surface containing $\{v\} \cup E_1 \cup E_2$.*

Proof. Let p and q be the points of intersection of E_1 and E_2, and c be the middle point of $[p, q]$. Denote by l the line through p and q, and by l_i the axis of affine symmetry of E_i which contains c and is distinct from l, $i = 1, 2$. Clearly, L_i contains $l_i \cup l$, $i = 1, 2$. Choose suitable coordinates ξ_1, ξ_2, ξ_3 in R^3 such that $c = o$, and l_1, l_2, l are, respectively, the coordinate ξ_1-, ξ_2-, and ξ_3-axes. A point $v = (\nu_1, \nu_1, \nu_3)$ belongs to $\mathrm{R}^3 \setminus (L_1 \cup L_2)$ if and only if $\nu_1 \nu_2 \neq 0$.

Re-scaling the unit vectors along the coordinate axes, we may suppose that $p = (0, 0, 1)$, $q = (0, 0, -1)$, and E_i are given by one of the following equations.

1. If E_i is an ellipse, then

$$\xi_i^2 + \xi_3^2 - 2\sigma_i \xi_i - 1 = 0, \quad \xi_j = 0, \text{ where } \sigma_i > 0, \ i, j \in \{1, 2\}, \ i \neq j. \tag{7}$$

2. If E_i is a parabola, then

$$\xi_3^2 - \xi_i - 1 = 0, \quad \xi_j = 0, \text{ where } i, j \in \{1, 2\}, \ i \neq j. \tag{8}$$

3. If E_i is a hyperbola, then

$$\xi_3^2 - \xi_i^2 - 2\sqrt{2}\, \xi_i - 1 = 0, \quad \xi_j = 0, \text{ where } i, j \in \{1, 2\}, \ i \neq j. \tag{9}$$

By symmetry, the may assume that E_1 and E_2 are combined as follows.

(a) Both E_1 and E_2 are ellipses, given by (7), with $i = 1, j = 2$ and $i = 2, j = 1$, respectively. Then the quadric surface containing $\{v\} \cup E_1 \cup E_2$ is described by the equation

$$\xi_1^2 + \xi_2^2 + \xi_3^2 + \mu \xi_1 \xi_2 - 2\sigma_1 \xi_1 - 2\sigma_2 \xi_2 - 1 = 0,$$

where $\mu = (2\sigma_1 \nu_1 + 2\sigma_2 \nu_2 - \nu_1^2 - \nu_2^2 - \nu_3^2 + 1)/(\nu_1 \nu_2)$.

(b) E_1 is an ellipse given by (7), with $i = 1, j = 2$, and E_2 is a parabola given by (8), with $i = 2, j = 1$. Then the quadric surface containing $\{v\} \cup E_1 \cup E_2$ is described by the equation

$$\xi_1^2 + \xi_3^2 + \mu \xi_1 \xi_2 - 2\sigma_1 \xi_1 - \xi_2 - 1 = 0,$$

where $\mu = (2\sigma_1 \nu_1 + \nu_2 - \nu_1^2 - \nu_3^2 + 1)/(\nu_1 \nu_2)$.

(c) E_1 is an ellipse given by (7), with $i = 1, j = 2$, and E_2 is a hyperbola given by (9), with $i = 2, j = 1$. Then the quadric surface containing $\{v\} \cup E_1 \cup E_2$ is described by the equation

$$\xi_1^2 - \xi_2^2 + \xi_3^2 + \mu\xi_1\xi_2 - 2\sigma_1\xi_1 - 2\sqrt{2}\,\xi_2 - 1 = 0,$$

where $\mu = (2\sigma_1\nu_1 + 2\sqrt{2}\,\nu_2 - \nu_1^2 + \nu_2^2 - \nu_3^2 + 1)/(\nu_1\nu_2)$.

(d) Both E_1 and E_2 are parabolas, given by (8), with $i = 1, j = 2$ and $i = 2, j = 1$, respectively. Then the quadric surface containing $\{v\} \cup E_1 \cup E_2$ is described by the equation

$$\xi_3^2 + \mu\xi_1\xi_2 - \xi_1 - \xi_2 - 1 = 0,$$

where $\mu = (\nu_1 + \nu_2 - \nu_3^2 + 1)/(\nu_1\nu_2)$.

(e) E_1 is a parabola given by (8), with $i = 1, j = 2$, and E_2 is a hyperbola given by (9), with $i = 2, j = 1$. Then the quadric surface containing $\{v\} \cup E_1 \cup E_2$ is described by the equation

$$\xi_3^2 - \xi_2^2 + \mu\xi_1\xi_2 - \xi_1 - 2\sqrt{2}\,\xi_2 - 1 = 0,$$

where $\mu = (\nu_1 + 2\sqrt{2}\,\nu_2 + \nu_2^2 - \nu_3^2 + 1)/(\nu_1\nu_2)$.

(f) Both E_1 and E_2 are hyperbolas, given by (9), with $i = 1, j = 2$ and $i = 2, j = 1$, respectively. Then the quadric surface containing $\{v\} \cup E_1 \cup E_2$ is described by the equation

$$\xi_3^2 - \xi_1^2 - \xi_2^2 + \mu\xi_1\xi_2 - 2\sqrt{2}\,\xi_1 - 2\sqrt{2}\,\xi_2 - 1 = 0,$$

where $\mu = (2\sqrt{2}\,\nu_1 + 2\sqrt{2}\,\nu_2 + \nu_1^2 + \nu_2^2 - \nu_3^2 + 1)/(\nu_1\nu_2)$.

The proof of Theorem 4 below uses the following well-known fact of analytical geometry. Since the determinant

$$\det Q = \begin{vmatrix} a_{11} & a_{12} & b_1 \\ a_{12} & a_{22} & b_2 \\ b_1 & b_2 & c \end{vmatrix}$$

is an invariant of a quadratic form

$$f(\xi, \eta) \equiv a_{11}\xi^2 + 2a_{12}\xi\eta + a_{22}\eta^2 + 2b_1\xi + 2b_2\eta + c = 0 \tag{10}$$

with respect to orthogonal transformations of the plane, the classification of quadratic curves implies that the real quadric $Q \subset \mathrm{R}^2$ described by (10) is non-degenerate (that is, does not lie in the union of two lines) if and only if $\det Q \neq 0$.

Theorem 4. *Any five distinct points in the plane belong to a quadric curve. This curve is uniquely determined by the points if and only if no four of them belong to a line.*

Proof. Let $p_i = (\xi_1, \eta_i)$, $1 \le i \le 5$, be any five points in the coordinate plane. A quadric curve, given by an Eq. (10) contains the set $P = \{p_1, \ldots, p_5\}$ if and only if $a_{11}, a_{12}, a_{22}, b_1, b_2, c$ satisfy the system of five linear homogenous equations

$$a_{11}\xi_i^2 + 2a_{12}\xi_i\eta_i + a_{22}\eta_i^2 + 2b_1\xi_i + 2b_2\eta_i + c = 0, \quad 1 \le i \le 5. \tag{11}$$

Since the number of variables here is greater than the number of equations, the system (11) has a nontrivial solution. If at least one of the scalars a_{11}, a_{12}, a_{22} is not zero, then (10) describes a quadric curve, and if $a_{11} = a_{12} = a_{22} = 0$, then the quadratic equation $(2b_1\xi + 2b_2\eta + c)^2 = 0$ gives a desired quadric curve.

If some four points from P belong to a line l, given by an equation $\alpha\xi + \beta\eta + \gamma = 0$, then, multiplying its left hand side by a linear polynomial vanishing on the fifth point of P, we obtain a family of quadratic equations which describe an infinite family of distinct quadric curves through P.

Suppose that no four points from P belong to a line. Choose a quadric curve Q containing P. If some three points from P belong to a line, then from the classification of quadric curves it follows that Q must be the union of two lines, say, l_1 and l_2. Let, for example, $p_1, p_2, p_3 \in l_1$ and $p_4, p_5 \in l_2$. It is easy to see that l_1 and l_2 give the only way to cover P by the union of two lines. Hence $Q = l_1 \cup l_2$ is the only quadric curve containing P.

Finally, let no three points of P belong to a line. Then any quadric curve containing P is non-degenerate. Assume, for contradiction, the existence of distinct quadric curves, Q and Q', both containing P. Let Q be given by (10), and Q' be given by an equation

$$g(\xi, \eta) \equiv a'_{11}\xi^2 + 2a'_{12}\xi\eta + a'_{22}\eta^2 + 2b'_1\xi + 2b'_2\eta + c' = 0. \tag{12}$$

Since $Q \neq Q'$, the polynomial $f(\xi, \eta) + tg(\xi, \eta)$ is not identically zero in ξ, η for any choice of $t \in \mathrm{R}$. Hence the equation $f(\xi, \eta) + tg(\xi, \eta) = 0$ in ξ, η determines a quadric curve $Q(t)$ for any choice of t in R.

Clearly, $P \subset Q(t)$, which shows that $Q(t)$ is non-degenerate for all $t \in \mathrm{R}$. On the other hand, the determinant

$$\det Q(t) = \begin{vmatrix} a_{11} + ta'_{11} & a_{12} + ta'_{12} & b_1 + tb'_1 \\ a_{12} + ta'_{12} & a_{22} + ta'_{22} & b_2 + tb'_2 \\ b_1 + tb'_1 & b_2 + tb'_2 & c + tc' \end{vmatrix}$$

is a polynomial of degree 3 in t, whose leading coefficient equals $\det Q'$ $(\neq 0)$. Hence the equation $\det Q(t) = 0$ has a real solution $t = t_0$, which shows that the quadric curve $Q(t_0)$ is degenerate. The obtained contradiction implies the uniqueness of Q.

Theorem 5. *For any set $P = \{p_0, p_1, p_2, p_3\}$ of four distinct points in the plane and a line l through p_0, there is a quadric curve which contains P and has l as a tangent line at p_0. This curve is unique if and only if the set $P \cap l$ consists of at most two points.*

Proof. Consider a quadric Q in the plane, expressed by (10). It is well-known that the tangent line of Q at a regular point $u = (p, q)$ is given by the linear equation

$$(a_{11}p + a_{12}q + b_1)\xi + (a_{12}p + a_{22}q + b_2)\eta + (b_1 p + b_2 q + c) = 0. \qquad (13)$$

Choosing suitable Cartesian coordinates ξ, η in the plane, we may assume that $p_0 = (0, 0)$ and l is the ξ-axis. Let $p_i = (\xi_i, \eta_i)$, $i = 1, 2, 3$. From (13) it follows that l is the tangent line of Q at p_0 if and only if $c = 0$ and $b_1 = 0$. Hence

$$f(\xi, \eta) \equiv a_{11}\xi^2 + 2a_{12}\xi\eta + a_{22}\eta^2 + 2b_2\eta = 0.$$

Furthermore, Q contains $\{p_1, p_2, p_3\}$ if and only the coefficients $a_{11}, a_{12}, a_{22}, b_2$ in (10) satisfy the system of three linear homogenous equations

$$a_{11}\xi_i^2 + 2a_{12}\xi_i\eta_i + a_{22}\eta_i^2 + 2b_2\eta_i = 0, \quad i = 1, 2, 3. \qquad (14)$$

Since the number of variables here is greater than the number of equations, (14) has a nontrivial solution. Hence a quadric curve Q satisfying theorem's conditions exists.

Assume first that Q is degenerate. Then Q is the union of two lines: l and, say, l'. The line l' is uniquely determined by P if and only if it contains at least two point of P. Hence Q is uniquely determined by P if and only if $P \cap l$ consists of at most two points. Furthermore, any other quadric T satisfying theorem's condition should also be degenerate. Indeed, if at least one of the points, p_1, p_2, p_3, say, p_1 belongs to l, then $l \subset T$. If none of p_1, p_2, p_3 is in l, then $\{p_1, p_2, p_3\}$ should lie in another line, m (because Q is degenerate), implying the inclusion $m \subset T$. In either case, T is degenerate.

Suppose now that Q is non-degenerate. Then $P \cap l = \{p_0\}$, and no three points of P belong to a line. In particular, $l \not\subset Q$. Furthermore, $a_{11} \neq 0$, since otherwise the polynomial

$$f(\xi, \eta) = (2a_{12}\xi + a_{22}\eta + 2b_2)\eta$$

would describe a degenerate quadric.

Assume for a moment the existence of another quadric curve Q' which contains P and has l as the tangent line at p_0. As shown above, any such a quadric is non-degenerate. Furthermore, Q' can be expressed by an equation

$$g(\xi, \eta) \equiv a'_{11}\xi^2 + 2a'_{12}\xi\eta + a'_{22}\eta^2 + 2b'_2\eta = 0,$$

where $a'_{11} \neq 0$. Since $Q \neq Q'$, the polynomial

$$f(\xi,\eta) + tg(\xi,\eta) = (a_{11} + ta'_{11})\xi^2 + 2(a_{12} + ta'_{12})\xi\eta + (a_{22} + ta'_{22})\eta^2 + 2(b_1 + tb'_2)\eta$$

is not identically zero in ξ, η for any choice of $t \in \mathrm{R}$. Hence the equation $f(\xi,\eta) + tg(\xi,\eta) = 0$ in ξ, η determines a quadric curve $Q(t)$ for any choice of t in R. Clearly, $Q(t)$ contains P and has l as the tangent line at p_0. By the proved above, $Q(t)$ should be a non-degenerate quadric for all $t \in \mathrm{R}$. On the other hand, the quadric $Q(t_0)$, with $t_0 = -a_{11}/a'_{11}$, described by the polynomial

$$f(\xi,\eta) + t_0 g(\xi,\eta) = \big(2(a_{12} + t_0 a'_{12})\xi + (a_{22} + t_0 a'_{22})\eta + 2(b_1 + t_0 b'_2)\big)\eta,$$

is degenerate. The obtained contradiction implies the uniqueness of Q.

3 Convex Quadrics and Their Plane Sections

In what follows, by *convex solid* in R^n, $n \geq 2$, we mean an n-dimensional closed convex sets, distinct from the whole space and, possibly, unbounded (*convex bodies* are compact convex solids). As usual, $\operatorname{bd} K$ and $\operatorname{int} K$ denote, respectively, the boundary and the interior of a convex solid $K \subset \mathrm{R}^n$.

A *convex hypersurface* in R^n is the boundary of a convex solid. This definition includes a hyperplane or a pair of parallel hyperplanes. There are different ways to define convex quadrics in R^n (see, e.g., a discussion in [42]). The most general one is given by the following definition.

Definition 1 ([39]). A convex hypersurface $S \subset \mathrm{R}^n$ is called *convex quadric* provided there is a quadric hypersurface $Q \subset \mathrm{R}^n$ and a component U of $\mathrm{R}^n \setminus Q$ such that U is a convex set and $S = \operatorname{bd} U$.

The following classification of convex quadrics is provided in [39].

Theorem 6 ([39]). *A convex hypersurface $S \subset \mathrm{R}^n$ is a convex quadric if and only if there are suitable Cartesian coordinates $\xi_1, \dots, \xi_n$ in R^n such that S can be expressed by one of the equations:*

$$a_1\xi_1^2 + \cdots + a_k\xi_k^2 = 1, \qquad 1 \leq k \leq n,$$

$$a_1\xi_1^2 - a_2\xi_2^2 - \cdots - a_k\xi_k^2 = 1,\ \xi_1 \geq 0, \qquad 2 \leq k \leq n,$$

$$a_1\xi_1^2 = 0,$$

$$a_1\xi_1^2 - a_2\xi_2^2 - \cdots - a_k\xi_k^2 = 0,\ \xi_1 \geq 0, \qquad 2 \leq k \leq n,$$

$$a_1\xi_1^2 + \cdots + a_{k-1}\xi_{k-1}^2 = \xi_k, \qquad 2 \leq k \leq n,$$

where all scalars a_i involved are positive.

In particular, convex quadrics in R^n which contain no lines can be expressed in suitable Cartesian coordinates $\xi_1, \dots, \xi_n$ by one of the equations:

$$a_1\xi_1^2 + \cdots + a_n\xi_n^2 = 1, \qquad \text{(ellipsoid)}$$

$$a_1\xi_1^2 - a_2\xi_2^2 - \cdots - a_n\xi_n^2 = 1, \ \xi_1 \geq 0, \qquad \text{(sheet of elliptic hyperboloid of two sheets)}$$

$$a_1\xi_1^2 - a_2\xi_2^2 - \cdots - a_n\xi_n^2 = 0, \ \xi_1 \geq 0, \qquad \text{(sheet of elliptic cone)}$$

$$a_1\xi_1^2 + \cdots + a_{n-1}\xi_{n-1}^2 = \xi_n, \qquad \text{(elliptic paraboloid)}$$

where all scalars $a_1, \dots, a_n$ are positive.

A recursive description of convex quadratics in R^n is given as follows.

1. Convex quadrics in R^2 are ellipses, branches of hyperbolas, parabolas, convex cones, lines, and pairs of parallel lines.
2. Convex quadrics in R^n, $n \geq 3$, are ellipsoids, sheets of elliptic hyperboloids of two sheets, sheets of elliptic cones, elliptic paraboloids, and cylinders based on convex quadrics in R^{n-1}.

In what follows, we will need the following definitions. Given a convex solid $K \subset \mathrm{R}^n$, we say that a point $x \in \mathrm{bd}\, K$ is *regular* provided there is a unique hyperplane through x supporting K. Furthermore, K is *regular* if all its boundary points are regular. The convex solid K is called *strictly convex* if its boundary does not contain segments. If $M \subset \mathrm{R}^n$ is a closed convex set, then $\mathrm{rbd}\, M$ and $\mathrm{rint}\, M$ mean, respectively, the relative boundary and the relative interior of M with respect to the smallest plane containing M (see, e.g., [46] for general references on convex sets).

We recall that the *recession cone* of a convex solid $K \subset \mathrm{R}^n$ is defined by

$$\mathrm{rec}\, K = \{y \in \mathrm{R}^n : x + ay \in K \text{ whenever } x \in K \text{ and } a \geq 0\}.$$

It is known that $\mathrm{rec}\, K$ is a closed convex cone with apex o, the origin of R^n; furthermore, $\mathrm{rec}\, K$ is distinct from $\{o\}$ if and only if K is unbounded. The subset $\mathrm{S}^{n-1} \setminus (\mathrm{rec}\, K \cup -\mathrm{rec}\, K)$ of the unit sphere $\mathrm{S}^{n-1} \subset \mathrm{R}^n$ consists of the *non-recessional* unit vectors for K. Equivalently, a unit vector $e \in \mathrm{R}^n$ is non-recessional for K if and only if the intersection of K with any line parallel to the one-dimensional subspace $l = \{ae : a \geq 0\}$ is either bounded or empty. The convex solid K has non-recessional unit vectors if and only if K is distinct from a closed halfspace of R^n.

The *linearity space* of a convex solid $K \subset \mathrm{R}^n$ is defined by

$$\mathrm{lin}\, K = \mathrm{rec}\, K \cap (-\mathrm{rec}\, K).$$

If $L \subset \mathrm{R}^n$ is a plane complementary to lin K, then K can be expressed as the direct sum

$$K = \operatorname{lin} K \oplus (K \cap L),$$

and $K \cap L$ is a closed convex set containing no lines.

The following characteristic properties of convex quadrics in terms of plane quadric sections will be used below (see also [42]). We will say that a plane $L \subset \mathrm{R}^n$ *properly* meets a convex solid $K \subset \mathrm{R}^n$ provided L meets both bd K and int K.

Theorem 7 ([37, 42]). *Let $K \subset \mathrm{R}^n$, $n \geq 3$, be a convex solid and p a point in K such that all proper sections of* bd K *by two-dimensional planes through p are convex quadric curves. Then* bd K *is a convex quadric or a convex cone with apex p.*

Theorem 8 ([38]). *If $K \subset \mathrm{R}^n$, $n \geq 3$, is a line-free convex solid and p a point in R^n, then the set* bd $K \setminus \big((p + \operatorname{rec} K) \cup (p - \operatorname{rec} K)\big)$ *lies in a convex quadric if and only if all proper bounded sections of* bd K *by two-dimensional planes through p are ellipses.*

It is interesting to compare Theorems 7 and 8 with similar generic results. For example, Lenz [25], using methods of projective geometry, proved the following theorem.

Theorem 9 ([25]). *Assume that a connected non-planar piece of a surface $S \subset \mathrm{R}^3$ is covered by an open family $\mathscr{C}$ of planes. If each section $S \cap P$, $P \in \mathscr{C}$, is a piece of a quadric curve, then S is a piece of a quadric surface.*

An open family of planes in Theorem 9 is defined as follows. Let $\mathscr{C} = \{P(e_\alpha, \gamma_\alpha)\}$ be a family of planes in R^3, each expressed as $P(e_\alpha, \gamma_\alpha) = \{x \in \mathrm{R}^3 : x \cdot e_\alpha = \gamma_\alpha\}$, where $\{e_\alpha\}$ are unit vectors and $\{\gamma_\alpha\}$ scalars. We say that $\mathscr{C}$ is *open* provided for any $P(e_\alpha, \gamma_\alpha) \in \mathscr{C}$ there is an $\varepsilon > 0$ such that $P(e, \gamma) \in \mathscr{C}$ for all unit vectors e and scalars γ satisfying the inequalities $\|e - e_\alpha\| < \varepsilon$ and $|\gamma - \gamma_\alpha| < \varepsilon$.

Clearly, Theorems 7 and 8 do not follow from Theorem 9 (even for the case $n = 3$) because the families of planes in these theorems are not open.

The following new result refines Theorem 7.

Theorem 10. *Let $K \subset \mathrm{R}^n$, $n \geq 3$, be a convex solid, p a point in R^n such that a proper section of* bd K *by a certain two-dimensional plane through p is not a branch of hyperbola. Then the following conditions are equivalent.*

1) bd K *is a convex quadric or a convex cone with apex p.*
2) *All proper sections of* bd K *by two-dimensional planes through p are convex quadric curves.*

Proof. Clearly, we need to show only that 2) $\Rightarrow$ 1). Since the cases $p \in \operatorname{int} K$ and $p \in \operatorname{bd} K$ are considered in [37] and [42], respectively, we assume that $p \in \mathrm{R}^n \setminus K$.

We proceed by induction on n (≥ 3). Let $n = 3$. If K contains a line m and L is a plane through p complementary to m, then bd K is a cylindric surface based on the convex quadric curve $L \cap \operatorname{bd} K$. Assume that K contains no lines. Let L_0 be a plane through p properly meeting K such that the curve $\Gamma_0 = L_0 \cap \operatorname{bd} K$ is not a branch of hyperbola. Then Γ_0 is either an ellipse, a parabola, or a convex cone.

I. Assume first the existence of a line $l \subset L_0$ through p meeting $\operatorname{int} K$ such that $K \cap l$ is a segment, $[u, z]$. Choose a pair of distinct two-dimensional planes L_1 and L_2 both containing l such that the sets $L_1 \cap K$ and $L_2 \cap K$ are bounded. By the assumption, $E_1 = L_1 \cap \operatorname{bd} K$ and $E_2 = L_2 \cap \operatorname{bd} K$ are convex quadric curves, whence they are ellipses. Choose a point $v \in \operatorname{bd} K \setminus (L_1 \cup L_2)$ so close to u that a certain two-dimensional plane L through the line $\langle p, v\rangle$ meets K along a segment $[v, w]$ and each of the sets $E_1 \cap L$, $E_2 \cap L$ has precisely two points. Clearly, v can be chosen such that $[v, w]$ meets $\operatorname{int} K$. As above, $L \cap \operatorname{bd} K$ is an ellipse. By Theorem 3, there is a quadric surface Q containing $\{v\} \cup E_1 \cup E_2$.

We state that $L \cap \operatorname{bd} K \subset Q$. Indeed, since both convex quadric curves $L \cap \operatorname{bd} K$ and $L \cap Q$ contain the five-point set $\{v\} \cup (L \cap E_1) \cup (L \cap E_2)$, which does not belong to a line, Theorem 4 implies that $L \cap \operatorname{bd} K = L \cap Q \subset Q$.

Slightly rotating L about the line $\langle p, v\rangle$, we obtain a family of ellipses $L \cap \operatorname{bd} K$ which cover an open subset V of $\operatorname{bd} K$. As above, $V \subset Q$. To show the inclusion $\operatorname{bd} K \subset Q$, choose a point $q \in V$ such that $\langle p, q\rangle$ meets $\operatorname{int} K$. Let $x \in \operatorname{bd} K \setminus \{q\}$, and denote by N the two-dimensional plane containing $\{p, q, x\}$. Since the quadric curves $N \cap \operatorname{bd} K$ and $N \cap Q$ coincide along the non-linear arc $N \cap V$, they must coincide: $N \cap \operatorname{bd} K = N \cap Q$. Hence $\operatorname{bd} K \subset Q$. Because $\operatorname{int} K$ is a convex component of $\mathrm{R}^n \setminus Q$, the surface $\operatorname{bd} K$ is a convex quadric.

II. Assume now that no line $l \subset L_0$ through p meets K along a segment. Since Γ_0 is not a branch of hyperbola, the latter happens only if the set $M = L_0 \cap K$ is a solid convex cone with an apex q such that p belongs to the symmetric cone $2q - M$. Denote by h_1 and h_2 the boundary halflines of M. Since both h_1 and h_2 belong to $\operatorname{bd} K$, there are planes H_1 and H_2 supporting K such that $h_1 \subset K \cap H_1$ and $h_2 \subset K \cap H_2$. If L is a plane through the line $l_0 = \langle p, q\rangle$ distinct from L_0, then the lines $L \cap H_1$ and $L \cap H_2$ bound the plane convex solid $L \cap K$, which shows that q is a singular point of $L \cap K$. Because $L \cap \operatorname{bd} K$ is a convex quadric, it must be a convex cone with apex q. Rotating L around l_0, we obtain that $\operatorname{bd} K$ is covered by a family of convex cones with apex q, implying that $\operatorname{bd} K$ is a convex cone with apex q.

Now, choose three planes L_1, L_2, L_3 through p such that neither curve $C_i = L_i \cap \operatorname{bd} K$, $i = 1, 2, 3$, is a convex cone and the component of $\operatorname{bd} K \setminus (C_1 \cup C_2 \cup C_3)$ containing q is bounded. Being unbounded, each of the convex quadrics C_1, C_2, C_3 is either a parabola or a branch of hyperbola. Since any parabola has only one recessional direction, and since each set $L_i \cap K$ has a two-dimensional recessional cone, none of C_i, $i = 1, 2, 3$, may be a parabola. Hence C_1, C_2, and C_3 are branches of hyperbolas. It is easy to see that the set $D_i = \cup([q, z\rangle : z \in C_i)$ is a piece of an elliptic cone. Since D_1, D_2, D_3 pairwise meet and cover the whole $\operatorname{bd} K$, we conclude that $\operatorname{bd} K$ is a sheet of an elliptic cone.

Let $n \geq 4$. Choose a two-dimensional plane L_0 through p such that $L_0 \cap \text{bd}\, K$ is not a branch of hyperbola. Let r be a point in $L_0 \cap \text{int}\, K$, and N be any two-dimensional plane through r. Consider a three-dimensional plane S through $\{p\} \cup N$ and the three-dimensional closed convex set $P = K \cap S$. If $L \subset S$ is a two-dimensional plane through p properly meeting P, then from $L \cap \text{rbd}\, P = L \cap \text{bd}\, K$ it follows that L meets rbd P along a convex quadric curve. By the proved above (the case $n = 3$), rbd P is a convex quadric in S. Hence $N \cap \text{bd}\, K = N \cap \text{rbd}\, P$ is a convex quadric curve, and Theorem 7 shows that bd K is a convex quadric.

Theorem 10 gives an additional argument to reiterate the following problem from [39, 42]: Is it true that the boundary of a convex solid $K \subset \mathbb{R}^n$, $n \geq 3$, is a convex quadric if and only if there is a point $p \in \mathbb{R}^n \setminus K$ such that all proper sections of bd K by two-dimensional planes through p are convex quadric curves? A combination of Theorems 7 and 10 reduces this problem to the following case.

Problem 1. Let $K \subset \mathbb{R}^n$, $n \geq 3$, be a convex solid and p a point in $\mathbb{R}^n \setminus K$ such that all proper sections of bd K by two-dimensional planes through p are branches of hyperbola. Is it true that bd K is a sheet of an elliptic hyperboloid of two sheets?

The next new result refines Theorem 5 from [40], proved there for the case $l \cap \text{int}\, K \neq \emptyset$. Given a line $l \subset \mathbb{R}^n$ and a scalar $\delta > 0$, denote by $C_\delta(l)$ the open circular cylinder of radius δ centered about the line l, and by $\mathcal{P}_\delta(l)$ the family of all two-dimensional planes which are parallel to l and whose distance from l is less than δ.

Theorem 11 ([40]). *Let $K \subset \mathbb{R}^n$, $n \geq 3$, be a convex solid, $l \subset \mathbb{R}^n$ a non-recessional for K line, and δ a positive scalar. The following conditions are equivalent.*

1) bd K *is a convex quadric.*
2) *For each two-dimensional plane $L \in \mathcal{P}_\delta(l)$ properly meeting K, the section $L \cap \text{bd}\, K$ is a convex quadric curve.*

Proof. Clearly, we need to show only that 2) $\Rightarrow$ 1). Since the case $l \cap \text{int}\, K \neq \emptyset$ is proved in [40], one can assume that $l \cap \text{int}\, K = \emptyset$. Furthermore, the case when $C_\delta(l) \cap \text{int}\, K \neq \emptyset$ can be reduced to the previous one. Indeed, choosing a suitable scalar $\varepsilon \in (0, \delta)$ and a line $l' \subset \text{int}\, C_\varepsilon(l)$ with the property $l' \cap \text{int}\, K \neq \emptyset$, we see that l' and $\delta' = \delta - \varepsilon$ satisfy condition 2). Hence we may suppose that $C_\delta(l) \cap \text{int}\, K = \emptyset$.

We further proceed by induction on n (≥ 3). Let $n = 3$. If K contains a line m and L is a two-dimensional plane through l which is complementary to m and properly meets K, then bd K is a cylindric surface based on the convex quadric curve $L \cap \text{bd}\, K$. Suppose that K contains no lines.

I. Assume first the existence of a proper section of bd K by a plane $L_0 \in \mathcal{P}_\delta(l)$ which is not a convex cone. Choose a line $l_0 \subset L_0$ parallel to l and meeting int K. Continuously rotating a plane L about l_0 from the initial position $L = L_0$ on a small angle of size $\varepsilon > 0$, we obtain a family $\mathcal{C}$ of planes L from $\mathcal{P}_\delta(l)$ properly meeting K. Since the curve $L_0 \cap \text{bd}\, K$ is not a convex cone, ε can be chosen so small that each section $L \cap \operatorname{bd} K$, with $L \in \mathscr{C}$, is a not a convex. Choose any distinct planes $L_1, L_2 \in \mathscr{C}$ and put $E_i = L \cap \operatorname{bd} K$, $i = 1, 2$. Since l is non-recessional for K, the set $K \cap l_0$ is a segment, say, $[p, q]$.

Let M be a plane through l which does not contain l_0 and meets both planes L_1 and L_2. Choose a point $v \in \operatorname{bd} K$ in the open triangular prism bounded by the planes L_1, L_2, M so close to p that the line l_v through v parallel to l meets $\operatorname{int} K$. By Theorem 3, there is a unique quadric surface $Q \subset \mathrm{R}^3$ containing $\{v\} \cup E_1 \cup E_2$.

We state that $\operatorname{bd} K \subset Q$. Indeed, choose a plane $L \in C_\delta(l)$ through v which meets the set $\{v\} \cup E_1 \cup E_2$ at five distinct points. By Theorem 4, the convex quadric curves $L \cap \operatorname{bd} K$ and $L \cap Q$ coincide. Hence $L \cap \operatorname{bd} K = L \cap Q \subset Q$. Continuously rotating L about l_v on a small angle such that L remains in $C_\delta(l)$, we obtain a family of convex quadrics $L \cap \operatorname{bd} K$ through v whose union covers an open piece, V_0, of $\operatorname{bd} K$. By the argument above, $V_0 \subset Q$. Denote by l_1 and l_2 the lines in $C_\delta(l) \cap M$ which lie at a distance $\delta/2$ from l on the opposite sides of l. Considering all sections of $\operatorname{bd} K$ by planes through l_1 which meet V_0, we enlarge V_0 to a new open piece V_1 of $\operatorname{bd} K$ also lying in Q. Similarly, the union of all sections of $\operatorname{bd} K$ by planes through l_2 which meet V_1 is a new open piece V_2 of $\operatorname{bd} K$ also lying in Q. Performing these enlargements (alternatively using planes through l_1 and l_2), we obtain an increasing sequence of open subsets $V_0 \subset V_1 \subset V_2 \subset \dots$ of $\operatorname{bd} K$ whose union covers $\operatorname{bd} K$ and lies in Q. Hence $\operatorname{bd} K \subset Q$. Since $\operatorname{int} K$ is a convex component of $\mathrm{R}^3 \setminus Q$, the set $\operatorname{bd} K$ is a convex quadric.

II. Next, assume that all proper sections of $\operatorname{bd} K$ by planes from $\mathscr{P}_\delta(l)$ are convex cones. We are going to show that this case is impossible. Indeed, choose any plane L_0 through l properly meeting K, and consider the convex cone $C_0 = L_0 \cap \operatorname{bd} K$. Let a_0 be the apex of C_0, and h_1 and h_2 its boundary halflines. Let M be a plane through l which does not contain C_0, and denote by l_1, l_2 the lines in $C_\delta(l) \cap M$ which lie at a distance $\delta/2$ from l on the opposite sides from l. By the assumption, the planes L through $\{x\} \cup l_i$, $x \in h_1$, meet $\operatorname{bd} K$ along convex cones, $i = 1, 2$. By a convexity argument, this is possible only if a_0 belongs to an open segment $(c_1, e_1) \subset \operatorname{bd} K$ such that $\operatorname{bd} K$ contains two unbounded 3-gonal regions $V_0, W_0 \subset \operatorname{bd} K$ based on $[c_1, e_1]$ and containing h_1, h_2, respectively. Similarly, considering the planes L through $\{x\} \cup l_i$, $x \in V_0$, we enlarge the regions V_0, W_0, to V_1, W_1, respectively, such that V_1 and W_1 are based on a segment $[c_2, e_2]$ properly containing $[c_1, e_1]$. Continuing the procedure, we obtain that $\operatorname{bd} K$ is the union of two closed halfplanes with the common boundary line through c_1 and e_1. In the latter case, K contains a line, contrary to the assumption above. Hence Case II is impossible.

Let $n \geq 4$. Choose a line l_0 which is parallel to l and meets $\operatorname{int} K$. Let L_0 be any two-dimensional plane through l_0 and M a three-dimensional plane through $l \cup L_0$. Then $C_\delta(l) \cap M$ is a cylinder of radius δ in M centered about l. Choose any two-dimensional plane $L \subset M$ which is parallel to l and whose distance from l is less than δ such that L properly meets K. According to condition 2), $L \cap \operatorname{bd} K$ is a convex quadric curve. Therefore, by the proved above ($n = 3$) the surface $M \cap \operatorname{bd} K$

is a convex quadric in M. Hence $L_0 \cap \operatorname{bd} K = L_0 \cap (M \cap \operatorname{bd} K)$ is a convex quadric. By Theorem 5 from [40], bd K is a convex quadric.

Problem 2. Is it true that Theorem 11 holds for any choice of the line l in R^n?

4 Convex Hypersurfaces with Hyperplanar Midsurfaces

In 1842, Bertrand [4] observed, with a sketch of proof, that any curve in the plane such that the middle points of every family of parallel chords of the curve belong to a line is necessarily a quadric curve. Bertrand's proof uses the following two arguments: (i) any five points in general position in the plane belong to a unique quadric curve, (ii) a convergent sequence of quadric curves tends to a quadric curve. Treating Bertrand's argument more analytically, Blaschke [7] (see §§ 7, 9, and 35) showed that a twice differentiable curve of constant curvature is a quadric curve (not necessarily convex) provided the middle points of any family of parallel chords of the curve belong to a line.

In 1889, Brunn [12, Chapter IV], using a technique of conjugate diameters, showed that a bounded convex curve C in the plane is an ellipse provided the middle points of every family of parallel chords of C belong to a line. For the same purpose, Blaschke [5] (see also [6, pp. 158–159]) uses the idea of affine symmetry. Namely, he chooses a pair of affine reflections ϕ and ϕ_n of C onto itself such that their composition Φ_n is an affine rotation with period 2^n. Applying a suitable affine transformation f, one can make Φ_n a usual rotation Φ_n^* on an angle $2\pi/2^n$ which maps $f(C)$ onto itself. When n tends to infinity, $f(C)$ remains invariant with respect to a rotation on any angle of size $2\pi m/2^n$, where $m, n \geq 1$. This argument shows that $f(C)$ is a circle, and whence C is an ellipse.

For some other proofs of this characteristic property, see, for example, Nakajima [29], Berger [3], Kneser [23], and Süss, Viet, Berger [44].

Grünbaum [20, p. 82] mentioned without proof the following two results about a bounded convex curve C in the plane: (a) C is an ellipse provided there is a scalar $\varepsilon > 0$ such that the intersection of each midcurve of C with an ε-neighborhood of C consists of two segments, (b) C is an ellipse provided it has infinitely many straight midcurves. We observe that statement (a) is extendable to higher dimension (see Theorem 14 below), while statement (b) cannot be generalized even to the case of unbounded convex curves in the plane. Indeed, let K denote the convex hull of the set $P = \{(k, k^2) : k \in \mathrm{Z}\}$. Then the convex curve $C = \operatorname{bd} K$ has infinitely many line midcurves: each of them is generated by the chords of C which are parallel to the line tangent to the parabola $y = x^2$ at (k, k^2).

Brunn's result was generalized to higher dimension by Blaschke [6, p. 159] for $n = 3$ and Busemann [14, p. 92] for all $n \geq 2$ (see also Grinberg [17] and Thompson [45, Section 3.4] for the case of a centrally symmetric convex body).

Theorem 12 ([14]). *The boundary of a convex body $K \subset \mathrm{R}^n$, $n \geq 2$, is an ellipsoid if and only if the middle points of each family of parallel chords of K belong to a hyperplane.*

The proof of Theorem 12 uses Blashke's method to characterize ellipses, followed by the following statement: the boundary of a convex body $K \subset \mathrm{R}^n$, $n \geq 3$, is an ellipsoid provided all sections of bd K by two-dimensional planes through a given point $p \in \mathrm{int}\, K$ are ellipses. See, e.g., [42] for various references on quadratic sections of convex solids.

Kubota [24] showed that the boundary of a convex body K in the plane is an ellipse provided for each family $\mathscr{F}$ of chords of K in the same direction, there is a scalar $\lambda_{\mathscr{F}} \in (0, 1)$ such that the locus of points that divide all chords from $\mathscr{F}$ in the ratio $\lambda_{\mathscr{F}}$ belongs to a line. (We say that, given a nonzero vector $e \in \mathrm{R}^n$, a chord $[x, z]$ (also a line $\langle x, z\rangle$) has direction e provided $z - x$ is a positive multiple of e. Furthermore, a point y divides $[x, z]$ in a ratio $\lambda \in [0, 1]$ provided $y = (1-\lambda)x+\lambda z$; obviously, $\|x - y\| = \lambda\|x - z\|$.)

Kubota's result was extended in [37] to the case of convex quadrics in R^n, as follows.

Theorem 13 ([37]). *If $K \subset \mathrm{R}^n$, $n \geq 2$, is a convex solid distinct from a halfspace, then the following conditions are equivalent.*

1) bd K *is a convex quadric surface.*
2) *The middle points of every family of parallel chords of K belong to a hyperplane.*
3) *For every family $\mathscr{F}$ of parallel chords of K in the same direction, there is a scalar $\lambda_{\mathscr{F}} \in (0, 1)$ such that the locus of points that divide the chords from $\mathscr{F}$ in the ratio $\lambda_{\mathscr{F}}$ lies in a hyperplane.*

Theorem 13 immediately follows from a sharper statement below, which considers families of parallel chords in a small neighborhood of bd K. We will say that a line $l \subset \mathrm{R}^n$ is *non-recessional* for a convex solid K if l is a translate of a one-dimensional non-recessional subspace for K. For a scalar $\delta > 0$ and a non-recessional line l which *supports* K, denote by $K_\delta(l)$ the set of points in K whose distance from l is at most δ. If l has a certain positive direction, then let $\mathscr{F}_\delta(l)$ be the family of chords of K lying in $K_\delta(l)$ and having the same direction as l.

A two-dimensional convex solid $M \subset \mathrm{R}^2$ (as well as its boundary curve bd M) will be called *δ-polygonal* provided it is locally polygonal, and for any non-recessional line l supporting M, the interior of $M_\delta(l)$ contains at most one vertex of M. Obviously, any convex polygon in R^2 is δ-polygonal for a suitable $\delta > 0$.

Theorem 14 ([37]). *For a convex solid $K \subset \mathrm{R}^n$, $n \geq 2$, distinct from a halfspace, the following conditions are equivalent.*

1) bd K *is a convex quadric or K is a direct sum of a subspace and a line-free closed convex set C of dimension m, $2 \leq m \leq 3$, such that C is a simplicial cone if $m = 3$, or C is δ-polygonal for a suitable $\delta > 0$ if $m = 2$.*

2) *There is a scalar* $\delta > 0$ *such that for each non-recessional directed line* l *supporting* K *one can find a scalar* $\lambda(l) \in [0, 1]$ *with the following property: the points which divide all chords* $m \in \mathscr{F}_\delta(l)$ *in the ratio* $\lambda(l)$ *belong to a hyperplane.*

Corollary 1. *For a convex solid* $K \subset \mathrm{R}^n$, $n \geq 2$, *distinct from a halfspace, the following conditions are equivalent.*

1) bd K *is a convex quadric or* K *is a direct sum of a subspace and a line-free closed convex set* C *of dimension* m, $2 \leq m \leq 3$, *such that* C *is a simplicial cone if* $m = 3$, *or* C *is a convex cone or a triangle if* $m = 2$.
2) *For each non-recessional directed line* l, *one can find a scalar* $\lambda(l) \in [0, 1]$ *with the following property: the points which divide in the ratio* $\lambda(l)$ *all chords of* K *in direction* l *belong to a hyperplane.*

The proof of Theorem 14 is organized in distinct steps. The statement 1) $\Rightarrow$ 2) follows from Theorem 1, with $\lambda(l) = 1/2$ if bd K is a convex quadric, and from the standard properties of triangles and three-dimensional simplicial cones, with $\lambda(l) = 0$ or $\lambda(l) = 1$.

The opposite statement 2) $\Rightarrow$ 1) is first considered for the case $n = 2$. The major steps here are as follows.

1. If the convex solid $K \subset \mathrm{R}^2$ is neither regular nor strictly convex, then K is δ-polygonal.
2. If the convex solid $K \subset \mathrm{R}^2$ is regular and strictly convex, then for any directed line supporting K the respective scalar $\lambda(l)$ satisfying condition 2) of the theorem equals 1/2.
3. If the convex solid $K \subset \mathrm{R}^2$ is regular and strictly convex, then for any tangent line of K, the part of bd K lying in $K_\delta(l)$ is an arc of a convex quadric curve.

For $n \geq 3$, the proof of 2) $\Rightarrow$ 1) uses the following arguments.

1. If there is a point $p \in$ int K such that each section of bd K by a two-dimensional plane through p is either a convex quadric curve or a locally polygonal line, then bd K is either a convex quadric hypersurface or a locally polyhedral surface.
2. If K is locally polyhedral and satisfies condition 2) of the theorem, then K is a direct sum of a subspace and a line-free closed convex set C of dimension m, $2 \leq m \leq 3$, such that C is a simplicial cone if $m = 3$, or C is δ-polygonal for a suitable $\delta > 0$ if $m = 2$.

Gruber [18, 19] gave the following refinements of Theorem 12 (see also Montejano and Morales [28] for the case when K in Theorem 15 is centrally symmetric and $\lambda(e) = 1/2$ for all $e \in \mathrm{S}^{n-1}$ in a small neighborhood of a given point).

Theorem 15 ([18]). *The boundary of a convex body* $K \subset \mathrm{R}^n$, $n \geq 2$, *is an ellipsoid provided* K *has the following property: there is a subset* T *of the unit sphere* $\mathrm{S}^{n-1} \subset \mathrm{R}^n$ *having nonempty interior with respect to* S^{n-1} *such that for each vector* $e \in T$ *one can find a scalar* $\lambda(e) \in (0, 1)$ *and a hyperplane* $H(e)$ *so that for any chord* $[x, z]$ *of* K *in direction* e, *the point* $(1 - \lambda(e))\, x + \lambda(e)\, z$ *belongs to* $H(e)$.

Theorem 16 ([18]). *The boundary of a convex body $K \subset \mathbb{R}^n$, $n \geq 2$, is an ellipsoid provided K has the following property: there is a convex subset T of the unit sphere $\mathbb{S}^{n-1} \subset \mathbb{R}^n$ having nonempty interior with respect to $\mathbb{S}^{n-1}$ and $\operatorname{cl} T$ containing a pair of opposite vectors of $\mathbb{S}^{n-1}$ such that for each vector $e \in T$ one can find a scalar $\lambda(e) \in [0,1]$ and a hyperplane $H(e)$ so that for any chord $[x,z]$ of K in direction e, the point $(1-\lambda(e))\,x + \lambda(e)\,z$ belongs to $H(e)$.*

Theorem 17 ([19]). *There are $(n-1)$-dimensional subspaces $L_1, \dots, L_4 \subset \mathbb{R}^n$ with the following property: the boundary of a convex body $K \subset \mathbb{R}^n$, $n \geq 2$, is an ellipsoid provided for each one-dimensional ordered subspace $l \subset L_1 \cup \dots \cup L_4$ one can find a hyperplane $H(l)$ and a scalar $\lambda(l) \in [0,1]$ so that for any chord $[x,z]$ of K in direction l, the point $(1-\lambda(l))\,x + \lambda(l)\,z$ belongs to $H(l)$.*

Based on Theorem 15, we prove the following result (a similar statement, with $\lambda(e) = 1/2$ for all $e \in T$ and $n \geq 2$, is given in [40]).

Theorem 18. *Given a line-free convex solid $K \subset \mathbb{R}^n$, $n \geq 3$, and an open nonempty subset T of $\mathbb{S}^{n-1} \setminus (\operatorname{rec} K \cup -\operatorname{rec} K)$, the following conditions are equivalent.*

1) bd K *is a convex quadric.*
2) *For each vector $e \in T$ one can find a scalar $\lambda(e) \in (0,1)$ and a hyperplane $H(e)$ so that for any chord $[x,z]$ of K in direction e, the point $(1-\lambda(e))x + \lambda(e)z$ belongs to $H(e)$.*

Proof. 1) $\Rightarrow$ 2) due to Theorem 1, with $\lambda(e) = 1/2$ for all $e \in T$.
2) $\Rightarrow$ 1) This part of the proof is organized by induction on $n \geq 3$.

Let $n = 3$. Translating K on a suitable vector, we assume that $o \in \operatorname{int} K$. Choose any vector $e \in T$ and denote by l the one-dimensional subspace containing e. Since K is line-free, there are distinct planes L_1 and L_2 both containing l such that the sets $L_1 \cap K$ and $L_2 \cap K$ are bounded. Clearly, $L_i \cap T$ is a nonempty open subset of

$$(L_i \cap \mathbb{S}^2) \setminus \big(\operatorname{rec}(L_i \cap K) \cup -\operatorname{rec}(L_i \cap K)\big), \quad i = 1, 2.$$

Choose any vector $u \in L_i \cap T$. By condition 2) of the theorem, there is a scalar $\lambda_i(u) \in (0,1)$ and a plane $H_i(u)$ so that for any chord $[x,z]$ of $L_i \cap K$ in direction u, the point $(1-\lambda(u))x + \lambda(u)z$ belongs to $L_i \cap H_i(u)$. Since $L_i \cap H_i(u)$ is a line in L_i, Theorem 15 (with $n = 2$) implies that both sections $E_1 = L_1 \cap \operatorname{bd} K$ and $E_2 = L_2 \cap \operatorname{bd} K$ are ellipses.

Choose a point $v \in \operatorname{bd} K \setminus (L_1 \cup L_2)$ so close to l that $v/\|v\| \in T$, a certain two-dimensional plane L through the line $\langle o, v\rangle$ meets K along a bounded set, and each of the sets $E_1 \cap L$, $E_2 \cap L$ has precisely two points. As above, $L \cap \operatorname{bd} K$ is an ellipse. By Theorem 3, there is a unique quadric surface $Q \subset \mathbb{R}^3$ containing $\{v\} \cup E_1 \cup E_2$. We observe that $L \cap \operatorname{bd} K \subset Q$. Indeed, Theorem 4 implies that the ellipse $L \cap \operatorname{bd} K$ is a unique quadric curve containing the five-point set $X = \{v\} \cup (E_1 \cap L) \cup (E_2 \cap L)$. Since $L \cap Q$ also is a quadric curve containing X, one has $L \cap \operatorname{bd} K = L \cap Q \subset Q$.

Slightly rotating L about the line $\langle o, v\rangle$, we obtain a family of ellipses $L \cap \operatorname{bd} K$ which cover an open subset V_0 of $\operatorname{bd} K$ consisting of two open "lenses" with a common endpoint v. As above, $V_0 \subset Q$. Similarly, if $\{x + L_1\}$ is the family of translates of the plane L_1 meeting V_0, then each section $(x + L_1) \cap \operatorname{bd} K$ is an ellipse lying in Q. Clearly, the union of these ellipses covers a larger than V_0 open subset V_1 of $\operatorname{bd} K$ enclosed by a pair of planes parallel to L_1. Performing next a similar procedure on V_1, with L_2 instead of L_1, we enlarge V_1 to another open subset V_2 of $\operatorname{bd} K$ which lies in Q and is enclosed by a pair of planes parallel to L_2. Alternatively repeating this enlargement procedures, we obtain a sequence of sets $V_0 \subset V_1 \subset V_2 \subset V_3 \subset V_4 \subset \ldots$ whose union covers $\operatorname{bd} K$ and lies in Q. Hence $\operatorname{bd} K \subset Q$. Since $\operatorname{int} K$ is a convex component of $\mathrm{R}^3 \setminus Q$, the surface $\operatorname{bd} K$ is a convex quadric.

Let $n \geq 4$. As above, we assume that $o \in \operatorname{int} K$. To prove that $\operatorname{bd} K$ is a convex quadric in R^n, it suffices to show that the intersection of $\operatorname{bd} K$ with any two-dimensional subspace $L \subset \mathrm{R}^n$ is a convex quadric curve (see Theorem 7). Choose a vector $e \in T \setminus L$ and put $M = \operatorname{span}(e \cup L)$. Then M is a three-dimensional subspace of R^n. Clearly, $M \cap T$ is a nonempty open subset of

$$(M \cap \mathrm{S}^{n-1}) \setminus \big(\operatorname{rec}(M \cap K) \cup -\operatorname{rec}(M \cap K)\big).$$

Choose any vector $u \in M \cap T$. By condition 2) of the theorem, there is a scalar $\lambda(u) \in (0,1)$ and a hyperplane $H(u)$ so that for any chord $[x,z]$ of $M \cap K$ in direction u, the point $(1-\lambda(u))x + \lambda(u)z$ belongs to $M \cap H(u)$. Since $M \cap H(u)$ is a plane in M, from the case $n = 3$ above it follows that $M \cap \operatorname{bd} K$ is a three-dimensional convex quadric. Hence $L \cap \operatorname{bd} K$ $(= L \cap M \cap \operatorname{bd} K)$ is a convex quadric curve. Therefore $\operatorname{bd} K$ is a convex quadric.

The question whether the statement of Theorem 18 holds in the case $n = 2$ remains open (see [40]). The following more general problem, if confirmed, will give an affirmative answer to this case.

Problem 3. Let f and g be, respectively, a convex and a concave functions on a closed segment $[a,b]$ such that $f(x) < g(x)$ for all $x \in [a,b]$. Furthermore, suppose the existence of a scalar $\varepsilon > 0$ such that for any directed line $l \subset \mathrm{R}^2$ forming with the y-axis of R^2 an angle of size at most ε there is a line $H = H(l)$ and a scalar $\lambda = \lambda(l) \in (0,1)$ so that the following property holds: if a translate of l meets the graphs of f and g at points u and v, respectively, then the point $z = (1-\lambda)u + \lambda v$ belongs to H. Is it true that the graphs of f and g are pieces of a convex quadric curve?

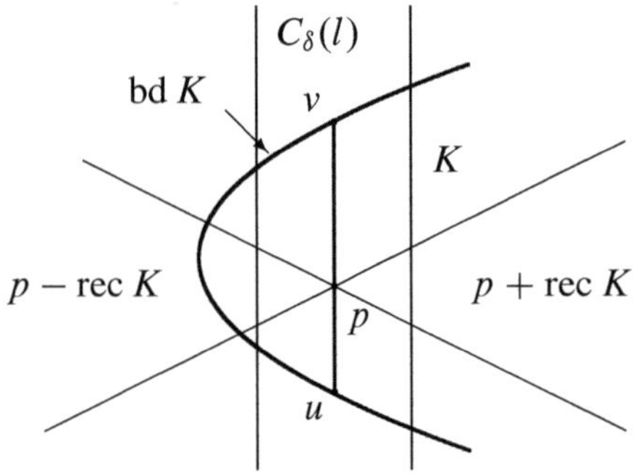

We need some definitions and notation to formulate one more result on local hyperplanarity of midsurfaces. Let $K \subset \mathbb{R}^n$ be a convex solid and p a point in $\mathbb{R}^n$ with the property that a certain line through p meets K along a segment $[u, v]$. Given a positive scalar δ, denote by $C_\delta(l)$ the closed circular cylinder of radius δ centered about the line $l = \langle u, v \rangle$, and by $\mathscr{F}_\delta(l)$ the family of all chords of K which are parallel to l and lie in $C_\delta(l)$. Furthermore, let

$$\Omega_\delta(p) = \cup\, (C_\delta(l) \cap \text{bd}\, K),$$

where the union is taken over all non-recessional lines of K which containing p (put $\Omega_\delta(p) = \emptyset$ if no such a line exists). Clearly, $\Omega_\delta(p)$ is a closed neighborhood of $\text{bd}\, K \setminus \big((p + \text{rec}\, K) \cup (p - \text{rec}\, K)\big)$ in bd K (see the figure above).

The following theorem (proved in [40] for the particular case $p \in \text{int}\, K$), addresses a question of Erwin Lutwak: Is it true that a convex body $K \subset \mathbb{R}^n$ is a solid ellipsoid provided there is a point $p \in \text{int}\, K$ and a scalar $\delta > 0$ such that, for every chord $[u, v]$ of K through p, the middle points of all chords of K which are parallel to $[u, v]$ and lie at a distance δ or less from $[u, v]$ belong to a hyperplane?

Theorem 19. *Given a convex solid $K \subset \mathbb{R}^n$, $n \geq 2$, a point $p \in \mathbb{R}^n$, and a scalar $\delta > 0$, the following conditions are equivalent.*

1) *The set $\Omega_\delta(p)$ lies in a convex quadric.*
2) *For each non-recessional line l of K which contains p and meets K, the middle points of all chords from $\mathscr{F}_\delta(l)$ belong to a hyperplane.*

Proof. 1) $\Rightarrow$ 2) Translating K on $-p$, we may assume that $p = o$. Choose a non-recessional line l of K which contains o and meets K. Then l is parallel to a unit vector $e \in S^{n-1} \setminus (\text{rec}\, K \cup -\text{rec}\, K)$. If $\Omega_\delta(o)$ is the neighborhood of $\text{bd}\, K \setminus (\text{rec}\, K \cup -\text{rec}\, K)$ in bd K that lies in a convex quadric, Q, then the cylinder $C_\delta(l)$ meets bd K within Q. By Theorem 1, the middle points of chords from $\mathscr{F}_\delta(l)$ belong to a hyperplane.
2) $\Rightarrow$ 1) As above, we assume that $p = o$. Furthermore, we may suppose that K is line-free. Indeed, let $\dim(\text{lin}\, K) \geq 1$. Choose a non-recessional line l of K which contains o. Let $M \subset \mathbb{R}^n$ be a subspace complementary to lin K and containing l. Put $K' = M \cap K$. Clearly, $\text{lin}\, K' = M \cap \text{lin}\, K = \{o\}$. If H is

a hyperplane that contains the middle points of chords from $\mathscr{F}_\delta(l)$, then $M \cap H$ contains the middle points of those chords from $\mathscr{F}_\delta(l)$ which lie in M. So, if we prove the existence of the neighborhood $\Omega'_\delta(o)$ of the set $\operatorname{rbd} K' \setminus (\operatorname{rec} K' \cup -\operatorname{rec} K')$ in $\operatorname{rbd} K'$ which lies in a convex quadric $Q' \subset M$, then, due to the equality $\operatorname{bd} K = \operatorname{rbd} K' \oplus \operatorname{lin} K$, we will conclude that the neighborhood $\Omega_\delta(o)$ of $\operatorname{bd} K \setminus (\operatorname{rec} K \cup -\operatorname{rec} K)$ in $\operatorname{bd} K$ lies in the convex quadric $Q' \oplus \operatorname{lin} K$.

First, we consider the case $n = 2$. Choose a non-recessional line l of K containing o. Put $[p_0, q_0] = K \cap l$, and denote by e_0 the unit vector which is a positive scalar of $q_0 - p_0$. We may choose δ so small that both boundary lines of the slab $C_\delta(l)$ meet $\operatorname{int} K$. Denote by e_m, $m \geq 1$, the unit vector forming with e_0 an angle of positive size π/m (according to counterclockwise bypass of $\operatorname{bd} K$). Clearly, there is a positive integer m_0 with the following property: for any $m \geq m_0$, there are points, denoted $p_{-1}(m)$ and $q_1(m)$, lying in $C_\delta(l) \cap \operatorname{bd} K$ such that both chords $[p_0, q_1(m)]$ and $[p_{-1}(m), q_0]$ have direction e_m.

Denote by $p_1(m)$, $m \geq m_0$, the point in $C_\delta(l) \cap \operatorname{bd} K$ such that $[p_1(m), q_1(m)]$ has directions e_0. By condition 1), there is a line $H(e_0)$ containing the middle points of $[p_0, q_0]$ and $[p_1(m), q_1(m)]$. Similarly, there is a line $H(e_m)$ containing the middle points of $[p_{-1}, q_0(m)]$ and $[p_0, q_1(m)]$. Since the set

$$Y_5(m) = \{p_0, q_0, p_1(m), q_1(m), p_{-1}(m)\}$$

does not belong to a line, there is a unique quadric curve $Q(m)$ containing $Y_5(m)$ (see Theorem 4).

If a point $q_k(m), k \geq 2$, is chosen in $C_\delta(l) \cap \operatorname{bd} K$ and the line through $q_k(m)$ in direction e_0 meets $H(e_0) \cap K$, then let $p_k(m)$ be the point in $C_\delta(l) \cap \operatorname{bd} K$ for which the segment $[p_k(m), q_k(m)]$ has direction e_0. If a point $p_k(m), k \geq 2$, is chosen in $C_\delta(l) \cap \operatorname{bd} K$ and the line through $p_k(m)$ in direction e_m meets both $H(e_m) \cap K$ and $C_\delta(l) \cap \operatorname{bd} K$, then denote by $q_{k+1}(m)$ the point in $C_\delta(l) \cap \operatorname{bd} K$ for which $[p_k(m), q_{k+1}(m)]$ has direction e_m.

Similarly, if a point $p_{-k}(m), k \geq 1$, is chosen in $C_\delta(l) \cap \operatorname{bd} K$ and the line through $p_{-k}(m)$ in direction e_0 meets $H(e_0) \cap K$, then denote by $q_{-k}(m)$ the point in $C_\delta(l) \cap \operatorname{bd} K$ for which the segment $[p_{-k}(m), q_{-k}(m)]$ has direction e_0. If a point $q_{-k}(m), k \geq 1$, is chosen in $C_\delta(l) \cap \operatorname{bd} K$ and the line through $q_{-k}(m)$ in direction e_m meets both $H(e_m) \cap K$ and $C_\delta(l) \cap \operatorname{bd} K$, then denote by $p_{-k-1}(m)$ the point in $C_\delta(l) \cap \operatorname{bd} K$ for which $[p_{-k-1}(m), q_{-k}(m)]$ has direction e_m.

A combination of condition 2) and Theorem 1 shows that the set

$$\begin{aligned} Y_{2k+2}(m) = \{&p_0, q_0, p_1(m), q_1(m), \dots, p_k(m), q_k(m), \\ &p_{-1}(m), q_{-1}(m), \dots, p_{-k}(m), q_{-k}(m)\} \end{aligned}$$

belongs to $Q(m) \cap C_\delta(l) \cap \operatorname{bd} K$. Clearly, there is an increasing sequence of positive integers $k(m)$, $m \geq m_0$, such that $Y_{2k(m)+2}(m)$ exists for all $m \geq m_0$, and the sets

$$Y_{2k(m_0)+2}(m_0), Y_{2k(m_0+1)+2}(m_0 + 1), \dots,$$

tend to a dense subset of $C_\delta(l) \cap \text{bd}\, K$. Hence the arcs of the quadratic curves

$$C_\delta(l) \cap Q(m_0),\ C_\delta(l) \cap Q(m_0 + 1), \dots$$

converge to $C_\delta(l) \cap \text{bd}\, K$, which shows that $C_\delta(l) \cap \text{bd}\, K$ consists of two arcs of the same quadric curve. Continuously rotating l about o, we cover $\text{bd}\, K \setminus (\text{rec}\, K \cup -\text{rec}\, K)$ with the family of overlapping pieces $C_\delta(l) \cap \text{bd}\, K$ of the same quadric curve. Hence the neighborhood $\Omega_\delta(o)$ of $\text{bd}\, K \setminus (\text{rec}\, K \cup -\text{rec}\, K)$ in $\text{bd}\, K$ lies in a convex quadric curve.

Let $n \geq 3$. Choose any two-dimensional subspace L such that $L \cap K$ is bounded (this is possible since K is line-free). Then $\text{rec}\,(L \cap K) = \{o\}$. If l is a non-recessional line through o meeting $L \cap K$, and if $H \subset \mathrm{R}^n$ is a hyperplane containing the middle points of all chords from $\mathscr{F}_\delta(l)$, then $L \cap C_\delta(l)$ is a slab of width 2δ centered about l and $L \cap H$ is a line that contains the middle points of chords of $L \cap K$ which belong to $\mathscr{F}_\delta(l)$. Hence $L \cap K$ satisfies condition 1) of the theorem (with L instead of R^n). By the proved above (see the case $n = 2$), $\text{rbd}\,(L \cap K)$ is a convex quadric; so, it is an ellipse because $L \cap K$ is bounded. Theorem 8 shows that $\text{bd}\, K \setminus (\text{rec}\, K \cup -\text{rec}\, K)$ lies in a convex quadric Q.

If K is bounded, then $\text{rec}\, K = \{o\}$ and the whole hypersurface $\text{bd}\, K$ is a convex quadric. Assume that K is unbounded and choose a halfline h with endpoint o that lies in $\text{int}\, K$. Then (see the case $n = 2$) for any two-dimensional subspace $L \subset \mathrm{R}^n$ containing h, the neighborhood $\Omega_\delta(o)$ of $(L \cap \text{bd}\, K) \setminus (\text{rec}\, K \cup -\text{rec}\, K)$ in $\text{rbd}\,(L \cap K)$ lies in $L \cap Q$. Therefore, the neighborhood $\Omega_\delta(o)$ of $\text{bd}\, K \setminus (\text{rec}\, K \cup -\text{rec}\, K)$ in $\text{bd}\, K$ lies in Q.

If $K \subset \mathrm{R}^n$ is a convex body, then $\text{rec}\, K = \{o\}$ and the set $\Omega_\delta(p)$ in Theorem 19 coincides with $\text{bd}\, K$ for any choice of the point $p \in \mathrm{R}^n$. This argument implies the following corollary.

Corollary 2. *Given a convex body $K \subset \mathrm{R}^n$, $n \geq 2$, a point $p \in \mathrm{R}^n$, and a scalar $\delta > 0$, the following conditions are equivalent.*

1) $\text{bd}\, K$ *is an ellipsoid.*
2) *For each line l which contains p and meets K, the middle points of all chords from $\mathscr{F}_\delta(l)$ belong to a hyperplane.*

We conclude this section with a joint characterization of solid ellipsoids and convex polyhedra by means of λ-surfaces.

Theorem 20 ([35]). *For a convex body $K \subset \mathrm{R}^n$, $n \geq 2$, the following conditions are equivalent.*

1) *K is either a solid ellipsoid or a convex polytope.*
2) *For each ordered line l in R^n there is a scalar $\lambda = \lambda(l) \in [0, 1)$ such that the set of points dividing in the ratio λ all chords of K in direction l lies within a polyhedral hypersurface.*

The interval [0, 1) in Theorem 20 cannot be replaced with [0, 1] (clearly, instead of [0, 1) one can consider (0, 1]). Indeed, with polar coordinates (ρ, φ) in the plane R^2, let $X = \{v_0, v_1, \ldots\}$, where $v_0 = (1, 0)$, $v_k = (1, \pi/k)$, $k \geq 1$. Since X is compact, its convex hull $K = \operatorname{conv} X$ is a convex body in R^2, which is neither an ellipse nor a polygon. At the same time, for any direction l in R^2 one of the λ-curves of K, corresponding to $\lambda = 0$ or $\lambda = 1$, in direction l is a polygonal line.

5 Convex Hypersurfaces with Hyperplanar Shadow-Boundaries

Given a convex solid $K \subset \mathrm{R}^n$ and a line $l \subset \mathrm{R}^n$, the *shadow-boundary* of K with respect to l, denoted $S_l(K)$, is the set of points in $\operatorname{bd} K$ at which the translates of l support K. This terminology comes from the concept of illumination of K by a family of rays which are parallel to a given direction (see, e.g., the survey of Martini and Soltan [27]). Since any two parallel lines determine the same shadow-boundary of K, we consider, in what follows, the shadow-boundaries generated by one-dimensional subspaces of R^n. If l is a one-dimensional subspace of R^n, then

$$S_l(K) = \operatorname{bd} K \cap \operatorname{bd}(K + l),$$

where $K + l$ is the vector sum of K and l (equivalently, $K + l$ is the union of all translates of l meeting K).

Blaschke ([5] and [6, p. 157–159], see also Blaschke and Hessenberg [9]) proved that a strictly convex body $K \subset \mathrm{R}^3$ with regular boundary is a solid ellipsoid if every shadow-boundary of K is a plane curve. Alexandrov [1], based on the work of Jitomirskii [21], obtained a far-reaching local version of Blaschke's result, which states that a non-planar bounded piece T of the boundary of a convex solid $K \subset \mathrm{R}^3$ lies in a convex quadric or in the boundary of a convex cone provided for each shadow-boundary $S_l(K)$ of K that meets T there is a plane H such that $S_l(K) \cap T \subset H$ (see also Blaschke [5] and [7, p. 119] for similar statements concerning regular non-convex surfaces).

Refining Blaschke's argument, Busemann [14, p. 93] proved the following statement (see also Borodin [11] and Šaĭdenko [32]).

Theorem 21 ([14]). *A convex body $K \subset \mathrm{R}^n$, $n \geq 3$, is a solid ellipsoid if every shadow-boundary of K lies in a hyperplane.*

Monejano and Morales-Amaya [28] proved the following result: A convex body $K \subset \mathrm{R}^n$ about the origin o is a solid ellipsoid if there is a hyperplane $H \subset \mathrm{R}^n$ through o such that for every one-dimensional subspace l sufficiently close to H the shadow-boundary $S_l(K)$ lies in a hyperplane. Another variation of Theorem 21 is mentioned by Rudin and Smith [31]: If $K \subset \mathrm{R}^n$ is a convex body centered at the origin o of R^n and $1 \leq r \leq n - 2$ an integer such that for each r-dimensional subspace $L \subset \mathrm{R}^n$, the set of points at which translates of L support K lies in a plane of dimension $n-r$, then K is an ellipsoid (see also Borodin [10] for a variety of similar conditions). Schneider [34] (respectively, Schwenk [33]) characterized ellipsoids as affine $(n-1)$-dimensional spheres in R^n which have at least $n+1$ (respectively, at least one) hyperplanar shadow-boundary resulted from parallel projection.

Marchaud [26] showed that a convex body $K \subset \mathrm{R}^3$ is a solid ellipsoid provided for any one-dimensional subspace $l \subset \mathrm{R}^n$ there is a plane H meeting $\operatorname{int} K$ such that

$$H \cap \operatorname{bd} K \subset S_l(K).$$

Although Marchaud's statement does not mention the condition $H \cap \operatorname{int} K \neq \emptyset$, his proof is essentially using it. Clearly, this condition cannot be omitted. Indeed, if K is a convex polytope in R^3, then for any one-dimensional subspace $l \subset \mathrm{R}^3$ there is a plane H which is not parallel to H and supports K along an edge lying in $S_l(K)$.

Gruber [19], refining Marchaud's argument, proved the following statement.

Theorem 22 ([19]). *There are $(n-1)$-dimensional subspaces $L_1, \dots, L_4 \subset \mathrm{R}^n$ with the following property: the boundary of a convex body $K \subset \mathrm{R}^n$, $n \geq 2$, is an ellipsoid provided for each one-dimensional subspace $l \subset L_1 \cup \dots \cup L_4$ one can find a hyperplane H satisfying the inclusion*

$$H \cap \operatorname{bd}(K+l) \subset S_l(K). \tag{15}$$

The following two lemmas clarify various planarity conditions used by various authors to characterize ellipsoids. We say that a one-dimensional subspace $l \subset \mathrm{R}^n$ is *sharp* for K if every line parallel to l and supporting K has precisely one point in K.

Lemma 1. *For a convex body $K \subset \mathrm{R}^n$, a one-dimensional subspace $l \subset \mathrm{R}^n$, and a hyperplane $H \subset \mathrm{R}^n$, the following conditions are equivalent:*

1) $S_l(K) \subset H$, 2) $S_l(K) \subset H \cap \operatorname{bd} K$,

3) $S_l(K) \subset H \cap \operatorname{bd}(K+l)$, 4) $S_l(K) = H \cap \operatorname{bd}(K+l)$.

Any of conditions 1)–4) *implies that l and H are not parallel and l is sharp for K.*

Proof. Since any line l' supporting K and parallel to l contains a point from $S_l(K)$, the set $S_l(K)$ cannot lie in a hyperplane parallel to l.

1) $\Leftrightarrow$ 2) If $S_l(K) \subset H$, then

$$\begin{aligned} S_l(K) &= \operatorname{bd} K \cap \operatorname{bd}(K+l) = (\operatorname{bd} K \cap \operatorname{bd}(K+l)) \cap \operatorname{bd} K \\ &= S_l(K) \cap \operatorname{bd} K \subset H \cap \operatorname{bd} K. \end{aligned}$$

Conversely, if $S_l(K) \subset H \cap \operatorname{bd} K$, then $S_l(K) \subset H$.

1) $\Leftrightarrow$ 3) If $S_l(K) \subset H$, then

$$S_l(K) = \operatorname{bd} K \cap \operatorname{bd}(K + l) = (\operatorname{bd} K \cap \operatorname{bd}(K + l)) \cap \operatorname{bd}(K + l)$$
$$= S_l(K) \cap \operatorname{bd}(K + l) \subset H \cap \operatorname{bd}(K + l).$$

Conversely, if $S_l(K) \subset H \cap \operatorname{bd}(K + l)$, then $S_l(K) \subset H$.

1) $\Leftrightarrow$ 4) Due to the proved above, it suffices to show that 1) $\Rightarrow$ 4); moreover, that $H \cap \operatorname{bd}(K + l) \subset S_l(K)$ provided condition 1) holds. Let $x \in H \cap \operatorname{bd}(K + l)$. Then there is a point $z \in \operatorname{bd} K$ such that the line through z contains x. Clearly, $z \in \operatorname{bd}(K + l)$, which shows that $z \in S_l(K) \subset H$. Since l is not parallel to H, x is the only point in $l \cap H$. Hence $x = z \in S_l(K)$, which proves the inclusion $H \cap \operatorname{bd}(K + l) \subset S_l(K)$.

Finally, assuming that l is not sharp for K, one can find a segment $[u, v] \subset \operatorname{bd} K$ parallel to l. Since $[u, v] \subset S_l(K)$, we obtain that $S_l(K)$ lies in H, while H is not parallel to l. The obtained contradiction shows that l is sharp for K.

Remark 1. The inclusion $S_l(K) \subset H \cap \operatorname{bd} K$ in Lemma 1 may be proper. Indeed, let $K = \{(\xi, \eta) : \xi^2 + \eta^2 \leq 1,\ \eta \geq 0\}$, and l be the η-axis of R^2. Then $S_l(K) = \{(-1, 0), (1, 0)\}$, and the only line H containing $S_l(K)$ is the ξ-axis. On the other hand, $H \cap \operatorname{bd} K$ is the segment with endpoints $(-1, 0)$ and $(1, 0)$.

Lemma 2. *For a convex body $K \subset \mathrm{R}^n$, a one-dimensional subspace $l \subset \mathrm{R}^n$, and a hyperplane $H \subset \mathrm{R}^n$, conditions* 1) *and* 2) *below are equivalent. If, additionally, H meets* int K, *then condition* 3) *becomes equivalent to both* 1) *and* 2).

1) $H \cap \operatorname{bd}(K + l) \subset S_l(K)$, 2) $H \cap K + l = K + l$, 3) $H \cap \operatorname{bd} K \subset S_l(K)$.

Each of conditions 1) *and* 2) *implies that l and H are not parallel.*

Proof. First, we observe that l and H are not parallel under condition 1), since otherwise $H \cap \operatorname{bd}(K + l)$ would contain a line, contrary to the compactness of $S_l(K)$. Similarly, under condition 2), if l and H were parallel, then the sum $H \cap K + l$ would be $(n - 1)$-dimensional, contrary to 2).

1) $\Rightarrow$ 2) Since $H \cap K + l \subset K + l$, it suffices to prove the opposite inclusion, which is equivalent to $\operatorname{bd}(K + l) \subset H \cap \operatorname{bd} K + l$. Choose any point $x \in \operatorname{bd}(K + l)$. Let l' be the line through x parallel to l. Since l and H are not parallel, they meet at a unique point, z. Clearly, $l' = z + l$ and $z \in \operatorname{bd}(K + l)$. By condition 1), $z \in S_l(K)$, whence $z \in \operatorname{bd} K$. Summing up, $x \in z + l \subset H \cap \operatorname{bd} K + l$, which shows the inclusion $\operatorname{bd}(K + l) \subset H \cap \operatorname{bd} K + l$.

2) $\Rightarrow$ 1) The equality $H \cap K + l = K + l$ immediately implies that $H \cap \operatorname{bd}(K + l) \subset H \cap \operatorname{bd} K$. Thus

$$H \cap \operatorname{bd}(K + l) = (H \cap \operatorname{bd}(K + l)) \cap \operatorname{bd} K = H \cap S_l(K) \subset S_l(K).$$

2) $\Leftrightarrow$ 3) If H meets $\text{int}\, K$, then condition 2) becomes equivalent to the equality $H \cap \text{bd}\,(K + l) = H \cap \text{bd}\, K$. Therefore,

$$H \cap \text{bd}\, K = (H \cap \text{bd}\, K) \cap H \cap \text{bd}\,(K + l) = H \cap S_l(K) \subset S_l(K).$$

Conversely, if condition 3) holds, then $H \cap \text{bd}\, K = H \cap \text{bd}\,(K + l)$, which gives $H \cap K + l = K + l$.

Remark 2. It is easy to see that condition 2) from Lemma 2 can be replaced by any of the following:

$$H \cap K = H \cap (K + l) \quad \text{and} \quad \text{bd}\,(K + l) = \text{rbd}\,(H \cap K) + l.$$

The next result, proved in [41], extends Theorem 22 to the case of convex solids.

Theorem 23 ([41]). *Given a convex solid $K \subset \mathrm{R}^n$, $n \geq 3$, the following conditions are equivalent.*

1) *For each one-dimensional subspace $l \subset \mathrm{R}^n$, there is a hyperplane $H \subset \mathrm{R}^n$ meeting $K + l$ such that the inclusion* (15) *holds.*
2) *For each one-dimensional non-recessional for K subspace $l \subset \mathrm{R}^n$, there is a hyperplane $H \subset \mathrm{R}^n$ meeting $K + l$ such that the inclusion* (15) *holds.*
3) *For each one-dimensional sharp for K subspace $l \subset \mathrm{R}^n$, there is a hyperplane $H \subset \mathrm{R}^n$ such that $H \cap \text{bd}\,(K + l) = S_l(K)$.*
4) *K has one of the following shapes:*

   (a) bd K *is a convex quadric,*
   (b) dim (lin K) $= n - 2$ *and K is the direct sum of* lin K *and a two-dimensional line-free closed convex set,*
   (c) dim (lin K) $= n - 3$ *and K is the direct sum of* lin K *and a three-dimensional line-free closed convex cone.*

The shapes (a)–(c) in condition 4) of Theorem 23 are not mutually exclusive: a cylinder based on a two-dimensional line-free convex quadric is a particular case of (b), and a cylinder based on a sheet of a three-dimensional elliptic cone is a particular case of (c).

Remark 3. Condition 1) of Theorem 23 (respectively, condition 1) of Corollary 3) is weaker than condition 2) of Corollary 1 (respectively, condition 3) of Theorem 13). Indeed, if for a given non-recessional directed line l, there is a scalar $\lambda(l) \in [0, 1]$ such that the points dividing in the ratio $\lambda(l)$ all chords a convex solid $K \subset \mathrm{R}^n$ in direction l belong to a hyperplane H, then $H \cap \text{bd}\,(K + l) \subset S_l(K)$.

The proof of Theorem 23 is organized by induction on $n \geq 3$. The case $n = 3$ uses the following result of Alexandrov [1].

Lemma 3 ([1]). *Let $K \subset \mathrm{R}^3$ be a convex solid and T a non-planar, bounded, open, and simply connected piece of* bd K. *If for any shadow-boundary $S_l(K)$ of K*

meeting T there is a plane H such that $S_l(K) \cap T \subset H$, then T is a piece of a line-free convex quadric or a piece of the boundary of a strictly convex cone.

We note that Lemma 3 deals with shadow-boundaries corresponding to all (possibly, non-sharp, or even recessional for K) one-dimensional subspaces l, and the plane H is allowed to be parallel to l. Furthermore, Lemma 3 refines Alexandrov's original conclusion "T is a piece of a convex quadric or a piece of the boundary of a convex cone." For $n \geq 4$, the proof of Theorem 23 uses Theorem 8.

Corollary 3 ([41]). *Given a convex solid $K \subset \mathrm{R}^n$, $n \geq 3$, the following conditions are equivalent.*

1) *For any one-dimensional non-recessional for K subspace $l \subset \mathrm{R}^n$, there is a hyperplane $H \subset \mathrm{R}^n$ such that $S_l(K) \subset H$.*
2) *K has one of the following shapes:*
   (a) bd K *is a convex quadric,*
   (b) dim (lin K) $= n-2$ *and K is the direct sum of* lin K *and a two-dimensional line-free closed convex set which is either unbounded or bounded and strictly convex,*
   (c) dim (lin K) $= n-3$ *and K is the direct sum of* lin K *and a three-dimensional line-free closed strictly convex cone.*

Kakutani [22] stated without proof (mistakenly attributing the result to Blaschke [6]) the following "dual" version of the line-to-hyperplane shadow-boundary characterization of ellipsoids in R^3: A regular convex body $K \subset \mathrm{R}^3$ symmetric about the origin $o \in \operatorname{int} K$ is an ellipsoid provided for each plane $H \subset \mathrm{R}^3$ through o there is a one-dimensional subspace $l \subset \mathrm{R}^3$ so that $H \cap \operatorname{bd} K \subset S_l(K)$. Kakutani's statement is a geometric interpretation of the following fact (with K being the unit ball of a three-dimensional normed space E^3): a norm in E^3 is Euclidean provided for each two-dimensional subspace H of E^3 there is a linear projection on H of norm 1.

Amir [2, pp. 99–100] (see also Borodin [10] for another method) proved Kakutani's statement in the geometric form described above, based on the following characterization of ellipsoids by Brunn [12, Chapter IV]: a bounded convex surface $S \subset \mathrm{R}^3$ is an ellipsoid provided all planar sections of S are centrally symmetric. Phillips [30] (see also Montejano and Morales-Amaya [28]) showed that the requirement on regularity and central symmetry of the convex body $K \subset \mathrm{R}^3$ can be omitted.

Using a polarity argument, Gruber [19] deduced from Theorem 22 the following statement.

Theorem 24 ([19]). *There are points $p_1, \dots, p_4 \in \mathrm{R}^n$ such that a convex body $K \subset \mathrm{R}^n$ with $o \in \operatorname{int} K$ is a solid ellipsoid centered at o provided for each $(n-1)$-dimensional subspace $H \subset \mathrm{R}^n$ with $H \cap \{p_1, \dots, p_4\} \neq \emptyset$, there is a one-dimensional subspace l satisfying the condition $H \cap K = H \cap (K + l)$.*

The points $p_1, \ldots, p_4$ in Theorem 24 can be chosen on any line not containing o and placed close to each other. Choosing them in a small neighborhood of a given one-dimensional subspace $L \subset \mathrm{R}^n$, one can deduce from Theorem 24 the following statement of Montejano and Morales-Amaya [28]: If $K \subset \mathrm{R}^n$ a convex body symmetric about the origin o and L a line through o such that for each hyperplane H through o sufficiently close to l there is a line $l(H)$ through o satisfying the inclusion rbd $(H \cap K) \subset S_{l(H)}(K)$, then K is an ellipsoid.

In the next theorem of Borodin [10], $C_M(S)$ means the union of all planes parallel to a given subspace $M \subset \mathrm{R}^n$ and supporting a bounded convex hypersurface $S \subset \mathrm{R}^n$.

Theorem 25 ([10]). *A bounded convex hypersurface $S \subset \mathrm{R}^n$ symmetric about the origin o is an ellipsoid if and only if any of the following conditions holds.*

1) *For each r-dimensional subspace $L \subset \mathrm{R}^n$, $r \geq 2$, there is a subspace $M \subset \mathrm{R}^n$ such that $C_M(S) \cap S = L \cap S$.*
2) *For each r-dimensional subspace $L \subset \mathrm{R}^n$, $r \geq 2$, there is a subspace $M \subset \mathrm{R}^n$ such that $C_M(S) \cap S \subset L \cap S$.*
3) *For each r-dimensional subspace $L \subset \mathrm{R}^n$, $r \geq 2$, there is a subspace $M \subset \mathrm{R}^n$ such that $C_M(S) \cap S \supset L \cap S$.*

In regard of Theorem 24 we put the following problem.

Problem 4. Let $K \subset \mathrm{R}^n$ be a convex body satisfying the property: for each $(n-1)$-dimensional subspace $L \subset \mathrm{R}^n$ there is a one-dimensional subspace $l \subset \mathrm{R}^n$ and a translate H of L such that $H \cap K = H \cap (K + l)$. Is it true that K is a solid ellipsoid?

6 Orthogonal Projections of Convex Quadrics

The main result of this section (see Theorem 26) makes use of a characteristic property of ellipsoids in terms of shadow-boundaries.

Blaschke and Hessenberg [9] observed without proof that a convex body $K \subset \mathrm{R}^3$ is a solid ellipsoid provided all orthogonal projections of K on two-dimensional planes are solid ellipses. Later Süss [43] proved a slightly weaker result: a convex body $K \subset \mathrm{R}^3$ is a solid ellipsoid provided all parallel projections of K on two-dimensional planes are solid ellipses. Blaschke [8] (also Lenz [25], using a

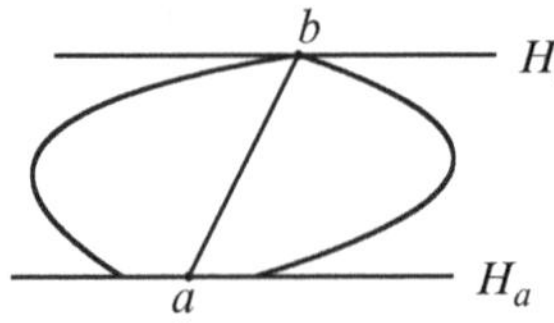

Fig. 2 Affine diameter of a convex body

projective technique) slightly generalized the result of Süss by proving that a regular convex body $K \subset \mathrm{R}^3$ is a solid ellipsoid provided all its parallel projections on two-dimensional planes are bounded by R-curves (that is, by centrally symmetric convex curves with the property that each affine diameter has a conjugate). We recall that a chord $[a, b]$ of a convex body $K \subset \mathrm{R}^n$ is an *affine diameter* of K (Fig. 2) provided there are two parallel, distinct hyperplanes H_a and H_b both supporting K such that $a \in H_a$ and $b \in H_b$ (see, e.g., [36] for a survey on various properties of affine diameters).

The argument of Süss was generalized by Chakerian [15] for all $n \geq 3$ in terms of projections on hyperplanes, and later expanded by Gardner [16, p. 102] in terms of projections on r-dimensional planes, $2 \leq r \leq n-1$. The following theorem goes back to the original observation of Blaschke and Hessenberg and provides a new method of proof.

Theorem 26. *For a convex body $K \subset \mathrm{R}^n$, $n \geq 3$, and an integer $2 \leq r \leq n-1$, the following conditions are equivalent.*

1) *K is a solid ellipsoid.*
2) *All orthogonal projections of K on r-dimensional planes of R^n are solid r-dimensional ellipsoids.*

Proof. 1) $\Rightarrow$ 2) First, assume that $r = n-1$. Let $H \subset \mathrm{R}^n$ be a hyperplane and l the one-dimensional subspace orthogonal to H. The orthogonal projection, M, of K on H can be expressed as the intersection of H with the cylinder $K + l$. By Theorem 2, the shadow-boundary $S_l(K)$ of K lies within a certain hyperplane G. This argument immediately shows that $K + l = (K \cap G) + l$. Since $K \cap G$ is an $(n-1)$-dimensional solid ellipsoid, $K + l$ is bounded by the elliptic cylinder $\mathrm{rbd}\,(K \cap G) + l$. Hence $M = (K + l) \cap H$ is a solid $(n-1)$-dimensional ellipsoid.

Let $2 \leq r < n-1$ (which is possible if $n \geq 4$). Choose an r-dimensional plane $L \subset \mathrm{R}^n$. The orthogonal projection $\pi : \mathrm{R}^n \to L$ onto L can be decomposed into a sequence of orthogonal projections

$$\mathrm{R}^n \mapsto L_1 \mapsto L_2 \mapsto \ldots \mapsto L_{n-r} = L,$$

where $L_1 \supset L_2 \supset \ldots \supset L_{n-r}$ is a nested sequence of planes such that $\dim L_j = n - j$ for all $j = 1, \ldots, n-r$. By the argument above, $\pi(K)$ is an r-dimensional solid ellipsoid.

2) $\Rightarrow$ 1) First, assume that $r = n-1$. We observe that K is strictly convex. Indeed, suppose for a moment that $\mathrm{bd}\,K$ contains a segment $[x, z]$. Choose a hyperplane G supporting K such that $[x, z] \subset K \cap G$ and a line $l \subset G$ which is not parallel to $[x, z]$. Then the relative boundary of the orthogonal projection, K', of K on the $(n-1)$-dimensional subspace perpendicular to l contains a segment (which is a projection of $[x, z]$), contrary to the condition that K' is a solid $(n-1)$-dimensional ellipsoid.

Now, choose any line $l \subset \mathrm{R}^n$. Then K has an affine diameter $[a, a']$ which is parallel to l. Let H and H' be distinct parallel hyperplanes both supporting K such that $a \in K \cap H$ and $a' \in K \cap H'$ (see, e.g., [36]). Since K is strictly convex, one has $K \cap H = \{a\}$ and $K \cap H' = \{a'\}$. Denote by b the middle point of $[a, a']$, and let H_0 be the hyperplane through b parallel to both H and H'.

We state that the shadow-boundary $S_l(K)$ of K lies within H_0. Indeed, choose a line l_1 parallel to l and supporting K at a certain point c_1. Since l_1 does not meet int K, there is a hyperplane G_1 containing l_1 and supporting K. Clearly, $K \cap G_1 = \{c_1\}$ because K is strictly convex. Let m be a one-dimensional subspace parallel to the $(n-2)$-dimensional plane $H \cap G_1$, and denote by M the orthogonal complement of m. Let $\pi : \mathrm{R}^n \to M$ be the orthogonal projection onto M. By condition 2), the set $\pi(K)$ is a solid $(n-1)$-dimensional ellipsoid. Clearly, both $(n-2)$-dimensional planes $M \cap H$ and $M \cap H'$ support $\pi(K)$ such that

$$(M \cap H) \cap \pi(K) = \{\pi(a)\} \quad \text{and} \quad (M \cap H') \cap \pi(K) = \{\pi(a')\}.$$

This argument shows that $\pi(K)$ is symmetric about the point $\pi(b)$.

The $(n-2)$-dimensional plane $G_1 \cap M$ supports $\pi(K)$ at a unique point c_1' which belongs to $H_0 \cap G_1$. Since c_1' is the orthogonal projection of c_1 on M, we conclude that c_1 also belongs to H_0. Summing up, $S_l(K) \subset H_0$. Theorem 21 implies that K is a solid ellipsoid. Hence the case $r = n - 1$ is proved.

Suppose that condition 2) holds for a given integer r, with $2 \le r < n - 1$ (this is possible if $n \ge 4$). We state that 2) holds for $r + 1$. Indeed, let $L \subset \mathrm{R}^n$ be a plane of dimension $r + 1$ and $\pi : \mathrm{R}^n \to L$ the orthogonal projection onto L. Then the orthogonal projection $\varphi : \mathrm{R}^n \to N$ onto an r-dimensional plane $N \subset L$ can be expressed as the composition $\varphi = \psi \circ \pi$, where ψ is the orthogonal projection of L onto N. Hence $\varphi(K)$ is the orthogonal projection of $\pi(K)$. By the assumption, $\varphi(K)$ is an r-dimensional solid for any choice of an r-dimensional plane L in N. Hence, by the proved above (with L instead of R^n), $\pi(K)$ is a solid $(r + 1)$-dimensional ellipsoid.

Consecutively incrementing the value of r, we obtain that condition 2) holds for $r = n - 1$. Hence K is a solid ellipsoid by the proved above.

Corollary 4. *A convex body $K \subset \mathrm{R}^n$, $n \ge 3$, is a solid ellipsoid if and only if there is a plane $L \subset \mathrm{R}^n$ of certain dimension s, $0 \le s \le n - 3$, and an integer r, with $s + 2 \le r \le n - 1$, such that all orthogonal projections of K on r-dimensional planes containing L are r-dimensional solid ellipsoids.*

Proof. Clearly, we have to verify the "only if" part. Let $L \subset \mathrm{R}^n$ be an s-dimensional plane and p a point in L. Choose a two-dimensional plane N through p. Since $s + 2 \le r$, there is an r-dimensional plane $M \subset \mathrm{R}^n$ containing $L \cup N$. By the assumption, the projection, K', of K on M is an r-dimensional ellipsoid. Hence the orthogonal projection of K on N (which is the same as the orthogonal projection of K' on N) is a solid ellipse. By Theorem 26 (with $r = 2$), K is a solid ellipsoid.

Problem 5. Let $K \subset \mathrm{R}^n$, $n \ge 3$, be a convex solid distinct from a halfspace such that for each vector $e \in S^{n-1} \setminus (\operatorname{rec} K \cup -\operatorname{rec} K)$ the orthogonal projection of K on the $(n-1)$-dimensional subspace $H(e) = \{x \in \mathrm{R}^n : x \cdot e = 0\}$ is an $(n-1)$-

dimensional closed convex set bounded by a convex quadric in $H(e)$. Is it true that bd K is a convex quadric (or, additionally, K is a convex cone if $n = 3$)?

We conclude this section by mentioning the following result of Burton [13]: A convex body $K \subset \mathrm{R}^n$, $n \geq 4$, is the sum of a polytope and a solid n-dimensional ellipsoid if and only if every orthogonal projection of K on a three-dimensional plane is the sum of a polytope and a three-dimensional ellipsoid.

Acknowledgements The authors thanks the referee for helpful comments on an earlier draft of this article.

References

1. Alexandrov, A. D. (1939). On convex surfaces with plane shadow-boundaries. (Russian) *Matematicheskii Sbornik, 5*, 309–316.
2. Amir, D. (1986). *Characterizations of inner product spaces*. Basel: Birkhäuser.
3. Berger, K. H. (1936). *Eilinien mit perspektiv liegenden Tangenten-und Sehnendreiecken* (vol. 4, pp. 1–11). S.-B. Heidelberg: Akad. Wiss.
4. Bertrand, J. (1842). Démonstration d'un théoreme de géométrie. *Journal de Mathématiques Pures et Appliquées, 7*, 215–216.
5. Blaschke, W. (1916). Räumliche Variationsprobleme mit symmetrischen Transversalitätsbedingungen. *Ber. Math. Phys. Kl. Königl. Sächs. Ges. Wiss. Leipzig, 68*, 50–55.
6. Blaschke, W. (1916). *Kreis und Kugel*. Leipzig: Viet.
7. Blaschke, W. (1923). *Vorlesungen über Differentialgeometrie II. Affine Differentialgeometrie*. Berlin: Springer.
8. Blaschke, W. (1956). Zur Affingeometrie der Eilinien und Elfächen. *Mathematische Nachrichten, 15*, 258–264.
9. Blaschke, W., & Hessenberg, G. (1917). Lehrsätze über konvexe Körper. Jahresber. *Deutsche Mathematiker-Vereinigung, 26*, 215–220.
10. Borodin, P. A. (1997). Quasi-orthogonal sets and conditions for the Hilbert property of a Banach space. (Russian) *Matematicheskii Sbornik, 188*, 63–74. English translation in: *Matematicheskii Sbornik, 188*, 1171–1182.
11. Borodin, P. A. (2003). A new proof of Blaschke's ellipsoid theorem. (Russian) *Vestnik Moskov. Univ. Ser. I Mat. Mekh., 3*, 17–22. English translation in: *Moscow University Mathematics Bulletin, 58*(3), 6–10.
12. Brunn, H. (1889). *Über Kurven ohne Wendepunkte*. Habilitationschrift. München: Ackermann.
13. Burton, G. R. (1976). On the sum of a zonotope and an ellipsoid. *Commentarii Mathematici Helvetici, 51*, 369–387.
14. Busemann, H. (1955). *The geometry of geodesics*. New York: Academic Press.
15. Chakerian, G. D. (1965). The affine image of a convex body of constant breadth. *Israel Journal of Mathematics, 3*, 19–22.
16. Gardner, R. J. (1995). *Geometric tomography*. New York: Cambridge University Press.
17. Grinberg, E. L. (1991). Isoperimetric inequalities for k-dimensional cross-sections and projections of convex bodies. *Mathematische Annalen, 291*, 75–87.
18. Gruber, P. M. (1974). Über kennzeichende Eigenschaften von eucklidischen Raumen und Ellipsoiden. I. *Journal für die reine und Angewandte Mathematik, 265*, 61–83.
19. Gruber, P. M. (1974). Über kennzeichende Eigenschaften von eucklidischen Raumen und Ellipsoiden. III. *Monatshefte für Mathematik, 78*, 311–340.
20. Grünbaum, B. (1972). *Arrangements and spreads*. Conference Board of the Mathematical Sciences, No. 10. Providence, RI: American Mathematical Society.

21. Jitomirskii, O. K. (1938). On surfaces with plane shadow-boundaries. (Russian) *Matematicheskii Sbornik, 3*, 347–352.
22. Kakutani, S. (1939). Some characterizations of Euclidean space. *Japanese Journal of Mathematics, 15*, 93–97.
23. Kneser, M. (1949). Eibereiche mit geraden Schwerlinien. *Math.-Phys. Semesterber, 1*, 97–98.
24. Kubota, T. (1916). On a characteristic property of the ellipse. *Tohoku Mathematical Journal, 9*, 148–151.
25. Lenz, H. (1958). Einige Anwendungen der projektiven Geometrie auf Fragen der Flächentheorie. *Mathematische Nachrichten, 18*, 346–359.
26. Marchaud, A. (1959). Un théoréme sur les corps convexes. *Annales Scientifiques de lÉcole Normale Supérieure, 76*, 283–304.
27. Martini, H., & Soltan, V. (1999). Combiatorial problems on the illumination of convex bodies. *Aequationes Mathematicae, 57*, 121–152.
28. Montejano, L., & Morales-Amaya, E. (2007). Variations of classic characterizations of ellipsoids and a short proof of the false centre theorem. *Mathematika, 54*, 35–40.
29. Nakajima, S. (1928). Eilinien mit geraden Schwerlinien. *Japanese Journal of Mathematics, 5*, 81–84.
30. Phillips, R. S. (1940). A characterization of Euclidean spaces. *Bulletin of the American Mathematical Society, 46*, 930–933.
31. Rudin, W., & Smith, K. T. (1961). Linearity of best approximation: a characterization of ellipsoids. *Indagationes Mathematicae, 23*, 97–103.
32. Šaĭdenko, A. V. (1980). Some characteristic properties of an ellipsoid. (Russian) *Sibirskii Matematicheskii Zhurnal, 21*, 232–234.
33. Schwenk, A. (1985). *Affinsphären mit ebenen schattengrenzen.* Lecture Notes in Math. vol. 1156, pp. 296–315. Berlin: Springer.
34. Schneider, R. (1967). Zur affinen Differentialgeometrie im Grossen. I. *Mathematische Zeitschrift, 101*, 375–406.
35. Soltan, V. (1995). Convex bodies with polyhedral midhypersurfaces. *Archiv der Mathematik, 65*, 336–341.
36. Soltan, V. (2005). Affine diameters of convex bodies–a survey. *Expositiones Mathematicae, 23*, 47–63.
37. Soltan, V. (2008). Convex solids with planar midsurfaces. *Proceedings of the American Mathematical Society, 136*, 1071–1081.
38. Soltan, V. (2009). Convex solids with homothetic sections through given points. *Journal of Convex Analysis, 16*, 473–486.
39. Soltan, V. (2010). Convex quadrics. *Buletinul Academiei de Stiinte a Republicii Moldova Matematica, 3*, 94–106.
40. Soltan, V. (2011). Convex solids with hyperplanar midsurfaces for restricted families of chords. *Buletinul Academiei de Stiinte a Republicii Moldova Matematica, 2*, 23–40.
41. Soltan, V. (2012). Convex solids with hyperplanar shadow-boundaries. *Journal of Convex Analysis, 19*, 591–607.
42. Soltan, V. (2012). Convex quadrics and their characterizations by means of plane sections. In B. Toni, et al. (Eds.), *Bridging mathematics, statistics, engineering and technology* (pp. 131–145). Springer Proceedings in Mathematics & Statistics, vol. 24.
43. Süss, W. (1953). Eine elementare kennzeichnende Eigenschaft des Ellipsoids. *Math.-Phys. Semesterber., 3*, 57–58.
44. Süss, W., Viet, U., & Berger, K. H. (1971). Konvexe Figuren. In H. Behnke, K. Fladt, & H. Kunle (Eds.) *Grundzüge der mathematik.* Bd. II: Geometrie. Teil B, (pp. 361–381). Göttingen: Vanderhoeck and Ruprecht.
45. Thompson, A. C. (1996). *Minkowski geometry*. Cambridge: Cambridge University Press.
46. Webster, R. J. (1994). *Convexity*. Oxford: Oxford University Press.

Classifying Normal, Nevus, and Primary Melanoma Skin Samples Using Penalized Ordinal Regression

Kellie J. Archer, Jiayi Hou, and André A.A. Williams

Abstract Many investigators conducting translational research are developing multigenic classifiers using data from high-throughput genomic experiments. While often the class to be predicted is nominal, sometimes it may be inherently ordinal. For example, tissue samples may be collected with the goal of classifying them as normal < pre-malignant < malignant. In this case, molecular features monotonically associated with the ordinal response may be important to disease development. While one can apply nominal response classification methods to ordinal response data, in so doing some information is lost that may improve the predictive performance of the classifier. We developed an `R` package, `glmpathcr`, capable of fitting a penalized continuation ratio model when the outcome to be predicted is ordinal. We demonstrate application of our method by predicting progression to melanoma using microarray gene expression data.

1 Introduction

It is estimated that 76,690 people in the United States will be diagnosed with melanoma and 9,480 will die from melanoma in 2013 [15]. Current methods of early diagnosis rely on visual assessment of existing moles following the ABCDE rule to note the presence of **A**ssymetry, irregular **B**order, variegated **C**olor, larger **D**iameter,

K.J. Archer (✉) • J. Hou
Department of Biostatistics, Virginia Commonwealth University,
Richmond, VA 23298-0032, USA
e-mail: kjarcher@vcu.edu; houj2@mymail.vcu.edu

A.A.A. Williams
Division of Biostatistics and Bioinformatics, National Jewish Health,
Denver, Colorado, 80206, USA
e-mail: WilliamsA@NJHealth.org

B. Toni (ed.), *New Frontiers of Multidisciplinary Research in STEAM-H (Science, Technology, Engineering, Agriculture, Mathematics, and Health)*,
Springer Proceedings in Mathematics & Statistics 90, DOI 10.1007/978-3-319-07755-0_5,

and **E**volution over time. Because stage of melanoma is directly linked to probability of survival, with 5-year relative survival estimated to be 98.3 %, 62.4 %, and 16.0 % for localized, regional, and distant stages, respectively [15], early diagnosis is important. Specifically, at an early stage, melanoma cells have not penetrated deep enough into the skin to reach blood vessels, so early stage melanoma is unlikely to metastasize to other areas such as the brain, liver, bones, central nervous system, or lymph nodes. Due to the poor clinical outcome of patients diagnosed with later stage metastatic melanoma, improved markers for early diagnosis are needed. Markers useful for early diagnosis may reduce time to treatment and possibly indicate novel therapeutic targets, and thereby yield improved patient outcomes. Therefore we sought to develop a multigenic classifier derived using gene expression microarray data capable of differentiating among normal, nevus, and primary melanoma skin samples.

Apart from our goal to classify normal, nevus, and melanoma skin samples using gene expression data, it is often of interest to develop a multigenic classifier to predict phenotype using high-throughput genomic data. In many biomedical settings where histopathological or health status data are collected, phenotypic variables are recorded on an ordinal scale. Examples include grading of adverse events (none < mild < moderate < severe) and tumor-node-metastasis stage (I < II < III < IV). Extensive simulation studies have demonstrated that when ordinal methods are appropriate, they result in lower classification error rates compared to traditional nominal response methods [25]. However, ordinal modeling methods are lacking for situations when the number of predictors exceed the number of observations, as in high-throughput genomic data. For our illustrative dataset, 70 Affymetrix HG-U133Av2 GeneChips were available for seven normal, 18 nevus, and 45 primary melanoma skin samples. Note that the sample size, $n = 70$, is much smaller than the number of available predictors, $p = 22,215$. Unfortunately, traditional statistical models cannot be estimated when $p > n$. Even if filtering is performed on this dataset using an F-test, there are still 13,365 probe sets remaining using a false discovery rate of 5 %. In this chapter, we describe a software package available in the R programming environment for fitting an L_1 penalized constrained continuation ratio model that can be applied to high-dimensional datasets for predicting an ordinal response. We apply our method to the melanoma dataset and discuss our findings. We also provide the code used in performing the analyses in the Appendix.

2 Regression Models

Before introducing our L_1 penalized constrained continuation ratio model, we first review linear regression and various penalized methods for the linear regression model. We then present the logistic regression model because it serves as the foundation for related ordinal response methods. We then review the continuation ratio model as a specific ordinal regression method.

2.1 Linear Regression

The simple linear regression model is given by,

$$y_i = \beta_0 + \beta_1 x_i + \epsilon_i \tag{1}$$

where y_i is the response or dependent variable for observations $i = 1, \ldots, n$, β_0 is the intercept, β_1 is the slope coefficient, x_i is the independent or predictor variable, and ϵ_i is the error where $E(\epsilon_i) = 0$ and $\mathrm{var}(\epsilon_i) = \sigma^2$ [16]. Note that the expected value of the response is

$$E(y_i) = E(\beta_0 + \beta_1 x_i + \epsilon_i) = E(\beta_0) + E(\beta_1 x_i) + E(\epsilon_i) = \beta_0 + \beta_1 x_i. \tag{2}$$

Estimation of the parameters in this model is solved by the method of least squares, which considers the deviation of y_i from its expected value $\beta_0 + \beta_1 x_i$. To fit a line such that the vertical distances from the fitted values to the datapoints are minimized, we minimize the residual sum of squared errors (RSS),

$$RSS = \sum_{i=1}^{n} (y_i - \beta_0 - \beta_1 x_i)^2. \tag{3}$$

To find the intercept and slope that minimize the RSS, the derivative of the RSS is taken with respect to the model parameters (β_0 and β_1),

$$\frac{\delta RSS}{\delta \beta_0} = -2 \sum_{i=1}^{n} (y_i - \beta_0 - \beta_1 x_i) \tag{4}$$

and

$$\frac{\delta RSS}{\delta \beta_1} = -2 \sum_{i=1}^{n} x_i (y_i - \beta_0 - \beta_1 x_i). \tag{5}$$

Setting both equations equal to 0 and solving for β_0 and β_1 yields the ordinary least squares (OLS) solution,

$$\hat{\beta}_1 = \frac{\sum_{i=1}^{n} (x_i - \bar{x})(y_i - \bar{y})}{\sum_{i=1}^{n} (x_i - \bar{x})^2} = \frac{\sum x_i y_i - \frac{\sum x_i \sum y_i}{n}}{\sum x_i^2 - \frac{(\sum x_i)^2}{n}} \tag{6}$$

and

$$\hat{\beta}_0 = \bar{y} - \hat{\beta}_1 \bar{x}. \tag{7}$$

Often there is more than one predictor variable so it is helpful to be familiar with the analogous matrix formulation of the least squares regression model, where the components of the model include

$$\mathbf{Y} = \begin{pmatrix} y_1 \\ y_2 \\ \vdots \\ y_n \end{pmatrix}, \tag{8}$$

$$\mathbf{X} = \begin{pmatrix} 1 & x_{11} & x_{12} & \cdots & x_{1p} \\ 1 & x_{21} & x_{22} & \cdots & x_{2p} \\ \vdots & \vdots & \vdots & \cdots & \vdots \\ 1 & x_{n1} & x_{n2} & \cdots & x_{np} \end{pmatrix}, \tag{9}$$

$$\mathbf{b} = \begin{pmatrix} \beta_0 \\ \beta_1 \\ \vdots \\ \beta_p \end{pmatrix}, \tag{10}$$

and

$$\mathbf{e} = \begin{pmatrix} e_1 \\ e_2 \\ \vdots \\ e_n \end{pmatrix} \tag{11}$$

The normal equation in matrix terms is

$$\mathbf{X}^T\mathbf{X}\mathbf{b} = \mathbf{X}^T\mathbf{Y} \tag{12}$$

and premultiplying both sides of the equation by $(\mathbf{X}^T\mathbf{X})^{-1}$

$$(\mathbf{X}^T\mathbf{X})^{-1}\mathbf{X}^T\mathbf{X}\mathbf{b} = (\mathbf{X}^T\mathbf{X})^{-1}\mathbf{X}^T\mathbf{Y} \tag{13}$$

yields the least squares solution,

$$\hat{\mathbf{b}} = (\mathbf{X}^T\mathbf{X})^{-1}\mathbf{X}^T\mathbf{Y} \tag{14}$$

and the regression model is $\mathbf{Y} = E(\mathbf{X}) + \mathbf{e}$ which is simply $\mathbf{Y} = \mathbf{X}\mathbf{b} + \mathbf{e}$. Note that the solution is the same when the model includes more than one predictor; the only modification that was necessary is that an additional column for each additional predictor is appended to **X** and an additional element for each additional predictor is appended to **b**.

2.2 *Penalization Methods*

When interest lies in modeling a continuous outcome y given p predictor variables measured for n samples, forward stepwise, backward stepwise, and best subsets methods are commonly used to obtain a parsimonious model [12]. For datasets where the number of covariates (p) exceeds the sample size (n), the backwards stepwise procedure cannot be undertaken. When p is large, the best subset procedure is computationally prohibitive. Moreover, β_j's estimated using these model selection procedures can exhibit extremely large variances when covariates are collinear. Penalized or regularization methods such as ridge regression, least absolute shrinkage and selection operator (LASSO), and elastic net are alternatives to traditional statistical methods that can be used to estimate a regression model when $p > n$ [12, 29, 34].

2.2.1 Ridge Regression

Solutions based on penalizing the size of the β_j estimates so that the bias introduced is traded for reduced variance of the estimates have been proposed. A specific method called ridge regression can be expressed as

$$\hat{\beta}^{ridge} = \arg\min_{\beta}\left(\sum_{i=1}^{n}(y_i - \beta_0 - \sum_{j=1}^{p} x_{ij}\beta_j)^2 + \lambda\sum_{j=1}^{p}\beta_j^2\right). \tag{15}$$

Here λ is the tuning parameter that controls the amount of shrinkage: if $\lambda = 0$ the solution is the OLS estimates; as λ increases, the amount of shrinkage of the parameter estimates increases. The ridge estimates differ depending upon the scaling of the covariates used, so typically covariates are standardized prior to model fitting. Referring to Eq. (15), note that β_0 is not included in the penalty term as penalizing the intercept would make the procedure dependent upon the origin chosen for y. For centered inputs, we estimate β_0 by $\frac{1}{n}\sum_{i=1}^{n} y_i$ and the remaining coefficients can be estimated using ridge regression without the intercept. Therefore, we have the matrix of covariates **X** of dimension $n \times p$. The penalized residual sum of squares (PRSS) is a modification of Eq. (3) written in matrix form as

$$PRSS(\lambda) = (\mathbf{y} - \mathbf{X}\boldsymbol{\beta})^T(\mathbf{y} - \mathbf{X}\boldsymbol{\beta}) + \lambda\boldsymbol{\beta}^T\boldsymbol{\beta}. \tag{16}$$

Taking the derivative yields

$$\frac{\delta PRSS(\lambda)}{\delta \boldsymbol{\beta}} = -2\mathbf{X}^T(\mathbf{y} - \mathbf{X}\boldsymbol{\beta}) + 2\lambda\boldsymbol{\beta} \quad (17)$$

and setting it equal to zero yields

$$-2\mathbf{X}^T(\mathbf{y} - \mathbf{X}\boldsymbol{\beta}) + 2\lambda\boldsymbol{\beta} = 0 \quad (18)$$

$$-2\mathbf{X}^T\mathbf{y} + 2\mathbf{X}^T\mathbf{X}\boldsymbol{\beta} + 2\lambda\boldsymbol{\beta} = 0 \quad (19)$$

$$\mathbf{X}^T\mathbf{X}\boldsymbol{\beta} + \lambda\boldsymbol{\beta} = \mathbf{X}^T\mathbf{y} \quad (20)$$

$$(\mathbf{X}^T\mathbf{X} + \lambda\mathbf{I})\boldsymbol{\beta} = \mathbf{X}^T\mathbf{y} \quad (21)$$

so that the ridge coefficients estimates are given by

$$\hat{\boldsymbol{\beta}}^{ridge} = (\mathbf{X}^{\mathbf{T}}\mathbf{X} + \lambda\mathbf{I})^{-1}\mathbf{X}^{\mathbf{T}}\mathbf{y}. \quad (22)$$

Adding $\lambda\mathbf{I}$ to $\mathbf{X}^T\mathbf{X}$ yields a non-singular matrix, allowing a solution to be obtained regardless if $p > n$.

Ridge coefficient estimates, though biased, have smaller variance and tend to have improved performance (lower test set error) when compared to OLS estimates. However, the ridge solution does not yield a parsimonious model because all p predictors will have non-zero coefficient estimates. For gene expression microarray data, this makes the final model lack interpretability.

2.2.2 Least Absolute Shrinkage and Selection Operator

The Least Absolute Shrinkage and Selection Operator (LASSO) [29] imposes a constraint similar to ridge regression, namely,

$$\hat{\beta}^{lasso} = \arg\min_{\beta} \left(\sum_{i=1}^{n} (y_i - \beta_0 - \sum_{j=1}^{p} x_{ij}\beta_j)^2 + \lambda \sum_{j=1}^{p} |\beta_j| \right). \quad (23)$$

Again, λ is the tuning parameter that controls the amount of shrinkage such that if $\lambda = 0$, the solution is the OLS estimate and as λ increases, the amount of shrinkage of the parameter estimates increases. Because the LASSO penalty is based on a multiplicative factor times the sum of the absolute values of the coefficient estimates, it is also referred to as an L_1 penalty or called an L_1 penalized model. Additionally, because the L_1 penalty shrinks some coefficients to be exactly zero, it is better than ridge regression in terms of model parsimony and interpretability.

Due to the L_1 constraint, there is not a closed form solution to Eq. (23). Various numerical methods such as the incremental forward stagewise method [11], least angle regression [7], and coordinate descent [20] have been used to obtain a LASSO solution.

2.2.3 Elastic Net Penalty

A generalized expression for penalized models is,

$$\tilde{\beta} = \arg\min_{\beta} \left(\sum_{i=1}^{n} (y_i - \beta_0 - \sum_{j=1}^{p} x_{ij}\beta_j)^2 + \lambda \sum_{j=1}^{p} |\beta_j|^q \right) \tag{24}$$

for $q \geq 0$. For the LASSO model, let $q = 1$; for ridge regression, let $q = 2$. Values of $q \in (1, 2)$ provide a compromise between the LASSO and ridge regression. However, users should note that when $q > 1$, coefficients are no longer set exactly equal to 0. Therefore the elastic net penalty was introduced to maintain benefits from both ridge and LASSO and is given by

$$\lambda \sum_{j=1}^{p} (\gamma\beta_j^2 + (1 - \gamma)|\beta_j|) \tag{25}$$

where γ weights the amount of the penalty placed on the ridge and LASSO components.

2.3 *Logistic Regression*

A binary (or dichotomous) response takes on one of only two possible values such as disease status (case/control) or response to therapy (responded/failed to respond). Dichotomous responses are conventionally coded as 0 and 1. It is not useful to fit a linear regression model to dichotomous response data since the response probabilities are confined to a [0,1] scale whereas regression models could predict off-scale values that are either below 0 or above 1. Instead, a model should be fit on a scale that preserves the range of response probabilities. Logistic regression is used to model the relationship between a dichotomous outcome variable and a set of predictor variables. Traditionally, logistic regression assumes that the observations are a random sample from a population where the model is expressed as

$$y_i = \pi(\mathbf{x}_i) + \epsilon_i \tag{26}$$

where y_i represents the dichotomous dependent or outcome variable, $\pi(\mathbf{x}_i)$ represents the conditional probability of experiencing the event given the independent predictor variables $\mathbf{x}_i$, and ϵ_i represents the binomial random error term. More formally, the conditional probability, $\pi(\mathbf{x}_i)$ is a function of the independent covariates

$$\pi(\mathbf{x}_i) = P(Y_i = 1|\mathbf{x}_i) = \frac{\exp(\mathbf{x}_i^T \boldsymbol{\beta})}{1 + \exp(\mathbf{x}_i^T \boldsymbol{\beta})} \tag{27}$$

where the matrix of independent variables and vector of model parameters are the same as those given in Eqs. (9) and (10) [14]. A convenient way to express the contribution to the likelihood function for observation i is

$$\psi(\mathbf{x}_i) = \pi(\mathbf{x}_i)^{y_i} (1 - \pi(\mathbf{x}_i))^{1-y_i}. \tag{28}$$

Since the observations are assumed to be independent, the likelihood function is simply the product of the n independent terms given in Eq. (28), or

$$L(\boldsymbol{\beta}|\mathbf{y}, \mathbf{x}_i) = \prod_{i=1}^{n} \psi(\mathbf{x}_i) = \prod_{i=1}^{n} \pi(\mathbf{x}_i)^{y_i} (1 - \pi(\mathbf{x}_i))^{1-y_i}. \tag{29}$$

Mathematically it is easier to maximize the log-likelihood which is given by

$$\log(L(\boldsymbol{\beta}|\mathbf{y}, \mathbf{x}_i)) = \sum_{i=1}^{n} (y_i \log(\pi(\mathbf{x}_i)) + (1 - y_i) \log(1 - \pi(\mathbf{x}_i))). \tag{30}$$

Penalized logistic regression models can also be used for variable selection and shrinkage where the log-likelihood is modified by subtracting the penalty term. For an L_1 penalized logistic regression model, this is expressed as

$$\log(L(\boldsymbol{\beta}|\mathbf{y}, \mathbf{x}_i)) - \lambda \sum_{j=1}^{p} |\beta_j|. \tag{31}$$

2.4 Ordinal Regression

For observations $i = 1, \ldots, n$, let the response y_i belong to one of K ordinal classes such that $k = 1, \ldots, K$. Let $\mathbf{x}_i$ represent a p-length vector of covariates for observation i. The backward formulation of the continuation ratio models the logit as

$$\texttt{logit}\,(P(y = k|y \leq k, \mathbf{X} = \mathbf{x})) = \alpha_k + \boldsymbol{\beta}_k^{\top} \mathbf{x} \tag{32}$$

whereas the forward formulation models the logit as

$$\texttt{logit}\,(P(y = k|y \geq k, \mathbf{X} = \mathbf{x})) = \alpha_{k\cdot} + \boldsymbol{\beta}_{k\cdot}^{\top}\mathbf{x}. \tag{33}$$

The different use of subscripts highlights the fact that the forward and backward formulations result in different coefficient estimates. Rather than describe both formulations in detail, here we present the backward formulation, which is commonly used when progression through disease states from none, mild, moderate, severe is represented by increasing integer values, and interest lies in estimating the odds of more severe disease compared to less severe disease [3]. Let $\mathbf{y}_i$ be a length K indicator vector for observation i representing ordinal class membership, such that $y_{ik} = 1$ if the response for observation i is in category k and 0 otherwise, such that $n_i = \sum_{k=1}^{K} y_{ik} = 1$. Using the logit link equation (34) represents the conditional probability for class k

$$\delta_k(\mathbf{x}) = P(y = k|y \leq k, \mathbf{X} = \mathbf{x}) = \frac{\exp(\alpha_k + \boldsymbol{\beta}_k^{\top}\mathbf{x})}{1 + \exp(\alpha_k + \boldsymbol{\beta}_k^{\top}\mathbf{x})}. \tag{34}$$

The likelihood for the continuation ratio model is then the product of conditionally independent binomial terms [6], given by

$$L(\boldsymbol{\beta}|\mathbf{y}, \mathbf{x}) = \prod_{i=1}^{n} \delta_2^{y_{i2}}(1 - \delta_2)^{1-\sum_{k=2}^{K} y_{ik}} \times \cdots \times \delta_K^{y_{iK}}(1 - \delta_K)^{1-y_{iK}}. \tag{35}$$

where here we have simplified our notation by not explicitly including the dependence of the conditional probability δ_k on $\mathbf{x}$. Further, simplifying our notation to let $\boldsymbol{\beta}$ represent the vector containing both the thresholds $(\alpha_2, \ldots, \alpha_K)$ and the log odds $(\beta_1, \ldots, \beta_p)$ for all $K - 1$ logits, the full parameter vector is

$$\boldsymbol{\beta} = (\alpha_2, \beta_{21}, \beta_{22}, \ldots, \beta_{2p}, \ldots, \alpha_K, \beta_{K,1}, \beta_{K,2}, \ldots, \beta_{K,p})^{\top} \tag{36}$$

which is of length $(K - 1)(p + 1)$. As can be seen from Eq. (35), the likelihood can be factored into $K - 1$ independent likelihoods, so that maximization of the independent likelihoods will lead to an overall maximum likelihood estimate for all terms in the model [3]. A model consisting of $K - 1$ different $\boldsymbol{\beta}$ vectors may be overparameterized so to simplify, one commonly fits a constrained continuation model, which includes the $K - 1$ thresholds $(\alpha_2, \ldots, \alpha_K)$ and one common set of p slope parameters, $(\beta_1, \ldots, \beta_p)$.

To accommodate situations where $p > n$, either an L_1, L_2, or elastic net penalty can be used for variable selection and shrinkage [10]. When using the elastic net penalty, the log likelihood is then taken to be

$$\log L(\boldsymbol{\beta}|\mathbf{y}, \mathbf{x}) - \lambda \sum_{j=1}^{p} (\gamma\beta_j^2 + (1 - \gamma)|\beta_j|). \tag{37}$$

3 Methods

To fit a constrained continuation ratio model, the original dataset can be restructured by forming $K - 1$ subsets, where for classes $k = 2, \ldots, K$, the subset contains those observations in the original dataset up to class k. Additionally, for the kth subset, the outcome is dichotomized as $y = 1$ if the ordinal class is k and $y = 0$ otherwise. Furthermore, an indicator is constructed for each subset representing subset membership. Thereafter the $K - 1$ subsets are appended to form the restructured dataset, which represents the $K - 1$ conditionally independent datasets in Eq. (35). Applying a logistic regression model to this restructured dataset yields an L_1 penalized constrained continuation ratio model.

We developed the `glmpathcr` package for the `R` programming environment [23] to fit a penalized constrained continuation ratio model. Our `glmpathcr` package depends on the `glmpath` package [20] that can fit penalized logistic regression models using coordinate descent. Specifically, the `glmpath.cr` function fits either a forward or backward (default) penalized constrained continuation ratio model by specifying `method="forward"` or `method="backward"` (default) in the `glmpath.cr` call. The `glmpath.cr` function first restructures the dataset to represent the $K-1$ conditionally independent likelihoods needed in Eq. (35) [3] and then fits the penalized continuation ratio model using the `glmpath` algorithm [20]. This allows fitting a penalized model for situations where the number of covariates p exceeds the sample size n. In addition, functions for returning class probabilities, the predicted class, coefficient estimates, and the Akaike Information Criterion (AIC) and Bayesian Information Criterion (BIC) for the continuation ratio model are provided. The `print` and `plot` methods from `glmpath` were also adapted for the returned `glmpath.cr` object. In this package, the nomenclature for these cutpoints is to use "cp$k$" where $k = 1, \ldots, K - 1$. In this dataset, $K = 3$ so the cutpoints are `cp1` and `cp2` with the `Intercept` being an offset.

We downloaded GSE3189 from Gene Expression Omnibus. In this dataset, Affymetrix HG-U133Av2 GeneChips were available for seven normal, 18 nevus, and 45 primary melanoma skin samples [28]. The MAS5 probe set expression summary method was used to summarize probe level data. Prior to analysis the control probe sets were removed, leaving 22,215 probe sets for statistical analyses. Subsequently, the MAS5 expression data were $\log_2$ transformed. When fitting the L_1 penalized constrained continuation ratio model using our `glmpathcr` package, the final model was selecting using the AIC. All `R` code used for this analysis appears in the Appendix to demonstrate usage of the `glmpathcr` package.

4 Results

Table 1 provides the cross-tabulation of the observed and predicted classes. The model had excellent performance as the re-substitution error rate was 0 % (100 % accuracy). We also used tenfold cross-validation (CV) as a means to assess generalization error; the tenfold CV error was 5.7 % (94.3 % accuracy).

Table 1 Cross-tabulation of observed and predicted class using the L_1 penalized constrained continuation ratio model

Predicted Class	Normal	Nevus	Melanoma
Normal	7	0	0
Nevus	0	18	0
Melanoma	0	0	45

Table 2 List of probe sets included in the final L_1 penalized constrained continuation ratio model

Probe Set	Entrez ID	Gene Symbol	Chromosome	$\hat{\beta}$
200755_s_at	813	CALU	7	0.0869
201022_s_at	11034	DSTN	20	−0.7493
201393_s_at	3482	IGF2R	6	0.0090
201591_s_at	11188	NISCH	3	−0.2533
201672_s_at	9097	USP14	18	0.6451
202022_at	230	ALDOC	17	−0.0937
204731_at	7049	TGFBR3	1	−0.5015
205236_x_at	6649	SOD3	4	−0.0886
205681_at	597	BCL2A1	15	0.0325
205883_at	7704	ZBTB16	11	−0.4180
207144_s_at	4435	CITED1	X	0.0141
208710_s_at	8943	AP3D1	19	0.0152
211762_s_at	3838	KPNA2	17	0.2661
212862_at	8760	CDS2	20	0.2269
213002_at	4082	MARCKS	6	0.2290
213029_at	4781	NFIB	9	−0.4125
213330_s_at	10963	STIP1	11	0.0467
216037_x_at	6934	TCF7L2	10	−0.1612
218692_at	55638	SYBU	8	−0.0387
219476_at	79098	C1orf116	1	−0.0622

Twenty probe sets were included in the final model (Table 2). The probe set having the largest absolute coefficient estimate was designed to interrogate destrin (actin depolymerizing factor) (DSTN). Expression of DSTN decreases as one moves from normal to nevus to melanoma (Fig. 1). Although DSTN has been studied in association with other cancers [33], no publications have previously reported an association between DSTN and melanoma. The two probe sets having the largest absolute coefficient estimates were considered to be the two most important predictors, namely DSTN and ubiquitin specific peptidase 14 (USP14). USP14 has been described as a tumor-promoting factor and over-expression of USP14 was associated with shorter overall survival in lung adenocarcinoma patients [32]. In this dataset, USP14 was also over-expressed in melanoma samples compared to normal and nevus samples. A scatterplot of the $\log_2$ expression values for these two probe sets reveals the three classes are almost linearly separable (Fig. 2).

With respect to the other genes included in the final model, several studies have identified ZBTB16 as under-expressed in primary and malignant melanoma [9]. When examining prognosis in melanoma patients, subjects with $\leq$ 10,000 ZBTB16 copies μg total tumor RNA had significantly worse survival compared to subjects

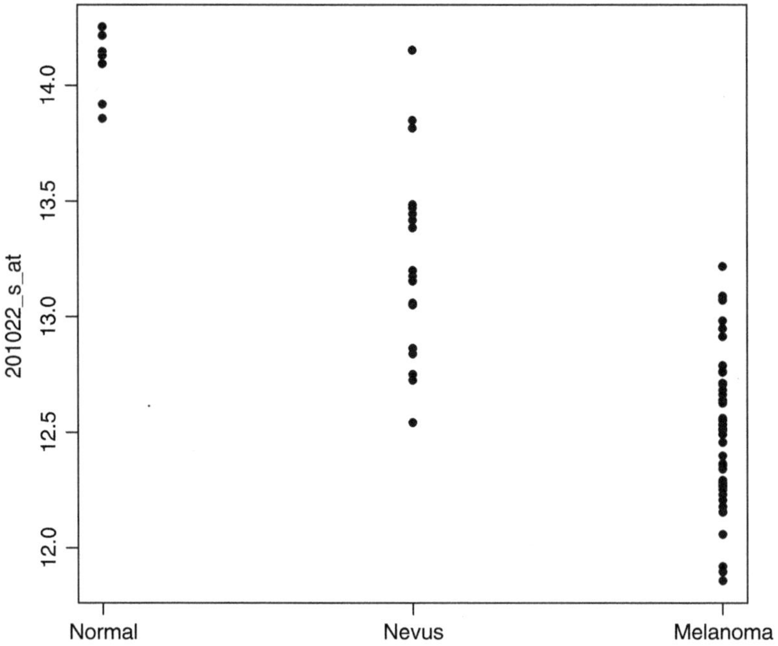

Fig. 1 Dotchart of 201022_s_at (DSTN) $\log_2$ expression by sample type (normal, nevus, and melanoma)

with > 10,000 ZBTB16 copies μg total tumor RNA (P = 0.0429) [4]. In a case study examining topical diphencyprone applied to a patient with in-transit metastatic melanoma of the scalp, ZBTB16 was not expressed before treatment but increased expression was observed after treatment [19]. IGF2R expression was inversely associated with miR-211 expression, suggesting a role of IGF2R in melanoma invasion or metastasis [17]. TGFBR3 is located on the short arm of chromosome 1 (1p22) and deletions in 1p22 occur in 17 % of melanomas [30]. BCL2A1 was over-expressed in 72 % of primary melanoma samples but had low expression in 70 % of nevi; its genomic region was also amplified in 31.8 % of melanoma samples evaluated [13]. In this same study, melanoma growth was significantly reduced by knocking down BCL2A1 in cell lines by siRNA and mouse xenografts by shRNA, indicating BCL2A1 is a melanoma oncogene [13]. In another study, BLC2A1 was significantly over-expressed in metastatic melanoma samples compared to primary melanoma samples [24]. This study also identified CITED1 as having lower expression in primary and malignant melanoma samples in comparison to normal human epithelial melanocytes [24]. In a different study, CITED1 had lower expression in more-aggressive compared to less-aggressive primary melanoma samples [26]. Gene expression for KPNA2 was significantly associated with 4-year distant metastasis-free survival; these results were confirmed using a larger sample when studying KPNA2 protein expression [31]. Finally, the role of MARCKS in

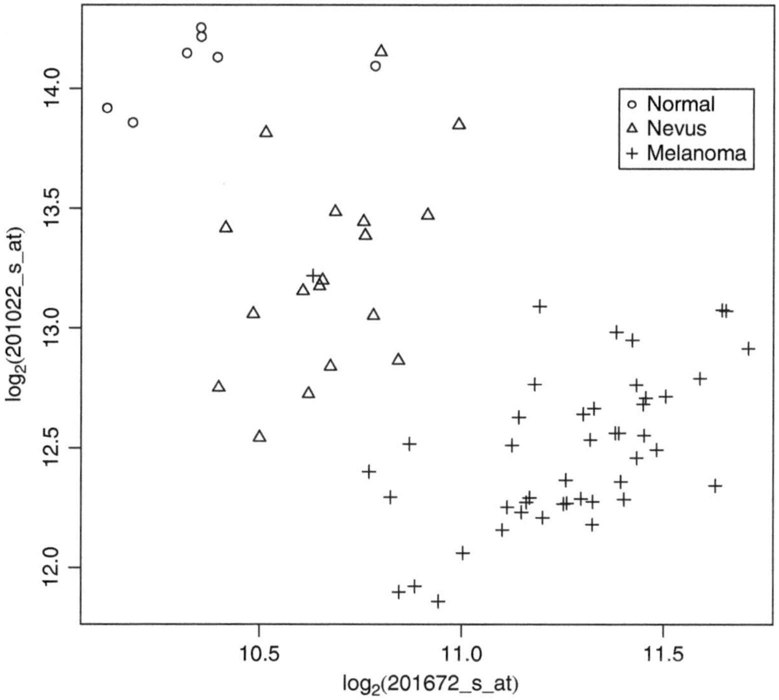

Fig. 2 Scatterplot of $\log_2$ expression for 201022_s_at (DSTN) against 201672_s_at (USP14) with plotting symbol indicating sample type (normal, nevus, and melanoma)

cell adhesion has also been studied in melanoma [8]. Other genes in our final model have been associated with other cancers including NISCH [2], SOD3 [27], AP3D1 [22], NFIB [21], STIP1 [5], and TCF7L2 [18]. We conclude that the genes identified by our L_1 penalized model appear to be relevant for the clinical question at hand.

5 Conclusion and Future Work

Using gene expression microarray data, we identified a twenty probe set classifier having 100 % accuracy on the training dataset and 94.3 % accuracy when using cross-validation as a means of assessing generalization error. Genes identified had important links to melanoma development and progression including ZBTB16, IGF2R, TGFBR3, BCL2A1, CITED1, KPNA2, and MARCKS while other genes have been associated with other cancers (DSTN, USP14, NISCH, SOD3, AP3D1, NFIB, STIP1, and TCF7L2). Multigenic tests consisting of a small number of genes may prove useful in surveillance strategies for detecting melanoma among at-risk patients.

A variety of statistical modeling procedures, namely, proportional odds, adjacent category, stereotype logit, and continuation ratio models can be used to predict an ordinal response. In this paper, we focused attention to the continuation ratio model because its likelihood can be easily re-expressed such that existing software can be readily adapted and used for model fitting. Herein we have described the `glmpathcr` package which works in conjunction with the `glmpath` package in the `R` programming environment. The package provides methods for fitting either a forward or backward penalized continuation ratio model. Moreover, the likelihood-based penalized constrained continuation ratios models have been demonstrated to have good performance in simulation studies and when applied to microarray gene expression datasets [1]. A similar package, `glmnetcr`, which uses the `glmnet` fitting algorithm for fitting a penalized constrained continuation ratio model has also been developed and is available for download from the Comprehensive `R` Archive Network. Functions for extracting coefficients, extracting non-zero coefficients, and obtaining fitted probabilities and predicted class in the `glmnetcr` package are similar to those in `glmpathcr` and both packages have similar performance [1]. Therefore either the `glmnetcr` or `glmpathcr` package should be helpful when predicting an ordinal response for datasets where the number of covariates exceeds the number of available samples. Our current research is to expand ordinal response modeling for high-dimensional datasets by modifying the generalized monotone incremental forward stagewise method for the cumulative logit, adjacent category, and stereotype logit models.

Acknowledgements Research reported in this publication was supported by the National Library Of Medicine of the National Institutes of Health under Award Number R01LM011169. The content is solely the responsibility of the authors and does not necessarily represent the official views of the National Institutes of Health.

Appendix

All analyses were performed in the R programming environment. We have provided the code used in performing the analyses in this appendix. R is a freely available programming environment structured after S. You can download R by going to http://cran.r-project.org/. To download a version that will run on Windows 95 or a later version, click on the '**Download R for Windows**' link located under **Download and Install R**. Then click on the **base** subdirectory or on the **install R for the first time** link and subsequently click on the **Download R-3.0.1 for Windows** link to save the R-3.0.1-win.exe file to your hard drive. Note that as new versions are released, which occurs every six months, the name of the executable file will change. To install R, simply go to <Start> <Run> and then browse for the R-3.0.1-win3.exe file and follow the prompts. Downloads are also available for Mac OS X and Linux operating systems.

Installing User Contributed Packages from CRAN

The Comprehensive R Archive Network (CRAN) makes hundreds of user contributed R packages publicly available for download. Once R has been installed, open R by double clicking on the icon or accessing it through the Start menu. To install user contributed R packages in a Windows environment, select '**Packages**' from the Toolbar, then select '**Select repositories**' and choose the appropriate repository (**CRAN** or **CRAN extras**). The command

```
R> setRepositories()
```

can also be issued at the command line to accomplish the same thing. Thereafter, select '**Packages**' from the Toolbar, then '**Install package(s)**.' A list of packages available for download will appear.

Installing Bioconductor

Once R has been installed, open R by double clicking on the icon or accessing it through the Start menu. Install the *biocLite* script which will install a subset of the most frequently used Bioconductor packages. From the R prompt,

```
R> source("http://www.bioconductor.org/biocLite.R")
```

then

```
R> biocLite()
```

To install additional Bioconductor R add-on packages, select '**Packages**' from the Toolbar, then select '**Select repositories**' and choose '**BioC software**'. Thereafter, select '**Packages**' from the Toolbar, then '**Install package(s)**.' A list of packages will appear. Install the additional packages needed by highlighting them (multiple packages may be installed by highlighting the desired packages while holding the <Ctrl> key). For further instructions see http://www.bioconductor.org/download. Download and installation instructions differ from those provided for a Unix or Mac OS platform. The BioConductor package `GEOquery` was used to download GSE3189 and will need to be installed to replicate the analyses.

R Code Example

Prior to analyzing the data, the control probe sets were removed. To simplify coding notation, the transpose of the gene expression data matrix was extracted as `x` and the ordinal factor was extracted and stored as `y`.

```
R> library(GEOquery)
R> library(Biobase)
R> GSE3189 <- getGEO("GSE3189")[[1]]
R> control.probes <- grep("AFFX", featureNames(GSE3189))
R> GSE3189 <- GSE3189[-control.probes,]
R> x <- t(exprs(GSE3189))
R> y <- factor(pData(GSE3189)$characteristics_ch1,
ordered = TRUE, levels = c("Normal", "Nevus", "Melanoma"))
```

Subsequently we $\log_2$ transformed the MAS5 expression summaries.

```
R> x <- log(x, 2)
```

The `glmpath.cr` function is in the `glmpathcr` package which is available from CRAN. Download the `glmpathcr` package and load it prior to model fitting.

```
R> library(glmpathcr)
```

The code for fitting a backward (default) continuation ratio model is given by

```
R> fit <- glmpath.cr(x, y)
```

As with `glmpath` model objects, methods such as `print` and `plot` can be applied to `glmpath.cr` model objects, which are helpful for selecting the step at which to select the final model from the solution path. For example, `plot` can be used to identify a more parsimonious model having an AIC close to the minimum AIC (Fig. 3).

```
R> plot(fit, xvar = "step", type = "aic")
```

The `plot` function can also be used for graphing the path of coefficient estimates (Fig. 4). The `par` function is merely used to provide more room in the margins for the probe set labels and to shrink the fonts using the character expansion parameter (`cex`).

```
R> par(mar = c(4, 4, 2, 5), cex=0.7)
R> plot(fit, xvar = "step", type = "coefficients")
```

For `plot`, the horizontal axis can be `"norm"`, `"lambda"`, or `"step"`. However extractor functions for `glmpath.cr` generally require the step to be selected, so we have selected `xvar = "step"` in these examples. The vertical axis can be `"coefficients"`, `"aic"`, or `"bic"`. The `model.select` function identifies the best fitting models using commonly used criterion, where the `which` parameter allows one to select either AIC or by default, BIC.

```
R> BIC.step <- model.select(fit, which="BIC")
R> AIC.step <- model.select(fit, which="AIC")
```

In this example, Step 11 corresponds to the model attaining the minimum BIC while Step 60 corresponds to the model attaining the minimum AIC. When extracting the model using the AIC criterion, there were 20 probe sets having a non-zero coefficient and two cutpoints in the final model. The two cutpoints result from having three ordinal classes. The `coef` function returns all estimated coefficients

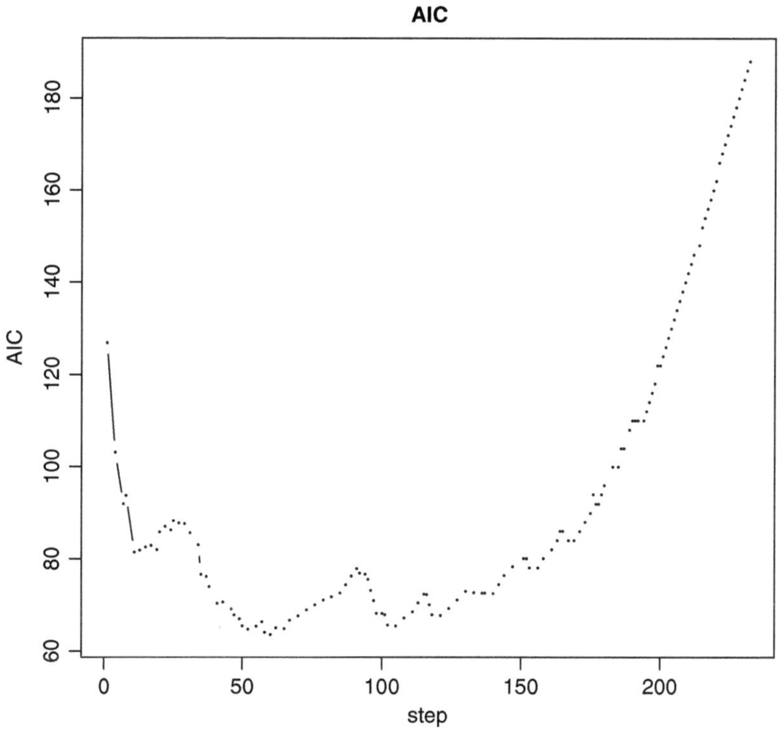

Fig. 3 Plot of Akaike Information Criteria (AIC) across the regularization path for the fitted `glmpath.cr` object using the GSE3189 melanoma data

for a `glmpath.cr` fitted model, including the intercept which is returned as the first element of the coefficient vector as well as the estimated slope and cutpoints. The $K - 1$ ordinal thresholds are given by the sum of the `Intercept` and the `cp1`, ..., `cpK-1` cutpoints, where `cp1`,...,`cpK-1` are the last $K - 1$ elements of the `coefficients` vector. The coefficient estimates are returned for a specific step of the regularization path by specifying the step number, `s`, to extract. The `nonzero.coef` function returns only those non-zero coefficient estimates for a selected model. This latter function is useful when the number of predictor variables is large.

```
R> coefficients<-coef(fit, s = AIC.step)
R> sum(coefficients != 0)
[1] 23
R>  nonzero.coef(fit, s = AIC.step)
  Intercept  200755_s_at  201022_s_at
15.470931606  0.086942969 -0.749283676
 201393_s_at  201591_s_at  201672_s_at
 0.009031235 -0.253298826  0.645105325
```

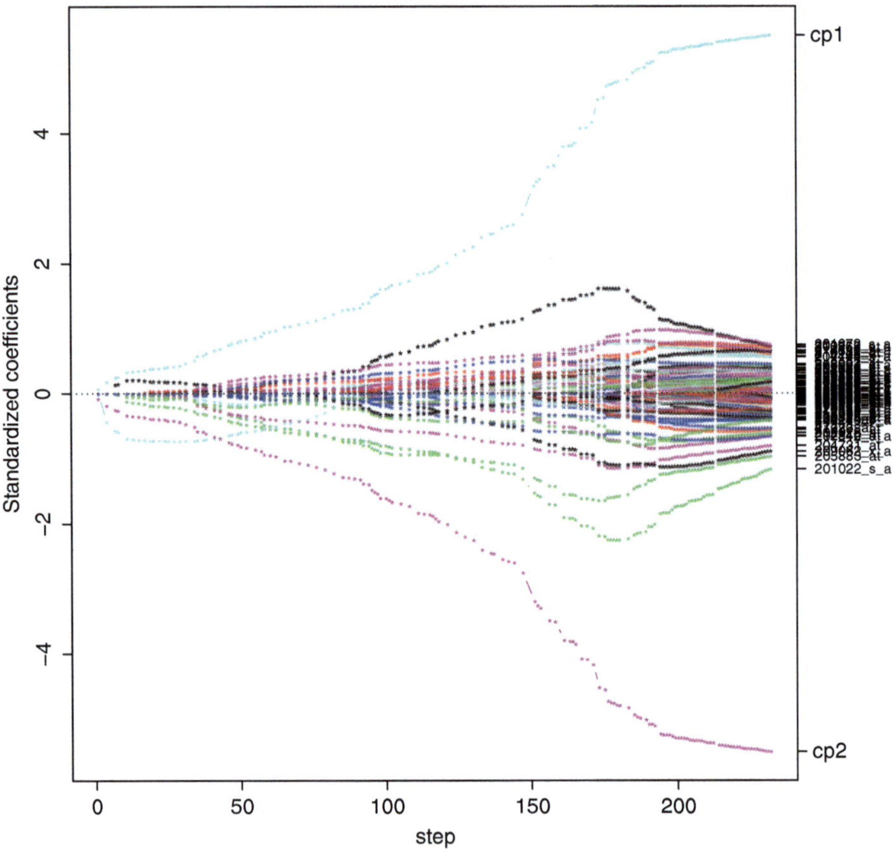

Fig. 4 Plot of estimated coefficients across the regularization path for the fitted `glmpath.cr` object using the GSE3189 melanoma data

```
   202022_at     204731_at  205236_x_at
-0.093681576 -0.501528045 -0.088600752
   205681_at     205883_at  207144_s_at
 0.032457082 -0.418007790  0.014087636
 208710_s_at   211762_s_at     212862_at
 0.015218824  0.266115729  0.226948228
   213002_at     213029_at  213330_s_at
 0.229035567 -0.412523980  0.046718171
 216037_x_at     218692_at     219476_at
-0.161153093 -0.038697412 -0.062208492
         cp1          cp2
 2.150935518 -2.150935517
```

Taking absolute values, the probe set having the largest non-zero coefficient is 201022_s_at. The dotchart in Fig. 1 was produced using the following code.

```
R> stripchart( x[, grep( "201022_s_at", dimnames(x)[[2]])]~ y,
   vertical = TRUE, pch = 16, ylab = "201022_s_at")
```

Using the annotate and hgu133a.db Bioconductor packages, the probe sets having a non-zero coefficient estimate correspond to the following Entrez IDs, gene symbols, and chromosomes:

```
R> library(annotate)
R> library(hgu133a.db)
R> beta <-nonzero.coef(fit, s = AIC.step)
R> beta <- beta[-c(1,22,23)]
R> EntrezID <- getEG(names(beta),
     "hgu133a.db")
R> EntrezID
200755_s_at 201022_s_at 201393_s_at 201591_s_at
      "813"     "11034"      "3482"     "11188"
201672_s_at   202022_at   204731_at 205236_x_at
     "9097"       "230"      "7049"      "6649"
  205681_at   205883_at 207144_s_at 208710_s_at
      "597"      "7704"      "4435"      "8943"
211762_s_at   212862_at   213002_at   213029_at
     "3838"      "8760"      "4082"      "4781"
213330_s_at 216037_x_at   218692_at   219476_at
    "10963"      "6934"     "55638"     "79098"
R> Gene.Symbol <- getSYMBOL(names(beta),
     "hgu133a.db")
R> Gene.Symbol
200755_s_at 201022_s_at 201393_s_at 201591_s_at
     "CALU"      "DSTN"     "IGF2R"     "NISCH"
201672_s_at   202022_at   204731_at 205236_x_at
    "USP14"     "ALDOC"    "TGFBR3"      "SOD3"
  205681_at   205883_at 207144_s_at 208710_s_at
   "BCL2A1"    "ZBTB16"    "CITED1"     "AP3D1"
211762_s_at   212862_at   213002_at   213029_at
    "KPNA2"      "CDS2"    "MARCKS"      "NFIB"
213330_s_at 216037_x_at   218692_at   219476_at
    "STIP1"    "TCF7L2"      "SYBU"  "C1orf116"
R> Chr <- unlist(mget(names(beta),env=hgu133aCHR))
R> Chr
200755_s_at 201022_s_at 201393_s_at
        "7"        "20"         "6"
201591_s_at 201672_s_at   202022_at
        "3"        "18"        "17"
  204731_at 205236_x_at   205681_at
        "1"         "4"        "15"
```

```
205883_at 207144_s_at 208710_s_at
     "11"          "X"         "19"
211762_s_at   212862_at   213002_at
     "17"         "20"          "6"
213029_at 213330_s_at 216037_x_at
      "9"         "11"         "10"
218692_at   219476_at
      "8"          "1"
```

R Code for producing the scatterplot of the probe set expression values for the two probe sets having the largest absolute coefficient estimates (Fig. 2) was produced using

```
R> plot(x[, grep( "201672_s_at", dimnames(x)[[2]])],
     x[, grep( "201022_s_at", dimnames(x)[[2]])],
     pch = c(1, 2, 3)[y],
     xlab = expression( log[2]("201672_s_at") ),
     ylab = expression( log[2]("201022_s_at")))
R> legend( 11.4, 14.0,
     legend = c( "Normal" ,"Nevus", "Melanoma"),
     pch = c(1, 2, 3), cex = 0.8)
```

Continuation ratio models model conditional probabilities so a new method to extract the class probabilities and predicted class was created [1]. The `predict` function returns the AIC, BIC, predicted class, and the class probabilities for the K classes for all steps along the regularization path. By default the training data is used to obtain model predictions, though predicted class and fitted probabilities can be obtained for a test dataset by specifying a different dataset using the `newx` parameter. The `predict` function extracts the predicted class when `type="class"` or the fitted probabilities when `type="probs"` for the K classes for a specific step (e.g., `which="AIC"`).

```
R> hat<-predict(fit, which="AIC", type="class")
R> table(hat, y)
            y
hat          Normal Nevus Melanoma
  Melanoma        0     0       45
  Nevus           0    18        0
  Normal          7     0        0
```

When there are small sample sizes in one or more groups (e.g., the normal group (N = 7)), cross-validation (CV) methods may not perform well as a means to estimate generalization error due to the random inclusion of samples into each of the folds. That is, multiple folds may include few if any subjects from the small classes. Nevertheless, we have provided the R code for those interested in using cross-validation on their own datasets. The `bootstrap` package is available from CRAN and must be downloaded for applying the cross-validation method. Here we have demonstrated 10-fold CV. Note that we set the random seed only so that our results from the cross-validation procedure can be reproduced.

```
R> library(bootstrap)
R> fit.cv<-function(x,y) {glmpath.cr(x,y) }
R> predict.cv<-function(fit,x) {
       predict(fit, newx=x, which="AIC", type="class")
       }
R> set.seed(12)
R> cr.cv<-crossval(x, y, fit.cv, predict.cv, ngroup=10)
R> table(cr.cv$cv.fit, y)
            y
             Normal Nevus Melanoma
  Melanoma        0     0       43
  Nevus           2    18        2
  Normal          5     0        0
```

which yields a generalized misclassification rate of 5.7 %.

References

1. Archer, K. J., & Williams, A. A. A. (2012). L_1 penalized continuation ratio models for ordinal response prediction using high-dimensional datasets. *Statistics in Medicine, 31*, 1464–1474.
2. Baranwal, S., Wang, Y., Rathinam, R., Lee, J., Jin, L., McGoey, R., et al. (2011). Molecular characterization of the tumor-suppressive function of nischarin in breast cancer. *Journal of the National Cancer Institute, 103*, 1513–1528.
3. Bender, R., & Benner, A. (2000). Calculating ordinal regression models in SAS and S-PLUS. *Biometrical Journal, 42*, 677–699.
4. Brunner, G., Reitz, M., Schwipper, V., Tilkorn, H., Lippold, A., Biess, B., et al. (2008). Increased expression of the tumor suppressor PLZF is a continuous predictor of long-term survival in malignant melanoma patients. *Cancer Biotherapy & Radiopharmaceuticals, 23*, 451–459.
5. Chao, A., Lai, C. H., Tsia, C. L., Hsueh, S., Hsueh, C., Lin, C. Y., et al. (2013). Tumor stress-induced phosphoprotein1 (STIP1) as a prognostic biomarker in ovarian cancer. *PLoS ONE, 8*, e57084.
6. Cox, D. R. (1975). Partial likelihood. *Biometrika, 62*, 269–276.
7. Efron, B., Hastie, T., Johnstone, I., & Tibshirani, R. (2004). Least angle regression. *The Annals of Statistics, 32*, 407–499.
8. Estrada-Bernai, A., Gatlin, J. C., Sunpaweravong, S., & Pfenninger, K. H. (2009). Dynamic adhesions and MARCKS in melanoma cells. *Journal of Cell Science, 122*, 2300–2310.
9. Felicetti, F., Bottero, L., Felli, N., Mattia, G., Labbaye, C., Alvino, E., et al. (2004). Role of PLZF in melanoma progression. *Oncogene, 23*, 4567–4576.
10. Friedman, J., Hastie, T., & Tibshirani, R. (2010). Regularization paths for generalized linear models via coordinate descent. *Journal of Statistical Software, 33*, 1–22.
11. Hastie, T., Taylor, J., Tibshirani, R., & Walther, G. (2007). Forward stagewise regression and the monotone lasso. *Electronic Journal of Statistics, 1*, 1–29.
12. Hastie, T., Tibshirani R., & Friedman, J. (2009). *The Elements of Statistical Learning: Data Mining, Inference, and Predition* (2nd Ed.). New York: Springer.

13. Haq, R., Yokoyama, S., Hawryluk, E. B., Jönsson, G. B., Frederick, D. T., McHenry, K., et al. (2013). BCL2A1 is a lineage-specific antiapoptotic melanoma oncogene that confers resistance to BRAF inhibition. *PNAS, 110*, 4321–4326.
14. Hosmer, D. W., & Lemeshow, S. (2000). *Applied Logistic Regression*, (2nd Ed.). Hoboken, New Jersey: John Wiley & Sons.
15. Howlader, N., Noone, A. M., Krapcho, M., Garshell, J., Neyman, N., Altekruse, S. F., et al. (1975–2010). *SEER cancer statistics review*. Bethesda, MD: National Cancer Institute.
16. Kutner, M. H., Nachtsheim C. J., Neter, J., & Li, W. (2005). *Applied Linear Statistical Models* (5th Ed.). New York: McGraw-Hill/Irwin.
17. Levy, C., Khaled, M., Iliopoulos, D., Janas, M. M., Schubert, S., & Pinner, S. (2010). Intronic miR-211 assumes the tumor suppressive function of its host gene in melanoma. *Molecular Cell, 40*, 841–849.
18. Ling, Q., Dong, F., Geng, L., Liu, Z., Xie, H., Xu, X., et al. (2013). Impacts of TCF7L2 gene polymorphisms on the susceptibility of hepatogenous diabetes and hepatocellular carcinoma in cirrhotic patients. *Gene, 522*, 214–218.
19. Martiniuk, F., Damian, D. L., Thompson, J. F., Scolyer, R. A., Tchou-Wong, K. M., & Levis, W. R. (2010). TH17 is involved in the remarkable regression of metastatic malignant melanoma to topical diphencyprone. *Journal of Drugs in Dermatology, 9*, 1368–1372.
20. Park, M. Y., & Hastie, T. (2007). L1-regularization path algorithm for generalized linear models. *Journal of the Royal Statistical Society, Series B, 69*, 659–677.
21. Persson, M., Andrén, Y., Moskaluk, C. A., Frierson Jr., H. F., Cooke, S. L., Futreal, P. A., et al. (2012). Clinically significant copy number alterations and complex rearrangements of MYB and NFIB in head and neck adenoid cystic carcinoma. *Genes, Chromosomes & Cancer, 51*, 805–817.
22. Petrenko, A. A., Pavlova, L. S., Karseladze, A. I., Kisseljov, F. L., & Kisseljova, N. P. (2006). Downregulation of genes encoding for subunits of adaptor complex-3 in cervical carcinomas. *Biochemistry (Mosc.), 71*, 1153–1160.
23. R Core Team. (2013). R: A Language and Environment for Statistical Computing. Available via R Foundation for Statistical Computing, Vienna, Austria. http://www.R-project.org. Cited30May2013.
24. Riker, A. I., Enkemann, S. A., Fodstad, O., Liu, S., Ren, S., Morris, C., et al. (2008). The gene expression profiles of primary and metastatic melanoma yields a transition point of tumor progression and metastasis. *BMC Medical Genomics, 1*, 13.
25. Rudolfer, S. M., Watson, P. C., & Lesaffre, E. (1995). Are ordinal models useful for classification? A revised analysis. *Journal of Statistical Computation and Simulation, 52*, 105–132.
26. Ryu, B., Kim, D. S., DeLuca, A. M., & Alani, R. M. (2007). Comprehensive expression profiling of tumor cell lines identifies molecular signatures of melanoma progression. *PLoS ONE, 7*, e594.
27. Singh, B., & Bhat, H. K. (2012). Superoxide dismutase 3 is induced by antioxidants, inhibits oxidative DNA damage and is associated with inhibition of estrogen-induced breast cancer. *Carcinogenesis, 33*, 2601–2610.
28. Talantov, D., Mazumder, A., Yu, J. X., Briggs, T., Jiang, Y., Backus, J., et al. (2005). Novel genes associated with malignant melanoma but not benign melanocytic lesions. *Clinical Cancer Research, 11*, 7234–7242.
29. Tibshirani, R. (1996). Regression shrinkage and selection via the Lasso. *Journal of the Royal Statistical Society, B, 58*, 267–288.
30. Walker, G. J., Indsto, J. O., Sood, R., Faruque, M. U., Hu, P., Pollock, P. M., et al. (2004). Deletion mapping suggests that the 1p22 melanoma susceptibility gene is a tumor suppressor localized to a 9-Mb interval. *Genes, Chromosomes & Cancer, 41*, 56–64.
31. Winnepenninckx, V., Lazar, V., Michiels, S., Dessen, P., Stas, M., Alonso, S. R., et al. (2006). Gene expression profiling of primary cutaneous melanoma and clinical outcome. *Journal of the National Cancer Institute, 98*, 472–482.

32. Wu, N., Liu, C., Bai, C., Han, Y. P., Cho, W. C. S., & Li, Q. (2013). Over-expression of deubiquitinating enzyme USP14 in lung adenocarcinoma promotes proliferation through the accumulation of β-catenin. *International Journal of Molecular Sciences, 14*, 10749–10760.
33. Yu, J., Liang, Q. Y., Wang, J., Cheng, Y., Wang, S., Poon, T. C. W., et al. (2013). Zinc-finger protein 331, a novel putative tumor suppressor, suppresses growth and invasiveness of gastric cancer. *Oncogene, 32*, 307–317.
34. Zhu, J., & Hastie, T. (2004). Classification of gene microarrays by penalized logistic regression. *Biostatistics, 5*, 427–443.

Structure–Activity Relationship Analysis of 7-Deazaadenosines as Anticancer Agents

Josue A. Nava-Bello, Ewa Wasilewski, Angelica M. Bello, and Alejandro A. Nava-Ocampo

Abstract The complex process to develop a successful therapeutic drug is lengthy and costly. In order to accelerate this process, molecular modeling has become a key component of drug design. Methods used in computational chemistry vary from Ab initio quantum chemistry methods, to semi-empirical calculations and molecular mechanics. A study of the anticancer activity of a series of 7-aryl- and 7-hetaryl-7-deazaadenosines published by Bourderioux (2011) showed that nucleosides with 5-member heterocycles at the position 7 were more potent in vitro cytostatic agents against hematological and solid tumor cell lines than molecules with 6-member heterocycles. We decided to conduct a quantitative structure–activity relationship (QSAR) analysis of these chemical moieties in order to have a better understanding of their structural properties and identify their molecular

J.A. Nava-Bello • E. Wasilewski
Center for Molecular Design and Preformulations, Toronto General Research Institute, University Health Network, TMDT/MaRS Center, 101 College St 5-357, Toronto, ON, Canada M5G 1L7
e-mail: bello.ja@live.com; epoduch@uhnres.utoronto.ca

A.M. Bello (✉)
Center for Molecular Design and Preformulations, Toronto General Research Institute, University Health Network, TMDT/MaRS Center, 101 College St 5-357, Toronto, ON, Canada M5G 1L7

University of Toronto, Leslie Dan Faculty of Pharmacy, Toronto, ON, Canada
e-mail: abello@uhnres.utoronto.ca

A.A. Nava-Ocampo
PharmaReasons – Pharmacological Research & Applied Solutions, 46 Lambertlodge Ave., Toronto, ON, Canada M6G 3Y8

Department of Pharmacology and Toxicology, Faculty of Medicine, University of Toronto, Toronto, ON, Canada

Division of Clinical Pharmacology & Toxicology, Hospital for Sick Children, Toronto, ON, Canada
e-mail: alex.nava@pharmareasons.com

B. Toni (ed.), *New Frontiers of Multidisciplinary Research in STEAM-H (Science, Technology, Engineering, Agriculture, Mathematics, and Health)*, Springer Proceedings in Mathematics & Statistics 90, DOI 10.1007/978-3-319-07755-0_6,

descriptors explaining their biological activities. We found that 5-member cyclic structures have three energy molecular descriptors that were negatively correlated to their biological activity, in particular, compounds with higher energies had higher biological potency represented by lower IC_{50} values. CLogP, a parameter of lipophilicity, was also found to be positively correlated to their biological activity, i.e. compounds with lower CLogP values had higher biological potency represented by lower concentrations inhibiting the growth of cancer cells by 50 %. Qualitatively, 5-member-ring heterocycles of 7-deazaadenosine had lower steric hindrance, i.e. were structurally smaller, than their 6-member counterparts. In this context, a QSAR analysis could be extraordinarily helpful in studying the mode of action of molecules with potential pharmacological relevance.

1 Introduction

The complex process to develop a successful therapeutic drug typically involves the synthesis of a series of chemicals moieties that undergo various in vitro and in vivo biological tests. During this process, a single chemical compound displays an interesting pharmacological profile and is selected as a potential therapeutic drug that will be carefully tested in humans before it can be authorized for clinical use.

In order to accelerate this process, molecular modeling has become a key component of drug design. Molecular structures can now be modeled, and their geometry, energies, and physical, electronic and spectroscopic properties can be calculated by using appropriate computer software [11]. This process includes the analysis of structural variations of the chemical moieties of interest, and can be complemented with the analysis of the interactions between the proposed moieties and their target enzymes or proteins. The ultimate goal is to optimise the synthesis of the new chemical entities with improved biological activities [7, 11].

Methods used in computational chemistry vary from Ab initio quantum chemistry methods, to semi-empirical calculations and molecular mechanics. Molecular surfaces, atoms, bonds and properties like hydrophilicity and lipophilicity can be visualized through graphic representations for better understanding of the molecular properties and interactions. The most common representation is the ball-and-spring model, which allows the visualization of flexible bonds (springs) between atoms (balls), volumes, and atoms types and interactions, allowing the chemist to interpret the results when calculating the lowest energy model for the molecule. The preferred molecular properties, or molecular descriptors, are those calculated by the computational chemistry software when the minimum energy geometry of the molecule, and consequently its most stable structure, has been reached.

In this context, the molecular descriptors are the result of a logic mathematical procedure that transforms chemical information encoded within a symbolic representation of a molecule into quantitative data representing standardized molecular conjectures. Molecular modeling has become rapidly popular, and is now part of the initial considerations when planning the synthesis of a series of chemical moieties with potential biological activity. Its use has been extended to other areas, and this has been reflected by the growing number of journals publishing studies in the broad application of computational chemistry. Of special consideration is the study of the relationship between the molecular descriptors and either qualitative or quantitative indicators of biological activity by statistical methods. This type of analyses is known as quantitative structure–activity relationship (QSAR) analysis.

We previously conducted several QSAR analyses where we used semi-empirical molecular mechanics to compute the molecular descriptors that were correlated to the biological activity of the molecules analyzed. For example, we found that alkaloids isolated from *Aconitum* roots were structurally more stable if they had an aroyl/aroyloxy group at R_{14} position than those with the aroyloxy group at R_4, and this characteristic explained their different local anesthetic activities [4]. Taking into consideration the previously reported enzymatic activity of chloroperoxidase for oxidizing organophosphorus pesticides, we used a similar QSAR analysis to propose a mechanism of the biological oxidation of these pesticides by the hepatic cytochrome P_{450} which could explain the generation of toxic metabolites [2]. We also used a QSAR analysis with semi-empirical molecular mechanics to identify that the anticonvulsant activity of valproic acid metabolites was mainly related to their lipophilicity, probably reflecting their ability to cross the blood–brain barrier [3].

These three examples of studies conducted by our group are cited to illustrate the use of semi-empirical methods available in commercial software, which in combination with simple statistical methods such as the Pearson correlation analysis and Student *t* test allowed us to conduct meaningful QSAR analyses. Recently, a study of the anticancer activity of a series of 7-aryl- and 7-hetaryl-7-deazaadenosines showed that nucleosides with 5-member heterocycles at the position 7 were more potent in vitro cytostatic agents against hematological and solid tumor cell lines than molecules with 6-member heterocycles [5]. Apart from being carefully conducted, the study attracted our attention for simultaneously containing detailed information of the chemical moieties and quantitative data of their anticancer activity in cell lines of breast, lung and colon cancer, which are of major interests under a public health perspective. According to the most recent statistics in Canada, altogether, these were the main causes of cancer in women, and represented 51 % of the new cases, after excluding skin cancers [6]. In men, lung and colon cancers were the second and third causes of cancer, only surpassed by prostate cancer, and together, they represented 27.6 % of the new cases.

Therefore, we decided to conduct a QSAR analysis of this series of 7-aryl- and 7-hetaryl-7-deazaadenosines in order to have a better understanding of their structural properties and to identify their molecular descriptors explaining their biological activities (Fig. 1). We used a similar approach to our previous studies [2–4], briefly described above, in order to illustrate the broad applicability of this procedure.

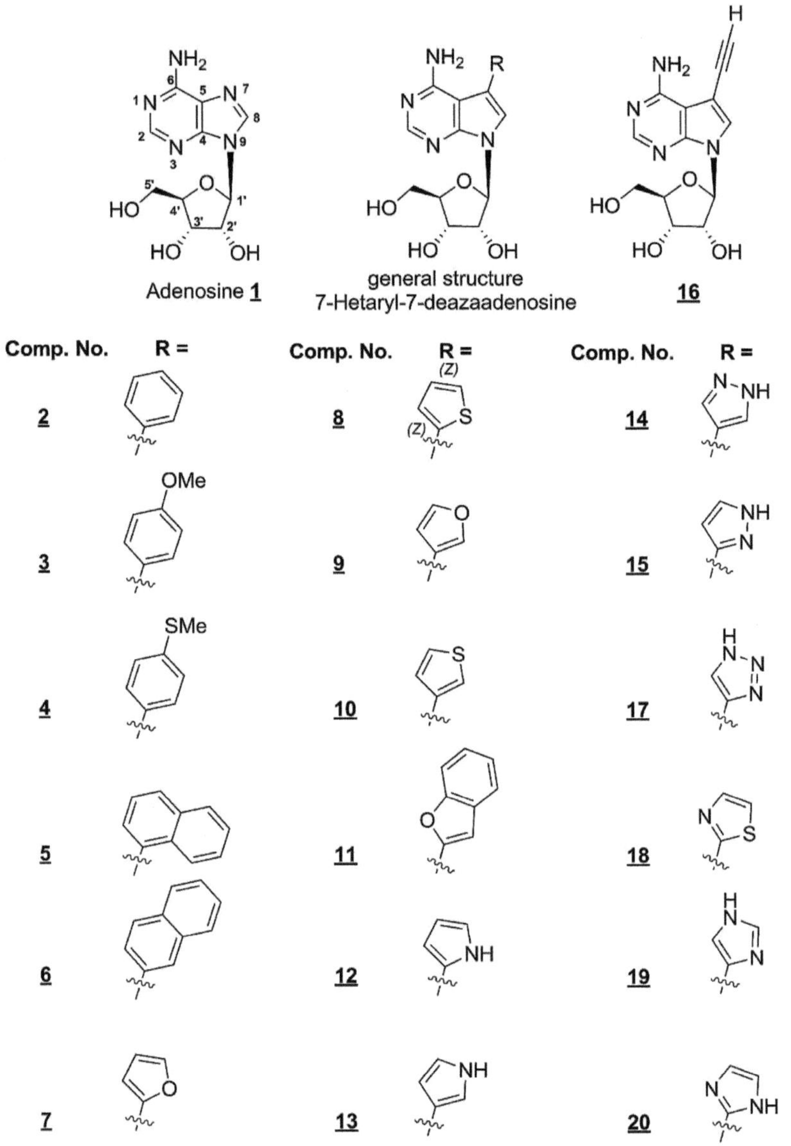

Fig. 1 Structures of the 7-aryl and 7-hetaryl-7-deazaadenosines synthesized and tested for anticancer effects by Bourderioux et al. [5], and used for the QSAR analysis in the present study

2 Methods

2.1 Molecular Modeling

We conducted a semi-empirical molecular modeling analysis of 5-member-ring ($n = 13$) and 6-member-ring ($n = 5$) heterocycles of 7- deazaadenosine previously published elsewhere by Bourderioux et al. [5]. The molecular modeling analysis was performed using the Computer-Aided Chemistry (CAChe) software v. 3.2 (Oxford Molecular Group, London, United Kingdom) as follows. The structure of a given chemical moiety is refined by means of pre-optimization calculations in Molecular Mechanics using Augmented MM3. To optimize the structure, the CAChe adjusts the initial geometry (bond lengths and angles) of the molecule in order to achieve minimal potential energy of the structure. The hybridization and bond lengths are established according with the atomic radii, the type of bond, and electronegativity of each atom. Once the structural potential energy has been calculated, the atomic positions are readjusted by changing the dihedral angles between the bonds and looking for positive interactions such as intramolecular hydrogen bonds and π interactions. The potential energy of the new structure is recalculated in every readjustment. This is an iterative process where the program moves the atoms and recalculates the energy, until a minimum energy is reached. The limit of change in energy between iterations was 0.001 kcal/mol. After the molecular geometry was optimized, the CAChe program was used to calculate the different forces that act within and around the molecule, which affect the intermolecular attractions and thus control its lipophilicity and hydrophilicity of the molecule. These forces include hydrogen bonds, dipole–dipole forces (electrostatic forces) and London dispersion forces.

2.2 Molecular Descriptors

The molecular descriptors generated by the CAChe program and used in the present study, were the following:

(a) The electrostatic force, or electrostatic energy, is a function of the charge on the non-bonded atoms, their interatomic distance, and a molecular dielectric expression that accounts for the attenuation of electrostatic interaction by the environment (the solvent or the molecule itself).
(b) The van der Waals energy is the energy of the sum of the attractive or repulsive forces between molecules, or between parts of the same molecule, when the distance between interacting atoms becomes less than the sum of their contact radii.

(c) The hydrogen bonds energy is the stabilization energy that corresponds to the intramolecular hydrogen bonds formation (an electrostatic bond between a hydrogen atom in a covalent bond and an electronegative atom, such as oxygen).
(d) The steric energy is the sum of the following net energies acting on each atom in the molecule: bond stretch energy, bond angle energy, dihedral angle energy, torsion energy, torsion stretch energy, bend–bend energy, van der Waals energy, electrostatic energy and hydrogen bond energy.
(e) The heat of formation (Kcal/mol) is the amount of heat absorbed in a reaction in which one mole of the substance in a specified state is formed from its elements in their standard states. It is considered a measurement of the stability of a compound.

In addition, the lipophilicity (CLogP) and topological polar surface area (TPSA) were calculated using Chem Draw Ultra v. 12.0 (Cambridge Soft Corporation, Perkin Elmer Informatics, Waltham, MA, USA). The calculation methods in this software can handle molecules containing carbon, hydrogen, oxygen, nitrogen, sulfur, halogens and phosphorus. Lipophilicity (CLogP) is the ability of a compound to dissolve in fat or lipids found in the human tissues and cell membranes, and hydrophilicity is the ability for a compound to dissolve in water. Water is a major component of the human body especially in the blood stream. Hence, the lipophilicity of a molecule would affect its ability to cross cellular membranes to reach the active site.

The topological polar surface area (TPSA), expressed in square angstroms, is given by the sum of the surface contribution of polar atoms, usually oxygen, nitrogen and their attached hydrogen, in a molecule. In addition to complementing the information provided by CLogP, TPSA can also be used to study receptor–ligand interactions. For example, this molecular descriptor was found to correlate negatively with activity data for marine pyridoacridine anticancer alkaloids, melatonin MT_1 and MT_2 agonists, monoamine oxidase-B and tumor necrosis factor-α inhibitors, as well as to correlate positively with inhibitory activity data for telomerase, phosphodiesterase-5, glycogen synthase kinase-3, DNA-dependent protein kinase, aromatase, malaria, trypanosomatids and cannabinoid CB_2 agonists [10].

2.3 Biological Activity

Of the different biological activities, we selected the concentrations reported to inhibit the cell growth by 50 % (IC_{50}) of NCI-H23 cell line for lung cancer, HCT116 cell line for colorectal carcinoma, and Hs578 cell line for breast cancer [5]. In order to facilitate the understanding of our QSAR analysis, it is important to note that lower IC_{50} values correspond to compounds that are more potent since less concentration is required in order to exert their anticancer activity.

2.4 Data Analysis

The biological activities and structural descriptors were summarized by their median and range values. For comparisons between the structural descriptors of the 5-member-ring and those of the 6-member-ring chemical structures, we used the Mann–Whitney *U* test due to the differences in the number of chemical moieties included in each group (n = 13 and 5, respectively). For the analysis of the relationship between the structural descriptors and the biological activity, we used the Pearson correlation analysis or linear regression analysis, since both the descriptors (independent or *x*-variables) and the biological effects (dependent or *y*-variables) were continuous variables. In the linear regression analysis, data were modeled using linear predictor functions, and unknown model parameters for a straight line model ($Y = mx + b$) were calculated from the data by the least squares method which minimises the sum of the squares of the errors associated with each *Y* point by differentiation. This error is the difference between the observed *Y* point and the *Y* point predicted by the regression equation. The slope, intercept and coefficient of correlation (*m*, *b* and r values, respectively) were computed. All the statistical analyses were conducted using StatsDirect v. 2.8.0 (StatsDirect Ltd., Cheshire, UK), and a $P < 0.05$ was considered as the significant limit in every analysis.

3 Results

3.1 QSAR

Apart from the differences in the biological activity of the 5-member-ring 7-deazaadenosines and those derivatives with a 6-member-ring, the two groups also had significant differences in four of the seven structural descriptors analyzed including lower van der Waals forces (kcal/mol) lower CLogP, higher steric energy (kcal/mol) and higer TPSA (Table 1).

In addition, of these three structural parameters, the structure energy and CLogP were consistently correlated to the IC_{50} (μM) of the 7-aryl- and 7-hetaryl-7-deazaadenosines derivatives in the colon, breast and lung cancer cell lines included in the analysis (Table 2). Although we analyzed data from the 5-member ring derivatives separate from the 6-member-ring derivatives, the small number of chemical moieties in the second group (n = 5) limited the statistical power of the Pearson correlation analysis, and therefore none of the correlation coefficients reached a significant level. However, the differences found between the correlation analyses when all the 7-deazaadenosines derivatives (n = 18) were included and those limited to the 5-member ring chemical moieties (n = 13), indicate that correlation between the 7-aryl- and 7-hetaryl-7-deazaadenosines and their IC_{50} in colon, breast and lung cancer cell lines was mainly driven by 5-member ring derivatives.

Table 1 Biological activities and structural properties of a series of 7-aryl- and 7-hetaryl-7-deazaadenosines

Parameter	5-Member-ring (n = 13)	6-Member-ring (n = 5)	P-value
Biological activity			
IC_{50} (μM)—lung cancer	0.035 (0.002; 5.4)	5.4 (0.39; >20)	**0.0044**
IC_{50} (μM)—colon cancer	0.015 (0.002; 3.74)	5.7 (0.63; >20)	**0.0016**
IC_{50} (μM)—breast cancer	0.015 (0.001; 6.73)	2.2 (0.81; >20)	**0.0016**
Structural descriptors			
Electrostatic forces (kcal/mol)	20.7 (19.2; 22.9)	20.7 (15.5; 22.2)	0.44
Van der Waals forces (kcal/mol)	13.9 (12.7; 15.9)	15.3 (13.8; 19.8)	**0.032**
Hydrogen bonds forces (kcal/mol)	4.6 (3.8; 6.9)	4.6 (4.3; 4.9)	0.91
Steric energy (kcal/mol)	95.2 (88.4; 100.0)	74.7 (69.6; 78.5)	**0.0002**
Heat of formation (kcal/mol)	−87.3 (−118.7; −64.3)	−89.7 (−128.8; −73.1)	0.55
CLogP	−0.91 (−1.5; 1.3)	1.0 (0.40; 1.66)	**0.0016**
TPSA ($Å^2$)	135.9 (123.9; −160.6)	123.9 (123.9; 133.1)	**0.015**

The biological activities for each chemical moiety were retrieved from the study conducted by Bourderioux et al. [5]. Data are summarized as the median and, in parenthesis, minimum and maximum values. Significant P-values are shown in *bold fonts*

3.2 *Qualitative Analysis of the 5-Member-Ring Substitutions*

The four most active 5-member-ring 7-deazaadenosines compounds for lung cancer were compounds number **19** > **8** > **17** > **18**, for colorectal carcinoma **20** > **8** > **19** > **17**, and for breast cancer **20** > **19** > **17** > **15**, arranged according to their anticancer activity. Compound **19**, which was reported to be the most active chemical moiety in lung cancer cells, consists of two nitrogen substitutions in positions 2 and 4 of the 5-member-ring. Nitrogen atoms as ring-members create a higher polar surface for interaction than a carbon atom in the ring. Compound **20** shares very similar structural characteristics than compound **19**, and therefore is not surprising that in the cell lines for colorectal carcinoma and breast cancer, the former is only slightly more potent than the latter chemical moiety.

Compound **8** which was reported to be the second most potent in lung and breast cancer cell lines has a very electronegative sulphur bonded at position 2 which would allow for both hydrogen bonding (since sulphur is a hydrogen bond acceptor) as well as stronger dipole–dipole moments as a result of the sulphur that is delta negative in comparison to the hydrogen atoms and carbon atoms present in the rest of the ring.

Compound **17** exhibited good biological activity in the three reported cell lines. This compound possesses three nitrogen atoms as substitutions in the 5-member-ring. However, the nitrogen atom in position 4 is bonded to a hydrogen atom thus making that area slightly more positive than the two other very/highly electronegative nitrogen atoms. This gives the compound a larger dipole moment as well as a greater ability to form hydrogen bonds; an amine group (–NH–) has the dual property of being hydrogen-bond donor and hydrogen-bond acceptor. However, the ability of 7-aryl- or 7-hetaryl-7-deazaadenosines to form hydrogen bonds does not seem to be directly affecting the activity of the compound in its receptor site.

Table 2 Linear regression equations ($y = mx + b$) between structural parameters (x) of 7-aryl- and 7-hetaryl-7-deazaadenosines derivatives that were significantly related to the IC_{50} (μM) in three types of cancer cell lines

Parameter	Slope (95 % CI)[a]	Intercept	R	P-value
HCT116 colon cancer cells				
5- and 6-member-ring (n = 18)				
Steric energy (kcal/mol)	−0.30 (−0.53; −0.069)	28.9	−0.57	0.014
CLogP	2.3 (0.18; 4.5)	2.8	0.50	0.034
5-member-ring (n = 13)				
Van der Waals forces (kcal/mol)	0.72 (0.075; 1.37)	−9.7	0.59	0.032
Steric energy (kcal/mol)	−0.22 (−0.42; −0.025)	21.4	−0.60	0.031
CLogP	0.72 (0.073; 1.4)	0.85	0.59	0.032
Hs578 breast cancer cells				
5- and 6-member-ring (n = 18)				
Steric energy (kcal/mol)	−0.28 (−0.52; −0.045)	27.5	−0.53	0.023
CLogP	2.3 (0.15; 4.5)	2.7	0.49	0.037
5-member-ring (n = 13)				
Van der Waals forces (kcal/mol)	1.5 (0.43; 2.5)	−20.2	0.68	0.0098
Steric energy (kcal/mol)	−0.44 (−0.77; −0.12)	42.6	−0.67	0.012
CLogP	1.5 (0.39; 2.5)	1.5	0.67	0.012
NCI-H23 lung cancer cells				
5- and 6-member-ring (n = 18)				
Steric energy (kcal/mol)	−0.28 (−0.52; −0.049)	27.9	−0.54	0.021
CLogP	2.4 (0.29; 4.5)	3.0	0.52	0.028
5-member-ring (n = 13)				
Electrostatic forces (kcal/mol)	−0.71 (−1.39; −0.030)	15.6	−0.57	0.042
Steric energy (kcal/mol)	−0.44 (−0.69; −0.18)	42.0	−0.76	0.0028
CLogP	1.34 (0.45; 2.23)	1.57	0.71	0.007

[a]Value and, in parenthesis, the 95 % confidence interval. None of the structural parameters of 6-member ring 7-deazaadenosine derivatives was significantly correlated to the IC_{50} (μM) in none of the cancer cells analyzed herein, more likely due to the limited number of chemical moieties included in the group (n = 5)

3.3 *Qualitative Analysis of the 6-Member-Ring Substitutions*

These compounds were significantly less active than those with 5-member-ring derivatives of 7-deazaadenosine. Therefore, we limited our qualitative analysis to emphasize that lower van der Waals and steric energies as well as lower structure energies were qualities in compounds **3**, **4**, and **6**, the three most active compounds in this group. Compound **6** contains a pair of fused-benzene rings at position 7 of the 7-deazapurine ring, whereas compounds **3** and **4** contain a single aromatic ring mono-substituted with an electronegative atom attached to a methyl group.

Hence, it appears that a bigger structure does not permit 7-deazaadenosine derivatives to have a proper interaction with their site of action. This can also be appreciated by the perpendicular orientation of the 6-member ring in relation to the 7-deazaadenosine, whereas the 5-member ring is oriented in the same plane to the nucleic base (Fig. 2).

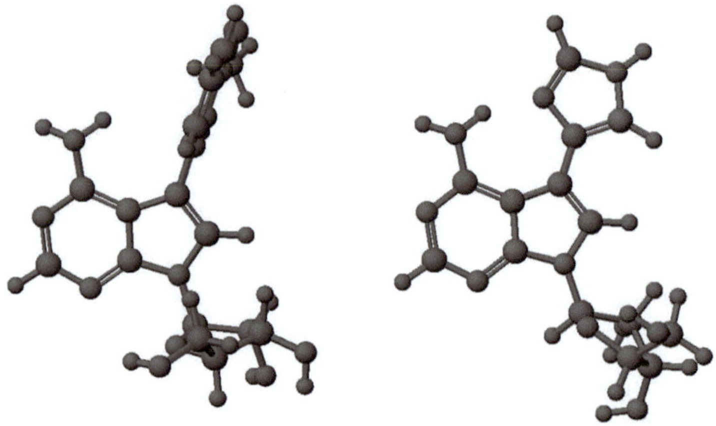

Fig. 2 Compound **4** (*left*) has a six member ring substituent which is positioned in a perpendicular plane to that of the 7-deazaadenosine, whereas compound **19** (*right*) has a five member ring substituent which stands in the same plane as the nucleic base. In the study by Bourderioux et al. [5], the latter exhibited more potent anticancer properties than the former

4 Commentary

Our QSAR analysis of the anticancer activity of 5-member-ring and 6-member-ring heterocycles of 7-deazaadenosine showed that 5-member cyclic structures reported elsewhere [5], have three energy molecular descriptors that where negatively correlated to their biological activity, i.e. compounds with higher energies had higher biological potency represented by lower IC_{50} values. These descriptors were the steric energy and electrostatic forces, which appear to be related to the stability of chemical moieties [3], and the van der Waals forces which represent steric interactions. Of these three energy molecular descriptors, electrostatic forces appear to be the less relevant parameter, since no statistical differences were observed when 5-member-ring heterocycles were compared to the 6-member-ring heterocycles (Table 1).

CLogP, a parameter of lipophilicity, was also found to be positively correlated to their biological activity, specifically compounds with lower CLogP values had higher biological potency represented by lower IC_{50} values. Since drugs need to cross cell membranes in order to enter into the cell, be adsorbed, or distributed, lipophilic chemical moieties are expected to cross better through these biological barriers. The optimum CLogP, however, may vary from one drug family to another. For example, we previously showed that the rate of transfer of local anesthetics from the central nervous system to the plasma is parabolic, with the slowest rate occurring at CLogP of 3.0 [8]

Qualitatively, 5-member-ring heterocycles of 7-deazaadenosine, which were reported to be biologically more active [5], had lower steric hindrance, i.e. were structurally smaller, than their 6-member counterparts. Together with the quantitative analysis, these findings suggest that bigger 7-deazaadenosine derivatives would have more difficulties interacting with their site of action. This can be supported by the findings reported in a study of a series of 7-substituted adenine derivatives where chemical moieties with high steric hindrance exhibited low anticancer activities, i.e. from 6.7 to $>100\ \mu M$ [12].

Other nucleosides consisting of pyrimidine or purine attached to either a ribose or deoxyribose have shown relevant cytostatic activity to treat solid and haematological malignancies [1, 9]. Cytostatic agents are able to halt cellular growth and multiplication, and have been widely used in chemotherapy for decades. Paradoxically, cytostatic agents have also been categorized as carcinogenic, mutagenic and teratogenic compounds, triggering concerns when treating cancer patients. Therefore, there is a continuous interest in developing cytostatic agents with high specificity for affecting malignant cells. In this context, a QSAR analysis could be extraordinarily helpful in studying the mode of action of molecules with potential pharmacological relevance in order to plan the synthesis of chemical moieties with a better pharmacological profile, as well as to understand and anticipate their potential toxicity when used in the clinical setting.

Acknowledgments Josue A. Nava-Bello contributed as a student of the International Baccalaureate Program at St. Robert Catholic High School, Thornhill, Ontario, Canada. He is currently a student at the École d'Ingénierie des Sciences Aérospatiales (ELISA), Saint Quentin, France.

References

1. Bello, A. M., Konforte, D., Poduch, E., et al. (2009). Structure–activity relationships of orotidine-5′-monophosphate decarboxylase inhibitors as anticancer agents. *Journal of Medicinal Chemistry, 52*, 1648–1658.
2. Bello-Ramírez, A. M., Carreón-Garabito, B. Y., & Nava-Ocampo, A. A. (2000). A theoretical approach to the mechanism of biological oxidation of organophosphorus pesticides. *Toxicology, 149*(2–3), 63–68.
3. Bello-Ramirez, A. M., Carreon-Garabito, B. Y., & Nava-Ocampo, A. A. (2002). Do structural properties explain the anticonvulsant activity of valproate metabolites? A QSAR analysis. *Epilepsia, 43*(5), 475–481.
4. Bello-Ramirez, A. M., & Nava-Ocampo, A. A. (2004). The local anesthetic activity of aconitum alkaloids can be explained by their structural properties: a QSAR analysis. *Fundamental and Clinical Pharmacology, 18*(2), 157–161.
5. Bourderioux, A., Naus, P., Perlikova, P., et al. (2011). Synthesis and significant cytostatic activity of 7-hetaryl-7-deazaadenosines. *Journal of Medicinal Chemistry, 54*, 5498–5507.
6. Canadian Cancer Society's Advisory Committee on Cancer Statistics (2013) *Canadian cancer statistics*. Toronto, ON. http://www.cancer.ca/statistics. Accessed 10.2.14. Robins RK, Revankar GR (1985) Purine analogs and related nucleosides and nucleotides as antitumor agents. Med Res Rev *S*:273–296
7. Lewards, E. (2003). *Computational chemistry. An introduction to the theory and applications of molecular and quantum mechanics*. New York: Kluwer Academic Publishers.
8. Nava-Ocampo, A. A., & Bello-Ramírez, A. M. (2004). Lipophilicity affects the pharmacokinetics and toxicity of local anaesthetic agents administered by caudal block. *Clinical and Experimental Pharmacology and Physiology, 31*, 116–118.

9. Parker, W. B. (2009). Enzymology of purine and pyrimidine antimetabolites used in the treatment of cancer. *Chemical Reviews, 109*, 2880–2893.
10. Prasanna, S., & Doerksen, R. J. (2009). Topological polar surface area: a useful descriptor in 2D-QSAR. *Current Medicinal Chemistry, 16*, 21–41.
11. Richon, A. B. (1994). An introduction to molecular modeling. *Mathematech, 1*, 83–100.
12. Tuncbilek, M., Guven, E. B., Onder, T., et al. (2012). Synthesis of novel 6-(4-substituted piperazine-1-yl0-9-(β-D-ribofuranosyl0purine derivatives, which lead to senescence-induced cell death in liver cancer cells. *Journal of Medicinal Chemistry, 55*, 3058–3065.

More than an African American Facilitator and a Prayer: Integrating Culture and Community into HIV Prevention Programs for African American Girls

Faye Z. Belgrave, Jasmine Abrams, Sarah Javier, and Morgan Maxwell

Abstract When integrating culture into HIV prevention and intervention efforts, for African American groups, programs need more than an African American facilitator and a "prayer." Understanding culture, cultural competency, and cultural integration is important for programs to be most effective. One research approach that seems particularly suited for culturally integrated interventions is Community Based Participatory Research, or CBPR. This chapter provides a discussion of community involvement in HIV prevention programs implemented by the authors. The chapter also discusses cultural attributes for African American girls including ethnic identity, relational orientation, and gender role beliefs. The chapter reviews why these cultural attributes are important and some of the authors' work implementing these attributes in prevention programs for African American girls. The chapter also discusses the importance of technology use in prevention interventions.

Sexually active African American adolescent females are at a heightened risk for contracting sexually transmitted infections including HIV/AIDS [15]. As such, there is a need for prevention and intervention programs to address this health disparity within a culturally sensitive and developmentally appropriate framework. Research has shown that culturally integrated interventions can be effective at reducing HIV risk [18, 19, 32]. The goals of this chapter are to: (1) define culture, cultural competency, and cultural integration; (2) discuss community integration in HIV prevention programs; and (3) discuss ways in which culture can be attended to and integrated in prevention and intervention efforts. The chapter addresses each goal in order, beginning with an overview of relevant concepts.

F.Z. Belgrave (✉) • J. Abrams • S. Javier • M. Maxwell
Department of Psychology, Virginia Commonwealth University, Richmond, VA, USA
e-mail: fzbelgra@vcu.edu

B. Toni (ed.), *New Frontiers of Multidisciplinary Research in STEAM-H (Science, Technology, Engineering, Agriculture, Mathematics, and Health)*, Springer Proceedings in Mathematics & Statistics 90, DOI 10.1007/978-3-319-07755-0_7,

1 Defining Culture, Cultural Competency, and Cultural Integration

When integrating culture into prevention and intervention efforts, for African American groups, programs need more than an African American facilitator and a "prayer." Understanding culture, cultural competency, and cultural integration is important for programs to be most effective. Culture is defined as "integrated patterns of human behavior that include the language, thoughts, communications, actions, customs, beliefs, values, and institutions of racial, ethnic, religious, or social groups" [42]. Cultural competence consists of congruent behaviors, attitudes, practices, and policies that come together in a system, agency or among professionals and enables that system, agency, or those professionals to work effectively in cross-cultural situations. Cultural competence facilitates culturally-appropriate prevention and treatment strategies that are (1) based on the cultural values of the targeted group; (2) reflect the subjective cultural characteristics of members of the targeted group; and (3) reflect the behavioral preferences and expectations of members of the group [28, 42].

Cultural dimensions for African American female adolescents include gender (i.e. being female), ethnicity (i.e. being African American), and age (i.e. being between the ages of 12–18). Other aspects of culture might include neighborhood, period in time, and group networks. Cultural integration involves attending to aspects of culture in both the format and the content of the program or activities that are being implemented [25].

The first step of culturally integrated interventions should be to understand the culture of the target population. To do so, one must take into account the unique and diverse attributes of the target population in research and programmatic activities including: development of research questions, execution of research, application of research findings, selection of the intervention program, identification of participants, recruitment and retention of participants, and evaluation of the intervention program. One research approach that seems particularly suited for culturally integrated interventions is Community Based Participatory Research, or CBPR.

2 CBPR and CBPAR: Community-Based Approaches to Culturally Competent Research

CBPR is a method through which communities can become involved in intervention efforts. The Kellogg Foundation has defined CBPR as a "*collaborative approach to research that equitably involves all partners in the research process and recognizes the unique strengths that each brings. CBPR begins with a research topic of importance to the community, has the aim of combining knowledge with action and achieving social change to improve health outcomes and eliminate health disparities*" [16]. In CBPR, community members are involved in all aspects of the research project, from conception to dissemination and sustainability. Community members work in close collaboration with the researchers and may be involved in planning and carrying out program activities as well as observing the proceedings and analyzing any data that may come out of the program. Use of CBPR has been encouraged by such agencies as the National Institutes of Health, Centers for the Disease Control and Prevention (CDC), Research! America, and the Public Health Foundation [16].

Similar to CBPR, Community Based Participatory Action Research (CBPAR) is seen as a culturally appropriate approach for health research and promotion [46]. Some characteristics of CBPAR include: (1) facilitating collaborative non-hierarchical partnerships and encouraging co-learning and capacity building amid partners; (2) highlighting community perspectives and indigenous perspectives on public health problems that identify and address multiple contributors to health and disease; (3) integrating and accomplishing a balance between research and action for the shared benefit of partners; (4) identifying and building on assets and resources within the community; and (5) involving partners in disseminating knowledge gained [22, 31]. Similar to CBPR, CBPAR is built on community participation and seeks to involve community stakeholders and members in every step of the research process [43].

The nature of the relationship between researchers and community members is very important in the participatory research approach. Generally, the first step in establishing a mutually beneficial and respectful relationship is to identify "stakeholders" who are invested in the program or research. Stakeholders can consist of program sponsors, or organizations that initiate and fund a program, the targets of the program themselves, and staff members trained to carry out the intervention [39].

The advantages of using community-based approaches are many and include (1) having a panel of stakeholders with diverse backgrounds and expertise; (2) improved strategies for recruitment and retention; (3) increased quality and validity of research by tailoring the program to fit the needs of the community; (4) increased trust and rapport; (5) increased resources to communities; and (6) increased potential to develop further interventions or to maintain the already-existing program [33].

When developing interventions aimed at reducing risk for contracting HIV/AIDS among African American girls, the above factors are important to consider. While all factors should be addressed to the extent possible, we believe three of the most important factors are (1) having diverse stakeholders; (2) having effective recruitment and retention strategies; and (3) implementing practices that promote the maintenance or sustainability of the program. Intervention sustainability will be revisited at the end of the chapter.

2.1 Having Diverse Stakeholders

HIV is typically viewed as a public health problem that results from individual-level behavioral practices (e.g., having unprotected sex with an HIV positive person). In order to understand HIV risk behaviors, a panel of diverse individuals with different levels of expertise is needed. For example, community partners could include participants from a non-profit health clinic, academicians from various backgrounds (e.g. Psychology, Public Health, African American studies, Education), professionals from school and after-school programs (e.g., teachers, administrators, prevention specialists), and residents from housing communities. African American adolescents and their parents are also stakeholders. Having a diverse panel of stakeholders will garner fresh and varied perspectives and insights. Stakeholders in our HIV prevention program for girls have included prevention specialists who served on an advisory council, staff and teachers from after school programs and local schools, parents, and representatives from teen pregnancy prevention programs, among others. Adolescent girls have also been involved stakeholders as members of a junior advisory board.

2.2 Recruiting and Retaining Participants

Recruitment and retention are essential aspects of ensuring the success of an intervention. It is impossible to obtain the benefits of a program unless participants are retained for the duration. A recommended recruitment and retention strategy is to provide transportation to and from the project site or hold the program activities in communities where people live. Accessibility may be important when working with African American youth who may live in underresourced communities, where transportation may not be accessible. Issues of security for program participants must also be considered if the site is in a high risk neighborhood.

Another recruitment and retention strategy is to work with existing groups of youth (e.g., church youth groups, community center groups, after-school program groups, sports teams). Participants from already intact groups typically attend more program sessions and are more likely to be available for follow up. We also recommend that staff members of these groups be utilized (in a paid position) if necessary to support recruitment and retention efforts. In our HIV prevention programs, we implemented the program within the community that girls lived and also at after-school sites.

Overall, both CBPR and CBPAR are useful approaches to conducting culturally competent research.

3 The Importance of Community Sustainability

Previously, we discussed the importance of CBPR. CBPR recognizes the importance of the community integration in HIV prevention efforts. It also places priority on programs being sustained by the community. When conducted correctly and efficiently, programs aimed at increasing HIV protective behaviors may be sustainable over the long-term. This is both in terms of the program itself and also with a permanent change in attitudes and behaviors among girls. In order to achieve sustainability, it is important to train community partners to be able to carry out the program when the researchers or program staff leaves. Several strategies can be used to assist in program sustainability. These include follow-up meetings or booster sessions, observations and critique of sessions, and training specifically on how to implement HIV prevention programs with fidelity. In programs that we have implemented in after-school and school settings, prevention specialists, and other professionals were trained in how to implement the curriculums. In other programs we have encouraged parents to continue to meet as a parenting support group to continue to realize the goals of the program.

4 Integrating Culture into Programming

Integrating culture into intervention programming should be a thoughtful process, occurring during program creation and implementation. At this stage, cultural dimensions should be integrated into both program content and format. There are several aspects of the culture of African American girls that could be addressed and incorporated. We describe next a cultural program we developed for African American girls. The program called, "Sisters of Nia," can be used as a stand alone cultural enrichment program or in combination with programs that target life skills (e.g. HIV and substance abuse prevention). We then focus on three culturally salient aspects of identity related to sexual risk taking for African American female adolescents: (1) ethnic identity, (2)) relational orientation, and (3) gender role beliefs. These cultural attributes have been incorporated into a number of effective HIV risk reduction interventions for African American girls and women, including Sister Informing Healing Learning and Empowering (SIHLE), and Project SAFE (Sexual Awareness for Everyone).

Sisters of Nia, developed for girls in early adolescence (11–14) uses a small group format that convenes 8–12 girls [6]. Fifteen one and a half hour sessions focus on the cultures of being female and of African descent. Girls are exposed to African American female intervention staff called "*mzees*" (Kiswahili for respected elders) that serve as role models of what females can do. The 15 sessions in Sisters of Nia are listed in Table 1 below:

Table 1 Sisters of Nia sessions

Session	Topic
1	Orientation
2	Jamaa building
3	Introduction to relationships
4	Relationships continued
5	Introduction to Africa and African Culture
6	African culture continued
7	Appearances: Judging others
8	Appearances: healing the hurt
9	Personal hygiene
10	Critical consciousness
11	Creativity
12	Leadership: African American women
13	Education awareness
14	Life course
15	Moving on: closing ceremony

An evaluation of Sisters of Nia showed an increase in ethnic identity and androgynous gender roles and a decrease in relational aggression among girls who had participated in the intervention [7].

We next provide examples of how we attend to ethnic identity, a relational orientation, and gender roles in interventions we have implemented and discuss how these three cultural elements can be incorporated into HIV intervention programing for African American girls. We also provide a brief discussion of research that supports why these attributes are important to attend to when working with African American girls.

4.1 Ethnic Identity

Ethnic identity refers to the degree to which one identifies with their racial, ethnic, and/or cultural sharing group. African Americans with high ethnic identity associate positive feelings with being a member of their ethnic group [35]. They are proud to be African American, desire to be around other African Americans, and engage in behaviors that support these desires (i.e., shopping in Black owned stores). African Americans with low ethnic identity tend to affiliate with other ethnic groups and have negative feelings about being African American.

For African American girls, high ethnic identity is associated with positive psychological and behavioral outcomes [17, 41]. Girls with high ethnic identity are less likely to experience sexual onset at a young age and more likely to protect themselves from sexually transmitted infections and pregnancy [5]. Further, they use fewer drugs and have more intolerant attitudes toward drugs compared to girls with low ethnic identity [13]. Given the relevance of these behaviors to decreasing

HIV risk, culturally integrated intervention programs for African American girls should aim to increase ethnic identity.

4.1.1 Integrating Ethnic Identity into Program Content

Ethnic identity can be incorporated into program content through sessions on ethnic pride and African culture that provide information on accomplishments and the diversity of people of African ancestry. For example, activities can include information about past and present queens of Africa, Africa being the birthplace of civilization, and various African countries that reflect rich natural resources, urban development, and cultural diversity. In addition, programs may benefit from providing intervention participants with opportunities to learn about successful African American women and encouraging girls that they have a long and rich history of which they can be proud.

Ethnic identity can also be addressed in sessions on self-image, with a focus on critically examining media portrayals of African American women. African American women are often portrayed as hypersexual in popular media. Addressing such stereotypical images and having discussions about what it means to be a person of African descent can help to correct myths about African American culture and increase ethnic identity [3].

Including sessions that focus on enhancing ethnic identity at the beginning of an intervention can help to establish a desirable programmatic climate. Topics related to ethnic identity should not be restricted to one or two sessions but integrated in all sessions.

4.1.2 Integrating Ethnic Identity into Program Format

Integrating ethnic identity into program format can be accomplished in a number of ways. One of the most obvious methods is to have African American female facilitators. This strategy is used in several HIV interventions for African American girls. Employing the use of cultural titles such as Sista or Mzee (respected female elder) is also a common technique for promoting positive ethnic identity. The use of cultural titles also emphasizes respect for elders.

Ethnic identity can also be attended to by employing ritualistic openings and closings of sessions with African American poetry, music, unity circles, prayers, call and response phrases, and/or libations. Creating opportunities for girls to participate in cultural activities is also a useful strategy for promoting ethnic identity. Examples of these strategies include having girls to read poetry to open or close a session, recite African proverbs, learn hip-hop or African dance, and attend cultural celebrations (e.g., Kwanzaa, Black History programs, Juneteenth celebrations, etc.).

It is also a good practice to create an atmosphere that reflects an appreciation of African culture. This can be done by decorating the intervention space with African art (paintings, pictures, and/or statues) and playing uplifting music that is popular in African American culture (socially conscious hip hop, R & B, and neo-soul). Finally, it is important to consider that people of African ancestry may think and behave within a communalistic worldview, which is closely related to having a relational orientation.

4.2 Relational Orientation

Mutually dependent relationships are important to adolescent girls. Such relationships, particularly those with adult females, shape the identity of girls and often serve as exemplars for behavioral expectations. While relationships hold significance for all girls, they may be especially important to African American girls given the relational values of both people of African ancestry and females [3].

As relationships are especially important to girls during early adolescence, significant others have the potential to positively or negatively influence girls' perception of self. A positive sense of self is associated with girls feeling valued by significant others (i.e., mothers, fathers, grandparents, teachers, mentors). A negative sense of self is associated with the opposite. Girls who do not feel valued in their relationships with significant others may have unmet relational needs and may attempt to seek fulfillment within other more damaging and negative relationships. Girls with positive family relationships are less likely to engage in risky behaviors when compared to girls without positive family relationships. Among girls, lower academic achievement, drug use, and risky sex are associated with poor family relationships [21, 29].

The relational orientation of African American girls can offer additional benefits through social support. Risky sexual behavior is less likely to occur among African American adolescents who report having more social support from parents and others [2, 40]. The types of individuals with whom social relationships are established can also influence sexual risk taking. For example, African American adolescents who reported having peers who engage in fewer risky sexual behaviors also reported fewer risky sexual behaviors [9, 30, 45, 47].

Creating a supportive environment is associated with increased intervention effectiveness among young African American females, as desired intervention outcomes are greater for participants who feel supported by their fellow group members than when they are not [4]. In one study, Belgrave and colleagues found that higher perceived support from group members in an HIV prevention intervention was associated with less sexual risk (e.g. higher condom negotiation efficacy and higher condom use efficacy). Research suggests that community based interventions may be most successful when the program allows for peer bonding and the fostering of friendships [27].

The value that girls place on relationships can also have negative impacts if girls are caring for significant others at the expense of their personal values, desires, and needs. If a girl values her partner more that she values herself, this mindset has the potential to manifest in risky sexual behaviors. Teaching the importance of self-value is essential to helping African American girls make safer sexual decisions.

4.2.1 Integrating a Relational Orientation into Program Content

Interventions that address the relational orientation of African American girls should include activities that help girls identify positive characteristics of relationships and emphasize the importance of positive relationship building. Activities should be structured in a way that encourages interpersonal connections with adult women and with co-participants.

The following strategies can assist with achieving high levels of group support, and positive relationships. Sessions should include teaching techniques such as active listening, how to provide positive affirmations, how to provide constructive criticism, and how to let others know that they are appreciated and respected. Sessions might also incorporate information to participants on how to access and gain support from others such as appropriate levels of disclosure, how to ask for help, and how to take constructive feedback [4].

4.2.2 Integrating Relational Orientation into Program Format

Interventions can provide an environment that is conducive to relationship building and garnering social support from peers. Ideally, this environment would allow for girls to support and encourage one another in safer sex decision-making. Providing ample opportunities for girls to interact with facilitators and work in small groups to discuss topics, solve problems, and carry out other activities promote collaborations and a sense of community. Within these small groups, girls learn to look out for and to be responsible for each other. Having adult African American female facilitators model positive collaborative interactions also helps to promote relationship building. Facilitators should encourage genuineness, sincerity, and respect in all interactions. It is also important to keep in mind that the dynamic of the group must be closely monitored by facilitators in order to avoid negative consequences of peer interactions.

Social support networks established from relationship building opportunities in interventions may assist with establishing sustainability of intervention effects. Facebook groups and group texting/chatting networks established during the intervention can offer girls support from their peers after the intervention has concluded.

4.3 Gender Role Beliefs

Gender role beliefs are socially constructed views that ascribe particular behavioral/role expectations to individuals based on their identity as a male or female.

Gender role beliefs affects how we interact with others [14], and guide behavior in a variety of situations including personal decisions, sexual behavior, family decision making, and human behavior in general [1, 8, 20, 38]. The development and expression of individual characteristics such as self-concept, identity, interpersonal skills, mental health, sexual behaviors, and attitudes are influenced by gender role beliefs [1]. Females who have both high levels of instrumental (masculine) and expressive and nurturing (feminine) gender roles are androgynous [12]. Being androgynous may serve as a protective factor for girls. Girls who are androgynous have higher self-esteem and self-worth and engage in less risky behaviors than those who are not [37].

African American girls are often socialized to simultaneously assume both masculine and feminine roles, including those as providers, assertive communicators, and self-reliant individuals while being a nurturer and a caregiver at the same time [10]. African American girls with androgynous gender role beliefs have better psychological functioning, engage in less risky behaviors, have higher ethnic identity, and increased levels of self-worth and self-esteem than those who do not [3, 12, 37]. African American girls may be raised to believe that part of womanhood is being independent, assertive, emotionally resilient, and economically capable of sustaining self [24]. These beliefs may operate as protective factors by promoting adaptive behaviors and effective coping skills especially in the face of stress or adversity.

4.3.1 Integrating Androgynous Gender Role Beliefs into Program Content

In HIV prevention interventions for girls, the focus should be on promoting androgynous gender role beliefs. Opportunities should be provided to encourage girls to express and internalize both expressive and instrumental roles. These can be encouraged through sessions that focus on developing general decision making skills, sex refusal efficacy, condom negotiation skills, condom use efficacy, and assertive communication skills. Knowledge and skills gained in sessions should reinforce the notion of them taking an active role in their sexual decision-making. They should be provided with the information and given opportunities to develop the skills needed to effectively communicate their desires and how to practice safer sexual activity. At the same time, sessions should emphasize the importance of being sensitive when interacting in a sexual situation.

Even with proper preparation, sexual situations may not go as planned. Thus, simulation and role playing sessions might demonstrate how to behave in certain sexual situations. For example, girls could take turns practicing what to say when in a situation with a partner who does not want to use a condom, then offer critiques and suggestions.

4.3.2 Integrating Gender Role Beliefs into Program Format

In order to promote androgynous gender roles, girls should be afforded opportunities to express both instrumental and expressive gender roles. Opportunities to increase instrumental gender roles can be facilitated through leadership positions. For example, girls can be given the opportunity to be responsible for opening and/or closing sessions, leading activities, or being responsible for calling order to a rambunctious environment. To increase expressive gender roles, program staff should acknowledge behavior that is seen as caring and responsive to others [3]. For example, in one program we implemented, we had girls identify (anonymously by putting a name in a box), one participant who had demonstrated kind and caring behavior. The person with the most names was recognized for that week. Girls can also be given homework assignments that encourage the adoption of androgynous gender roles. These may include asking girls to visit a senior citizen home, develop an activity for a younger sibling (to strengthen expressive/feminine gender roles) or initiate a conversation about leadership opportunities within her school and/or place of worship (to strengthen instrumental gender roles).

5 Considering Technology

Technology can be used to promote or inhibit positive developmental outcomes among girls. Technology usage, particularly internet and cell phone use, is prevalent among African American adolescents. A recent study found that 72 % of African American youth (from low-income backgrounds) use the Internet at least once or twice a week to acquire information and learn (Whiteley et al. 2011). In addition, about 60 % use the internet for social networking (Whiteley et al. 2011). Furthermore, research suggests that increasing numbers of African American youth have constant internet access with about 45% accessing the internet from their personal cell phones [26].

HIV risk reduction interventions can use this technology for decreasing risk for two primary reasons: (1) there are numerous Internet sites that provide inaccurate information about sexual health and (2) youth have begun to use the Internet to identify sex partners. Although the relationship between HIV risk and online technology is understudied, research indicates that meeting sex partners online is associated with drug and alcohol use and unprotected sex among African American youth [44]. As such, interventions should discuss prevention strategies related to unique online social interactions with prospective sex partners. Programs may also want to provide African American girls with knowledge of and strategies for identifying appropriate internet resources (government health websites/factsheets) to gather information about their sexual health.

In addition, as increasing numbers of African American youth have personal cell phones, programs may benefit from using brief phone calls or text messaging as a prevention strategy. Indeed, African American adolescents are receptive to receiving phone based messages related to HIV prevention (DiClemente et al. 2009; St. Lawrence et al. 2009). In addition to being thoughtful about the content of the messages, it is important to consider timing and frequency. Focus groups conducted with African American adolescents revealed that they preferred to be texted during the hours of 4:00–6:00 pm and wanted to receive a maximum of three messages per day (St. Lawrence et al. 1995).

The incorporation of online and cell phone technology into intervention delivery is demonstrating successful trends in HIV prevention (DiClemente et al. 2009). Considering the prevalence of use and importance of these technologies would be useful in creating culturally integrated HIV prevention programs for African American adolescent girls.

6 Conclusion

In conclusion, culturally integrated interventions are essential in effective HIV prevention strategies. It is important to acknowledge that culturally integrated interventions for African American girls require more than an African American facilitator and a prayer. Health professionals interested in creating or implementing culturally appropriated programming should invest time in understanding the target population. For African American girls, the following cultural values should be attended to in culturally relevant HIV prevention programming: (1) ethnic identity, (2) relational orientation, and (3) gender role beliefs. It is also important to consider the prevalence of technology use and utilize strategies that allow for the incorporation of these into interventions. Further, community integration is essential for the maintenance of culturally integrated interventions program.

References

1. Amaro, H., Raj, A., & Reed, E. (2001). Women's sexual health: The need for feminist analyses in public health in the decade of behavior. *Psychology of Women Quarterly, 25*(4), 324–334.
2. Aronowitz, T., Rennells, R. E., & Todd, E. E. (2005). Heterosocial behaviors in early adolescent African American girls: The role of mother–daughter relationships. *Journal of Family Nursing, 11*(2), 122–139.
3. Belgrave, F. Z. (2009). *African American girls: Reframing perceptions and changing experiences*. London: Springer.
4. Belgrave, F. Z., Corneille, M., Hood, K., Foster-Woodson, J., & Fitzgerald, A. (2010). The impact of perceived group support on the effectiveness of an HIV prevention program for African American women. *Journal of Black Psychology, 36*(2), 127–143.
5. Belgrave, F. Z., Marin, B., & Chambers, D. B. (2000). Cultural, contextual, and intrapersonal predictors of risky sexual attitudes among urban African American girls in early adolescence. *Cultural Diversity & Ethnic Minority Psychology, 6*(3), 309–322.
6. Belgrave, F. Z., Rawls, V., Butler, D., & Townsend, T. (2008). *Sisters of nia: A cultural curriculum to empower African American girls*. Champaign: Research.

7. Belgrave, F. Z., Reed, M. C., Plybon, L. E., Butler, D. S., Allison, K. W., & Davis, T. (2004). Sisters of Nia: A cultural program for African American girls. *Journal of Black Psychology, 30*, 329–343.
8. Bielby, W. T., & Bielby, D. D. (1992). I will follow him: Family ties, gender-role beliefs, and reluctance to relocate for a better job. *American Journal of Sociology, 97*(5), 1241–1267.
9. Black, M. M., Ricardo, I. B., & Stanton, B. (1997). Social and psychological factors associated with AIDS risk behaviors among low-income, urban, African American adolescents. *Journal of Research on Adolescence, 7*(2), 173–195.
10. Boyd, K., Ashcraft, A., & Belgrave, F. (2006). The impact of mother–daughter and father–daughter relationships on drug refusal self-efficacy among African American adolescent girls in urban communities. *Journal of Black Psychology, 32*(1), 29–42.
11. Buckley, T., & Carter, R. (2005). Black adolescent girls: Do gender role and racial identity: Impact their self-esteem? *Sex Roles, 53*(9–10), 647–661.
12. Burlew, K., Neely, D. K., Johnson, C., Hucks, T., Purnell, B., Butler, J., et al. (2000). Drug attitudes, racial identity, and alcohol use among African American adolescents. *Journal of Black Psychology, 26*(4), 402–420.
13. Carli, L. L. (2001). Gender and social influence. *Journal of Social Issues, 57*(4), 725–741.
14. Centers for Disease Control and Prevention CDC. (2010). *Sexually transmitted disease surveillance*. Atlanta: Department of Health and Human Services.
15. Community-Campus Partnerships for Health (2013). *Community-based participatory research*. Retrieved on July 19, 2013 from http://depts.washington.edu/ccph/commbas.html
16. Corneille, M. A., & Belgrave, F. Z. (2007). Ethnic identity, neighborhood risk, and adolescent drug and sex attitudes and refusal efficacy: The urban African American girls' experience. *Journal of Drug Education, 37*(2), 177–190.
17. DiClemente, R. J., & Wingood, G. M. (1995). A randomized controlled trial of an HIV sexual risk reduction intervention for young African–American women. *Journal of the American Medical Association, 274*(16), 1271–1276.
18. DiClemente, R. J., Wingood, G. M., Harrington, K. F., Lang, D. L., Davies, S. L., & Hook, E. (2004). Efficacy of an HIV prevention intervention for African American adolescent girls: A randomized controlled trial. *JAMA, 292*(2), 171–179.
19. Eccles, J. S. (1987). Gender roles and women's achievement-related decisions. *Psychology of Women Quarterly, 11*(2), 135–172.
20. Hacker, D., Belgrave, F. Z., Grisham, J., Abrams, J., & Colson, D. (2013). Culturally integrated sex and prevention programs for substance abuse. Prevention programs for middle school students. In C. Clauss-Ehlers, Z. Serpell, & M. Weiss (Eds.), *Handbook of culturally responsive school mental health: advancing research, training, practice, and policy*. New York: Springer.
21. Israel, B. A., Schulz, A. J., Parker, E. A., & Becker, A. B. (2001). Community-based participatory research: Policy recommendations for promoting a partnership approach in health research. *Education for Health: Change in Learning & Practice, 14*(2), 182–197.
22. Kerrigan, D., Andrinopoulos, K., Johnson, R., Parham, P., Thomas, T., & Ellen, J. M. (2007). Staying strong: Gender ideologies among African–American adolescents and the implications for HIV/STI prevention. *Journal of Sex Research, 44*(2), 172–180.
23. Kreuter, M., Lukwago, S., Bucholtz, D., Clark, E., & Sanders-Thompson, V. (2003). Achieving cultural appropriateness in health promotion programs: Targeted and tailored approaches. *Health Education & Behavior, 30*, 133–146.
24. Lenhart, A., Purcell, L., Smith, A., & Zickuhr, K. (2010). Social media & mobile Internet use among teams and young adults. Millennials. Few Internet & American Life project.
25. Loder, T., & Hirsch, B. (2003). Inner-city youth development organizations: The salience of peer ties among early adolescent girls. *Applied Developmental Science, 7*, 2–12.
26. Marín, B. V. (2003). HIV prevention in the Hispanic community: Sex, culture, and empowerment. *Journal of Transcultural Nursing, 14*(3), 186–192.
27. Miller, K. S., Forehand, R., & Kotchick, B. A. (1999). Adolescent sexual behavior in two ethnic minority samples: The role of family variables. *Journal of Marriage and the Family, 61*(1), 85–98.

28. Millstein, S. G., & Moscicki, A. B. (1995). Sexually-transmitted disease in female adolescents: Effects of psychosocial factors and high risk behaviors. *Journal of Adolescent Health, 17*(2), 83–90.
29. Minkler, M., & Wallerstein, N. (2002). *Community-based participatory research for healthi.* From Process to outcomes. John Wiley & Sons.
30. Neumann, M. S., Johnson, W. D., Semaan, S., Flores, S. A., Peersman, G., Hedges, L. V., et al. (2002). Review and meta-analysis of HIV prevention intervention research for heterosexual adult populations in the United States. *Journal of Acquired Immune Deficiency Syndromes, 30,* 106–117.
31. Office of Behavioral and Social Sciences Research (2013). *Community-based participatory research.* Retrieved on July 19, 2013 from http://obssr.od.nih.gov/scientific_areas/methodology/community_based_participatory_research/
32. Phinney, J., & Kohatsu, E. (1997). Ethnic and racial identity development and mental health. In J. Schulenberg, J. Maggs, & K. Hurrelman (Eds.), *Health risks and developmental transitions in adolescence* (pp. 420–443). New York: Cambridge University Press.
33. Rose, A. J., & Montemayor, R. (1994). The relationship between gender role orientation and perceived self-competency in male and female adolescents. *Sex Roles, 31*(9–10), 579–595.
34. Rosenthal, L., Levy, S., & Earnshaw, V. A. (2012). Social dominance orientation relates to believing men should dominate sexually, sexual self-efficacy, and taking free female condoms among undergraduate women and men. *Sex Roles, 67*(11–12), 659–669.
35. Rossi, P. H., Lipsey, M. W., & Freeman, H. E. (2004). *Evaluation: A systematic approach.* Thousand Oaks: Sage Publications. New R.
36. St. Lawrence, J. S., Brasfield, T. L., Jefferson, K. W., Alleyne, E., O'Bannon, R. E., III, & Shirley, A. (1995). Cognitive-behavioral intervention to reduce African American adolescents' risk for HIV infection. *Journal of Consulting and Clinical Psychology, 63*(2), 221–237.
37. Townsend, T., & Belgrave, F. Z. (2000). The impact of personal identity and racial identity on drug attitudes and use among African American children. *Journal of Black Psychology, 26*(4), 21–36.
38. U.S. Department of Health and Human Services, Office of Minority Health. (2005). *What is cultural competency?* Retrieved July 10, 2011, from http://minorityhealth.hhs.gov/templates/browse.aspx?lvl=2&lvlID=11
39. Viswanathan, M., Ammerman, A., Eng, E., Gartlehner, G., Lohr, K.N., Griffith, D., et al. (2004). *Community-based participatory research: Assessing the evidence.* Evidence Report/Technology Assessment. No. 99 Publishing, Rockville
40. Whiteley, L. B., Brown, L. K., Swenson, R. R., Valois, R. F., Vanable, P. A., Carey, M. P., et al. (2011). African American adolescents meeting sex partners online: Closing the digital research divide in STI/HIV prevention. *Journal of Primary Prevention, 33*(1), 13–18.
41. Wolff, J. M., & Crockett, L. J. (2011). The role of deliberative decision making, parenting, and friends in adolescent risk behaviors. *Journal of Youth and Adolescence, 40*(12), 1607–1622.
42. Yancey, A. K., Kumanyika, S. K., Ponce, N. A., McCarthy, W. J., Fielding, J. E., Leslie, J. P., et al. (2004). Population-based interventions engaging communities of color in healthy eating and active living: A review. *Preventing Chronic Disease, 1,* 1–18.
43. Young, M. A., & Vazsonvi, A. T. (2011). Parents, peers, and risky sexual behaviors in rural African American adolescents. *The Journal of Genetic Psychology: Research and Theory on Human Development, 172*(1), 84–93.

Dynamics of Niche Construction in Models "Consumers–Renewable Resource" and "Consumers–Predators–Renewable Resource"

Faina S. Berezovskaya and Georgiy P. Karev

Abstract In this chapter a question of "how much over-consumption a renewable resource can tolerate" is addressed using mathematical models, where a consumer population compete for the common resource, can contribute to resource restoration, and is subject to attacks of predators. The bifurcation analysis of the systems shows that well-adapted predators can keep the system in a stable equilibrium even for "strong" prey over-consumption, when the initial system of resource–consumer goes to extinct. It means that predators may extend the domain of the total system coexistence.

1 Introduction: Description of Models

The identification of mechanisms responsible for the observed patterns of coexistence in populations whose survival is intimately connected to their ability to share a common resource is central to the study of ecological sustainability. The notion of niche construction provides but one way to organize and understand how populations can sustainably coexist with their resources. It is the goal of this paper to study this question using a simple resource-consumer framework.

The term "niche" was first introduced by Grinnell [7] in 1917 in his efforts to describe how an organism or a population responds to and competes for a common resource. The interactions of organisms or populations with available resources within their niche are not limited to consumption. Odling-Smee et al. [15]

F.S. Berezovskaya (✉)
Department of Mathematics, Howard University, Washington, DC 20059, USA
e-mail: fberezovskaya@howard.edu

G.P. Karev
National Centre for Biotechnology Information, Bethesda, MD 20894, USA
e-mail: karev@ncbi.nlm.nih.gov

B. Toni (ed.), *New Frontiers of Multidisciplinary Research in STEAM-H (Science, Technology, Engineering, Agriculture, Mathematics, and Health)*,
Springer Proceedings in Mathematics & Statistics 90, DOI 10.1007/978-3-319-07755-0_8,

referred to the notion of "niche construction" in situations where organisms not only adapt in response to environmental pressures (for example, consuming the resource in the most efficient manner) but in the process also modify the environment. These adaptive interactions of consumers with their environment re-shape the niche to the needs of the communities that share the resource consumer–producer (*N*–*R*) systems, where the individuals that compete for resources also contribute differentially to "increases in the size of the pie". The carrying capacity of *N*–*R* systems turns out to be a function of the adaptive interactions between resources and consumers. Niche is therefore not a static concept but is an adaptive system in itself, for example, equilibrium of the system. The ability to understand and predict possible directions in which the consumer–resource system may evolve is crucial in order to successfully achieve sustainable coexistence with common resources and to avoid "tragedy of common" (Hardin [8]).

It has been supposed that bifurcations in dynamical systems describing the *N*–*R* model can correspond to "tipping points" in complex adaptive systems, which in turn may signal upcoming crises.

For these purposes a simplified version of the model introduced by Krakauer et al. [13] was used, which is presented in the form

$$\begin{aligned} \frac{dN}{dt} &= N\left(c - \frac{bN}{R}\right) \\ \frac{dR}{dt} &= \gamma - \delta R + \frac{e(1-c)N}{R+N} \end{aligned} \tag{1}$$

In this model consumers $N(t)$ compete for common resource $R(t)$, which determines the carrying capacity of the population. A resource is suppose to be renewable in such a way that consumers not only consume the resource but also be able to contribute to its restoration, i.e., contribute to increase of the common population carrying capacity.

The per capita birth rate is equal to the rate c of the resource consumption. The per capita death rate is equal to $\frac{bN}{R}$, where b characterizes the efficiency of resource consumption. Resource $R(t)$ is restored naturally at constant rate $\gamma \geq 0$, deteriorates at the rate $\delta R (\delta > 0)$ and can be replenished by the activity of N. The rate of restoration of the common resource in response to the activity of individuals is modeled by the function $\frac{eN(1-c)}{R+N}$, where parameter $0 < e \leq 1$ denotes the proportion of total resource that is consumed or restored. As the number of consumers increases, the amount of resource R will increase or decrease depending on the value of the parameter $c \geq 0$.

The resource consumption/restoration parameter c is restricted to the interval $c \in [0, a], a \geq 1$, since within the frameworks of this model, the rate of niche-construction can neither be negative nor infinite. Letting $a = 1$ implies that the individuals in the population never consume more than they restore, making the population completely "altruistic". Letting $a > 1$ allows for the presence of over-consumers in the system, so $(1 - c)$ can take on negative values, which accounts for strictly consumerist behavior. The solutions to the equation for population growth always remain positive. Solutions for the equation for $\frac{dR}{dt}$ are positive when $c \in [0, 1 + \gamma/e)$.

Certainly, resource can be exhausted if individuals use it more than contribute to its restoration. A boundedness of a renewable resource is the principal distinction of the considering model of many others models, which describe consumer competition. The first problem of our interest is *what is the role of* ***over-consumption*** *in the dynamics of the system*. For this purpose in Sect. 2 we study the effects of the resource (over) consumption on the entire population, identifying all possible dynamical regimes that the population can go through with increasing of parameter c up to exhausting of its resources.

We evaluate what transitional regimes the population can go through over-consumption and extinction and describe the bifurcation boundaries; discussion of the problem how the boundaries can be used for forecasting of the system collapse are contained in [3, 10–12, 14].

Next, we "introduce" in the system predators ($P(t)$) which attack consumers–preys and study the common dynamics of the system "consumers–predators–renewable resource". The second problem of the interest is *what is the role of predators in the total system dynamics, and how predators can govern the coexistence of resource and consumers*. We study this problem in Sect. 3 using the model

$$\begin{cases} \frac{dN}{dt} = N\left(c - \frac{N}{R} - P\right) \equiv f_1(N, P, R), \\ \frac{dP}{dt} = \beta P(N - m) \equiv f_2(N, P, R), \\ \frac{dR}{dt} = \gamma - \delta R + \frac{e(1-c)N}{N+R} \equiv f_3(N, P, R) \end{cases} \quad (2)$$

Many works were devoted to studying predator–prey systems (see, for example, [3] and references herein). The second equation in (2) describing the predator dynamics is taken in the simplest (Volterra) form. It introduces to the model two additional positive parameters, β, m where $\beta \leq 1$ is the efficiency coefficient of transformation of preys to predators biomass, and βm is the mortality of predators. Notice that the parameter m characterizes the equilibrium level of coexistence of the predator–prey system. It was shown in [3] that in the predator–prey model with the logistic prey growth there exists such value $m = m^*$ that predators and preys can coexist in a stable equilibrium or stable oscillating for $m < m^*$, and predators get extinction if $m > m^*$. So, m is one of the most important model parameters in our analysis below, whereas β is fixed.

Interpretations and discussions of results are presented in the Sect. 4.

2 The model "Consumers–Renewable Resource"

2.1 *Simplification of Model (1); Equilibria*

System (1) has a singular point at the origin $O(N = 0, R = 0)$. To handle this singularity we consider the following system of equations, obtained from (1) via transformation

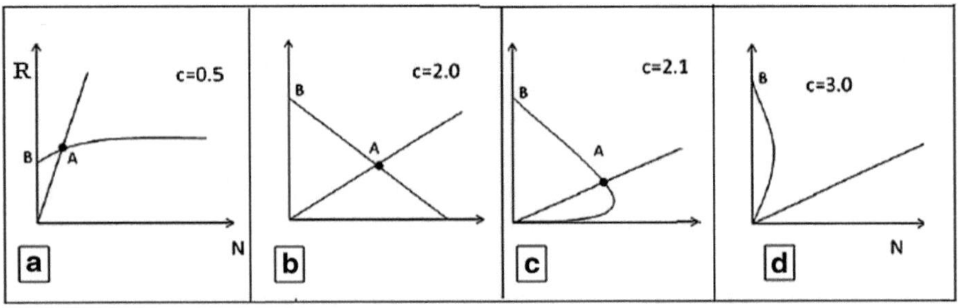

Fig. 1 Four types of placing of null-clines

$$R\,(N + R)\,d\tau = dt$$

$$\begin{cases} N' \equiv \frac{dN}{d\tau} = N\,(cR - N)\,(N + R) \equiv G_1\,(N, R)\,, \\ z' \equiv \frac{dR}{d\tau} = ((\gamma - \delta R)(\,N + R\,) + eN(\,1 - c\,))\,R \equiv G_2\,(N, R)\,, \end{cases} \tag{3}$$

The following statement holds.

Proposition 2.1 (1) *In the positive quadrant of* (N, R)*-plane systems* (3) *and* (1) *are topologically orbital equivalent everywhere except for the point* O*, which is an equilibrium point of system* (3) *for all parameter values but not for system* (1).

(2) *Trajectories of system* (3) *are bounded in the first quadrant.*

The statement (1) follows from [1], the statement (2) is proven in Sect. 2.7. Coordinates of equilibria of (3) satisfy the equations:

$$N\,(cR - N) = 0, \quad R\,((\gamma - \delta R)\,(N + R) + eN\,(1 - c)) = 0. \tag{4}$$

Proposition 2.2 *System* (3) *has*

(1) *Non-hyperbolic equilibrium point* $O(0, 0)$;
(2) *A saddle point* $B_2(N = 0, R = \gamma/\delta)$;
(3) *For* $0 \leq \frac{\gamma}{e} < \frac{c(c-1)}{c+1}$ *the system has nontrivial equilibrium*

$A_2\left(N = c\left(\frac{ce(1-c)}{(1+c)\delta} + \frac{\gamma}{\delta}\right), R = \frac{ce(1-c)}{(1+c)\delta} + \frac{\gamma}{\delta}\right)$ *(see Fig.* 1*).*

The point A_2 *is a stable topological node if* $\frac{\gamma}{e} > \frac{c(c-1)(c(c+1)+\delta(c+2))}{(c+1)^2(c+\delta)}$ *and unstable topological node if* $\frac{c(c-1)}{c+1} < \frac{\gamma}{e} < \frac{c(c-1)(c(c+1)+\delta(c+2))}{(c+1)^2(c+\delta)}$.

The Proof of Proposition is given in Sect. 2.5.

We use Proposition 2.2 to identify three parameter boundaries (see Fig. 4) that correspond to qualitatively different phase portraits of System (3):

$$O\gamma : \gamma = 0, \quad Nul : \frac{\gamma}{e} = \frac{c\,(c-1)}{c+1}, \quad H : \frac{\gamma}{e} = \frac{c\,(c-1)\,(c\,(c+1) + \delta\,(c+2))}{(c+1)^2\,(c+\delta)}$$

Crossing the boundary *Nul* from the bottom to the top is accompanied by the appearance of a positive node A_2; crossing the boundary $O\gamma$ leads to the appearance of a saddle B_2 in the first quadrant; the boundary H corresponds to changing of stability of equilibrium A_2, which is accompanied by appearance or disappearance of limit cycles in the phase plane (Andronov–Hopf bifurcations). Analysis of the model behavior in a neighborhood of equilibrium point A_2 with parameters close to the boundary H is performed in Sect. 2.3.

2.2 Structure of the Equilibrium Point at the Origin

The point $O(0,0)$ is the non-hyperbolic equilibrium of System (3), since both eigenvalues of the Jacobian matrix at the point O are equal to zero. We will apply the "blowing-up transformation" to analyze this point (for general aspects of this method see [4] and references within). We show that the orbit structures in a neighborhood of the point O depend on the parameters in the following way:

Proposition 2.3 *For any positive fixed values of parameters e and δ, the parameter half-plane ($\gamma \geq 0, c \geq 0$) of System* (3) *in a neighborhood of point O is divided into three domains of topologically different phase portraits* (*see Fig.* 2). *Boundaries between the domains are the lines $O\gamma : \gamma = 0$ and $K : \frac{\gamma}{e} = c - 1$ and $Nul : \frac{\gamma}{e} = \frac{c(c-1)}{c+1}$.*

Non-trivial asymptote of the orbits tending to O is

$$N = \frac{R}{\gamma + e\,(1-c)}\,(1 + o(1)), \quad \frac{\gamma}{e} > c - 1 > 0$$

The proof of the Proposition is given in Sect. 2.6.

2.3 Hopf Bifurcations and Separatrix Bifurcations in the Model

Let $H : \frac{\gamma}{e} = \frac{c(c-1)(c(c+1)+\delta(c+2))}{(c+1)^2(c+\delta)}$ be a curve in the parametric space (γ, c) at fixed (δ, e); if the point (γ, c) belongs to the curve H, then the equilibrium A_2 has coordinates $A_2\left(N = \frac{\gamma c}{c(1+c)+(2+c)\delta}, R = \frac{\gamma}{c(1+c)+(2+c)\delta}\right)$.

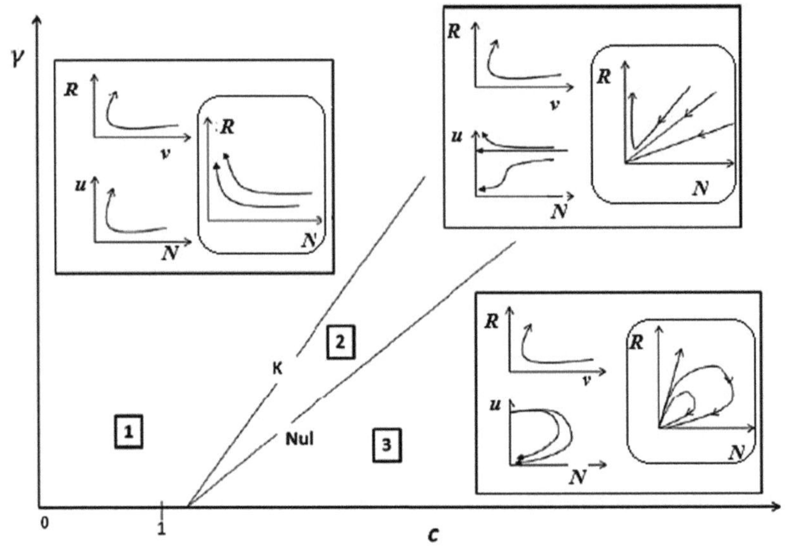

Fig. 2 Bifurcation diagram of the equilibrium point O, shown through blowing-up transformations where $u = R/N, v = N/R$

Proposition 2.4 *The line H is the Hopf boundary such that equilibrium A_2 changes stability when parametric point (γ, c) crosses the boundary H.*

(1) *Let $\delta > 5+\sqrt{24}$. Then there exists δ^* such that the "generalized Hopf" (Bautin) bifurcations of co-dimension 2, a change of stability of equilibrium A_2 with appearance of two limit cycles happens at parameter values*

$$\chi_\pm \left(\delta^*, c^*_\pm = \frac{\delta^*-1\pm\sqrt{1-10\delta^*+\delta^{*2}}}{2}, \right.$$
$$\left. \gamma^*_\pm = e^{\frac{c^*_\pm\left(c^*_\pm-1\right)\left(c^*_\pm\left(c^*_\pm+1\right)+\delta^*\left(c^*_\pm+2\right)\right)}{\left(c^*_\pm+1\right)^2\left(c^*_\pm+\delta^*\right)}} \right).$$

(2) *The supercritical Hopf bifurcation of co-dimension 1 is realized in the equilibrium A_2 with crossing H for $\delta > \delta^*$ if $\delta > 5+\sqrt{24}$ (see the line H^-, Fig.* 3).

(3) *The subcritical Hopf bifurcation of co-dimension 1 is realized in the equilibrium A_2 with crossing H for $5+\sqrt{24} < \delta < \delta^*$ or for any $\delta \in \left(0, 5+\sqrt{24}\right)$ (see the line H^+, Fig.* 3). *Schematic bifurcation diagram of Hopf bifurcations is shown in Figs.* 3c *and* 4.

The proof of the Proposition is performed in Sect. 2.5.

Due to the Bautin Theorem (see, for example, [14]) each of the parametric points χ_+ and χ_- has in its vicinity the line C_+ and C_- correspondingly of one more bifurcation of co-dimension 1 corresponding to a non-degenerate fold bifurcation of

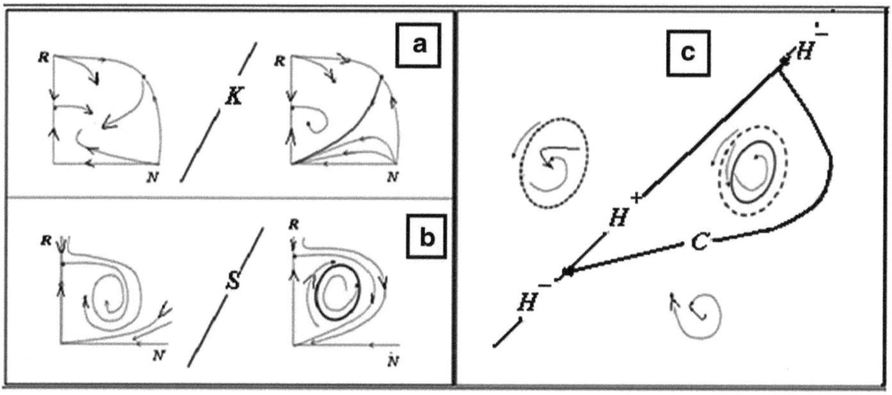

Fig. 3 Schematic bifurcation diagrams of heteroclinics and Hopf bifurcations in system (3): (**a**) appearance/disappearance of attractive parabolic sector in the vicinity of the origin; (**b**) appearance/disappearance of unstable limit cycle containing A_2 inside itself; (**c**) Hopf bifurcations

the cycles. Numerical analysis showed that the bifurcation lines C_+ and C_- merge and compose a unique curve C (see Fig. 3c). We call to the curve C as bifurcation boundary of *saddle-node cycles*. The parameter domain which is bounded by H^+ and C contains two limit cycles, unstable and stable, and unstable equilibrium point A_2. Crossing the boundary C limit cycles collide and disappear.

In Fig. 4 H^- is the boundary between Domains 3 and 4, it corresponds to subcritical Hopf bifurcation of equilibrium A_2, H^+ is the boundary between Domains 3 and 6, it corresponds to supercritical Hopf bifurcation of A_2. C is the boundary between Domains 6, which contains two limit cycles, and Domain 4 where the system has no limit cycles.

System (3) demonstrates *separatrix* bifurcations. One of such bifurcations corresponds to the appearance of an attracting parabolic sector in a positive neighborhood of equilibrium O_2 (see Fig. 2). It happens with parameter values belonging to the boundary K, for which the separatrix of the "infinite" equilibrium in the Poincaré coordinates $u = 1/N, v = R/N$ reaches O_2 (see Figs. 3a, and 4a, the boundary between Domains 1 and 2).

An unstable limit cycle that contains stable equilibrium point A_2 inside itself (see Fig. 4a, Domain 3) we were able to identify numerically. This cycle appears from the heteroclinics composed by the separatrices of the saddle point B_2 and the saddle-node point O_2 (see Fig. 3b). The curve S is the parameter boundary which corresponds to this bifurcation. In Fig. 4a curve S is the boundary between Domains 2 and 3.

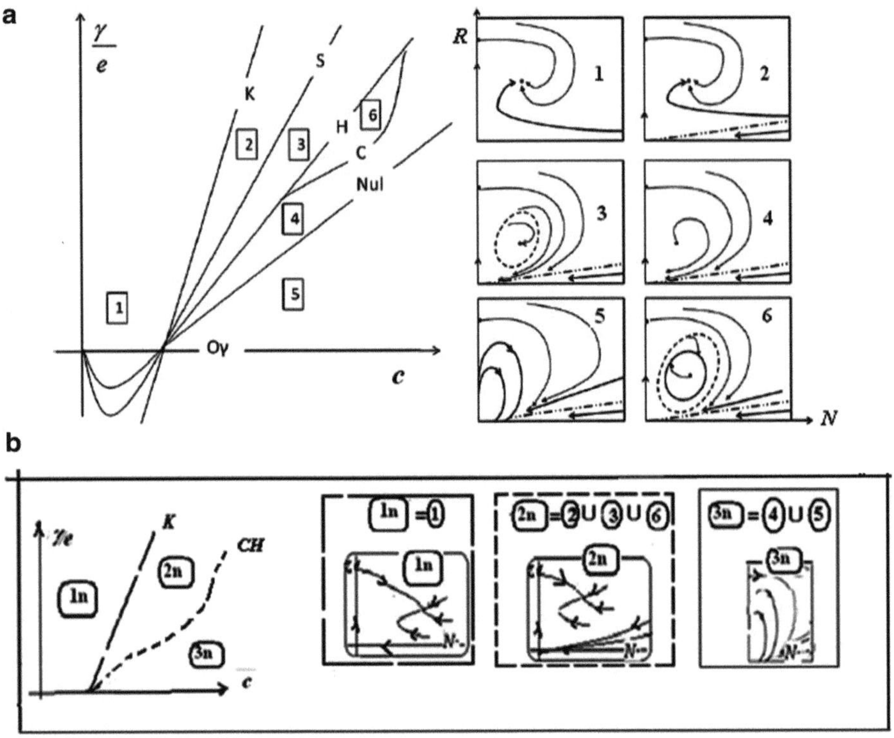

Fig. 4 Bifurcation diagram of system (1) for non-negative parameter values (c,γ) and phase variables (N,R) at fixed $e=1$ and positive δ. **a**: The partition of (c,γ)-parameter portrait into six Domains of topologically non-equivalent phase portraits (see Theorem 2.1 for details). The boundaries between Domains, $K, S, H^{+}, H^{-}, C, Nul$ are described in Table 2, attracting sets and boundaries of the basins in every Domain are described in Table 1 and Theorem 2.1. **b**: The partition of (c,γ)-parameter portrait into three Domains with different number of stable non-trivial modes: the single globally stable equilibrium A_2 in Domain 1n; the locally stable equilibrium A_2 or limit cycle in Domain 2n; there is no stable non-trivial equilibria or cycles in Domain 3n. Line K is the boundary between Domains 1n and 2n, $CH = H^{-} \cup C$ is the boundary between Domains 2n and 3n (see the lower row of Table 2)

2.4 Bifurcation Diagram of the System "Consumers–Renewable Resource"

The results of our analysis are summarized in the following statement.

Theorem 2.1 *For any fixed parameter* $0 < e \leq 1$ *the bifurcation diagram of system* (1) *consists of six Domains of qualitatively (topologically) different parameter-phase portraits, presented in Fig.* 4a, *namely*:

- *Domain of monostability 1, there exist a single non-trivial equilibrium* $A_2\left(N = c\left(\frac{ce(1-c)}{(1+c)\delta} + \frac{\gamma}{\delta}\right), R = \frac{ce(1-c)}{(1+c)\delta} + \frac{\gamma}{\delta}\right)$ *which is globally stable*;

Table 1 Domains of qualitatively different phase behaviors in system (3), attracting sets and boundaries of the basins

	Stability	Attracting sets	Boundary of basin
1	Monostability	Equilibrium A_2	Global stability
2	Bistability	Equilibrium A_2, origin	Separatrix $z = \frac{N}{c+\delta}$, $N >> 1$
3	Bistability	Equilibrium A_2, origin	Unstable limit cycle
4	Monostability	Origin	Global stability
5	Elliptic sector	Origin	Elliptic sector
6	Bistability	Stable limit cycle, origin	Unstable limit cycle, point A_2

- *Domains of bistability 2 and 3; in domain 2 equilibrium A_2 shares basins with the equilibrium O_2 and separatrix of O_2 serves as a boundary of their basins; in Domain 3 an unstable limit cycle containing inside equilibrium A_2 serves as a boundary of the basin of A_2;*
- *Domains of monostability 4, where only equilibrium O_2 is globally stable;*
- *Domain 5, where the system has no stable modes and an elliptic sector exists in a neighborhood of O_2*
- *Domain 6 of bistability, where unstable limit cycle is a common boundary of stable equilibrium O_2 and a stable limit cycle.*

Structures of basins are described in Table 1, boundaries between Domains, K, S, H^+, H^-, C, Nul, are described in Table 2. Schematically presented bifurcation diagram is shown in Fig. 4a. For any fixed value of parameter e parameter space (γ, c, δ) of the model is divided into six domains of different phase behaviors.

For further analysis and interpretations we schematically present the modified partition of (c, γ)-parameter portrait of system (3) in panel **b** of Fig. 4. This partition contains *three* domains corresponding to different numbers of stable non-trivial modes in (N, R)-phase portraits. Equilibrium A_2 is globally stable in Domain 1n of **b**-panel as well as in Domain 1 of **a**-panel. Domain 2n of **b**-panel is the union of Domains 2, 3 and 6 of **a**-panel; it is the region of bistability; there exist stable equilibrium O_2 and other stable manifold, equilibrium A_2 or a stable limit cycle, containing unstable point A_2. Domain 3n of **b**-panel is the union of Domains 4 and 5 of **a**-panel; there are no stable non-trivial equilibria or cycles. The boundaries of Domains 1n, 2n and 3n are shown in Fig. 4b, described the captions and presented in Table 2.

2.5 *Proof of Propositions 2.2 and 2.4*

1) Jacobian $J(x, z)$ for system (3) is of the form

$$J(N, R) = \begin{pmatrix} (cR - N)(N + R) + N(cR - 2N) & N((c-1)N + 2cR) \\ R(\gamma - \delta R + e(1-c)) & eN(1-c) + \gamma(N + 2R) - \delta R(2N + 3R) \end{pmatrix}$$

Table 2 Boundaries of Domains, corresponding to bifurcations in system (3)

Domains	Boundary	Bifurcation
Domains 1, 2	$\boldsymbol{K}$:$\gamma/e = c-1$; denote $c_b = 1+\gamma/e$ for fixed γ, e, δ	Appearance of stable parabolic sector in a positive neighborhood of equilibrium O
Domains 2, 3	$\boldsymbol{S}$ (no analytical description) corresponds to the heteroclinics composing by separatrices of equilibria B_2 and O_2	Appearance of unstable limit cycle containing stable equilibrium A_2 inside itself
Domains 3, 4	$\boldsymbol{H}^+ : \frac{\gamma}{e} - \frac{c(c-1)(c(c+1)+\delta(c+2))}{(c+1)^2(c+\delta)} = 0$; the 1st Lyapunov value is positive; δ is small $c = c_h$ is a root of the equation for that parameters	Subcritical Hopf bifurcation: Stable equilibrium A_2 becomes unstable due to merging with unstable limit cycle
Domains 3, 6	$\boldsymbol{H}^- : \frac{\gamma}{e} - \frac{c(c-1)\left(c(c+1)+\delta(c+2)\right)}{(c+1)^2(c+\delta)} = 0$ the 1st Lyapunov value is negative; δ is large $c = c_h$ is the root of the equation for that parameters.	Supercritical Hopf bifurcation: equilibrium A_2 loses stability producing stable limit cycle
Domains 6, 4	$\boldsymbol{C}$ (no analytical description) corresponds to merging and disappearing of stable and unstable limit cycles. c=c_c: for fixed e, δ (c_c,γ) $\in \boldsymbol{C}$	Saddle-node bifurcations of limit cycles
Domains 4, 5	$\boldsymbol{Nul}$: $\frac{\gamma}{e} - \frac{c(c-1)}{1+c} = 0, c \geq 1$,$c = c_0$ is the root of the equation	Merging of equilibria A_2 and O_2
Domains $2_n = 3 \cup 6$, 4	$\boldsymbol{CH}$: $\{c = c_{ch}, \gamma)$, where $c_{ch} = (c_0 \vert (c,\gamma) \in \boldsymbol{C}), c_{ch} = (c_h \vert (c,\gamma) \in \boldsymbol{H}^-)$	Disappearance of non-trivial stable modes

In $B_2(N = 0, R = \gamma/\delta)$ one has $J\ (0, \gamma/\delta) = \begin{pmatrix} c(\gamma/\delta)^2 & 0 \\ e\,(1-c)\,\gamma/\delta & -\gamma^2/\delta \end{pmatrix}$.

Thus, B_2 is a saddle whose eigenvalues are $\lambda_1 = c(\gamma/\delta)^2, \lambda_2 = -\gamma^2/\delta$.

In $A_2\left(c\left(\frac{ce(1-c)}{(1+c)\delta} + \frac{\gamma}{\delta}\right), \frac{ce(1-c)}{(1+c)\delta} + \frac{\gamma}{\delta}\right)$., where $N = cR$, one has

$$J\ (A_2) = \begin{pmatrix} -c\,(1+c)\,R^2 & c^2\,(1+c)\,R^2) \\ R\,(e\,(1-c)+\gamma-\delta R) & -R\,(-\gamma+(2+c)\,\delta R) \end{pmatrix}$$

So, $Det(J(A_2)) = c\delta(1+c)^2R^4 \geq 0$, $Trace(J(A_2)) = R(\gamma - (c(c+1)^2 + \delta\ (2+c)R)$.

The first inequality guarantees that equilibrium A_2 is a topological node. Eigenvalues of A_2 become imaginary when $Trace(J(A_2)) = 0$. A simple algebra shows that the line $H : \frac{\gamma}{e} = \frac{c(c-1)(c(c+1)+\delta(c+2))}{(c+1)^2(c+\delta)}$ is the boundary of the Hopf bifurcation in the system. For verifying a "direction" of Hopf bifurcation (sub- or super-critical one) with crossing the H-boundary we compute the first Lyapunov quantity L_1 [1, 2] with parameter values belonging to H. We found that (up to a positive factor)

$$L_1 \cong \frac{(ec\,(c-1))^6 \delta \left(c^2 - c\,(\delta - 1) + 2\delta\right)}{(c+1)^{10}(c+\delta)^5}.$$

Denote $\psi(c,\delta) \equiv c^2 + c(1-\delta) + 2\delta$. Evidently, $L_1 > 0$ if $\psi(c,\delta) > 0$, which corresponds to a sub-critical Hopf bifurcation, and $L_1 < 0$ if $\psi(c,\delta) < 0$, which corresponds to a super-critical Hopf bifurcation, under condition that $\frac{d(Trace(J(A_2)))}{d(\gamma/e)} \neq 0$ [14].

For the system considered

$$\frac{d\,(Trace\,(J\,(A_2)))}{d\,(\gamma/e)} = \frac{1}{\delta^2}\left(\frac{ce\left(c^3 + c(\delta-1) - 2\delta\right) - (c+1)^2(c+\delta)\gamma}{(c+1)} - (c+\delta)\left(ce\,(1-c) + \gamma\,(1+c)\right)\right),$$

which vanishes for $\frac{\gamma}{e} = \frac{c\left(2c^3 + c(\delta-2) - 3\delta + c^2\delta\right)}{2(c+1)^2(c+\delta)}$ if $(\gamma, c) \in H$.

One can see that $\psi(c,\delta)$ and consequently L_1 are positive for $\delta \in \left(0.5 + \sqrt{24}\right)$ and for $\delta > \delta* \in \left(5 + \sqrt{24}, \infty\right)$ if $c \in (c_+^*, \infty), \gamma > \gamma_+^*$ and is negative for $\delta > \delta* \in \left(5 + \sqrt{24}, \infty\right)$ if $c \in (c_-^*, c_+^*), \gamma_-^* < \gamma < \gamma_+^*$

The first Lyapunov value L_1 vanishes for $(c = c_-^*, \gamma = \gamma_-^*)$, $(\ c = c_+^*, \gamma = \gamma_+^*)$. Notice that

$$\left.\frac{dL_1}{dc}\right|_{\delta=\delta^*, c=c_\pm^*} = \frac{\left(e\,(c_\pm^* - 1)\,c_\pm^*\right)^6 \delta^* \sqrt{1 - 10\delta^* + \delta^{*2}}}{\left(c_\pm^* + 1\right)^{10}\left(c_\pm^* + \delta^*\right)^5}$$

does not vanish for the corresponding c, δ. The sign of the second Lyapunov value [1, 2, 14] for $c = c_\pm^*, \delta = \delta*, \gamma = \gamma_{\pm-}^*$ coincides with the sign of the function:

$$L_{2t} = -4 + 28\delta^* - 55\delta^{*2} + 7\delta^{*3} + \left(4 - 20\delta^* + 7\delta^{*2}\right)\sqrt{1 - 10\delta^* + \delta^{*2}}$$

which is always positive. All the conditions of Bautin theorem [2, 14] are fulfilled.

Propositions 2.2 *and* 2.4 *are proven.*

2.6 Structure of Non-hyperbolic Equilibrium in the Origin: Proof of Propositions 2.3

For studying the structure of non-hyperbolic point $O(0,0)$ we use blowing-up method developed in [4, 5].

System (3) has two integral manifolds $R=0, N\geq 0$ and $N=0, R\geq 0$. It was shown in [4] that two blowing-up transformations

$$(N,R)\to(N,u=R/N),\quad N\neq 0 \tag{5}$$

and

$$(N,R)\to(v=N/R,R),\quad R\neq 0 \tag{6}$$

allow us to reveal the structure of a neighborhood of the point $O(0,0)$ in the first quadrant (see Fig. 2). Applying (5) to System (3) and letting $d\tau\to Nd\tau$, we obtain the system

$$\begin{aligned} N' &\equiv \tfrac{dN}{d\tau} = -N^2\left(1+u\,(c-1)-cu^2\right)\\ u' &\equiv \tfrac{du}{d\tau} = u\,(e\,(1-c)+\gamma+\gamma u)+Nu\,(1-\delta+(c+\delta-1)u-cu^2) \end{aligned} \tag{7}$$

It has equilibria on u-axis, $O_1(N=0,u=0)$ with eigenvalues $\lambda_1=0$, $\lambda_2=(1-c)e+\gamma$ and

$O_u(N=0,u=(e(c-1)-\gamma)/\gamma$ with eigenvalues $\lambda_1\,(O_u) = \frac{(1-c)(ce+\gamma(1-c))}{\gamma^2}e$, $\lambda_2\,(O_u) = 0$. According to [1], in the first quadrant point O_1 has a positive saddle sector if $\frac{\gamma}{e}<c-1$and an attracting parabolic sector if $\frac{\gamma}{e}>c-1>0$; point O_u has positive u-coordinate only for $\frac{\gamma}{e}>c-1>0$; in this case it has a saddle sector in its positive neighborhood if $c-1<\frac{\gamma}{e}<\frac{c(c-1)}{c+1}$, and a repelling parabolic sector if $\frac{c(c-1)}{c+1}<\frac{\gamma}{e}$.

Applying (6) to System (3) and letting $d\tau\to Rd\tau$we obtain the system

$$\begin{aligned} v' &\equiv \tfrac{dv}{d\tau} = -v\,(\gamma+(e\,(1-c)+\gamma)v+R^2(\delta-c+((\delta-c+1)v+v^2)\\ R' &\equiv \tfrac{dR}{d\tau} = R\,(\gamma+(e\,(1-c)+\gamma)v)-\delta R^2\,(1+v) \end{aligned} \tag{8}$$

Equilibrium point $O_2(v=0,R=0)$ of system (8) for any $\gamma>0$ has a saddle sector for positive R. Combining obtained results we have

1) For $0<\frac{\gamma}{e}<c-1$ systems (7) and (8) have only one equilibrium point at the axis, O_1 and O_2 correspondingly; each of these points has a saddle sector in its positive neighborhood (see Fig. 2);
2) For $c-1<\frac{\gamma}{e}<\frac{c(c-1)}{c+1}$ equilibrium O_1 has a positive attracting parabolic sector, equilibrium O_u as well as equilibrium O_v have a positive saddle sector;

3) For $\frac{\gamma}{e} > \frac{c(c-1)}{c+1}$ equilibrium O_1 has a positive attracting parabolic sector, equilibrium O_u has a positive repelling parabolic sector, and O_v has a positive saddle sector.

Using the results of [4] we can state: in Domain 1, $0 < \frac{\gamma}{e} < c - 1$, equilibrium O is a saddle for positive R, N; in Domain 2, $c - 1 < \frac{\gamma}{e} < \frac{c(c-1)}{c+1}$, equilibrium O is a saddle-node in the first quadrant; in Domain 3, $\frac{\gamma}{e} > \frac{c(c-1)}{c+1}$, equilibrium O contains elliptic sector for positive R, N. The results of our analysis are summarized in Fig. 2 and completely prove all statements of the Proposition.

2.7 Equilibria "At Infinity": Proof of the Boundedness of the Model

For study the structure of equilibrium points "at infinity" we use the Poincaré sphere method [1, 2].

(a) The change of variables

$$(N, R) \to (u = 1/N, v = R/N), \quad N \neq 0 \tag{9}$$

and

$$dt = u^2 d\tau \tag{10}$$

transforms system (3) to

$$\begin{aligned} \frac{dw}{d\tau} &= -w(1+u)(-1+cu), \\ \frac{du}{d\tau} &= -u\left(-1+(c+\delta)u^2+(-1+c)ew-\gamma w+u(-1+c+\delta-\gamma w)\right) \end{aligned} \tag{11}$$

Non-negative equilibrium points of system (11) in the axis u are $(w=0, u=0)$, which an unstable node with eigenvalues is $\lambda_1 = \lambda_2 = 1$, and $(w=0, u=1/(c+\delta))$, which is a non-trivial saddle with the eigenvalues is $\lambda_1 = \delta(c_1+\delta+1)/(c_1+\delta)^2$, $\lambda_2 = -(c_1+\delta+1)/(c_1+\delta)$; the separatrix of the saddle stays in a bounded domain of the first quadrant.

(c) Making the transformation

$$(N, R) \to (u = 1/R, w = N/R), \quad R \neq 0 \tag{12}$$

and (10), we obtain the system of equations:

$$\begin{aligned} \frac{du}{d\tau} &= u\left(\delta\,(1+w) - \gamma u((1+w) - (ecw-1)\right), \\ \frac{dw}{d\tau} &= -u\left(u+u^2 - \delta\,(1+w) + \gamma u\left(1 + ecu + u - c\,(1+u+e\gamma cuw)\right)\right) \end{aligned} \tag{13}$$

All equilibria of system (13) with $w \neq 0$ are topologically equivalent to corresponding equilibria of system (11). The equilibrium $(u=0, w=0)$ has positive eigenvalues $\lambda_1 = \delta, \lambda_2 = c$, so it is an unstable node.

We got that at the equators of the Poincare sphere non-negative equilibria are repelling; axes N, R are integral manifolds of system (14). Thus every trajectory, which is beginning inside the first quadrant does not leave it with $t \to \infty$ (see Fig. 3a).

Statement (2) *of Proposition 2.1 is proven.*

3 Model "Consumers–Predators–Renewable Resource"

3.1 Simplifications of the Model

System (2) [as well as system (1)] has a singularity at $z = 0$. To handle this singularity, we consider the system of equations, obtained from (2) via transformation $dt \to (N+z)\,z\,d\tau$:

$$\begin{cases} \frac{dN}{d\tau} = N\left((c-P)\,R - N\right)(N+R) \equiv F_1\,(N,P,R)\,, \\ \frac{dP}{d\tau} = \beta P\,(N-m)\,R\,(N+R) \quad \equiv F_2\,(N,P,R)\,, \\ \frac{dR}{d\tau} = R\left((\gamma - \delta R)(N+R) + e(1-c)N\right) \equiv F_3\,(N,P,R)\,, \end{cases} \tag{14}$$
$$F_1 = (N+R)\,f_1\,(N,P,R)\,, F_2 = R\,(N+P)\,f_2\,(N,P,R)\,, F_3 = R\,(N+R)\,f_3\,(N,P,R)$$

where N, P, R are scaling amounts of preys–consumers, predators, and common resource, γ, c, m are "free" parameters and β, e, δ are "fixed" parameters of the system.

The following statement holds.

Proposition 3.1 *Systems* (14) *and* (2) *are topologically orbitally equivalent inside the positive* (N, P, R)*-octant everywhere except for the plane* $R = 0$, *which is the plane of non-isolated equilibria for system* (14) *with all parameter values.*

In what follows we study behaviors of system (14) depending on positive parameters (c, m) while the parameters β, δ, e are fixed. More exactly, we construct and study (c, m)-cuts of the complete (γ, m, c)-bifurcation diagram depending on values of γ. We divide the parameter space into domains of different numbers of stable non-trivial modes in (N, R)-phase portraits of system (14) (similar to the partition of the phase-parameter portrait of system (3), see Fig. 4b).

3.2 "Predator-Induced" and "Predator-Free" Equilibria of the Model (14): Bifurcation Diagram in (c, m)-Plane

Coordinates of equilibria of the system satisfy equations:

$$\begin{aligned} &N\left((c-P)\,R-N\right)=0,\\ &PR\,(N-m)=0,\\ &R\left((\gamma-\delta R)(N+R)+e(1-c)N\right)=0. \end{aligned} \tag{15}$$

Solving (15) we see that system (14) can have up to five equilibrium points. Three of them,

$O(0,\ 0,\ 0)$, $B\left(N=0, P=0, R=\frac{\gamma}{\delta}\right)$, $A\left(N=\left(\frac{c}{\delta}\left(\gamma+\frac{ce(1-c)}{1+c}\right), P=0,\right)\right.$ $\left.R=\frac{1}{\delta}\left(\gamma+\frac{ce(1-c)}{1+c}\right)\right)$ have the same N, R-coordinates as the points O_2, B_2, A_2 of system (3), i.e. have counterparts in the sense, that $N(B)=N(B_2)$, $R(B)=R(B_2)$, and $N(A)=N(A_2)$, $R(A)=R(A_2)$. In what follows we use the notations $N(A), P(A), R(A)$ for corresponding coordinates of the point $A(N, P, R)$

New, "predator-induced" equilibria of system (14), $C(N*, p*, z*)$, have coordinates $N*=m$, $P*\ =\ c-\frac{m}{z*}$, $R=z*$ where $z*$ satisfies the quadratic equation

$$\delta R^2-(\gamma-\delta m)\,R-m\,(e\,(1-c)+\gamma)=0. \tag{16}$$

Let us denote $z^+=\frac{\gamma-\delta m+\sqrt{D}}{2\delta}$, $z^-=\frac{\gamma-\delta m-\sqrt{D}}{2\delta}$ where $D\equiv(\gamma+\delta m)^2+4e\delta m$ $(1-c)$. Thus, system (14) can have up to two "predator-induced" equilibria $C^+(m, P^+, z^+), C^-(m, P^-, z^-)$ with $P^{\pm}\ =\ c-\frac{m}{z^{\pm}}$. The domain Δ where C-equilibria are defined is given by the condition $D\geq 0$. The boundary of the domain Δ is defined by the equation $D=0$ and consists of two branches, where m_Δ^+, m_Δ^- for fixed γ, e, δ are the curves in (c, m)-plane (see Fig. 5)

$$\begin{aligned} m_\Delta^+ : m &= \frac{2\sqrt{e\,(c-1)\,(e\,(c-1)-\gamma)}+(2e\,(c-1)-\gamma)}{\delta},\\ m_\Delta^- : m &= \frac{-2\sqrt{e\,(c-1)\,(e\,(c-1)-\gamma)}+(2e\,(c-1)-\gamma)}{\delta}. \end{aligned}$$

The domain Δ is bounded by the coordinate axes and the boundary $m_\Delta^+\cup m_\Delta^-$. We show that there exist "wide" sub-domains of the domain Δ where equilibria C^+, C^- have positive coordinates.

Let us define the curve $m_{N(A)} : m=\frac{c}{\delta}\left(\frac{ce(1-c)}{(1+c)}+\gamma\right)\equiv N\,(A_2)$ with $0\leq c\leq c_0$, where c_0 is a positive root of the equation $\frac{\gamma}{e}=\frac{c(c-1)}{1+c}$, which defines the boundary

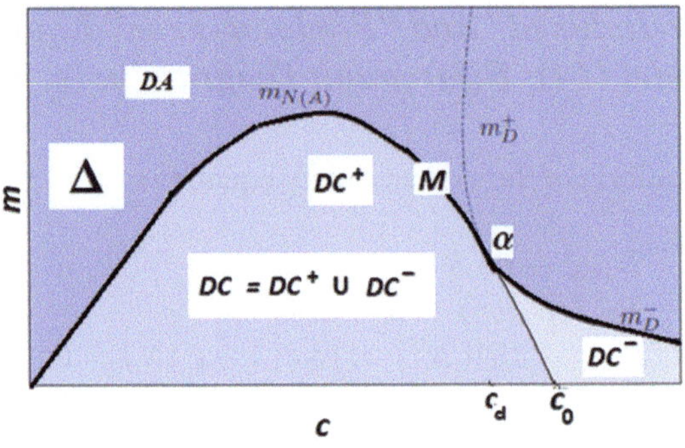

Fig. 5 For any positive γ: (**a**) the lines $m_\Delta^\pm$: $m = \frac{\pm 2\sqrt{(c-1)e((c-1)e-\gamma)}+(2(c-1)e-\gamma)}{\delta}$ are boundaries of the Domain Δ. The line $m_{N(A)}$: $m = \frac{c}{\delta}\left(\frac{ce(1-c)}{(1+c)} + \gamma\right), 0 < c < c_o$ and m_Δ^- have a common point $\alpha(c_d, m_d)$ where c_d is a positive root of the equation $\gamma/e = \frac{c_d(c_d-1)(c_d+2)}{(1+c_d)^2}$; (**b**) the line $M(c,m) = \begin{cases} m_{N(A)}, 0 \le c \le c_d, \\ m_\Delta^-, \quad c > c_d \end{cases}$ divides $0 \le (c,m)$-quadrant into domains *DA* and *DC* of qualitatively different phase behaviors of system (14), see Theorem 3.1, Propositions 3.2, 3.3

Nul in Fig. 4 (see Table 2). For $c = c_0$ the equilibria A_2 and O_2 of system (3) merge; A_2 leaves the positive quadrant if $c > c_0$ (Fig. 4a, D .5). Similarly the equilibrium $A\left(N = \left(\frac{c}{\delta}\left(\gamma + \frac{ce(1-c)}{1+c}\right), p = 0, R = \frac{1}{\delta}\left(\gamma + \frac{ce(1-c)}{1+c}\right)\right)\right.$ is positive only if $0 \le c \le c_0$.

Denote DC^+ the domain in positive (c,m)-quadrant bounded by the curve $m_{N(A)}$ and the interval $0 \le c \le c_0$ (Fig. 5). We have found that the curves $m_{N(A)}$ and the branch m_Δ^- of the boundary $D = 0$ have a common point $\alpha(c_d, m_d)$, where c_d is a positive root of the equation

$$\gamma/e - \frac{c\,(c-1)\,(c+2)}{(1+c)^2} = 0 \tag{17}$$

and the coordinate $m_d = \frac{c_d}{\delta}\left(\frac{c_d e\left(1-c_d\right)}{\left(1+c_d\right)} + \gamma\right)$ for any fixed β, γ, δ, e.

Next, let us define the continuous curve $M\,(c,m) = \begin{cases} m_{N(A)}, 0 \le c \le c_d, \\ \quad m_\Delta^-, \quad c > c_d \end{cases}$

Denote *DA*, *DC* the domains in non-negative (c,m)-quadrant that are placed, correspondingly, upper and lower than $M(c,m)$, $0 \le c < \infty$ (see Fig. 5). Evidently, $DC \subset \Delta$. We show below that the "predator-free" equilibrium A is stable in the domain *DA*, while the "predator-induced" equilibrium C^+ is stable in the domain *DC*.

In what follows we study behaviors of the model depending on positive parameters (c,m) while the parameters β, δ, e are fixed. More exactly, for any fixed γ we construct the phase-parametric portraits of model (14) and compare it with those of system (3). We pay the main attention to the analysis of *stable modes* of the model.

Denote DC^+ the domain bounded by the curve $m_{N(A)}$ and the interval $0 \le c \le c_0$, denote also $DC^- = DC \setminus DC^+$ (see Fig. 5). The following statements hold.

Proposition 3.2 *Let β, γ, δ, e be arbitrary fixed constants.*

1. *If $(c,m) \in DC$ then equilibrium $C^+(m, p^+, z^+)$ is positive;*
2. *Equilibrium $C^-(m, p^-, z^-)$ is positive if $(c,m) \in DC^-$ and $c > c_0$*

In the following statement we describe characteristics of N, R-coordinates of equilibrium C^+ and compare them with N, R-coordinates of equilibrium A.

Proposition 3.3 *Let $\gamma > 0$ and $(c,m) \in DC$ where equilibrium C^+ is stable. Then*

1) *the coordinate $p(C^+)$ is always positive and increases if the parameter c increases;*
2) *for any $0 < c < 1$ the coordinate $N(C^+) = m < N(A)$, $R(C^+) < R(A)$;*
3) *for $1 < c < c_d$ the coordinate $N(C^+) = m < N(A)$ but $R(C^+) > R(A)$;*
4) *if $c = c_d$ then $N(C^+) = N(A)$ and $R(C^+) = R(A)$;*
5) *if $c_d < c < c_0$ then two cases, $m = N(C^+) > N(A)$ and $m = N(C^+) \le N(A)$ are possible; in both cases $R(C^+) > R(A)$;*
6) *if $c \ge c_0$ then there is no predator-free equilibria.*

Propositions 3.2 and 3.3 are proven in Sect. 3.5.

Analysis of stability of equilibrium points of system (14) implies the following statement.

Theorem 3.2 (1) *Equilibrium $B\left(N = 0, P = 0, R = \frac{\gamma}{\delta}\right)$ is unstable; it has one positive and two negative eigenvalues: $\lambda_1(B) = \frac{c\gamma^2}{\delta^2}$, $\lambda_2(B) = -\frac{\beta m \gamma^2}{\delta^2}$, $\lambda_3(B) = -\frac{\gamma^2}{\delta}$;*

(2) *Equilibrium $A\left(N = \frac{c}{\delta}\left(\gamma + \frac{ce(1-c)}{1+c}\right), P = 0, R = \frac{1}{\delta}\left(\gamma + \frac{ce(1-c)}{1+c}\right)\right)$ is unstable for $(c,m) \in DC^+$ and stable for $(c,m) \in DA, 0 < c < c_{ch}$* (see Figs. 5 and 6);

(3) *Equilibrium $C^+(m, p^+, z^+)$ is positive and stable for $(c,m) \in DC$;*

(4) *If $(c,m) \in DC^-$ then the equilibrium $C^-(m, p^-, z^-)$ is positive and unstable; it has at least one positive eigenvalue;*

(5) *Equilibrium $O(0, 0, 0)$ is non-hyperbolic; the projection of its neighborhood to the plane $P = 0$ has the same local structure in (N, R)-plane as the local structure of the equilibrium $O_2(0, 0)$ of system* (3).

Theorem 3.2 is proven in Sect. 3.6.

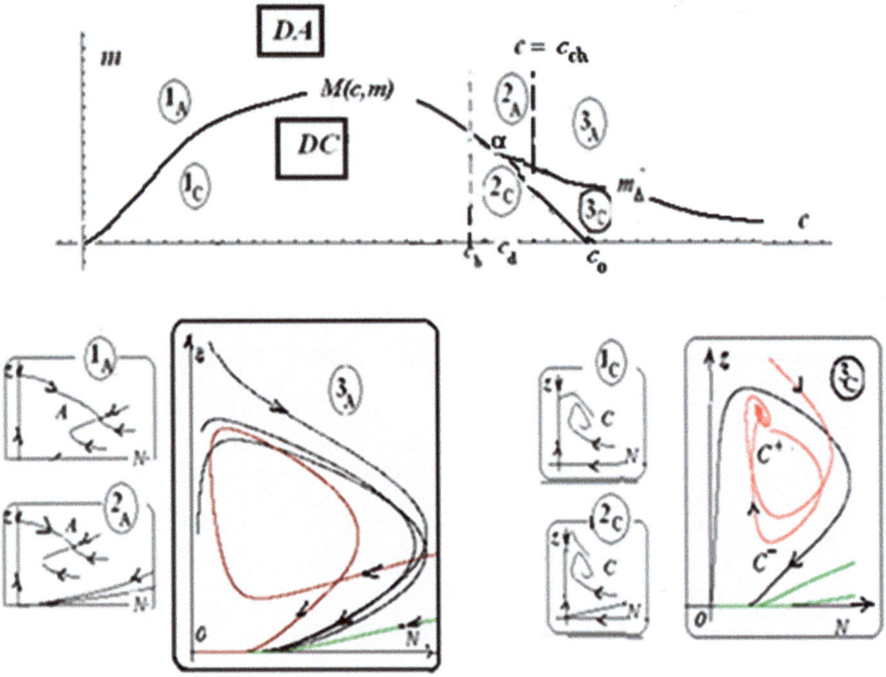

Fig. 6 Schematically presented (N,R)-cut of bifurcation diagram of system (14) in (m,c)-parameter plane (at fixed γ). The curve $M(c,\ m)$ divides the plane to the regions *DA* where predator-free equilibrium *A* is stable, and *DC* where predator-induced equilibrium C^+ is stable. Each region consists of three Domains with qualitatively different stable modes. The Domains are numerated by the integer and sub-index *A* in *DA*, sub-index *C* in *DC*. Domains 1A and 1C are divided by the curve $M(c,\ m)$ and bounded the lines $c=0, c=c_b$. Domain 2A is bounded by the curve $M(c,\ m)$ and the lines $c=c_b, c=c_{ch}$; $M(c,\ m)$ is the boundary of Domain 3A for $c>c_{ch}$. Domain 2C is bounded by the line $c=c_b$ and the curve $m_{N(A)}, c_d \le c \le c_0$, whereas Domain 3C is bounded by the curves $m_{N(A)}, c_d < c < c_0$ and $m_{\Delta^-}, c > c_0$ (see Table 2 for definitions). In model (14) (N,R)-coordinates of the equilibria A, C^+, C^- coincide if $c=c_d, m=m_d$. The (N,R)-phase portraits of the system are presented in the *lower panel*. Portraits 3A and 3C are constructed for $\beta=\gamma=\delta=e=1, c=3$, and $m=0.2, m=0.1$ correspondingly

3.3 Sketch of the Bifurcation Diagram of Model (14)

Theorem 3.1 together with the computer analysis of model (14) allows us to describe the stable modes of system (14) in the regions *DA* and *DC* under conditions that parameters β, δ, e are fixed (see Figs. 5 and 6)

Proposition 3.4 *For any fixed* $\gamma > 0$ *the parameter domain DA in* (c,m)*-plane is dividing into three subdomains* $1A, 2A, 3A$ *of qualitatively different phase behaviors of system* (14). *The boundaries of these Domains are the lines* $c=0, c=c_b(\gamma)c=c_{ch}(\gamma)$ *(see Fig. 5 and the captions). Specifically, the equilibrium A is globally stable in the domain* 1A *and shares basins of attraction with O in*

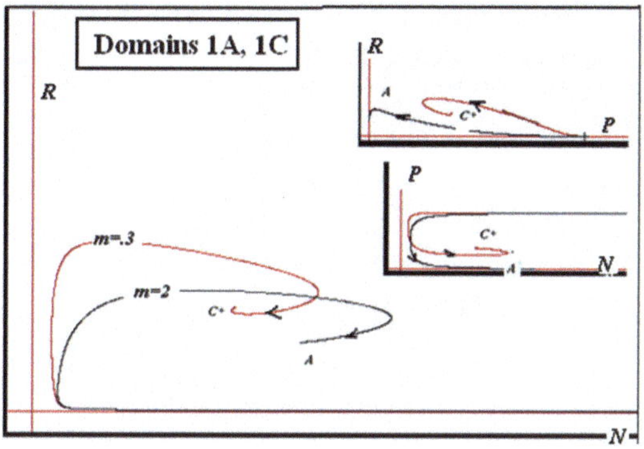

Fig. 7 Phase curves of system (14) in Domains 1*A* and 1*C* of Fig. 6. Parameters $\beta, \gamma, \delta, e = 1$, $c = 1.95$. In Domain 1A with $m = 2$ phase curves (*black*) tend to $A(0.72, 0, 0.37)$). In Domain 1C with $m = 0.15$ phase curves (*red*) tend to $C^{+}(0.15, 1.77, 0.86)$)

2A. *If the parameters c, m belong to Domain* 3*A then O is a single non-negative equilibrium point; O has attractive sector and may have elliptic sector in its positive neighborhood.*

Notice, that projections of phase curves corresponding to the parameter domain *DA* of three-dimension system (14) onto (*N*, *R*)-plane are qualitatively equivalent to the portraits of two-dimension system (3) in Domains 1n, 2n and 3n presented in Fig. 4b.

Proposition 3.5 *Non-trivial point* C^{+} *is stable equilibrium of system* (14) *in parameter domain* $DC \equiv DC^{+} \cup DC^{-}$ (Figs. 5 and 6). *This point is globally stable in the domain 1C bounded by the lines* $c = 0, c = c_b(\gamma)$ *and shares basins of attraction with equilibrium O in domains 2C and 3C.* (*N*, *R*)*-phase portraits of system* (14) *in the parameter Domains* 1*C and* 2*C are qualitatively equivalent to those in domains* 1*A*, 2*A*. (*N*, *R*)*-phase portrait in Domain* 3*C* (*see Fig*. 6) *contains the stable equilibrium* C^{+} *and equilibrium O, which has an attractive sector and may have elliptic sector for large c, in the last case basins of* C^{+} *and O are divided by separatrixes of equilibrium* C^{-}.

Proposition 3.6 *The boundary* $M(c, m)$ *between regions DA and DC corresponds to the following bifurcations in the system: the trans-critical bifurcation of changing stability of equilibria A and* C^{+} *for* $0 < c < c_d$ *and merging of equilibria* C^{+} *and* C^{-} *for* $c > c_d$.

Parameter-phase portrait of system (14) is schematically presented in Fig. 6. Phase portraits of the system in different parameter domains are shown in Figs. 7, 8, 9 and 10 and will be discussed below.

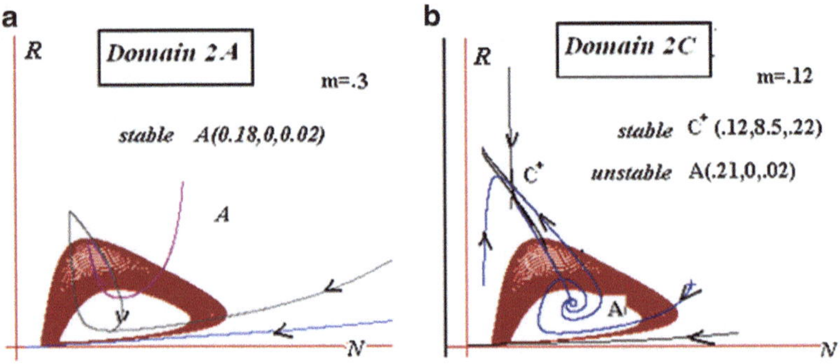

Fig. 8 Parameters $\beta = e = 1$, $\gamma = 7.74, \delta = 22, c = 9.033$; initial value $P(0) = 0.3$. (N, R)-projections of phase curves of system (14) are shown with different values of the parameter m (see Fig. 6); (**a**) $m = 0.3$; in Domain 2A phase curves tend to the stable equilibrium $A(0.18, 0, 0.02)$ (**b**) m=0.12; in Domain 2C phase curves tend to the stable equilibrium $C^+(0.12, 8.5, 0.22)$

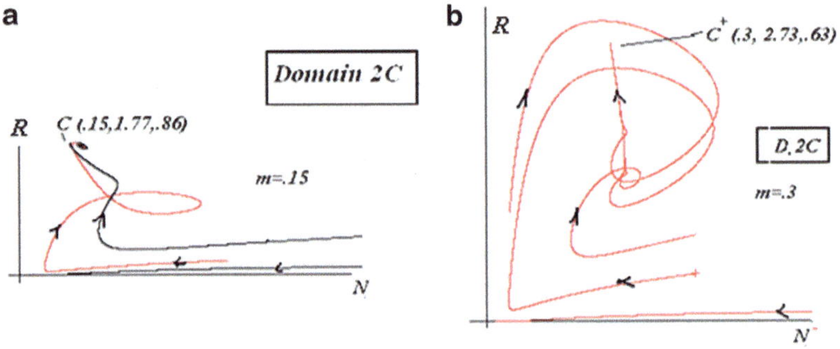

Fig. 9 Evolution of phase portraits of system (14) with increasing of parameter m in Domain 2C (Fig. 6) for fixed $c = 2.31, \beta = e = \gamma = \delta = 1, P(0) = 1, N(A) = 0.2$; (**a**) $m = 0.15$; phase curves tend to the origin or to the stable equilibrium $C^+(0.15, 1.77, 0.86)$; (**b**) $m = 0.3$; phase curves tend to the origin or to the stable equilibrium $C^+(0.3, 2.73, 0.63)$

3.4 Comparison the Bifurcation Diagram of Model (14) in Regions DA and DC

Let a parameter point $(c, m) \in 1C$, e.g., $0 < c < c_b(\gamma)$. Then the equilibrium $C^+(m, p^+, z^+)$ is stable while the equilibria A and O are unstable (see Proposition 3.4). If a parameter point $(c, m) \in 1A$, e.g., $0 < c < c_b(\gamma)$. Then the equilibrium A is stable while the equilibria C^+ and O are unstable (see Proposition 3.5). Thus, the phase portraits of system (14) in Domains 1A and 1C are equivalent. We, however, do not combine them into one because they have different "*biological*" *meanings*, namely, in Domain 1C the equilibrium amount of predators is non-zero while in Domain 1A there are "no" predators.

The same is true for Domains of bistability 2A and 2C, where $c_b(\gamma) < c < c_0(\gamma)$. Depending on initial values trajectories of the system tend either to O, or to

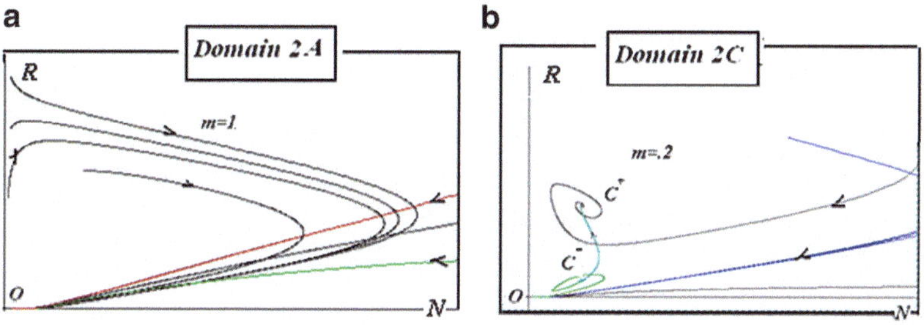

Fig. 10 Parameters $c = 2.42, \gamma = \delta = e = 1$, initial value $P(0) = 1$. (N, R)-projections of phase curves of system (14): (**a**) in Domains 3A for $m = 1$ and (**b**) in Domains 3C for $m = 0.2$ (see Fig. 2). Phase curves in Domain 3A, depending on initial values, tend to O for $t \to \infty$ or compose elliptic sector close to O. In Domain 3C phase curves tend to the stable equilibrium $C^+(0.2, 2.12, 0.68)$ or to O; system has also unstable equilibrium $C^-(0.2, 0.81, 0.12)$ whose separatrixes divide basins of C^+ and O

non-trivial equilibrium point (e.g., to A in Domain 2A and to C^+ in Domain 2C). Notice, that due to Proposition 3.3 in Domain 2C value $R(C) > R(A)$.

The most crucial difference of system dynamics is observed for parameters belonging to Domains 3A and 3C (see Fig. 7). For $(c, m) \in 3A$ e.g., $c > c_{ch}$, system (14) has no non-trivial stable equilibria. For $(c, m) \in 3C$ system (14) the equilibrium C^+ remains locally stable for any c. The (N, R)-phase portrait in domain 3C, Fig. 6 shows that the system has two attractive equilibria, C^+ and O, as well as "saddle-type" equilibrium C^- whose separatrices divide the basins of C^+ and O. Non-hyperbolic point O has a complex structure, in a neighborhood of O there exists an elliptic sector in (N, R)-plane (see Fig. 2). The results of computer studying of the system behaviors with different parameter values are shown in Figs. 7, 8, 9 and10.

So, the existence of non-trivial equilibrium C^+ in Domain 3C shows that predators are able to support and stabilize coexistence of the "consumers–predators–resource" system in "critical" situation of over-consumption when the system in absence of predators goes to extinction.

3.5 On Positivity of Predator-Induced Equilibria: Proofs of Propositions 3.2 and 3.3

Here we prove the following three statements:

1) *The point* $C^+(N = m, P = p^+, R = z^+)$ *is positive, i.e.* $p^+, z^+ > 0$, *if* $(c, m) \in DC^+ \cup DC^-$;

2) *The point* $C^+(N = m, P = p^-, R = z^-)$ *is positive, i.e.* $p^-, z^- > 0$, *if* $(c, m) \in DC^-$, where

$$z^{+} = R\left(C^{+}\right) = \frac{\gamma - \delta m + \sqrt{(\gamma + \delta m)^{2} + 4e\delta m\,(1-c)}}{2\delta}, p^{+} = P\left(C^{+}\right) = c - m/z^{+},$$

$$z^{-} = R\left(C^{-}\right) = \frac{\gamma - \delta m - \sqrt{(\gamma + \delta m)^{2} + 4e\delta m\,(1-c)}}{2\delta}, p^{-} = P\left(C^{-}\right) = c - m/z^{-}$$

3) $z^{+} = R(C^{+}) > R(A)$, if $c > c_d$ *where* c_d *is a positive root of the equation* $\gamma/e = \frac{c(c-1)(c+2)}{(1+c)^2}$ *and* $R(A) = \frac{ce(1-c)+(1+c)\gamma}{(1+c)\delta}$

To prove of the first statement we solve the system of inequalities

$$\begin{aligned} (\gamma + \delta m)^2 + 4e\delta m\,(1-c) \geq 0, \\ \sqrt{(\gamma + \delta m)^2 + 4e\delta m\,(1-c)} \geq -(\gamma - \delta m), \\ c\sqrt{(\gamma + \delta m)^2 + 4e\delta m\,(1-c))} \geq 2\delta m - c\,(\gamma - \delta m) \end{aligned} \tag{18}$$

The first inequality defines the domain Δ. The second inequality holds in Δ if $m \leq \gamma/\delta$ or if $\gamma/e \geq c-1$. The third inequality holds for $m \leq \frac{c\gamma}{\delta(c+2)}$ or $m \leq \frac{c}{\delta}\left(\gamma + \frac{ce(1-c)}{1+c}\right) \equiv m_{N(A)}$.

Let $\gamma = 0$. Then the domain Δ where z^{+} is positive has the boundaries $m = 0, m = 4e(c-1)/\delta$, thus $DC^{-} = \varnothing$. The domain Δ contains positive sub-domain DC^{+}, which is bounded by the curves $m = 0$ and $m_{N(A)} : m = \frac{c^2 e(1-c)}{\delta(1+c)}, 0 \leq c \leq 1$. Thus, C^{+} is positive in $DC^{+} \equiv DC$ (see Fig. 5).

If $\gamma > 0$ then the curves $m_{N(A)}$ and $m_{\Delta}^{-} : m = \frac{-2\sqrt{(c-1)e((c-1)e-\gamma)}+(2(c-1)e-\gamma)}{\delta}$ have a common "point" $\alpha(c_d, m_d)$ where the value c_d is a positive root of the equation $\gamma/e = \frac{c(c-1)(c+2)}{(1+c)^2}$, and the coordinate $m_d = \frac{c_d}{\delta}\left(\frac{c_d e\left(1-c_d\right)}{\left(1+c_d\right)} + \gamma\right)$.

Remark, that the point $\alpha(c_d, m_d)$ belongs also to the curve $m = \frac{c\gamma}{\delta(c+2)}$, so the parameter point α is the common for these three lines, m_{Δ}^{-}, $m_{N(A)}$ and $m = \frac{c\gamma}{\delta(c+2)}$. Domain DC^{-} (see Fig. 5), which is bounded by the curves m_{Δ}^{-} where $c_d < c < \infty$, $m_{N(A)}$ where $c_d < c \leq c_o$ and $m = 0, c > c_o$, belongs to the domain Δ and DC^{-} is not empty. The first statement is proven.

For proving the second statements we consider and study the system

$$\begin{aligned} (\gamma + \delta m)^2 + 4e\delta m\,(1-c) \geq 0, \\ \sqrt{(\gamma + \delta m)^2 + 4e\delta m\,(1-c)} \leq \gamma - \delta m, \\ c\,(\gamma - \delta m) - 2\delta m \geq \sqrt{(\gamma + \delta m)^2 + 4e\delta m\,(1-c)} \end{aligned} \tag{19}$$

similarly to the previous case. At last, the third system of inequalities can be reduced to the system

$$\begin{gathered}(\gamma+\delta m)^2+4e\delta m\,(1-c)\geq 0,\\ \sqrt{(\gamma+\delta m)^2+4e\delta m\,(1-c)}\geq-(\gamma-\delta m)\\ (1+c)\sqrt{(\gamma+\delta m)^2+4e\delta m\,(1-c)}>(1+c)(\gamma+\delta m)+2ce\,(1-c)\end{gathered}\qquad(20)$$

Solutions of (20) are

$$c>1\ and\ \tfrac{\gamma}{e}>\tfrac{c(c-1)(c+2)}{(1+c)^2} and\ m\leq\tfrac{c(ce(1-c)+\gamma(1+c))}{(1+c)\delta}\equiv N(A), or$$
$$\tfrac{\gamma}{e}<\tfrac{c(c-1)(c+2)}{(1+c)^2}\ and\ m\leq\frac{-2\sqrt{(c-1)e((c-1)e-\gamma)}+(2(c-1)e-\gamma)}{\delta}\equiv m_{\Delta}^{-}$$

This formulas mean that system (20) is satisfied if $m\leq m_{N(A)}$ for $1<c<c_d$ and if $m<m_{\Delta}^{-}$ for $c>c_d$. The statement is proven.

3.6 On Stability of Equilibria

We use the standard linearization method for analyzing stability of equilibrium points. Let $J\,(N,P,R)=\left(\frac{\partial F_i(N,P,R)}{\partial N}\ \frac{\partial F_i(N,P,R)}{\partial P}\ \frac{\partial F_i(N,P,R)}{\partial R}\right), i=1,2,3,$ be Jacobian of (14). Then

$$J\,(N,P,R)=$$
$$\begin{pmatrix}-3N^2+2N\,(-1+c-p)\,z & -NR\,(N+R) & N\,(N\,(-1+c-P)+2(c-P\,)R)\\ \quad+(c-p)\,z^2 & & \\ \beta PR\,(-m+2N+R) & \beta\,(-m+N)\,R\,(N+R) & \beta\,(-m+N)\,P\,(N+2R)\\ R\,(e\,(1-c)+\gamma-\delta R) & 0 & eN\,(1-c)+\gamma\,(N+2R)\\ & & \quad-\delta z\,(2N+3R)\end{pmatrix}$$

1) Substituting coordinates of B in $J(N,p,z)$ we get

$$J\Big(B_3\,(0,0,\gamma/\delta)=\begin{pmatrix}c(\gamma/\delta)^2 & 0 & 0\\ 0 & -\beta m(\gamma/\delta)^2 & 0\\ e\,(1-c)\,\gamma/\delta & 0 & -c\gamma^2/\delta\end{pmatrix}.$$

Thus, $\lambda_1(B)=\frac{c\gamma^2}{\delta^2}, \lambda_2(B)=-\frac{\beta m\gamma^2}{\delta^2}, \lambda_3(B)=-\frac{\gamma^2}{\delta}$, and statement 1) is proven.

2) Substituting $P=0$ to $J(N,P,R)$ we get

$$Det\,(J(A)-\lambda) = Det\begin{pmatrix} -3N^2+2N(-1+c)R +cR^2-\lambda & -NR(N+R) & N(N(-1+c)+2cR) \\ 0 & \beta(-m+N)R(N+R)-\lambda & 0 \\ R(e(1-c)+\gamma-\delta R) & 0 & eN(1-c)+\gamma(N+2R) -\delta z(2N+3R)-\lambda \end{pmatrix} =$$

$$(\beta(-m+N)R(N+R)-\lambda)\,Det\begin{pmatrix} -3N^2+2N(-1+c)R +cR^2-\lambda & N(N(-1+c)+2cR) \\ R(e(1-c)+\gamma-\delta R) & eN(1-c)+\gamma(N+2R) -\delta R(2N+3R)-\lambda \end{pmatrix}$$

Thus, in the point $A(N=N(A), p(A)=0, R=R(A))$ the characteristic polynomial is of the form $Det(J(A)-\lambda)=(\beta(-m+N)R(N+R)-\lambda)Det(J(A_2)-\lambda)$ where $Det(J(A_2)-\lambda)$ is the characteristic polynomial in the point A_2 (see Sect. 2.5).

From the latter formula we get $\lambda_3 = \beta(N(A)-m)R(N(A)+R)$. So, with positive R and β the eigenvalue $\lambda_3<0$ for $N(A)<m$, which is true for $(c,m)\in DA$ (see Fig. 5b).

3) Substituting coordinates of the points $C^{\pm}$ to $J(N,P,R)$ we get the matrix:

$$J\left(C^{\pm}\right) = \begin{pmatrix} -m(m+R) & -mR(m+R) & m(c-P)(m+R) \\ \beta PR(m+R) & 0 & 0 \\ R(e(1-c)+\gamma-\delta R) & 0 & R(\gamma-\delta R-\delta(m+R)) \end{pmatrix}$$

where $R=R(C^{\pm})=z^{\pm}$.

The characteristic polynomials in $C^{\pm}$ are

$$Q(\lambda)=\lambda^3-\lambda^2((-m^2-(1+\delta)mR+z(\gamma-2\delta R))-\lambda(m(m+R)(em(1-c) +(m+R)(\gamma+R(-2\delta+\beta(m-cR))))-\beta mR^2(m+R)^2(m-cR)(-\gamma+\delta(m+ 2R))$$

Let $\lambda_1, \lambda_2, \lambda_3$ be the roots of $Q(\lambda)$. Then

$$\begin{aligned} \lambda_1+\lambda_2+\lambda_3 &= -m^2-(1+\delta)mR+R(\gamma-2\delta R), \\ \lambda_1\lambda_2+\lambda_1\lambda_3+\lambda_2\lambda_3 &= -(m(m+R)(em(1-c)+(m+R), \\ &\quad (\gamma+r(-2\delta+\beta(m-cR)))\Big) \\ \lambda_1\lambda_2\lambda_3 &= \beta mR^2(m+R)^2(m-cR)(-\gamma+\delta(m+2R)) \end{aligned} \tag{21}$$

We consider only positive $z^{\pm} = R\left(C^{\pm}\right) = \frac{\gamma-\delta m\pm\sqrt{D}}{2\delta}$, $D \equiv (\gamma+\delta m)^2 + 4e\delta m(1-c) \geq 0$ and $P\left(C^{\pm}\right) = \frac{cz^{\pm}-m}{z^{\pm}}$.

In the third equation of (21) the factor $m - cz^{\pm} < 0$, the factor $-\gamma + \delta(m + 2R) = \pm\sqrt{D}$; in the first equation of (21)

$$-m^2 - (1+\delta)mR + z(\gamma - 2\delta r) = -\frac{D \pm \sqrt{D}(\gamma - \delta m + m) + m(\gamma - \delta m)}{2\delta}.$$

We see that for the point C^- the product $\lambda_1\lambda_2\lambda_3 > 0$. It is possible if at least one of the eigenvalues $\lambda_1, \lambda_2, \lambda_3$ is positive. So the equilibrium C^- is unstable.

For the point C^+ the product and the sum of eigenvalues are negative. Thus, all roots of the characteristic polynomial have negative real parts. So, the point C^+ is stable.

4 Discussion and Conclusion

In this work we consider two kinds of systems, "consumer–renewable resource" and "consumer–predator–renewable resource" and compare their dynamics. A common resource R determines the carrying capacity of the consumer population N. The main peculiarity of both systems is that the common resource is supposed to be renewable in such a way that consumers not only consume the resource but also are able to contribute to its restoration increasing the common population carrying capacity.

We suppose that the resource R is naturally restored at constant rate $\gamma \geq 0$ and deteriorated at the rate δR ($\delta > 0$) and can be replenished by the activity of consumers. We suppose also that the per capita birth rate of consumers is proportional to the rate c of resource consumption whereas the rate of resource restoration/utilization by the consumers–producers is proportional to $(1-c)$. The case $c > 1$ models over-consumption, when consumers utilize more resource than they restore; we call to them as *over-consumers*.

We use model (1) in order to investigate two questions: whether a largely consumerist population can nevertheless sustainably coexist with common renewable resources without exhausting it over time, and if it cannot, what is the critical level of over-consumption and what dynamical regimes it goes through before it collapses.

The dynamics of "consumer–renewable resource" model was described by means of bifurcation diagram in parameters (c, γ) (see Fig. 4), where parameter space is divided into six domains of qualitatively different (N, R)-phase portraits.

Our analysis indicates that even when the population consists of over-consumers, there is a threshold for system resistance to over-consumption, which is directly proportional to the natural growth rate of the resource γ and inversely proportional to individuals' efficiency of niche construction, modeled by the parameter e. Hence, the system can tolerate more over-consumers if it can restore itself quickly enough or if the individuals are not overly efficient in resource consumption.

As the consumption rate c increases, the population goes through a series of transitional regimes (see Fig. 4) before it collapses. When the value of c is small, consumers–producers and the resource can coexist in a stable non-trivial equilibrium point. As c increases, an unstable limit cycle appears around the stable equilibrium point. The system enters in the domain of bistability where it may either still coexist in a stable equilibrium or in stable oscillations or go to extinction depending on initial values of the consumers and resource $P(0), R(0)$.

Further increases in c drive the dynamics into the domain, where the non-trivial equilibrium point still exists but is unstable. Further increases in c lead to disappearance of the non-trivial equilibrium from the first quadrant, changing the point $O(0, 0)$ from a saddle-node to an elliptic sector, which corresponds to eventual extinction of both the resource and the population, although in infinite time.

The system has a non-hyperbolic singular point at the origin that, as the parameters are varied, changes its structure from a saddle to a stable saddle-node with a sector of trajectories tending to the origin, to an elliptic sector. As parameters are varied, the non-trivial equilibrium in the model changes its stability as a result of a "catastrophic" Andronov–Hopf bifurcation, yielding a parameter region, where an unstable limit cycle divides the basins of attraction of the nontrivial equilibrium and the origin. No stable oscillations can be observed, and the system eventually "dies out" because as c increases, an increasing number of trajectories tend towards the origin. It is the mutual placement of separatrixes that determines the structure of the phase portrait (or in other words, it is the ratio of the consumer to the resource that determines the existence of the attractive sector at the origin). Notably, the unstable limit cycle appears from heteroclinic orbits of the origin and the saddle point B_2. These are new type of dynamics compared to other models that have a complex equilibrium point at the origin [3, 4, 5, etc.]. With further increasing of c the system goes to total extinction.

The second model describes the dynamics of "consumer–predator–renewable resource" system; informally, we add predators to the first system such that consumers are **subject to attacks of predators**. The main problem of interest here is how predators P change the dynamics of the initial "consumer–renewable resource" system. We analyze the role of predators with the help of three-dimension model (2). This model has an additional important parameter m m, which characterizes the critical density of predator for non-trivial equilibrium coexistence with preys. The dynamics of "standard" predator–prey model (which correspond to the case of unlimited resource) is well known (see, e.g., [3]); it demonstrates only two types of behaviors depending on model parameters: either predators go to extinct and only preys survive in the system, or both populations coexist in a stable equilibrium. Model (2) can be considered as a generalization of that models; the carrying capacity of preys is now determined by a renewable resource R, which is governed by the third equation of model (2). Analysis of system (2) shows that for any fixed values of parameters β, γ, δ, e and any value of c there exists a boundary $M(c)$, such that consumers–producers, predators and resource coexist in a stable equilibrium for $m < M(c)$. The existence of that equilibrium reveals the "governing and stabilizing" role of predators, which increase a mortality rate of consumers and hence save more resources.

Our analysis of the "consumer–predator–renewable resource" model shows that predators can change the steady states and dynamics of the system. Predators don't change essentially the dynamics of the "consumer–renewable resource" system when the level of over-consumption is not too large. In contrast, predators are able to keep a stable equilibrium with non-zero amounts of the preys and resource even when the level of overconsumption is so large that the "predator free" consumers–resource system goes to extinction. The amount of predators in this equilibrium increases as the parameter c increases. The equilibrium point has a bounded basin and trajectories that start out of this basin tend to O. Notice that the level of over-consumption corresponding to this equilibrium can be arbitrary, and the amount of predators P^*, which keeps non-zero equilibrium of the system increases proportionally to the parameter c.

Computer experiments revealed that even small amount of predators in the system increases the life time of the system even in the case when in the "final" equilibrium the amount of predators is zero. We may conclude that the model reveals possible "positive" influence of predators which can increase sustainability of the "consumers–predators–renewable resource" system and prevent it from extinction.

It is our hope that the model and the results of its study may be interpreted in terms of socio-economics systems, but this is out of the scope of this work.

References

1. Andronov, A., Leontovich, E., Gordon, I., & Maier, A. (1973). *Qualitative theory of second-order dynamic systems*. N.-Y.-Toronto: Wiley.
2. Bautin, N., & Leontovich, E. (1976). *Methods and techniques for qualitative analysis of dynamical systems on the plane*. Moscow: Nauka.
3. Bazykin, A.D. (2000). Nonlinear dynamics of interacting populations. *World scientific series on Nonlinear Science, Ser. A* (Vol. 11). Singapore: Word Scientific.
4. Berezovskaya, F. (1979). *Non-hyperbolic equilibrium of a planar system of ordinary differential equation*. Analysis by Newton diagram method. PhD Dissertation. Institute of Mathematics, Ukraine Acad. Sci.
5. Berezovskaya, F. S., Novozhilov, A. S., & Karev, G. P. (2007). Population models with singular equilibrium. *Mathematical Biosciences, 208*(1), 270–299.
6. Brauer, F., & Castillo-Chavez, C. (2001). *Mathematical models in population biology and epidemiology*. New York: Berlin, Heidelberg, Hong Kong, London, Milan, Paris, Tokyo: Springer.
7. Grinnell, J. (1917). The niche-relationships of the California thrasher. *The Auk, 34*(4), 427.
8. Hardin, G. (1968). The tragedy of the commons. *Science, 162*(5364), 1243–8.
9. Hooper, D. U., Chapin, F. S., III, Ewel, J. J., Hector, A., Inchausti, P., Lavorel, S., et al. (2005). Effects of biodiversity on ecosystem functioning: A consensus of current knowledge. *Ecological Monographs, 75*(1), 3–35.
10. Kareva, I., Berezovskaya, F., & Castillo-Chavez, C. (2012). Transitional regimes as early warning signals in resource dependent competition models. *Mathematical Biosciences, 240*, 114–123.
11. Kareva, I., Berezovskaya, F., & Karev, G. (2013). Mixed strategies and natural selection in resource allocation. *Mathematical Biosciences and Engineering, 10*(5&6), 1561–1586.

12. Kareva, I., Morin, B., & Karev, G. (2012). Preventing the tragedy of the commons through punishment of over-consumers and encouragement of under-consumers. *Bulletin of Mathematical Biology, 75*, 565–588.
13. Krakauer, D. C., Page, K. M., & Erwin, D. H. (2009). Diversity, dilemmas, and monopolies of niche construction. *The American Naturalist, 173*(1), 26–40.
14. Kuznetsov, Y. (1998). *Elements of applied bifurcation theory*. New York: Springer.
15. Odling-Smee, F. J., Laland, K. N., & Feldman, M. W. (2003). Niche construction: The neglected process in evolution, *Ser. Monographs in Population Biology, 37*. Princeton University Press.
16. Odum, E. (1971). *Fundamentals in ecology* (3rd ed.). Philadelphia: Saunders.
17. Post, D. M., Conners, M. E., & Goldberg, D. S. (2000). Prey preference by a top predator and the stability of linked food chains. *Ecology, 81*, 8–14.
18. Volterra, V. (1931). *Leçons sur la Th'eorie Math'ematique de la Lutte pour la Vie*. Paris: Gauthier-Villare.

Recent Advances in Approaches to the Study of Gene Locus Control Regions

Benjamin D. Ortiz

Abstract In the many decades of investigation into the regulation of gene transcription in vertebrates, the locus control region (LCR) has emerged as perhaps the most powerful *cis*-acting regulatory DNA element that one can envision. An LCR element is unique in that it supports both specific spatiotemporal regulation of transcription during development, and a poorly understood "insulation capacity" that prevents genomic interference with the gene regulatory program it would impose upon a linked transgene. As such, it represents a complete, compact and portable package of the DNA sequence information required to establish an independently and predictably regulated gene locus in native chromatin of a whole animal. Both in vivo and cell culture models have contributed significantly to building the field of LCRs. Nevertheless, the gold standard experimental approach to LCR study is transgenic mice, which has been dominant in the progress made in the field over the past 25 years. However, recent technological advances are resulting in a re-emergence of cell culture based approaches to LCR study, portending a coming era of more rapid progress in this significant but understudied field. The investigation of the unique and powerful gene regulatory activities supported by LCR elements offers unparalleled opportunities to gain insight into *cis*-mediated transcriptional regulation at the single gene locus level. Furthermore, such insights are critical to advancing the safety and efficacy of gene therapy.

B.D. Ortiz (✉)
Department of Biological Sciences, City University of New York, Hunter College and Graduate Center, 695 Park Avenue, Room 927 N, New York, NY 10065, USA
e-mail: ortiz@genectr.hunter.cuny.edu

B. Toni (ed.), *New Frontiers of Multidisciplinary Research in STEAM-H (Science, Technology, Engineering, Agriculture, Mathematics, and Health)*,
Springer Proceedings in Mathematics & Statistics 90, DOI 10.1007/978-3-319-07755-0_9,

1 Introduction

The processes regulating gene transcription in time and space are of paramount importance to organismal development and cell type-differentiation. Therapeutic interventions that harness and/or manipulate these processes are also on the cutting edge of novel approaches to treating disease. The molecular machinery of gene regulation is presumed to be directed to specific gene loci primarily by *cis*-acting DNA sequences that can be either proximal or distal to the transcription start site of a given target gene. Thus, such *cis*-acting DNA elements have been a major focus of study in the effort to understand the spatiotemporal control and gene locus selectivity of cell type-specific transcription programs. Of these DNA elements, transcriptional enhancers have received the bulk of experimental attention, as they can directly modulate linked gene promoter activity [6]. However, other classes of *cis*-acting DNA elements have been characterized with profound, though usually more indirect, impact on the subsequent activity of RNA polymerase bound to a promoter of a native gene locus [37]. These elements are thought to influence the localization and/or accessibility of DNA subsequences within a gene locus that would interact with *trans*-acting nuclear protein factors in chromatin. Arguably, the most powerful of these distinct *cis*-acting DNA elements is the locus control region (LCR).

An LCR is unique in its ability to virtually eliminate integration site-dependent position effects that would interfere with the expression of a linked transgene inserted at a random, ectopic location in the genome of mice [38]. Such position effects lead to unpredictability, or even absence, of transgene expression in some or all lines of mice that are transgenic for a given transcription unit [48]. While transcriptional enhancers can be important regulators of transcription in vivo [6], they generally cannot, on their own, suppress integration site-dependent position effects on transgene expression. In contrast, LCR-driven transgene expression is "integration site-independent" and, thus, consistently observed to be high-level across all lines of mice bearing the transgene. Furthermore, in an LCR-regulated system, mRNA expression levels will directly correlate with the number of transgene copies incorporated into the genome of a given transgenic mouse line. The specificity of transgene expression in time and space will be similar in all lines, and generally parallel the pattern of a given LCR's gene locus of origin. As such, an LCR represents a complete, compact and portable package of the DNA sequence information required to establish an independently and predictably regulated gene locus in native chromatin of a whole animal. The investigation of the gene regulatory activities supported by LCR elements offers unparalleled opportunities to gain insight into *cis*-mediated transcriptional regulation at the single gene locus level. Furthermore, such insights are critical to advancing the safety and efficacy of gene therapy.

2 LCRs Versus Insulator Elements

It has been speculated that the integration site-independence aspects of LCR activity are related to the function of vertebrate insulators [19]. There are two distinct classes of insulator elements identified in vertebrates: enhancer blockers [8] and chromatin barriers [52]. Distinct DNA-binding protein effectors, and modes of action, have been identified for each insulator type (reviewed in [1]). Evidence has accumulated in support of the view that LCRs can contain within them one or more sub-elements with activities reminiscent of *barrier type* insulators. A small handful of LCR sub-elements have been isolated and shown to provide a linked transgene with a degree of integration-site independence similar to that provided by *bona fide* barrier insulators [13, 26, 46, 44]. However it is important to note the dramatic distinctions between the activity of barrier insulators and LCRs. Barrier insulators do seem able to render most integration sites in a genome permissive for some expression of a linked gene. However, in contrast to LCRs, barrier insulators do not provide the copy number-dependent mRNA expression levels indicative of complete suppression of integration site-derived position effects [37]. Furthermore, unlike LCRs, the vertebrate barrier insulator elements identified to date seem not to bear any particular developmental and/or cell type specificity in their function, and their identified molecular effectors are ubiquitously expressed [11, 64]. In short, while the eventual discovery of clear molecular connections between the activity of barrier insulators and LCRs seems likely, it is clear that the LCR supports a considerably more complex (and comprehensive) function in terms of gene regulation.

3 The First Locus Control Region

The first LCR was discovered in the human β-globin gene locus. Early attempts to make human β-globin transgenic mice, containing well-characterized promoter and downstream enhancer elements failed to produce transgenic mice with predictable human β-globin gene expression [40, 60]. It was also known that disruption of non-coding DNA flanking the β-globin gene locus can cause human β-thalassemia by inactivating the locus [32]. These data taken together provided a strong rationale to search for hypothesized *cis*-acting elements located at a greater distance from the naturally occurring human β-globin gene. Initial evidence for such came from DNase I hypersensitivity studies that yielded candidate regulatory regions distant from the coding regions of the β-globin locus [16, 25, 61].

Using a somatic cell fusion approach [58], Forrester et al. provided important correlative evidence that the distant DNase hypersensitive regions actively regulated β-globin gene expression [15]. Briefly, these experiments utilized non-erythroid human cells, in which the β-globin gene, borne by chromosome 11, would be inactive. These cells were fused with a mouse erythroid cell line, in which the endogenous mouse β-globin gene was active. The hypothesis behind this experiment was that the mouse erythroid cell derived factors would activate in *trans* the dormant human β-globin locus de novo. Resulting human chromosome 11 bearing mouse/human hybrid cells indeed displayed both transcriptional activation of the locus, and concurrent formation of the developmentally stable DNase hypersensitive region flanking the human β-globin locus.

In 1987, Grosveld et al. used transgenic mice to test human β-globin locus derived transcription units that included the putative distant flanking control regions [24]. Unlike the integration site-dependent "position effects" that were routinely and generally observed to lead to significant line-to-line variability in transgene expression [48], these new transgenes displayed consistent activity across transgenic mouse lines (i.e. at multiple integration sites). The human β-globin transgene expression pattern paralleled that of the endogenous β-globin locus in time, space and even mRNA level per transgene copy. These astonishing results would (with a subsequent accord on nomenclature [37]) give birth to the field of locus control regions. It would also begin the dominance of the transgenic mouse approach to the study of LCR activity.

Following the above events, an LCR was discovered in the lymphocyte-specifically expressed human CD2 gene locus [36]. In 1989, both the β-globin and CD2 LCRs were reported to dominantly transfer the spatiotemporal expression patterns of their gene locus of origin to an unrelated transcription unit [3, 22] while maintaining the ability to confer integration site-independent and transgene copy number-dependent expression levels. These remarkable findings would help establish the definition of a *bona fide* LCR as a uniquely powerful *cis*-acting gene regulatory element distinct from classical transcriptional enhancers. At the same time, it augured high potential for the future application of LCR activity to the developing field of gene therapy, where vectors engineered to support robust and specifically targeted gene expression would be highly desirable.

4 Study of LCR Activity in Transgenic Mice

Since the late 1980s, transgenic mice have been utilized to identify LCRs in multiple gene loci expressed in various cell types [38]. Aside from yielding important information on the molecular mechanisms regulating their native gene loci, LCR-driven transgene systems began yielding more general insight into the nature of mammalian gene regulation in native chromatin, even before the term "histone code" became popularized [29]. Chief among these insights was the observation that LCRs can dominantly overcome the repressive impact of heterochromatin on transgene expression at an ectopic integration site [14, 42]. Furthermore, in this context, the LCR mode of action clearly involved *cis*-mediated, tissue-specific regulation of long-range chromatin structure [12, 44] and epigenetic modifications of histones [23] and DNA [54].

Despite the important contributions of LCR study to locus-specific, and more general, gene regulatory knowledge, the field has been relatively slow moving.

This state-of-affairs is most directly attributable to the field's dependence on transgenic mouse models, wherein the full range of LCR activities were first demonstrated. Transgenic mouse experiments are resource intensive, high cost and involve protracted timetables. While this approach has remained the most complete and rigorous model for assessing LCR activity, the slow pace of progress inherent in this technology inspired various attempts to develop assays for LCR activity that did not require transgenic mice. For many years, these efforts yielded only partial success, although, in the process, much was learned about the unanticipated requirements for establishing LCR activity in a cell.

5 The Search for Alternatives to Transgenic Mice

Since cell culture based technology yielded important data leading to the initial discovery of the β-globin LCR, it is surprising that the field struggled for well over a decade to validate assays for the complete range of LCR activities that were not dependent on transgenic mice. Somatic cell genetics experiments in erythroid cell lines began to indicate that the activity of the β-globin LCR on chromatin, in the context of its endogenous locus, would manifest itself differently from its apparent impact on a transgene at an ectopic site in the genome [53]. This notion would later be confirmed in vivo [2] and is likely due to native locus derived functional redundancy. These findings notwithstanding, at the time there was no a priori basis for doubting that a differentiated adult erythroid cell line would support the full activity of a β-globin LCR-driven transgene introduced de novo into its genome. Nevertheless, an informative "failure" in this regard was published in 1998. Skarpidi et al. discovered that the same mouse cell line that was able to "*trans*-activate" the endogenous human β-globin locus in human/mouse cell hybrids was unable to support integration site-independent β-globin LCR activity at an ectopic site [57]. The copy number-dependent transgene mRNA expression levels supported by the β-globin LCR in transgenic mice were not reproduced in the erythroleukemia cell clones stably transfected with similar reporter transgenes.

An important clue to explaining the puzzling inability of erythroid cell lines to support LCR activity was reported the following year [62]. In this work, β-globin LCR-driven reporter transgenes were first stably integrated into the genome of mouse fibroblasts. These transfected fibroblasts were then fused with the same erythroid cells used in the above experiments. The fusion resulted in activation of the ectopically integrated LCR-driven transgenes, much in the same manner as the endogenous human β-globin gene locus was activated by similar fusion experiments performed in the earliest days of the field. The copy number-dependence of transgene expression was restored in the hybrid cells. The interpretation of these experiments was that the first steps of LCR activation require it to be present in chromatin of an undifferentiated cell. If this were true, it would explain the failure of de novo introduction into the chromatin of already differentiated erythroid cells to support the establishment of an LCR's activity.

It might seem trivial to point out that during development, all cell type-specifically expressed gene loci will exist for some time in an undifferentiated chromatin environment in precursor cells, before they become activated upon subsequent cellular differentiation events. But the reasons why such pre-differentiation presence would be *required* for the later establishment of LCR activity were far less obvious at the time. Subsequent reports would discover pre-differentiation molecular "priming" events at the human β-globin locus occurring early in development [4, 5, 63]. While the system used in Vassilopoulos et al. [62] does not recapitulate normal erythroid cell development *per se*, the data supporting the significance of prior transcriptional priming to subsequent β-globin locus activation pointed to a potential mechanistic basis for their findings. Nevertheless, even after these efforts, the field remained without a validated cell culture model supporting the full range of LCR activities manifested at an ectopic genomic site. As it happened, the next steps forward in this effort would come from the study of LCRs that are active in a different cell lineage altogether, the T cells of the immune system. These recent advances would be enabled by a breakthrough procedure that supported the direct observation of the full trajectory of T cell development from undifferentiated precursors, without the use of transgenic mice.

6 A Method to Recapitulate T Cell Development In Vitro

Embryonic stem cells (ESC) are the pluripotent founder cells of mammalian embryos. ESCs can be isolated from zygotes and propagated in cell culture in way that maintains their undifferentiated state. These cells can also be directed to differentiate in vitro into a wide variety of specialized cell types, upon provision of the appropriate micro-environmental components (both soluble and cell-associated factors) in a sequence that mimics the normal developmental events that yield a given cell lineage in vivo [18]. Although a number of hematopoietic cell types had been successfully derived in vitro from ESC [7, 43], T cells remained recalcitrant to differentiation in cell culture. It was then discovered that input from the Notch signal transduction pathway shifted the fate of developing lymphocytes into the T lineage at the expense of B lineage cells in vivo [51]. This key information was used to develop a clever strategy to coax developing hematopoietic cells towards T cell fates in vitro (Fig. 1).

A bone marrow derived stromal cell line (called OP9 [43]) was transduced with a Notch receptor-triggering ligand [named Delta-like-1 (DL1)] to create the OP9-DL1 cell line [56]. With the provision of the appropriate cytokines, OP9-DL1 cells supported robust, quantitative differentiation of mouse ESCs into T cells within 3 weeks of co-culture [55]. OP9-DL1 cells are also able to support the direction of human embryonic stem cells and human induced pluripotent stem cells into the T lineage in cell culture [31]. DNA constructs can be readily introduced into mouse ESCs by electroporation, and other means. Thus, this system seemed to offer a way to assess the regulatory activity of reporter gene-linked DNA elements during hematopoietic differentiation of cloned, stable-transfected ESCs in vitro.

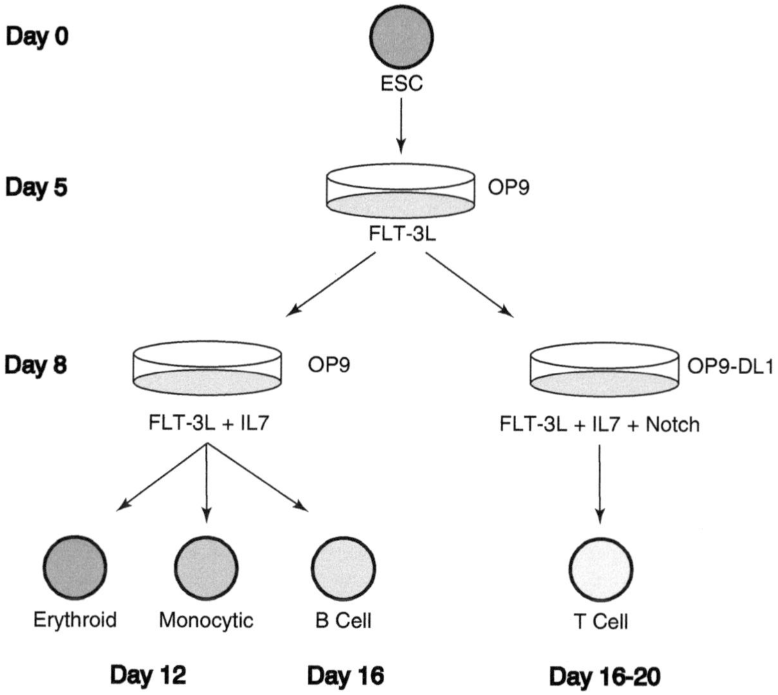

Fig. 1 The ESC-OP9 co-culture procedure for reproducing hematopoiesis in vitro [55]. ESCs are initially seeded on a monolayer of OP9 bone marrow derived stromal cells. On day 5 of co-culture, the cytokine fms-like-tyrosine kinase receptor-3 ligand (Flt-3 L) is included in the culture to derive hematopoietic progenitor cells by day 8. At that point, interleukin-7 (IL7) is included and developing progenitors are re-plated on either OP9 cells, to generate non T cell types or OP9-DL1 cells to produce T-lineage cells. The process in completed within 3 weeks of co-culture

7 A Step Forward: The Human Perforin Gene LCR

The idea to use in vitro T cell differentiation to examine an LCR was first tested in studies of selected aspects of human Perforin gene LCR activity [49]. The Perforin gene encodes a key cytotoxic effector molecule employed by natural killer (NK) and CD8-lineage T-killer cells during cell-mediated immune responses [30]. This gene becomes activated as part of a late-stage response to the triggering of mature T cells via their antigen receptor. This response coincides with the development of cytolytic effector function 2 days after initial T cell receptor stimulation [9, 34, 47].

The long, eighteen hypersensitive site region of the putative human Perforin LCR was first identified using microcell-mediated chromosome transfer (a variation of cell fusion). In this approach, a mouse T cell line capable of expressing the Perforin gene (after immune receptor triggering) receives a copy of the Perforin-bearing human chromosome 10 from microsomes derived from a donor fibroblast cell line [49]. The fusion moved the inactive, native human Perforin gene locus out of an undifferentiated cell, and into a differentiated cellular environment. Much like the experiments carried out 20 years earlier on the β-globin locus, this event enabled the de novo activation of the DNase hypersensitive region, and subsequent induction of Perforin gene expression.

In this study, in vitro ESC differentiation was used to test the cell-type specificity of this LCR's activity. A bacterial artificial chromosome (BAC) bearing the human Perforin gene, and its putative LCR, was used to transfect mouse ESCs. BAC bearing ESC clones were differentiated into various hematopoietic progeny in vitro. Among the generated progeny cell types, transgenic Perforin expression was only seen in cytolytic lymphocyte cell types that had been appropriately activated to simulate an immune response [49]. This pattern is what would have been predicted from the characteristics of endogenous Perforin gene expression.

In the above report on the initial identification of the Perforin LCR [49], multiple assays using both ESC and a cultured T cell line were combined to characterize its activities. Thus, this effort represented a step forward in the drive to develop alternatives to transgenic mice for LCR study. However, the Perforin LCR has yet to be isolated and shown to transfer its characteristics to a linked heterologous transgene. Furthermore, there have not yet been reports of its activity in vivo. Nevertheless, the successful, if limited, use of in vitro ESC differentiation in this work seemed to point the way toward the development of a single, complete assay for LCR activity that was not dependent on transgenic mice. A direct test of this idea would require the use of an isolated, *bona fide* LCR, the activity of which had already been fully characterized in vivo. Another LCR active in T cells would eventually provide the opportunity for this test.

8 The T Cell Receptor-α Gene LCR

The LCR residing in the mouse T cell receptor (TCR)-α gene locus (Fig. 2a) would become the third LCR to be discovered in a T cell expressed gene locus [10]. It was initially described as a series of nine DNase I hypersensitive sites spread over 13-kb of genomic DNA in between the TCRα constant region exons on the 5′-end, and the last exon of the essential Defender against Death (Dad)-1 gene on the 3′-end [28]. The study of this LCR has had significant impact on the field [1, 17, 37, 38]. Like the human β-globin and CD2 LCRs, the TCRα LCR has been well demonstrated to dominantly transfer its gene regulatory properties to a linked heterologous transgene [27, 33, 44]. Furthermore, the TCRα LCR is one of a small handful of LCRs for which significant structure-function information has been obtained [21, 26, 45] (Fig 2b). This LCR's activity seems to result from functional synergy between a 3′-"chromatin opening" property that is widely active in multiple tissues [46] and an adjacent 5′-DNA region that provides developmental timing- and cell type-specificity to this activity in a whole animal [44]. While the molecular bases for these functions are not completely known, there is considerable evidence

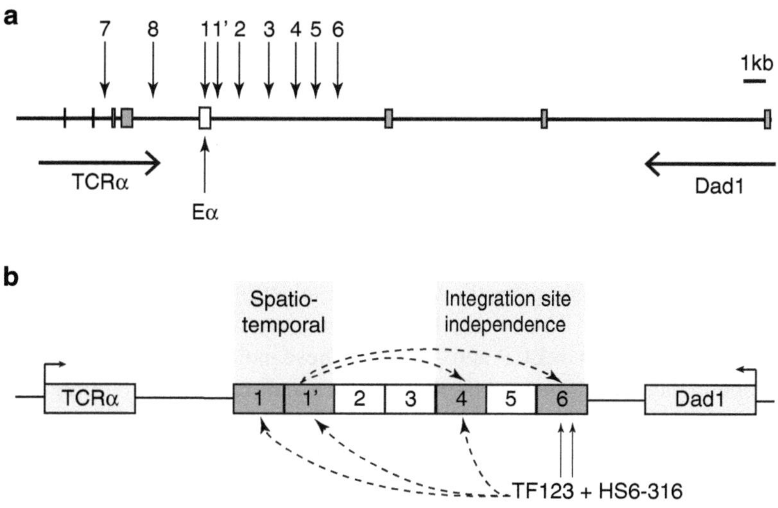

Fig. 2 The TCRα gene locus control region. (**a**) Scale diagram of the endogenous mouse TCRα/Dad 1 genomic locus showing the locations of the nine DNase I hypersensitive sites of its resident LCR (*numbers with arrows*). Exon sequences are marked with *filled boxes*. The *open box* represents the classical transcriptional enhancer of the locus (Eα). The transcriptional orientations of the TCRα and Dad1 genes are shown. Only the four TCRα constant region exons are shown here. The somatically rearranging variable region exons of the TCRα gene extend over 1-Mbp of DNA to the *left* (i.e. 5′) of what is depicted. (**b**) Depiction of TCRα LCR structure–function information obtained to date. The positions of the TCRα and Dad1 genes are shown as simple *boxes* with *arrows* indicating their transcriptional orientations. The *numbers* indicate the DNase hypersensitive sites (HS) of the core TCRα LCR [45] noting those known to be involved in the LCR's spatiotemporal specificity [44, 45] and its integration site-independence function [21, 26, 46]. TF123 and HS6-316 are functional sub-regions of HS6. *Dashed arrows* indicate known functional interactions between the numbered HS regions [26, 45, 46, 54]. *Note*: this depiction is not drawn to scale

that the cooperation between these two regions involves distinct LCR sub-elements that provide *cis*-mediated direction to *trans*-regulatory mechanisms that target both chromatin [44, 45] and DNA for cell type-specific epigenetic modification [54].

Aside from two reports identifying enhancer-blocking insulator activity within the TCRα LCR [39, 65], virtually all published data on the TCRα LCR are derived from transgenic mouse experiments. Because of the high cost and long timelines of such experiments, efforts were made to study TCRα LCR activity in cultured cell lines [26]. As in the β-globin case, TCRα LCR activity was incomplete after its direct introduction into chromatin of T cells [35]. These data made it clear that for a cell culture model to support full LCR activity, it would have to meet the apparent requirement (revealed in the β-globin LCR model [62]) for the LCR-driven reporter transgene to be present in the genome prior to differentiation and cell type specification. We hypothesized that an approach of introducing TCRα LCR driven reporter gene constructs into ESCs followed by their in vitro differentiation would satisfy this criterion, and yield T cells capable of supporting all the known aspects of TCRα LCR activity observed in vivo.

9 The TCRα LCR Manifests Full Activity in T Cells Derived In Vitro from Mouse ESC

Because the function of the TCRα LCR had been extensively tested and confirmed in vivo, it was an ideal candidate for testing the notion that committed cell types derived in vitro from transfected ESCs would support full LCR activity. An in vivo proven TCRα LCR linked human CD2 reporter gene construct [27] was introduced into mouse ESC by standard electroporation protocols. A co-transfected neomycin-G418 resistance gene allowed for the selection and propagation of stable-transfectant ESC clones. The in vitro ESC differentiation system allowed us to assay for all the key properties of LCR activity [35]. These include provision of integration site-independence, cell type-specificity, appropriate developmental timing and copy number-related mRNA production levels to a linked transcription unit.

In the absence of the LCR, the unlinked hCD2 reporter gene alone was not expressed in T cells derived from any of the transfected ESC clones ($n = 6$). In sharp contrast, we observed that every ESC clone bearing intact TCRα LCR linked hCD2 transgene integrants yielded T cells that expressed the hCD2 reporter gene robustly ($n > 12$). As the independent clones resulting from ESC transfection represent independent integration events, each presumably at varying random locations in the genome, these results provided strong evidence that the TCRα LCR-linked transgene can establish activity at any site of chromosomal integration in this assay [35], as it does in transgenic mice [27].

In addition to T cells, ESCs can be directed to differentiate into various blood cell types in the OP9 cell co-culture system including monocytic, erythroid and B-lineage lymphocytes [7, 43]. Thus, this technology enables some assessment of the cell type-specificity of TCRα LCR activity, something that generally is not possible in transfected, lineage-committed, cultured cell lines. In short, consistent, high level expression of the reporter gene was only observed in T-lineage cells derived from transfected ESC clones [35]. Furthermore, the entire course of T cell development (that normally takes place in the thymus) can be monitored in this co-culture system, using flow cytometry-based detection of key cell surface-expressed developmental marker proteins [20]. In such experiments, the temporal course of TCRα LCR activation generally parallels that of its gene locus of origin [35]. Thus the regulatory characteristics of this LCR in time and space are completely manifested in the in vitro ESC differentiation system.

In addition to the qualitative aspects of gene expression described above, there are important *quantitative* facets to LCR activity that are, generally speaking, not shared by other types of *cis*-acting gene regulatory DNA elements. That is to say, LCR-driven mRNA production levels per transgene copy only vary within a very narrow (<3-fold) range across multiple lines of transgenic mice [13]. This leads to a strong direct correlation between LCR-driven gene expression and integrated transgene copies. Neither classical enhancers, nor even insulator elements [37] have the capacity to achieve this degree of quantitative predictability of linked transgene activity. Previous work in transgenic mice has amply demonstrated, in assays that have utilized four different transcription units, that the TCRα LCR supports this key property [10, 27, 33, 44].

For the in vitro ESC differentiation system to serve as a true alterative to transgenic mice for LCR study, it was critical to determine if the copy number-dependence property of the TCRα LCR was supported in this co-culture system. Indeed, we observed copy number-related mRNA production levels from TCRα LCR driven transgenes in T cells derived in vitro from transfected ESCs [35]. Multiple independent sets of transfectants, each containing multiple independent transfectant clones were analyzed in real time, quantitative reverse transcriptase-mediated (qRT)-PCR to obtain this data. The range of reporter mRNA expression levels per transgene copy was a very tight 1.6-fold.

Interestingly, mRNA production levels from identical transgenes that were directly transfected into already differentiated T cell lines did not display the strong degree of copy number dependence described above. This is despite the fact that nearly all (10 of 11) of the T cell transfectant clones did express the transgene at some level indicating a high degree of integration site-independence. These results could be produced by a barrier-type insulator element, which is thought to prevent the spread of heterochromatin into a linked transcription unit [1]. It is possible that the TCRα LCR functions as a barrier insulator-like element in the directly transfected T cell lines, even though it is not displaying complete the LCR activity evident in T cells derived from transfected ESCs. Further work is needed to determine the molecular bases of the difference between complete LCR activity and the barrier insulator-like activity we observed in the T cell lines. Collectively, the data lead to a novel hypothesis that the copy number dependence property of LCRs arises from additional molecular mechanisms distinct from those that support integration site-independence.

10 Conclusion

The report of the above-described data on the TCRα LCR has validated the in vitro ESC to T cell differentiation system for the study of LCR activity [35]. Our report represented a major advance in that it demonstrated a complete assay for the full range of a *bona fide* LCR's activity at an ectopic genomic site that is not dependent on the use of transgenic mice [35]. The timeframe from ESC transfection with a reporter gene, through to the emergence of T cells from in vitro differentiation of isolated transfected cell clones, can be as short as 6 weeks. This is a much faster "time-to-data" scenario when compared with the timeline from pronuclear microinjection of a transgene into fertilized eggs, to the harvesting T cells from stably established transgenic mouse lines. Thus, wider adoption of this technology for the study of LCR elements that are active in hematopoietic cell types should greatly speed progress in the field of LCRs.

There is also very high potential for LCRs to be useful in the design of gene therapy vectors for use in stem cell transplantation. During genetically engineered stem cell differentiation in a transplant recipient, vector components derived from LCRs should provide predictable expression patterns and levels to a gene encoding a therapeutic protein. This idea has already yielded β-globin LCR containing vectors, versions of which can be used to treat Thalassemia patients [41]. Therapeutic genetic engineering of T cells has recently yielded very promising results in B cell leukemia patients [50]. The viral vectors used to transduce these cells have limited space for exogenous DNA sequences [59]. Thus, there is much merit to the continued structure/function study of LCRs to identify their key functional sequences. These key sequences will very likely have to be isolated and reassembled into a smaller form in order for them to be successfully accommodated by lentiviral vectors. These "miniaturized" LCR versions will have to be extensively tested to determine if they can reproduce the full function of the original LCR they are derived from. The in vitro ESC differentiation system we validated should enable these tests to be more readily undertaken, and more quickly completed in the future.

Acknowledgments The author thanks D. Loayza, J. Lovett, M. Kučerová-Levisohn and G.D. Raghupathy for helpful comments on the manuscript and A. Lahiji for the digital artwork in Figs. 1 and 2. Research in the author's laboratory is funded by National Institutes of Health (NIH) grant SC1-GM095402. The research infrastructure of Hunter College is supported in part by NIH grant MD007599.

References

1. Barkess, G., & West, A. G. (2012). Chromatin insulator elements: Establishing barriers to set heterochromatin boundaries. *Epigenomics, 4*(1), 67–80. doi:10.2217/epi.11.112.
2. Bender, M. A., Bulger, M., Close, J., & Groudine, M. (2000). Beta-globin gene switching and DNase I sensitivity of the endogenous beta-globin locus in mice do not require the locus control region. *Molecular Cell, 5*(2), 387–393.
3. Blom van Assendelft, G., Hanscombe, O., Grosveld, F., & Greaves, D. R. (1989). The beta-globin dominant control region activates homologous and heterologous promoters in a tissue-specific manner. *Cell, 56*(6), 969–977.
4. Bottardi, S., Aumont, A., Grosveld, F., & Milot, E. (2003). Developmental stage-specific epigenetic control of human beta-globin gene expression is potentiated in hematopoietic progenitor cells prior to their transcriptional activation. *Blood, 102*(12), 3989–3997. doi:10.1182/blood-2003-05-1540 2003-05-1540 [pii].
5. Bottardi, S., Ross, J., Pierre-Charles, N., Blank, V., & Milot, E. (2006). Lineage-specific activators affect beta-globin locus chromatin in multipotent hematopoietic progenitors. *EMBO Journal, 25*(15), 3586–3595. doi:10.1038/sj.emboj.7601232. 7601232 [pii].
6. Bulger, M., & Groudine, M. (2011). Functional and mechanistic diversity of distal transcription enhancers. *Cell, 144*(3), 327–339. doi:10.1016/j.cell.2011.01.024. S0092-8674(11)00063-8 [pii].
7. Cho, S. K., Webber, T. D., Carlyle, J. R., Nakano, T., Lewis, S. M., & Zuniga-Pflucker, J. C. (1999). Functional characterization of B lymphocytes generated in vitro from embryonic stem cells. *Proceedings of the National Academy of Sciences of the United States of America, 96*(17), 9797–9802.

8. Chung, J. H., Whiteley, M., & Felsenfeld, G. (1993). A 5′ element of the chicken beta-globin domain serves as an insulator in human erythroid cells and protects against position effect in Drosophila. *Cell, 74*(3), 505–514.
9. Crabtree, G. R. (1989). Contingent genetic regulatory events in T lymphocyte activation. *Science, 243*(4889), 355–361.
10. Diaz, P., Cado, D., & Winoto, A. (1994). A locus control region in the T cell receptor alpha/delta locus. *Immunity, 1*(3), 207–217.
11. Dickson, J., Gowher, H., Strogantsev, R., Gaszner, M., Hair, A., Felsenfeld, G., et al. (2010). VEZF1 elements mediate protection from DNA methylation. *PLoS Genetics, 6*(1), e1000804. doi:10.1371/journal.pgen.1000804.
12. Elefant, F., Su, Y., Liebhaber, S. A., & Cooke, N. E. (2000). Patterns of histone acetylation suggest dual pathways for gene activation by a bifunctional locus control region. *EMBO Journal, 19*(24), 6814–6822. doi:10.1093/emboj/19.24.6814.
13. Ellis, J., Tan-Un, K. C., Harper, A., Michalovich, D., Yannoutsos, N., Philipsen, S., et al. (1996). A dominant chromatin-opening activity in 5′ hypersensitive site 3 of the human beta-globin locus control region. *EMBO Journal, 15*(3), 562–568.
14. Festenstein, R., Tolaini, M., Corbella, P., Mamalaki, C., Parrington, J., Fox, M., et al. (1996). Locus control region function and heterochromatin-induced position effect variegation. *Science, 271*(5252), 1123–1125.
15. Forrester, W. C., Takegawa, S., Papayannopoulou, T., Stamatoyannopoulos, G., & Groudine, M. (1987). Evidence for a locus activation region: the formation of developmentally stable hypersensitive sites in globin-expressing hybrids. *Nucleic Acids Research, 15*(24), 10159–10177.
16. Forrester, W. C., Thompson, C., Elder, J. T., & Groudine, M. (1986). A developmentally stable chromatin structure in the human beta-globin gene cluster. *Proceedings of the National Academy of Sciences of the United States of America, 83*(5), 1359–1363.
17. Fraser, P., & Grosveld, F. (1998). Locus control regions, chromatin activation and transcription. *Current Opinion in Cell Biology, 10*(3), 361–365.
18. Fuchs, E., & Segre, J. A. (2000). Stem cells: A new lease on life. *Cell, 100*(1), 143–155. S0092-8674(00)81691-8 [pii].
19. Gaszner, M., & Felsenfeld, G. (2006). Insulators: Exploiting transcriptional and epigenetic mechanisms. *Nature Reviews Genetics, 7*(9), 703–713.
20. Germain, R. N. (2002). T-cell development and the CD4–CD8 lineage decision. *Nature Reviews Immunology, 2*(5), 309–322. doi:10.1038/nri798.
21. Gomos-Klein, J., Harrow, F., Alarcón, J., & Ortiz, B. D. (2007). CTCF-independent, but not CTCF-dependent, elements significantly contribute to TCRa locus control region activity. *Journal of Immunology, 179*(2), 1088–1095.
22. Greaves, D. R., Wilson, F. D., Lang, G., & Kioussis, D. (1989). Human CD2 3′-flanking sequences confer high-level, T cell-specific, position-independent gene expression in transgenic mice. *Cell, 56*(6), 979–986.
23. Gribnau, J., Diderich, K., Pruzina, S., Calzolari, R., & Fraser, P. (2000). Intergenic transcription and developmental remodeling of chromatin subdomains in the human beta-globin locus. *Molecular Cell, 5*(2), 377–386.
24. Grosveld, F., van Assendelft, G. B., Greaves, D. R., & Kollias, G. (1987). Position-independent, high-level expression of the human beta-globin gene in transgenic mice. *Cell, 51*(6), 975–985.
25. Groudine, M., Kohwi-Shigematsu, T., Gelinas, R., Stamatoyannopoulos, G., & Papayannopoulou, T. (1983). Human fetal to adult hemoglobin switching: changes in chromatin structure of the beta-globin gene locus. *Proceedings of the National Academy of Sciences of the United States of America, 80*(24), 7551–7555.
26. Harrow, F., Amuta, J. U., Hutchinson, S. R., Akwaa, F., & Ortiz, B. D. (2004). Factors binding a non-classical Cis-element prevent heterochromatin effects on locus control region activity. *Journal of Biological Chemistry, 279*(17), 17842–17849.

27. Harrow, F., & Ortiz, B. D. (2005). The TCRalpha locus control region specifies thymic, but not peripheral, patterns of TCRalpha gene expression. *Journal of Immunology, 175*(10), 6659–6667.
28. Hong, N. A., Cado, D., Mitchell, J., Ortiz, B. D., Hsieh, S. N., & Winoto, A. (1997). A targeted mutation at the T-cell receptor alpha/delta locus impairs T-cell development and reveals the presence of the nearby antiapoptosis gene Dad1. *Molecular Cell. Biology, 17*(4), 2151–2157.
29. Jenuwein, T., & Allis, C. D. (2001). Translating the histone code. *Science, 293*(5532), 1074–1080.
30. Kagi, D., Ledermann, B., Burki, K., Seiler, P., Odermatt, B., Olsen, K. J., et al. (1994). Cytotoxicity mediated by T cells and natural killer cells is greatly impaired in perforin-deficient mice. *Nature, 369*(6475), 31–37. doi:10.1038/369031a0.
31. Kennedy, M., Awong, G., Sturgeon, C. M., Ditadi, A., LaMotte-Mohs, R., Zuniga-Pflucker, J. C., et al. (2013). T lymphocyte potential marks the emergence of definitive hematopoietic progenitors in human pluripotent stem cell differentiation cultures. *Cell Reports, 2*(6), 1722–1735. doi:10.1016/j.celrep.2012.11.003. S2211-1247(12)00384-1 [pii].
32. Kioussis, D., Vanin, E., deLange, T., Flavell, R. A., & Grosveld, F. G. (1983). Beta-globin gene inactivation by DNA translocation in gamma beta-thalassaemia. *Nature, 306*(5944), 662–666.
33. Knirr, S., Gomos-Klein, J., Andino, B. E., Harrow, F., Erhard, K. F., Kovalovsky, D., et al. (2010). Ectopic T cell receptor-alpha locus control region activity in B cells is suppressed by direct linkage to two flanking genes at once. *PLoS One, 5*(11), e15527. doi:10.1371/journal.pone.0015527.
34. Krensky, A. M., Weiss, A., Crabtree, G., Davis, M. M., & Parham, P. (1990). T-lymphocyte-antigen interactions in transplant rejection. *New England Journal of Medicine, 322*(8), 510–517. doi:10.1056/NEJM199002223220805.
35. Lahiji, A., Kucerova-Levisohn, M., Lovett, J., Holmes, R., Zuniga-Pflucker, J. C., & Ortiz, B. D. (2013). Complete TCR-alpha gene locus control region activity in T cells derived in vitro from embryonic stem cells. *Journal of Immunology, 191*(1), 472–479. doi:10.4049/jimmunol.1300521. jimmunol.1300521 [pii].
36. Lang, G., Wotton, D., Owen, M. J., Sewell, W. A., Brown, M. H., Mason, D. Y., et al. (1988). The structure of the human CD2 gene and its expression in transgenic mice. *EMBO Journal, 7*(6), 1675–1682.
37. Li, Q., Harju, S., & Peterson, K. R. (1999). Locus control regions: Coming of age at a decade plus. *Trends in Genetics, 15*(10), 403–408.
38. Li, Q., Peterson, K. R., Fang, X., & Stamatoyannopoulos, G. (2002). Locus control regions. *Blood, 100*, 3077–3086.
39. Magdinier, F., Yusufzai, T. M., & Felsenfeld, G. (2004). Both CTCF-dependent and -independent insulators are found between the mouse T cell receptor alpha and Dad1 genes. *Journal of Biological Chemistry, 279*(24), 25381–25389.
40. Magram, J., Chada, K., & Costantini, F. (1985). Developmental regulation of a cloned adult beta-globin gene in transgenic mice. *Nature, 315*(6017), 338–340.
41. May, C., Rivella, S., Callegari, J., Heller, G., Gaensler, K. M., Luzzatto, L., et al. (2000). Therapeutic haemoglobin synthesis in beta-thalassaemic mice expressing lentivirus-encoded human beta-globin. *Nature, 406*(6791), 82–86. doi:10.1038/35017565.
42. Milot, E., Strouboulis, J., Trimborn, T., Wijgerde, M., de Boer, E., Langeveld, A., et al. (1996). Heterochromatin effects on the frequency and duration of LCR-mediated gene transcription. *Cell, 87*(1), 105–114.
43. Nakano, T., Kodama, H., & Honjo, T. (1994). Generation of lymphohematopoietic cells from embryonic stem cells in culture. *Science, 265*(5175), 1098–1101.
44. Ortiz, B. D., Cado, D., Chen, V., Diaz, P. W., & Winoto, A. (1997). Adjacent DNA elements dominantly restrict the ubiquitous activity of a novel chromatin-opening region to specific tissues. *EMBO Journal, 16*(16), 5037–5045.
45. Ortiz, B. D., Cado, D., & Winoto, A. (1999). A new element within the T-cell receptor alpha locus required for tissue-specific locus control region activity. *Molecular Cell Biology, 19*(3), 1901–1909.

46. Ortiz, B. D., Harrow, F., Cado, D., Santoso, B., & Winoto, A. (2001). Function and factor interactions of a locus control region element in the mouse T cell receptor-alpha/Dad1 gene locus. *Journal of Immunology, 167*(7), 3836–3845.
47. Ortiz, B. D., Nelson, P. J., & Krensky, A. M. (1997). Switching gears during T-cell maturation: RANTES and late transcription. *Immunology Today, 18*(10), 468–471.
48. Palmiter, R. D., & Brinster, R. L. (1986). Germ-line transformation of mice. *Annual Review of Genetics, 20*, 465–499.
49. Pipkin, M. E., Ljutic, B., Cruz-Guilloty, F., Nouzova, M., Rao, A., Zuniga-Pflucker, J. C., et al. (2007). Chromosome transfer activates and delineates a locus control region for perforin. *Immunity, 26*(1), 29–41.
50. Porter, D. L., Levine, B. L., Kalos, M., Bagg, A., & June, C. H. (2011). Chimeric antigen receptor-modified T cells in chronic lymphoid leukemia. *New England Journal of Medicine, 365*(8), 725–733. doi:10.1056/NEJMoa1103849.
51. Pui, J. C., Allman, D., Xu, L., DeRocco, S., Karnell, F. G., Bakkour, S., et al. (1999). Notch1 expression in early lymphopoiesis influences B versus T lineage determination [In Process Citation]. *Immunity, 11*(3), 299–308.
52. Recillas-Targa, F., Pikaart, M. J., Burgess-Beusse, B., Bell, A. C., Litt, M. D., West, A. G., et al. (2002). Position-effect protection and enhancer blocking by the chicken beta-globin insulator are separable activities. *Proceedings of the National Academy of Sciences of the United States of America, 99*, 6883–6888.
53. Reik, A., Telling, A., Zitnik, G., Cimbora, D., Epner, E., & Groudine, M. (1998). The locus control region is necessary for gene expression in the human beta-globin locus but not the maintenance of an open chromatin structure in erythroid cells. *Molecular Cell Biology, 18*(10), 5992–6000.
54. Santoso, B., Ortiz, B. D., & Winoto, A. (2000). Control of organ specific demethylation by an element of the T-cell receptor alpha locus control region. *Journal of Biological Chemistry, 275*, 1952–1958.
55. Schmitt, T. M., de Pooter, R. F., Gronski, M. A., Cho, S. K., Ohashi, P. S., & Zuniga-Pflucker, J. C. (2004). Induction of T cell development and establishment of T cell competence from embryonic stem cells differentiated in vitro. *Nature Immunology, 5*(4), 410–417. doi:10.1038/ni1055 ni1055 [pii].
56. Schmitt, T. M., & Zuniga-Pflucker, J. C. (2002). Induction of T cell development from hematopoietic progenitor cells by delta-like-1 in vitro. *Immunity, 17*(6), 749–756. doi:S1074761302004740 [pii].
57. Skarpidi, E., Vassilopoulos, G., Stamatoyannopoulos, G., & Li, Q. (1998). Comparison of expression of human globin genes transferred into mouse erythroleukemia cells and in transgenic mice. *Blood, 92*(9), 3416–3421.
58. Takegawa, S., Brice, M., Stamatoyannopoulos, G., & Papayannopoulou, T. (1986). Only adult hemoglobin is produced in fetal nonerythroid x MEL cell hybrids. *Blood, 68*(6), 1384–1388.
59. Thomas, C. E., Ehrhardt, A., & Kay, M. A. (2003). Progress and problems with the use of viral vectors for gene therapy. *Nature Reviews Genetics, 4*, 346–358.
60. Townes, T. M., Lingrel, J. B., Chen, H. Y., Brinster, R. L., & Palmiter, R. D. (1985). Erythroid-specific expression of human beta-globin genes in transgenic mice. *EMBO Journal, 4*(7), 1715–1723.
61. Tuan, D., Solomon, W., Li, Q., & London, I. M. (1985). The "beta-like-globin" gene domain in human erythroid cells. *Proceedings of the National Academy of Sciences of the United States of America, 82*(19), 6384–6388.
62. Vassilopoulos, G., Navas, P. A., Skarpidi, E., Peterson, K. R., Lowrey, C. H., Papayannopoulou, T., et al. (1999). Correct function of the locus control region may require passage through a nonerythroid cellular environment. *Blood, 93*(2), 703–712.
63. Vieira, K. F., Levings, P. P., Hill, M. A., Crusselle, V. J., Kang, S. H., Engel, J. D., et al. (2004). Recruitment of transcription complexes to the beta-globin gene locus in vivo and in vitro. *Journal of Biological Chemistry, 279*(48), 50350–50357. doi:10.1074/jbc.M408883200. M408883200 [pii].

64. West, A. G., Huang, S., Gaszner, M., Litt, M. D., & Felsenfeld, G. (2004). Recruitment of histone modifications by USF proteins at a vertebrate barrier element. *Molecular Cell, 16*(3), 453–463.
65. Zhong, X. P., & Krangel, M. S. (1999). Enhancer-blocking activity within the DNase I hypersensitive site 2 to 6 region between the TCR alpha and Dad1 genes. *Journal of Immunology, 163*(1), 295–300.

Dynamical Roles of Jacobian Feedback Loops and Qualitative Modeling

Bourama Toni

Abstract The chapter presents a mathematical methodology applicable to qualitative modeling of systems using the feedback loops encoded in the Jacobian matrix and described by the products of the Jacobian entries under cyclic permutations of the indices. The relation between these feedback loops and the Jacobian spectrum defines their dynamical properties. We determine the conditions of nondegeneracy and appearance of multiple equilibria in terms of feedback loops as well as the conditions of loop stability to induce the stability analysis of the systems. In particular we emphasize the applications to qualitative modeling in biological and biochemical sciences, economics. We present a complete loop analysis of the celebrated Lorenz and Rossler systems predicting their global dynamics.

The methodology is proved efficient to assert the possibility of multistationarity, periodicity, self-sustained oscillations, and chaos using strictly the qualitative relations and assumptions of the systems, to achieve primarily qualitative understanding rather than quantitative numerical prediction. We also show the Jacobian loops technique is easy to implement and could quickly demarcate both parameter and phase spaces into exciting regions (limit cycle, multiple equilibria, chaotic behavior), non-exciting ones (single stable fixed points), hard-instance regions (ergodic behavior). Therefore as such the technique could be useful in surveying dynamical responses of models simulating physico-chemical, biological, biochemical, economical systems and game theory.

AMS Subject Classification: 34C, 58F, 92B, 93D

B. Toni (✉)
Department of Mathematics and Computer Science, Virginia State University, Petersburg, Virginia, VA 23806, USA
e-mail: btoni@vsu.edu

B. Toni (ed.), *New Frontiers of Multidisciplinary Research in STEAM-H (Science, Technology, Engineering, Agriculture, Mathematics, and Health)*, Springer Proceedings in Mathematics & Statistics 90, DOI 10.1007/978-3-319-07755-0_10,

1 Introduction

Interest in qualitative mathematical methods has been expanding during the last decades, mostly due the wide applications in biosciences, social sciences including behavioral sciences and economics. In these mathematical models variables define parts of the modeled system, parameters designate factors that influence the system dynamic but are not usually influenced by it; the system dynamics are defined as relationships among system parts and between the parts and some extra systematic factors; the dynamics are described by the model equations which could be in the differential or difference format. For models representing complex systems in biological and behavioral sciences, it is usually impossible or infeasible to determine the quantitative value or the precise functional form of most of the interactions between system parts. However, it is often possible to determine the qualitative properties of these interactions; sometimes what can only be ascertained is that there is or there is not interaction between variables, which could be translated by yes or no, 0 or 1, e.g., in *Boolean* models, making qualitative modeling more appropriate in these sciences. For example within ecology, qualitative models are more easily ascertained in the attempt to estimate intrinsic growth rate, carrying capacity, competition coefficients. Economists trust more the sign and direction of interactions between major parts of the economy but doubt their functional form can be determined more precisely. In psychology, there is little expectation for a precise mathematical function to accurately represent human behavior as reflected in imprecise belief states or preferences of typical real-world agents. Indeed, in biosciences, physical-chemistry, economics and behavioral sciences, informations about the underlying dynamics often reside in the rules of construct of the system and not in the absolute quantitative values. The data and phenomena being studied are essentially qualitative. Therefore, absent the precise quantitative, qualitative modeling concerns what properties, in particular dynamical properties, can be derived from these qualitative relations between the model variables.

Results established by qualitative models, with less commitment to details, tend to achieve a greater generality. In addition this type of modeling allows an understanding of phenomena less susceptible to the drawbacks of the quantitative usual idealization methods. Simplifications are inherent to both quantitative and qualitative models; in the former, they are realized by decreasing specificity, whereas, in the latter, they usually involve unrealistic assumptions in order to use some precise and tractable mathematical equations with fewer or more easily estimable parameters, in the hope that these intentional misrepresentations will not distort the salient features of the system. Several qualitative methods have been proposed. See Levins, Puccia and Levins, Orzack and Sober [9, 10, 14]. However as claimed by Levins *Scientific modeling can maximize at most two of three virtues: generality, realism, and precision:*

1. Sacrifice generality for precise quantitative predictions about specific systems and maximize realism by representing as many system details as possible.

2. Sacrifice realism to make unrealistic assumptions so systems can be described with general mathematically tractable equations producing precise quantitative predictions.
3. Sacrifice precision to abandon quantitative accuracy for qualitative relations between variables for maximum generality and realism.

This chapter introduces a recent and progressive methodology in qualitative modeling, for instance, to showcase how much can be achieved about the structure and behavior of systems partially specified by using the sign of an interspecific interaction. In a series of studies we have developed an efficient tool in the qualitative study of systems described by differential or difference equations, namely, a tool based on the dynamical roles of Jacobian Feedback Loops. Such a tool intends to survey the dynamical response of models simulating physico-chemical, biological and economical systems by stressing qualitative understanding as the primary goal rather than numerical prediction [29, 30, 33–36].

Dynamical systems theory is mostly based on quantitative values of the Jacobian entries. But for some systems, mainly in biosciences such as biology and biochemistry, in economics and behavioral sciences, the relevant informations are of qualitative nature. Quantitative results are rare in studying interactions in a system of biochemical compounds: A gene X could be shown to be an activator (or a repressor) of the expression of a gene Y, but usually without knowing the strength of the interaction, the concentrations and their kinetics. In a conflict resolution model featuring the variables of Attitude (A), Behavior (B) and Contradiction (C), one can "accurately" determine if the variables mutually influence each other positively (+), negatively (−) or no influence (0) with no need to quantify the strength of the influences.

The theory of Jacobian loops is therefore the analysis of the dynamics (simple and complex) using solely the loop-pattern Jacobian matrix, that is, even when only the signs, not the magnitudes of the Jacobian terms, are known. Section 2 presents the preliminary concepts, definitions and examples. Section 3 discusses the Jacobian loops required for the existence of multiple equilibria (multistationarity). Section 4 addresses loop and qualitative stability, whereas Sect. 5 is devoted to the applications, in particular the complete loop analysis of the well-known Lorenz and Rossler systems. The last section presents directions for future research on qualitative modeling as an efficient tool to address the ever increasing complexity of system mathematical models.

2 Preliminaries

Consider the autonomous differential system

$$\begin{aligned} \dot{x}_i(t) &= \frac{dx_i(t)}{dt} = F_i(x, a) \\ x &= (x_1, \cdots, x_n) \in \mathbb{R}^n, \quad a = (a_1, a_2, \cdots, a_N) \in \mathbb{R}^N, \end{aligned} \tag{1}$$

describing a dynamical system with phase space in $\mathbb{R}^n$, and the parameter/control space in $\mathbb{R}^N$. The component functions F_i, $i = 1, 2, \cdots, n$ of $F(x, a)$ are assumed to be at least $C^1(U)$, that is, differentiable along with their first partial derivatives on U an open set of $\mathbb{R}^n$. The partial order relation $x \leq y \iff x_i \leq y_i, i = 1, \ldots, n$ defines the vector order in $\mathbb{R}^n$. The Jacobian matrix at $\bar{x} = (\bar{x}_1, \ldots, \bar{x}_n)$ is given by

$$J(\bar{x}) = DF(\bar{x}) = \left[\frac{\partial(F_1, \cdots, F_n)}{\partial(x_1, \cdots, x_n)}(\bar{x})\right] = \left[\frac{\partial F_i}{\partial x_j}(\bar{x})\right]_{1 \leq i,j \leq n} = [J_{ij}]_{1 \leq i,j \leq n}, \quad (2)$$

and in general depends on the state variables, except for linear systems [1, 38].

Remark 2.1. Actually the relation $\dot{x}_i(t) = F_i(x, a)$ shows how the rate of change in variable x_i is dependent on changes in any given variable x_j. Therefore the Jacobian entry $\frac{\partial F_i}{\partial x_j}(x) = J_{ij}$, for $1 \leq i, j \leq n$ describes the interaction between the variables x_i and x_j, as positive (respectively negative, no) interaction for $J_{ij} > 0$ (respectively $J_{i,j} < 0,\ J_{i,j} = 0$).

2.1 Jacobian Loops: Definitions and Notations

2.1.1 Permutations of Indices and Their Properties

Let $\mathscr{I}_n$ be the set of indices $1, \cdots, n$ and denote by $I_k = \{i_1, \cdots, i_k\}$ an ordered subset of k different elements of $\mathscr{I}_n$ and by $\tilde{I}_k = \pi_k(I_k) = \{j_1, \cdots, j_k\}$, with $\pi_k \in \Xi_k$ a permutation of I_k. Recall $Card(\Xi_k) = k!$, i.e., there are $k!$ permutations. Every permutation π_k may be factored into ν disjoint circular (cyclic) permutations $\sigma_i, i = 1, \cdots, \nu$, that is, $\pi_k = \sigma_1\sigma_2\cdots\sigma_\nu$. The signature of π_k, denoted $sg(\pi_k)$, is $(-1)^\eta$, η the number of inversions in π_k, that is, the number of pairs (j_m, j_n) with $j_m > j_n$ while $i_m < i_n$, for $j_m = \pi_k(i_m)$, and $j_n = \pi_k(i_n)$. The permutation π_k is even (resp. odd) for an even (resp. odd) η. There are exactly $\frac{k!}{2}$ even and exactly $\frac{k!}{2}$ odd permutations in Ξ_k. We denote Ξ_k^c (resp. Ξ_k^e, Ξ_k^o) the subset of circular (resp. even, odd) permutations [3]. The set Ξ_n is the classic symmetric group of permutations on the set of indices $\mathscr{I}_n$.

We have the following defining concepts [33, 34].

Definition 2.2. 1. The set of nonzero terms J_{ij}, $i \in I_k$, and $j \in \tilde{I}_k$, describes a *Jacobian loop* associated with the nonzero product

$$P(\pi_k, J) := \prod_{l=1}^{l=k} J_{i_l \pi_k(i_l)} = J_{i_1\pi_k(i_1)} J_{i_2\pi_k(i_2)} \ldots J_{i_k\pi_k(i_k)} \quad (3)$$

called a *loop product*.

2. The loop is called a k-order *simple Jacobian loop* L_k when the permutation π_k is a k-cycle$= (i_1, i_2, \cdots, i_k)$ with the loop product

$$p_k = J_{i_1 i_2} J_{i_2 i_3} \cdots J_{i_{k-1} i_k} J_{i_k i_1}. \quad (4)$$

3. Its sign $sgn(L_k)$ is that of the loop product $p_k = P(L_k) := P(\pi_k, J)$. Its length or dimension $l(L_k) = k$ is the number of loop factors $J_{i_l \pi_k(i_l)}$ involved, as well as the number of related system variables x_i.

Remark 2.3. 1. A simple Jacobian loop L_k is positive (resp. negative) for an even (resp. odd) number of negative loop factors $J_{i_l \pi_k(i_l)}$ in the loop product $P(L_k)$.
2. The loop L_k has the following representation called *loop graph* or *interaction graph* and denoted by $\mathbb{L}_k$: it consists of k distinct vertices given by the system variables $x_i, i = 1, \ldots, k$ and k edges $E_{ij} = (x_i, x_j, s_{ij})$ directed from j to i where $s_{ij} = sign(J_{ij})$ denotes the nature of the interaction between the variable x_j and x_i.
3. A positive (resp. negative) loop involving the variables $x_1, x_2, \ldots, x_k$ is also conveniently denoted $L^+_{x_1 x_2 \cdots x_k}$ (resp. $L^-_{x_1 x_2 \cdots x_k}$).
4. A loop graph $\mathbb{L}_k$ is *complete* if for every $i \neq j$ there is a directed polygonal line connecting x_i to x_j, that is, $\overline{x_i x_{k_1}}, \overline{x_{k_1} x_{k_2}}, \cdots, \overline{x_{k_r} x_j}$.

Definition 2.4. 1. A non-circular permutation π_k yields a *union* of simple Jacobian loops, called a *composite loop* $\mathscr{L}_k^\nu = \cup_{i=1}^{i=\nu} L_i = (L_1, \ldots, L_\nu)$ of dimension $l(\mathscr{L}_k^\nu) = k$ given by the sum of the lengths of its ν simple component loops, i.e., $k = \sum_{i=1}^{i=\nu} d(L_i) = 1 + \cdots + \nu$.
2. A proper composite loop $\mathscr{L}_k^\nu$ of *resonance* (ν, k) is a disjoint union of ν simple loops of total length k, that is, the component loops do not share a vertex.

Remark 2.5. We denote P_k^ν the loop product of a composite loop $\mathscr{L}_k^\nu$. The sign of a composite loop $\mathscr{L}_k^\nu$ is the sign of P_k^ν, or equivalently, $sign(\mathscr{L}_k^\nu) = \prod_{l=1}^{l=\nu} sign(L_i) = (-1)^{\nu_-}$, where ν_- is the number of negative simple loops in $\mathscr{L}_k^\nu$.
$\chi_k^\nu = (-1)^{\nu+1}$ is the characteristic of the proper composite loop of resonance (ν, k). Therefore a k-order proper composite loop has a negative (resp. positive) resonance, i.e., a negative (resp. positive) characteristic for ν even (resp. odd).

Definition 2.6. A k-order Feedback F_k is defined by

$$F_k = \sum_{all \nu} (-1)^{\nu+1} P_k^\nu, \tag{5}$$

where P_k^ν is the loop product of the proper composite loop $\mathscr{L}_k^\nu$.

Consequently we have the following

Lemma 2.7. *1. A composite loop $\mathscr{L}_k^\nu$ is positive (resp. negative) for an even (resp. odd) number of its negative simple loops.*
2. A proper composite loop $\mathscr{L}_k^\nu$ with all component simple loops negative has a negative resonance in the Feedback F_k as defined above.

Definition 2.8. 1. We call *qualitative matrix* S a matrix consisting exclusively of the signed entries $s_{ij} \in \{+,-,0\}$, that is, $S := [s_{ij}]_{1 \le i,j \le n}$. We denote S by A_q for a qualitative matrix associated with a matrix $A = [a_{ij}]$, that is, $s_{ij} = sign(a_{ij})$, for $1 \le i, j \le n$.
2. The *loop structure (or qualitative structure),* denoted $\mathbb{L}_{\mathscr{R}}$, corresponding to the region $\mathscr{R}$ in the phase space or to a sign-pattern is the set of all Jacobian loops (simple and proper composite) along with their signs.

2.2 Methodology Requirements

The Jacobian loop analysis required first the determination of the loop structure associated to the system in a given region of the phase space, either from the signed entries of a Jacobian matrix evaluated at equilibria or constant at some parameter values or solely from the qualitative evaluation of the interaction between the variables in terms of positive, negative or zero. Determining and analyzing the qualitative structure anywhere in the phase space, including around the steady states, if any, yields some understanding of the local and global dynamics of the system.

2.2.1 Some Terminologies and Notations

Consider a matrix $A = [a_{ij}]$, possibly a Jacobian, and its corresponding qualitative matrix $S = A_q$.

1. For $x = (x_1, \cdots, x_n) \in \mathbb{R}^n$ $x_q := (sign(x_i), i = 1, \cdots, n)$ is called a qualitative vector. The corresponding equivalence class is $[x] := \{y \in \mathbb{R}^n / y_q = x_q\}$.
2. A matrix $B = [b_{ij}]_{1 \le i,j \le n}$ is qualitatively or sign equivalent to $A = [a_{ij}]_{1 \le i,j \le n}$ in the region $\mathscr{R}$ if B has the same sign pattern as A, i.e., $A_q = B_q$. We denote $A \odot B$.
3. $\langle A \rangle$ denotes the qualitative equivalence class of matrix A represented by the *qualitative matrix* A_q.
4. The qualitative equivalence class $\langle\rangle$ is represented by a $n \times n$ array of nonzero $s_{ij} = +,-,0$ entries such that, for any matrix $A = [a_{ij}] \in \langle\rangle$, $sign(a_{ij}) = s_{ij}, 1 \le i, j \le n$.
5. A matrix $B = [b_{ij}]_{1 \le i,j \le n}$ is loop equivalent to $A = [a_{ij}]_{1 \le i,j \le n}$ in the region $\mathscr{R}$, if B yields the same loop structure $\mathbb{L}_{\mathscr{R}}$ as A. We denote $A \circlearrowleft B$.
6. $\rangle A \langle$ denotes the loop equivalence class of matrix A. The class $\rangle\langle = \mathbb{L}$ is a loop structure represented by a set of signed loops, such that any matrix $A \in \rangle\langle$ has the loop structure $\mathbb{L}$.

2.2.2 Examples

To illustrate consider in the 3-variables xyz-phase-space the following three distinct matrices:

$$A = \begin{pmatrix} -1 & -3 & -2 \\ 5 & 0 & 8 \\ 1 & -4 & 7 \end{pmatrix}; \quad B = \begin{pmatrix} -8 & -9 & -1 \\ 3 & 0 & 7 \\ 4 & -2 & 5 \end{pmatrix}; \quad M = \begin{pmatrix} 2 & 3 & -6 \\ -9 & 0 & -8 \\ 7 & 5 & -4 \end{pmatrix}.$$

The qualitative equivalence classes are represented by

$$\langle A \rangle = \langle B \rangle \equiv A_q = B_q = \begin{pmatrix} - & - & - \\ + & 0 & + \\ + & - & + \end{pmatrix}, \quad \langle M \rangle \equiv M_q = \begin{pmatrix} + & + & - \\ - & 0 & - \\ + & + & - \end{pmatrix},$$

where $\equiv$ indicates representation. Therefore

1. $A \odot B$, but A and M are not qualitative equivalent, as are not B and M.
2. $A \circlearrowleft B \circlearrowleft M$.
3. The common loop structure $\mathbb{L}$ contains the following simple loops and their composition:
   (a) Negative 1-loops given by either L_x^- or L_z^-; positive 1-loops given by L_z^+ or L_x^+
   (b) Three negative 2-loops given by L_{xy}^-; L_{yz}^-; L_{xz}^-
   (c) One positive 3-loop L_{xyz}^+, and one negative 3-loop L_{xzy}^-.

Remark 2.9. 1. Clearly the real vector spaces $\mathbb{R}^n$ and the space $\mathcal{M}_n$ of all real matrices are partitioned as

$$\mathbb{R}^n = \cup_{x \in \mathbb{R}}[x], \quad \mathcal{M}_n = \cup_{A \in \mathcal{M}} \langle A \rangle.$$

2. The classes $[x]$, $\langle A \rangle$, and $\rangle A \langle$ are convex cones respectively in $\mathbb{R}^n$ and $\mathcal{M}_n$, closed by addition and multiplication by a positive scalar. The cone $[x]$ is solid if $sign(x_i) \neq 0, i = 1, \ldots, n$. The set $\overline{[x]} := \{y \in \mathbb{R}^n | y_l = x_l, \quad or \quad 0\}$ is the closure of $[x]$. Similarly one defines the solid cone $\langle A \rangle$, and $\rangle A \langle$, and the closure $\overline{\langle A \rangle}$, and $\overline{\rangle A \langle}$.
3. The equation $Ax = b$ is *qualitatively or sign solvable* if $B \in \langle A \rangle$, $c \in [b]$, $By = c$ implies $y \in [x]$.
4. Denote by $\langle xAy \rangle$ the set of matrices which map the set $[x]$ into $[y]$. Of course an interesting question will be to characterize algebraically $\langle xAy \rangle$ [18, 21].

Some general observations based on the above definitions and examples and some results in matrix theory lead to the following lemma:

Lemma 2.10. *1. Qualitative equivalence obviously implies loop equivalence but not inversely.*

2. *Given two matrices A and B loop equivalent, the qualitative class $\langle B \rangle$ may be obtained from that of A by some combination of negation, transposition, permutation, and signature similarity.*

Proof. Note that the loop equivalence class yields the loop structure uncovered from a representative qualitative matrix Q. Therefore if $A \circlearrowleft B$ then they are sign-patterned in a way to provide the same loop structure, though they are not necessarily qualitatively equivalent in the sense $sign(a_{ij}) = sign(b_{ij})$. Thus matrix operations such as negation, transposition, permutations, and signature similarity allow to derive one sign pattern from the other. □

An immediate consequence is the following result:

Corollary 2.11. *Let Λ_A denotes the spectrum of matrix A, that is, the set of the eigenvalues of A. If $A \circlearrowleft B$ therefore $\Lambda_A = \Lambda_B$.*

Next we show how the loops and their combinations, i.e., composite loops and their Feedback, have their dynamical roles uncovered from the Jacobian characteristic equation.

2.3 Loops and Jacobian Spectrum

For a matrix A, Jacobian or otherwise, given by $A = [A_{ij}]_{1 \leq i,j \leq n}$ the characteristic polynomial is defined by the monic polynomial

$$\mathscr{C}_A(\lambda) = |\lambda I - A| = \lambda^n + c_1\lambda^{n-1} + \ldots + c_k\lambda^{n-k} + \ldots + c_{n-1}\lambda + c_n. \qquad (6)$$

From Linear Algebra [1,6] the coefficients may be expressed as

$$c_k = coefficient\,(\lambda^{n-k}) = \sum (-1)^k m_k, \quad k = 0, \cdots, n-1, \qquad (7)$$

where the sum extends over all kth order principal minors m_k of A. For instance we have

$$\begin{aligned} c_n &= (-1)^n det(A) = (-1)^n |A|, \quad \text{for k = n,} \\ c_1 &= -\sum A_{ii} = -Tr(A), \quad \text{where Tr(A) is the trace of A.} \end{aligned} \qquad (8)$$

From the theory of determinant and permutations we may write

$$m_k = \sum_{\pi_k \in \Xi_k} (-1)^\eta \prod_{i_l \in \mathscr{I}_k} A_{i_l \pi_k(i_l)} = \sum_{all\,\nu} (-1)^{k-\nu} \prod_{i=1}^{i=\nu} P(\sigma_i, A) = \sum_{all\,\nu} (-1)^{k-\nu} P(\mathscr{L}_k^\nu), \qquad (9)$$

where the permutation $\pi_k \in \Xi_k$ of the indices $1 \leq i_1 < i_2 < \cdots < i_k \leq n$ factors into the cyclic permutations $(\sigma_1, \cdots, \sigma_\nu)$ yielding the proper composite loop $\mathscr{L}_k^\nu = (\sigma_1, \cdots, \sigma_\nu)$ with loop product $P(\mathscr{L}_k^\nu) = P_k^\nu$ as defined above.

Consequently we obtain an expression of the characteristic coefficients c_k in terms of the proper composite loops. Importantly we obtain [20]

Lemma 2.12. *The kth order Feedback $\mathscr{F}_k$ involving all the proper composite loops $\mathscr{L}_k^\nu$ with $\nu = 1, \cdots, k$ may be expressed in terms of the coefficients of the characteristic polynomial by*

$$\mathscr{F}_k = c_k = \sum_{all\nu} (-1)^{\nu+1} P_k^\nu, \quad k = 1, \cdots, n \tag{10}$$

Proof. Immediate from the above formulas, the k-order principal subdeterminant D_k of matrix A is written as

$$D_k = \sum_{all\nu} (-1)^{k-\nu} P_k^\nu, \quad \text{giving}(-1)^{k+1} D_k = \sum_{all\nu} (-1)^{\nu+1} P_k^\nu = \mathscr{F}_k.$$

□

Remark 2.13. 1. First recall the zeros of the characteristic polynomial are the eigenvalues of the matrix A, that is, they are the elements of the spectrum Λ_A. They are of multiplicity m if $(z - \lambda)^m$ factorizes $\mathscr{C}_A(z)$. For $m = 1$ the corresponding eigenvalue is said to be simple, such as when A has n distinct eigenvalues.
2. Importantly the k-order Feedback $\mathscr{F}_k$ being the k-order coefficient of the characteristic polynomial entails that the loop factors $A_{i\sigma(i)}$ defined above are the only Jacobian entries contributing to the characteristic equation, and therefore, influence directly the eigenvalues of the matrix, and consequently the dynamics.
3. From the standard theory of equations it is also known that the coefficients $c_k = \mathscr{F}_k$ are related to the eigenvalues λ_i in a systematic way by the following Viete formulas: (See [1, 6, 13]).

$$\begin{aligned}
c_1 &= F_1 = -(\lambda_1 + \cdots + \lambda_n) = -\sum_{i=1}^{i=1} A_{ii}. \\
c_2 &= F_2 = \sum_{i,j=1,i<j} \lambda_i \lambda_j = \sum_{i,j=1,i<j} (A_{ii}A_{jj} - A_{ij}A_{ji}) \\
&= \lambda_1\lambda_2 + \lambda_1\lambda_3 + \cdots + \lambda_{n-1}\lambda_n. \\
c_3 &= F_3 = -\sum_{i,j,k=1,i<j<k} \lambda_i \lambda_j \lambda_k \\
&= -(\lambda_1\lambda_2\lambda_3 + \lambda_1\lambda_2\lambda_4 + \cdots + \lambda_{n-2}\lambda_{n-1}\lambda_n). \\
&\dots \\
c_n &= F_n = (-1)^n \lambda_1 \lambda_2 \cdots \lambda_n.
\end{aligned} \tag{11}$$

Immediate from the above definitions and the theory of determinants [1, 6], we have

Lemma 2.14.

$$
\begin{aligned}
det(A) = |A| &= \sum_{\pi_n \in \Xi_n} sg(\pi_n) P(\pi_n, A) \\
&= \sum_{\pi_n^e \in \Xi_n^e} P(\pi_n^e, A) - \sum_{\pi_n^o \in \Xi_n^o} P(\pi_n^o, A).
\end{aligned}
\tag{12}
$$

Proof. Every permutation $\pi_n = \sigma_1\sigma_1 \cdots \sigma_n$ is associated with a composite loop $\mathscr{L}_n = (L_1, L_2, \cdots L_n)$ with the simple loop L_i defined by the cyclic permutation σ_i. π_n^e (resp. π_n^o) denotes an even (resp odd) permutation. In the expression (12) of $|A|$ all the loop product $P(\pi_n^e, A)$ have the same sign opposite to that of $P(\pi_n^o, A)$. In fact if $\mathscr{L}_n^e$ (resp. $\mathscr{L}_n^o$) is the composite loop associated with π_n^e (resp. π_n^o) then it has an even (resp. odd) number ν of components L_i for n even, and it has an odd (resp. even) number ν of components L_i for n odd. □

A classic result therefore leads to the following theorem:

Theorem 2.15. *A necessary condition to have all eigenvalues with negative real parts $\mathscr{R}_e < 0$ is that all kth order Feedback $\mathscr{F}_k$ must be positive.*

Proof. The proof is straightforward from a classic lemma we recall (See for instance [3]): given a n-degree polynomial with real coefficients

$$
p(z) := z^n + a_1 z^{n-1} + \cdots + a_k z^{n-k} + \cdots + a_{n-1} z + a_n, \tag{13}
$$

a necessary condition for $p(z)$ to have all its zeros z_* with negative real part, i.e., $p(z)$ is a *strongly stable* polynomial, is that all the coefficients $a_k > 0, k = 1, \cdots, n$. Indeed, if all real parts are negative then we have either the form

$$
\begin{aligned}
p(z) &= \prod (z - (\alpha + i\beta)) \\
&= \prod (z^2 - 2\alpha z + \alpha^2 + \beta^2) \\
&= \prod (z^2 + az + b), \quad a > 0, \quad b > 0,
\end{aligned}
\tag{14}
$$

or

$$
p(z) = \prod (z - \alpha) = \prod (z + a), \quad a > 0. \tag{15}
$$

By successive multiplication we necessarily obtain $p(z)$ with $a_k > 0$. □

We also have

Lemma 2.16. *A proper composite loop $\mathscr{L}_k^\nu$ with all ν components simple loops negative has a negative resonance in the kth order Feedback, that is, its characteristic χ_k^ν is negative.*

Proof. Indeed the term $(-1)^{\nu+1}P(\mathscr{L}_k^{\nu})$ in the Feedback $\mathscr{F}_k$ has the sign $(-1)^{\nu}(-1)^{\nu+1} = -1$. Hence the claim. □

Theorem 2.17. *If there is no proper composite loop $\mathscr{L}_k$ of dimension $k \leq n$, then the characteristic coefficient $c_k = 0$.*

Moreover at least one proper composite loop $\mathscr{L}_n$ of the system dimension is necessary to have a nonsingular jacobian matrix.

Proof. From the above Viete formulas the characteristic coefficient c_k can be written as

$$c_k = \mathscr{F}_k = \sum(-1)^{\nu}\prod_{l=1}^{l=r_1} J_{i_l\pi_k(i_l)} \prod_{l=r_1+1}^{l=r_2} A_{i_l\pi_k(i_l)} \cdots \prod_{l=r_{\nu-1}}^{l=r_\nu} A_{i_l\pi_k(i_l)}. \tag{16}$$

Terms in the expression of c_k with one cyclic permutation correspond to $r_1 = k$, those with two cyclic permutations correspond to $r_1 < k,\ r_2 = k - r_1$, and so on. Therefore, if there is no proper composite loop $\mathscr{L}_k$ of dimension k, then each term of the sum is zero. For $k = n$ the system dimension, these formulas yield clearly $Det(A) = |A| = 0$. hence the claim. □

Definition 2.18. We say that the qualitative equivalent class $\langle A\rangle$ or the loop equivalence class $\rangle A\langle$ is *qualitatively nondegenerate* if every matrix in the class is nonsingular in the sense $|A|$ is nonzero.

We prove

Theorem 2.19. *If the loop structure $\mathbb{L}$ does contains a composite loop $\mathscr{L}_n$ of the dimension of the system, and all such loop $\mathscr{L}_n$ have the same sign, then the corresponding Jacobian determinant $|A|$ is nonzero.*

Proof. Indeed suppose all the composite loops $\mathscr{L}_n$ of the dimension of the system have the same sign. Then c_n, consisting of nonzero terms of the same sign, is therefore nonzero. Consequently, the Jacobian determinant is nonzero.

Moreover c_n is positive (resp. negative) if all $\mathscr{L}_n$ have an odd (resp. even) number ν_o (resp. ν_e) of simple loops L_i. □

We also prove

Theorem 2.20. *A positive simple loop in the loop equivalence class is a necessary condition for the Jacobian matrix to have a positive real eigenvalue.*

Proof. Recall the characteristic coefficients given in the Viete formulas, that is,

$$c_k = \sum_{\mathscr{L}_k=(L_1,\cdots,L_\nu)} (-1)^{\nu}P(\sigma_1, A)\cdot P(\sigma_2, A)\cdot \cdots \cdot P(\sigma_\nu, A), \tag{17}$$

where the simple Jacobian loops $L_i, i = 1,\cdots,\nu$ are defined by the cyclic permutations $\sigma_i, i = 1,\cdots,\nu$. Now assume that the region has a negative loop equivalence class, i.e., there is no positive simple loop in its loop structure. So every simple loop L_i defined by $\sigma_i \in \Xi_k^c$ is negative. Therefore the corresponding nonzero loop product $P(\sigma_i, A)$ is also negative. Then for a composite loop $\mathscr{L}_k = (L_1, \cdots, L_\nu)$ we have

$$sign(-1)^\nu P(\sigma_1, A) \cdot P(\sigma_2, A) \cdot \cdots \cdot P(\sigma_\nu, A) = sign((-1)^{2\nu}) = +. \quad (18)$$

Thus all the characteristic coefficients c_k are positive. This entails a characteristic polynomial of degree n with only positive coefficients. By Descartes' rules of sign it cannot have a positive real root. Hence the claim. □

3 Jacobian Loops for Multiple Equilibria

Equilibria or steady states of system (1) are solutions of the equations $F(x, a_0) = 0$ at the parameter value a_0. Together with closed orbits they are the simple dynamics or limit sets of a system, and sometimes they are "essentially" all that can occur, e.g., for gradient systems and planar systems. Variants of the qualitative study of the existence of multiple equilibria may also be found in other literatures. See for instance [11, 26, 27, 29, 30].

Assume that the elements of the Jacobian matrix J are constant in a region $\mathscr{D}$ (open convex) of the phase space not necessarily a neighborhood of a steady state, and that the equivalence class $\langle J \rangle$ is qualitatively nondegenerate. Set $\mathbb{L}_{\mathscr{R}}$ to be the corresponding loop structure. We prove

Theorem 3.1. *Assume the loop structure* $\mathbb{L}$ *in a region* $\mathscr{D}$ *contains a composite loop* $\mathscr{L}_n$ *of the dimension n of the system, and that all such loops have the same sign. Then there is a subregion* $\bar{\mathscr{D}} \subset \mathscr{D}$ *where the dynamical system cannot have more than one fixed point.*

The proof is based on the following lemmas.

Lemma 3.2 (n-dimensional Mean Value Theorem). *Assume f is a differentiable function in an open convex domain U of* $\mathbb{R}^n$. *Then For any* $a = (a_1, \cdots, a_n)$ *and* $b = (b_1, \cdots, b_n)$ *in U there exists* $c = (c_1, \cdots, c_n) \in]a, b[$, $c = (1-t)a + tb$, *for some* $t \in]0, 1[$ *such that*

$$f(b) - f(a) = \nabla f(c).(b - a), \quad (19)$$

where $\nabla f(c)$ *is the gradient of f at c, i.e.,* $\nabla f(c) = \sum_{j=1}^{j=n} \frac{\partial f}{\partial x_j}(c)$.

Consequently if $f(a) = f(b)$ then there exists $c = (c_1, \cdots, c_n) \in]a, b[$ such that $\nabla f(c)$ and $b - a$ are orthogonal. This lemma entails the following.

Lemma 3.3. *Let* $F = (F_1, \cdots, F_n) \in \mathscr{C}^1(U)$, U *open set in* $\mathbb{R}^n$. *Assume* $|J_F(x_0)| \neq 0$ *for some* $x_0 \in U$. *Then there exists a neighborhood* W_{x_0} *of* (x_0) *where* F *is one-to-one, that is, for* $a, b \in W_{x_0}$, $F(a) = F(b)$ *implies* $a = b$.

Indeed, a neighborhood being convex, from the n-dimensional Mean Value Theorem it follows that

$$0 = F_i(b) - F_i(a) = \nabla F_i(c_i).(b-a), \quad i = 1, \cdots, n \quad c_i \in]a, b[\subset W_{x_0}. \tag{20}$$

But this is a system of linear equations

$$\sum_{k=1}^{n} (y_k - x_k) \frac{\partial F_i}{\partial x_k}(c_i) = 0 \tag{21}$$

with a nonzero determinant. Hence $y_k - x_k = 0$ for every k.

Proof (Proof of Theorem 3.1). Indeed points in the region $\mathscr{D}$ have a nonzero determinant, and therefore, admit a neighborhood where the vector field F defining the dynamical system is one-to-one. Hence the claim. □

We now address the following questions: *What will be a necessary condition in terms of Jacobian loop to have more than one fixed point in a given region* $\mathscr{R}$?

We actually prove

Theorem 3.4. *A positive simple Jacobian loop is a necessary condition for multiple equilibria in a given region of the phase-space.*

Proof. We consider two $a = (a_1, a_2, \cdots, a_n) \in \mathscr{R} \subset \mathbb{R}^n$ and $b = (b_1, b_2, \cdots, b_n) \in \mathscr{R} \subset \mathbb{R}^n$. Then the line segment $]a, b[= \{x \in \mathbb{R}^n | \quad a < x < b\}$, with respect to the partial order $\mathscr{O}$ previously defined, is included in the open convex set $\mathscr{R}$. The components $F_i, i = 1, \cdots, n$ of the $\mathscr{C}^1$ vector field $F = (F_1, F_2, \cdots, F_n)$ are also $\mathscr{C}^1$ on $\mathscr{R}$. Therefore, from the n-dimensional Mean-value theorem, we obtain, for every $i = 1, \cdots, n$

$$F_i(b) - F_i(a) = \nabla F_i(c^i).(b-a), \quad \mathrm{for some c^i} = (\mathrm{c^i_1, c^i_2, \cdots, c^i_n}) \in]\mathrm{a, b}[. \tag{22}$$

We have $c^i = (1-s_i)a + s_i b$, for some $s^i \in]0, 1[$. That is c^i has components in the form $c^i_j = (1-s^i)a_j + s^i b_j, \quad j = 1, \cdots, n$. Therefore we may rewrite the preceding formula as

$$F_i(b) - F_i(a) = \sum_{j=1}^{j=n} \frac{\partial F_i}{\partial x_j}((1-s^i)a_1 + s^i b_1, (1-s^i)a_2 + s^i b_2, \cdots, (1-s^i)a_n + s^i b_n)(b_j - a_j). \tag{23}$$

We now assume $F(a) = F(b)$. (e.g. a and b are fixed points). Then (22) yields the homogeneous systems of n linear equations of n unknowns

$$0 = \sum_{j=1}^{j=n} \frac{\partial F_i}{\partial x_j}((1-s^i)a_1 + s^i b_1, (1-s^i)a_2 + s^i b_2, \cdots, (1-s^i)a_n + s^i b_n)\delta_j, \tag{24}$$

with $\delta_j = b_j - a_j$. This system admits the unique trivial solution $(\delta_1, \delta_2, \cdots, \delta_n) = (0, 0, \cdots, 0)$ if the corresponding Jacobian determinant is nonzero, for instance for a and b in the neighborhood of a point of nonzero Jacobian determinant. That is, in such a case, $F(a) = F(b)$ implies $a = b$.

Let now analyze the case $a \neq b$. With respect to the partial order, we may take $a \leq b$, i.e., $a_j \leq b_j, j = 1, \cdots n$. Denote $\Lambda_n = \{1, 2, \cdots, n\}$, $I_r = \{j \in \Lambda_n | a_j < a_j\}$, and $\bar{I}_r = \Lambda_n - I_r$. Without loss of generality we may take $I_r = \{1, 2, \cdots, r\}$, and $\bar{I}_r = \{r+1, r+2, \cdots, n\}$. Thus for $j \in I_r$, there exist a positive real δ_j such that $b_j = a_j + \delta_j$. And $a_j = b_j$ for $j \in \bar{I}_r$. Therefore we get from formula (24)

$$c^i = (a_1 + s^i\delta_1, a_2 + s^i\delta_2, \cdots, a_r + s^i\delta_r, a_{r+1}, \cdots, a_n), \quad \text{and } 0 = \sum_{j=1}^{j=r} \frac{\partial F_i}{\partial x_j}(c^i)\delta_j. \tag{25}$$

Set

$$J_F(c) = [\frac{\partial F_i}{\partial x_j}(c^i)]_{1 \leq i \leq n, 1 \leq j \leq r} = [J_{ij}(c^i)]_{1 \leq i \leq n, 1 \leq j \leq r} \tag{26}$$

the $n \times r$ corresponding matrix. The assumption $a \neq b$ requires $\delta = (\delta_1, \delta_2, \cdots, \delta_r)$ be a nontrivial solution. It is well-known that the system (24) of n linear equations of $1 \leq r \leq n$ unknowns admits a nontrivial solution $\delta = (\delta_1, \delta_2, \cdots, \delta_r) \neq (0, 0, \cdots, 0)$ if and only if every $r \times r$ determinant

$$Det_r = |J_{ij}(c^i)|_{1 \leq i,j \leq r} = |\frac{\partial F_i}{\partial x_j}(c^i)|_{1 \leq i,j \leq r} \tag{27}$$

formed by r rows is zero. We denote by J^r the corresponding matrix. From results about determinants in the previous section, we also know that every such determinant may be written in the form

$$Det_r = \sum_{\pi_r \in \Xi_r} sg(\pi_r)P(\pi_r, J_r) = \sum_{\pi_r^e \in \Xi_r^e} P(\pi_r^e, J_r) - \sum_{\pi_r^o \in \Xi_r^o} P(\pi_r^o, J_r). \tag{28}$$

Therefore

$$Det_r = 0 \iff \sum_{\pi_r^e \in \Xi_r^e} P(\pi_r^e, J_r) = \sum_{\pi_r^o \in \Xi_r^o} P(\pi_r^o, J_r). \tag{29}$$

Let now assume that all simple Jacobian loops L_i are negative in the system loop structure corresponding to the region where $F(a) = F(b) = 0$ with $a \neq b$. For an even (resp. odd) number r the associated compound loop $\mathscr{L}_r^e$ having an even (resp. odd) number ν of simple component loops L_i, and $\mathscr{L}_r^o$ an odd (resp. even) number ν of simple component loops L_i, yields all $P(\pi_r^e, J_r)$ of same sign but opposite to that of all $P(\pi_r^o, J_r)$. Therefore Det_r cannot be zero. Hence there must be a positive simple loop in the loop structure. □

Remark 3.5. The previous theorem (22) actually settles Thomas conjecture 1 in the circuits formalism as recalled below [27, 29, 32].

Let now assume that the matrix $J = [J_{ij}]_{1 \leq i,j \leq n}$ in (1) is such that its loop equivalence class $J_l \equiv \langle J \rangle$ has only positive simple Jacobian loops L_k, $k \geq 2$. We prove the following theorem.

Theorem 3.6. *Under the previous assumptions, the corresponding system (1) is linearly equivalent to the system*

$$\dot{y} = f(y) = (f_1(y), f_2(y), \cdots, f_n(y)), \quad with \frac{\partial f_i}{\partial x_j} \geq 0 \ for \ i \neq j. \tag{30}$$

Proof. Indeed there is a linear transformation $y = Tx$ converting the Jacobian J with a positive loop equivalence class to a Jacobian $A = [a_{ij}]_{1 \leq i,j \leq n}$ whose off-diagonal elements are nonnegative, and such that $x(t)$ is a solution of (1) if and only if $y(t) = Tx(t)$ is a solution of system (30). We recall here the algorithm to construct the transformation matrix $T = [T_{ij}]_{1 \leq i,j \leq n}$. T is the diagonal matrix defined by

$$\begin{aligned} &T_{11} = 1, \quad T_{ij} = 0, \quad i \neq j \\ &T_{ii} = 1, \quad i \neq 1, \quad \text{if } J_{1i} > 0 \text{ or } J_{i1} > 0. \\ &T_{ii} = -1, \quad i \neq 1, \quad \text{if } J_{1i} < 0 \text{ or } J_{i1} < 0. \\ &T_{ii} = \text{arbitrary if } \mathrm{J_{1i}} = \mathrm{J_{i1}} = 0. \end{aligned} \tag{31}$$

We also have $T^{-1} = T$, $T^2 = I_n$. The transformed vector field $f(y)$ and its Jacobian matrix A are obtained as

$$f(y) = TF(Ty), \quad y = (y_1, \cdots, y_n), \quad \text{with } A = [\frac{\partial f_i}{\partial x_j}]_{1 \leq i,j \leq n} = T \cdot J \cdot T. \tag{32}$$

Hence the theorem. See also [8]. □

In terms of Jacobian loops we get

Corollary 3.7. *Assume we have a loop equivalence class $\rangle J \langle$ whose simple Jacobian loops $L_k, k \geq 2$ are all positive. Then there exists a qualitative diagonal matrix T_q such that the qualitative matrix given by*

$$A_q = T_q \cdot J_q \cdot T_q \tag{33}$$

has only nonnegative signs s_{ij}, for $i \neq j$.

Indeed T_q is the qualitative matrix associated with T.

4 Loop and Qualitative Stability Analysis

The loop stability refers to the invariance of the Jacobian spectrum under any variation of entries that leave unchanged its loop structure. Jacobian loops and their combinations provide valuable information about the stability of a system even when only the signs, not the magnitudes of the Jacobian terms, are known.

Recall that Λ_A denotes the spectrum of matrix A, i.e., the set of all eigenvalues λ of A or zeros of the characteristic polynomial $\mathscr{C}_A(\lambda)$.

We summarize the classic characterizations of stability from Routh-Hurwitz and Lyapunov theories. For more details see [1, 6, 13, 15, 17].

Theorem 4.1 (Stability Criteria). *The necessary and sufficient conditions to have all real parts negative are given by:*

1. *(Lyapunov) There exist a positive definite symmetric matrix Q such that $QA + A^t Q$ is a negative definite matrix.*
2. *(Routh-Hurwitz) All the Hurwitz determinants H_i are positive, where*

$$H_1 = c_1,$$
$$\vdots$$
$$H_n = \begin{vmatrix} c_1 & c_3 & c_5 & \cdots & c_{2n-1} \\ 1 & c_2 & c_4 & \cdots & c_{2n-2} \\ 0 & c_1 & c_3 & \cdots & c_{2n-3} \\ \vdots & \vdots & \vdots & \cdots & \vdots \\ 0 & 0 & 0 & \cdots & c_n \end{vmatrix} = c_n H_{n-1}, \quad c_j = 0, \quad j > n. \tag{34}$$

3. *(Liénard-Chipart) For all $k = 1, \cdots, n$ $c_k > 0$, and the alternate Hurtwitz determinants up to order n are positive.*

Definition 4.2. 1. The matrix A is stable (resp. asymptotically stable) if its characteristic polynomial $\mathscr{C}_A(\lambda)$ is stable (resp. strongly stable), that is, for every eigenvalue λ of A we have $\mathscr{R}_e(\lambda) \leq 0$ $(resp. < 0)$.
2. A is unstable if it is not stable. In other words, there is at least one eigenvalue λ such that $\mathscr{R}_e(\lambda) > 0$.
3. The matrix A is of *saddle-type* if $\mathscr{R}_e(\lambda) > 0$ for some eigenvalues, and $\mathscr{R}_e(\lambda) < 0$ for the remaining.

Therefore a sink equilibrium has a asymptotically stable Jacobian, whereas a source has an unstable Jacobian.

Moreover, if $\mathscr{R}_e(\lambda) > 0$ for every $\lambda \in \Lambda_A$ the instability is said to be *strong*, e.g., a source is strongly unstable. A *weak instability* is characterized by $\mathscr{R}_e(\lambda) > 0$, for some eigenvalues, and $\mathscr{R}_e(\lambda) = 0$ for the remaining.

Recall that, as a function of the matrix, Λ_A is neither additive nor multiplicative. Moreover we have the followings:

1. For a nonsingular matrix A, i.e., $|A| = det(A) \neq 0$, $\Lambda_{A^{-1}} = \{\frac{1}{\lambda}, \quad \lambda \in \Lambda_A\}$. So A and A^{-1} must be stable simultaneously.
2. $\Lambda_A = \Lambda_{A^t}$, A^t denotes the transpose of matrix A. Thus A and A^t must be stable simultaneously.

Definition 4.3. 1. The matrix A is *loop stable (resp. asymptotically stable)* if every matrix in the loop equivalence class $\rangle A\langle$ is stable (resp. asymptotically stable).
2. A is *loop unstable* if every matrix in the loop equivalence class $\rangle A\langle$ is unstable.

Remark 4.4. A loop equivalence class $\rangle A\langle$ that is not stable is not necessarily unstable. Instead it does have at least one matrix that is unstable, i.e., with an eigenvalue of positive real part in its spectrum.

The loop analysis is addressed here for irreducible matrices. Recall a matrix $A = [a_{ij}]_{1\leq i,j\leq n}$ is *irreducible or indecomposable* if there is no simultaneous row-and-column permutation $P_r - P_c$ such that A is similar to

$$P_r A P_c = \begin{pmatrix} B & O \\ C & D \end{pmatrix}. \tag{35}$$

where P_r and P_c are respectively the row and column permutation matrices, and B, C are respectively a $p \times p$ and a $q \times q$ block such $p + q = n$, and O a $p \times q$ block of zeros. By a Laplace decomposition the spectrum of A is given by $\Lambda_A = \Lambda_B + \Lambda_D$, thus reducing its eigenvalue analysis to that of the individual diagonal block of lower dimension. Therefore we assume all matrices are irreducible without loss of generality. Actually a $P_r - P_c$ permutation amounts to a renumbering of the system variables, and renumbering should certainly not affect the properties of the system in general, and its asymptotic behavior in particular. [citations]

From the above stability criteria we readily derive

Theorem 4.5. *All kth order Feedback F_k positive is a necessary condition for a stable loop equivalence class.*

Moreover we obtain

Theorem 4.6. *If the loop structure $\mathbb{L}$ has a positive simple loop L_k then the corresponding loop equivalence class cannot be stable.*

Proof. Indeed it suffices to construct a representative matrix with k eigenvalues arbitrarily close to the kth roots of unity by continuity, and one of them is a simple positive eigenvalue. Therefore the matrix is unstable. Hence the claim. □

Similarly one can prove

Theorem 4.7. *The loop equivalence class is unstable if it has a composite loop $\mathscr{L}_n$ of the dimension of the system, positive for n even, and negative for n odd.*

Proof. Indeed for any matrix A in such a loop structure, the presence of a composite loop of the dimension of the system ensures a nonzero determinant such that the characteristic coefficient $c_n = F_n = \mathscr{C}_A(0)$ is > 0 for n even, and < 0 for n odd. This entails the characteristic polynomial of any matrix in the class has a positive root. The claim is proved. □

Combining the above definitions and properties results in the following [19,34,35].

Remark 4.8. 1. The Routh-Hurwitz stability criteria also imply for the k-order feedback loop $\mathscr{F}_k$

$$\begin{aligned} \mathscr{F}_k &> 0, \quad \forall k \\ \mathscr{F}_3 &> \mathscr{F}_1 \times \mathscr{F}_2. \end{aligned} \tag{36}$$

2. If $\mathscr{F}_1 = 0$ then $\mathscr{F}_{2r+1} = 0, \quad r = 1, \cdots, (n-1)$
3. Positive loops have a negative contribution to $\mathscr{F}_k$, and therefore hey tend to destabilize a steady state. Hence their presence promotes instability in the system
4. A steady state with $\mathscr{F}_k \neq 0$, but $\mathscr{F}_{k+1} = 0$ cannot be asymptotically stable.
5. To have $\mathscr{F}_1 < 0$ requires at least one negative 1-loop.
6. To have $\mathscr{F}_3 < 0$ requires the loop structure to contain at least one negative loop of order 2 or higher.
7. An important consequence of the last is that in the absence of negative feedback loops of at least order two, the system cannot have any stable periodic behavior such as stable limit cycles. See also [snoussi]. Therefore negative feedback loops promote oscillations in the system.
8. Whenever the positive and negative feedback loops fulfill their dynamical role, they are said to be *functional.*

5 Applications

5.1 *Qualitative Modeling in Biological and Biochemical Sciences*

5.1.1 Background

Many proteins are transcription factors binding to DNA to regulate the transcription of specific genes, synthesizing RNA from coding regions of chromosomal DNA. Regulation occurs during the complete process of gene expression. Identical DNA does not imply identical gene expression. Most genes are part of a gene network from which one can draw an interaction graph consisting of vertices (genes) and directed edges endowed with sign, positive to indicate activation, negative for inhibition and zero to indicate the absence of effect of a gene on another. In general the strength of the interactions between genes is unknown, that is, the lack of quantitative information. Therefore determining the dynamical properties of a gene network solely for the qualitative topology of the interaction has proved to be difficult; some methods were based on numerical simulation choosing realistic kinetic parameters. Other methods involve the study of the statistical properties of gene networks comparing the interaction graph with random ones. Decomposition of the graph into submodules of biological significance has also been tried.

Feedback loop patterns are important for many biological activities, such as the circadian rhythm of sleep-wake cycle generated by genes in a network of positive and negative feedback loops, or for the fates of cells in the transition of an egg to a multi-cellular organism.

Cell differentiation was suggested earlier on in the 40*s* e.g., Max Delbruck in [3], to be associated with distinct states of expression in the cell genetic regulatory networks. Differences transmissible from cell to cell in the absence of any genetic difference are called epigenetic differences, also involved in *cell differentiation*, and part of the more general process of the so-called *multistationarity* the display of multiple steady states. During cell differentiation a gene can be activated by the product of another gene and remains on after the disappearance of this product. Biologist for many years have recognized the stabilizing roles of negative feedback loops within networks of interacting genes and proteins: for instance the protein product of a gene could act to inhibit its synthesis, eventually turning it off as the protein concentration increases. As for the positive feedback loop, a gene expression could trigger further increase in its expression, leading in the absence of mitigating factors to an unbounded increase in various protein concentrations. As a result, with other factors coming into play, the cell could be switched from on stable condition to another. It has been known for some time there exists a positive feedback loop in all biological systems displaying *multistationarity* on which is based a cellular memory. Then Thomas in 1981 formally conjectured that the presence of positive feedback loop (termed circuit) is a necessary condition for *multistationarity*. Thomas also conjectured at the same time that negative feedback loops are necessary for the biological homeostasis and periodicity. While these conjectures could be easily stated in the biological context, a major challenge has been to state and prove them using the mathematical models of biological processes, which display essentially qualitative features. The Jacobian feedback loop methodology has indeed helped to settle these conjectures [2–4, 11, 22, 28, 30, 31].

It has been also shown that positive loops account for many features of memory *stricto sensu*(neural memory and mnesic evocation) and *largo sensu* (differentiation and immunological memory). The combination of positive and negative loops could provide some powerful regulatory modules, with enormous dynamical possibilities in neurobiology. An application of such combination has been given in problems of synchronization and desynchronization of a neural model for hippocampus memory evocation processes. See [2, 4, 11].

Note also that in biological networks, the occurrence of feedback loop is often in a coupled structure rather than in a single isolated form. Thus the importance of understanding the dynamics of coupled feedback loops, in particular dynamics not predictable by just the combinatorial dynamics of the individual component loops, as it is often the case. We see the dynamical role of coupled feedback loops below in the qualitative Lorenz and Rossler systems [16, 22, 23].

5.1.2 Thomas Conjectures

We restate Thomas's conjectures in the loop formalism [29–32].

1. Conjecture 1 [1981] The presence of a positive feedback loop (somewhere in the phase space) is necessary condition for multistationarity.
2. Conjecture 2 [1981] The presence of a negative feedback loop of length at least two (somewhere in phase space) is a necessary condition for stable periodicity.
3. Conjecture 3 [1999] Chaotic dynamics require both a positive feedback (*to allow multistationarity*) and a negative feedback loop (*to allow for permanent periodicity*).

Remark 5.1. 1. Soulé in 2003 presented a proof of conjecture 1. Some partial results were also given by Plahte et al (1995), Snouussi (1998), Gouzé (1998), Cinquin and Demongeot (2002). See [7, 19, 26, 27]
2. Under additional assumptions Snoussi and Gouzé also proved Conjecture 2. See [4, 26]
3. Conjecture 3 is also included in a more general conjecture by Toni et al in 1999, yet to be settled [34–36].

5.2 *Eisenfeld Qualitative Stability*

Eisenfeld et al also studied in [5] qualitative stability, that is, strictly from the sign patterns. The results could be easily derived as well from the above analysis using the qualitative equivalence terminology. For instance

Lemma 5.2. *If the loop structure of the n-dimensional system contains a composite loop of length n and all such composite loops have same sign then the Jacobian is nondegenerate, i.e., $det(J) \neq 0$.*

See details in [5]. Lemma that entails

Theorem 5.3 (Eisenfeld Stability). *Assume the loop structure has all composite n-loop of same sign. Then a necessary and sufficient condition for a stable sign equivalence class is that the loop structure has the following features*

1. *There is at least a 1-loop.*
2. *There are no positive 1-loops*
3. *There are no positive 2-loops*
4. *There are no k-loops with $k \geq 3$.*

And leads to

Corollary 5.4. *A necessary condition of the sign equivalence class to be unstable is that its loop structure contains no more than $(n-1)$ negative 1-loops.*

For a system to undergo a Hopf bifurcation, ensuring the existence of limit cycles, its Jacobian must admit an exchange of stabilities. Therefore we have

Corollary 5.5. *A Hopf bifurcation requires a sign equivalence class that is neither stable nor unstable.*

Remark 5.6. 1. An example of a sign equivalence class neither stable nor unstable is one associated with a loop structure containing n negative 1-loop $L^{-}_{x_i}, i = 1, \cdots, n$ i.e., at each vertex and at least one k-loop with $k \geq 3$.
2. Intuitively the negative loop may be seen as stabilizing whereas the positive loop could be seen as destabilizing.

In terms of composite loops and their sign as defined in the previous sections, we state

Theorem 5.7. *A sufficient condition for an unstable sign equivalence class (qualitatively unstable matrix) is that in the loop structure there is at least one integer k, ($1 \leq k \leq n$) such that either there is no composite k-loop or they are all positive (strong instability.)*

See details and proof in [5].

5.3 Loop Analysis in the Plane

Consider a square matrix A of order 2 given by

$$A = \begin{pmatrix} a_{11} & a_{12} \\ a_{21} & a_{22} \end{pmatrix} \tag{37}$$

with entries constant with respect to the state variables, possibly depending on some parameters. The above stability criteria translate into

$$\begin{aligned} Tr(A) &= a_{11} + a_{22} \leq 0 \\ |A| &= a_{11}a_{22} - a_{12}a_{21}. \end{aligned} \tag{38}$$

We obtain the following loop interpretation. There is a parabolic boundary line at $Tr(A)^2 - 4|A| = 0$ between real and complex eigenvalues. Complex eigenvalues, i.e., oscillations are possible only when $a_{12}a_{21} < 0$, which corresponds to a negative 2-loop. Crossing the boundaries of the second quadrant is the fundamental way to lose stability, leading to the appearance of a positive feedback loop rendering positive the real part of the eigenvalues. Thereby the stable periodicity is destabilized into a limit cycle for complex eigenvalues, or promoting multistationarity, i.e., saddle point for real eigenvalues.

Oscillations require the necessary condition $Tr(A)^2 - 4|A| < 0$, that is, $(a_{11} - a_{22})^2 < -4a_{12}a_{21}$. Therefore a negative 2-loop L^-_{xy} is necessary for any periodic behavior (center, stable focus, or limit cycle).

For a two-dimensional system, we also have the following results.

Theorem 5.8. *Any loop structure in the plane consisting of two 1-loop L_1 of opposite signs and a negative 2-loop cannot be loop stable or loop unstable.*

The proof is based on the following lemma:

Lemma 5.9. *The qualitative equivalence class given by*

$$\langle\rangle = \begin{pmatrix} + & - \\ + & - \end{pmatrix}, \tag{39}$$

cannot be stable and cannot be unstable, that is, there is a matrix $A \in \langle\rangle$ such that A is unstable or stable.

Proof. The equivalence class above is the so-called 1-striped sign pattern; as such, given any monic quadratic polynomial $q(x) = x^2 + bx + c$, there is a matrix M with characteristic polynomial $\mathscr{C}_A(\lambda) = q(\lambda)$. Therefore there is certainly one whose spectrum contains an eigenvalue with a positive (resp. negative) real part. Hence the claim. □

Proof (Proof of Theorem 5.8). The proof of theorem 5.8 follows immediately; indeed by combinations such as negation, transposition, permutation, signature similarity, one can construct a matrix M with the loop structure of two 1-loop of opposite signs and a negative 2-loop in a such way that the matrix M is in a class of the type in Lemma 5.9. □

Based on the Linear Stability Theory, we can classify the loop equivalence classes in the plane as follows. See also [12].

Theorem 5.10. *1. The loop structure consists of only a negative 2-loop L^-_2: The dynamic is that of a linear center, that is, a family of periodic orbits surrounding the origin.*
2. The loop structure $\mathbb{L}$ consists of a negative 2-loop L^-_2 and two 1-loop of same sign; the dynamic is that of a focus point, that is, the presence of sustained oscillations around the origin. If the system is bounded then there exists a limit cycle surrounding the origin.
3. The loop structure consists of only two 1-loops of opposite sign (resp. same sign). Then the origin is a saddle point (resp. a node stable for negative 1-loops, and unstable for positive 1-loops).
4. These loop equivalence classes are the only loop equivalence classes in the plane.

Proof. Immediate from Linear stability theory adapted in terms of sign patterns. Indeed every 2×2 matrix is amenable, via nonsingular linear transformations to one of the forms:

$$\begin{pmatrix} a & 0 \\ 0 & b \end{pmatrix}, \quad \begin{pmatrix} a & b \\ -b & a \end{pmatrix}. \tag{40}$$

The sign equivalence classes consist of the following sign patterns:

1. Absence of 2-loops.

$$\langle\rangle = \begin{pmatrix} \pm & 0 \\ 0 & \pm \end{pmatrix}, \quad \langle\rangle = \begin{pmatrix} \pm & 0 \\ 0 & \mp \end{pmatrix}, \tag{41}$$

2. Absence of 1-loops

$$\langle\rangle = \begin{pmatrix} 0 & \pm \\ \pm & 0 \end{pmatrix}, \quad \langle\rangle = \begin{pmatrix} 0 & \pm \\ \mp & 0 \end{pmatrix}, \tag{42}$$

3. Presence of 1-loops and 2-loops including of same or opposite sign.

$$\langle\rangle = \begin{pmatrix} \pm & \pm \\ \pm & \pm \end{pmatrix}, \tag{43}$$

□

Therefore we have formally, as indicated above

Corollary 5.11. *A negative 2-loop L_2^- is a necessary condition for any periodic behavior such as a center, a focus, or a limit cycle, i.e., an isolated periodic orbit.*

5.3.1 Biochemical Application: Two-Component Oscillators

Consider a two-component network of a chemical reaction system given by

$$\begin{aligned} \dot{x}_1 &= f_1(x_1, x_2) \\ \dot{x}_2 &= f_2(x_1, x_2), \end{aligned} \tag{44}$$

possibly with some dependence of f_1 and f_2 on the kinetic parameters. Recall that the Bendixon's negative criterion claims that if the divergence $div(f_1, f_2)$ is of constant sign in a region of the plane, then there can be no periodic solution in that region. For chemical reaction systems the diagonal entries of the Jacobian matrix are usually negative. If both a_{11} and a_{22} are always negative, that is, the existence of a constant negative 1-loop in the loop structure, then the trace never changes sign, and Hopf bifurcation cannot occur in such a system. So at least one of them must be positive, indicating autocatalysis. With the diagonal elements of opposite sign, in order to have the determinant positive, the off-diagonal elements must also be of opposite sign. The typical sign patterns for a Hopf bifurcation are given by the following.

$$\langle\rangle = \begin{pmatrix} + & + \\ - & - \end{pmatrix}, \tag{45}$$

representing the so-called substrate-depletion oscillator. [Tyson]. The production of x_1 is autocatalytic, and the reaction speeds up as x_1 increases, until the substrate x_2 is depleted to the extend that the reaction ceases. The matrix

$$\langle\rangle = \begin{pmatrix} + & - \\ + & - \end{pmatrix}, \tag{46}$$

corresponds to the *activator-inhibitor* models. Intuitively, when x_2 is rare, x_1 increases autocatalytically. The degradation of the x_2 is inhibited with their accumulation stimulated by abundant x_1, which feeds back to inhibit the production of x_1. After x_1 disappears, x_2 is also destroyed, and then x_1 can make a comeback.

Therefore a two-component biochemical reaction system can oscillate if there exists in its loop structure at least one positive 1-loop, that is, autocatalysis along with a negative 2-loop. *Autocalysis* represented by $a_{ii} > 0$ has a major role in biochemical oscillations, and usually occurs when a chemical decelerates the rate of its own destruction. $a_{ij} > 0$ together with $a_{ji} > 0$ indicates the x_i activates the production of x_j and vice versa, leading to a *feedback loop* generating an indirect autocatalysis. When $a_{ij}a_{ji} < 0$, there exist a negative 2-feedback loop indicating that x_i activates the production of x_j but x_j inhibits the production of x_i [35].

5.3.2 More Illustrative Planar Examples

1. Consider the sign equivalence class

$$\langle\rangle = \begin{pmatrix} - & + \\ - & - \end{pmatrix} \tag{47}$$

given by the system

$$\begin{aligned} \dot{x} &= -x^3 + y \\ \dot{y} &= -x - y \end{aligned} \tag{48}$$

The loop structure consists of two simple negative 1-loops, a simple negative 2-loop, and a proper composite loop of resonance $(2, 2)$ given by $\mathscr{L}_2^2 = L_x^- \cup L_y^-$. The absence of any positive loop excludes multistationarity; and indeed the system has a single steady state. From the previous theorem, this single steady is a stable focus as indicated by the presence of a negative 2-loop and two negative 1-loops. Such conclusion is also confirmed by the traditional stability analysis.

2. For the sign equivalence class given by

$$\langle\rangle = \begin{pmatrix} + & \pm \\ \pm & + \end{pmatrix} \tag{49}$$

note that all loops are positive, allowing for multistationary, that is, the existence of multiple steady states, none of which could be stable, due to the presence of the positive 1-loops. Indeed the traditional methods ensure the presence of two unstable nodes separated by a saddle point.

3. If the sign equivalence class is

$$\langle\rangle = \begin{pmatrix} - & - \\ - & - \end{pmatrix} \tag{50}$$

the loop structure has a simple positive 2-loop L_{xy}^{+} and two simple negative 1-loops L_x^- and L_y^-. This allows for multistationarity, with possibly the presence of stable steady states, actually two stable nodes and a saddle point from conventional methods.

5.3.3 Two-Dimensional Model for Electrochemical Corrosion

We present how the Jacobian feedback loop methodology is applied to the electrochemical corrosion model initially developed by Talbot and Oriani. See also [36]. That is, a metal M is dissolving in an electrolyte solution in such a way that any given point of the metal surface at any given time is either bare or covered with adsorbed MOH to passivate the underlying metal. The model system reproduces the dynamics observed during potentiostatic dissolution of copper in an acetate buffer. In terms of dimensionless variables the system is given by

$$\begin{aligned} \dot{x} &= p(1-y) - qx \\ \dot{y} &= x(1-y) - ye^{-\beta y} \end{aligned} \tag{51}$$

with the state variables confined to interval $[0, 1]$, for the positive parameters values $p,\ q$, and β. In the region of steady states, the Jacobian matrix is

$$J = \begin{pmatrix} -q & -p \\ 1-y & a_{22} \end{pmatrix}, \tag{52}$$

where a_{22} is a function of the parameters $p,\ q$, and β given by $a_{22} = \frac{p(y-1)}{qy}(\beta y^2 - \beta y + 1)$. The associated qualitative matrix is

$$J = \begin{pmatrix} - & - \\ + & sign(a_{22}) \end{pmatrix}.$$

Therefore the loop structure has a negative 1-loop L_x^- and a negative 2-loop L_{xy}^-, satisfying the necessary condition for periodic behavior. Moreover the only way to secure a positive loop to promote multistationarity and sustained oscillations is to get $a_{22} > 0$. This is realized for $\beta > 4$ and $y \in [Y_1, Y_2]$ with $Y_{1,2} = \frac{1}{2}(1 \mp \sqrt{1 - \frac{4}{\beta}})$. This implies there exists a region in the parameter space with a positive loop necessary for multistationarity. There is also the possibility of sustained oscillations due to the presence of the negative 2-loop L^-_{xy}.

Therefore the loop analysis allows to predict a region of coexistence of multistationarity and sustained oscillations, actually limit cycle. See more in details in [36]. We note that the sign of the single Jacobian term a_{22} was crucial for both multistationarity and limit cycle. We were able to predict the global dynamics without resorting to actual integration of the system.

5.4 *Loop Analysis for a Three-Dimensional System*

Consider the corresponding matrix

$$A = \begin{pmatrix} a_{11} & a_{12} & a_{13} \\ a_{21} & a_{22} & a_{23} \\ a_{31} & a_{32} & a_{33} \end{pmatrix}. \tag{53}$$

The characteristic polynomial $\mathscr{C}_A(\lambda) = \lambda^3 + c_1\lambda^2 + c_2\lambda + c_3$ has the coefficients

$$\begin{aligned} c_1 &= -(a_{11} + a_{22} + a_{33}) = -Tr(A) \\ c_2 &= -a_{12}a_{21} - a_{23}a_{32} - a_{31}a_{12} + a_{11}a_{22} + a_{22}a_{33} + a_{33}a_{11} \\ c_3 &= -a_{12}a_{23}a_{31} - a_{13}a_{32}a_{21} + a_{11}a_{23}a_{32} + a_{22}a_{13}a_{31} + a_{33}a_{12}a_{21} - a_{11}a_{22}a_{33} \end{aligned} \tag{54}$$

5.4.1 Saddle-Focus Loop Interpretation

In the associated 3-variable xyz systems, a *saddle-focus* may be described as follows in terms of loops

1. A repulsive periodicity in the xy-plane, attractive along the z-axis with the loop structure

$$\mathbb{L} = L^-_{xy} \cup L^+_{x/y} \cup L^-_z,$$

   where $L^+_{x/y}$ stands for a positive 1-loop at either x or y.
2. A steady state periodically attractive in the xz-plane and repulsive along the y-axis requires

$$\mathbb{L} = L^-_{xz} \cup L^-_{x/z} \cup L^+_y.$$

The presence of a positive 3-loop $L^{\pm}_{xyz}$ could destabilize these periodicities. Note there are only two 3-loop in the matrix A given respectively by $a_{12}a_{23}a_{31}$ and $a_{31}a_{32}a_{21}$. In addition wherever they exist together they are coupled to the three 2-loops $a_{12}a_{21}$, $a_{13}a_{31}$, and $a_{23}a_{32}$. Such a strong coupling might preclude the appearance of a chaotic attractor, leading to the conjecture

Theorem 5.12 (Conjecture). *There are no three-dimensional chaotic systems consisting of two coupled 3-loops.*

We also prove

Theorem 5.13. *Any three-dimensional loop structure containing the two negative 2-loops L^-_{xy} and L^-_{yz}, the positive 2-loop L^+_{xz}, in the absence of any positive 1-loop L^+_y and negative 1-loops L^-_x and L^-_z is destabilized by the two positive 3-loops whenever they are present.*

Proof. Indeed such a loop may be represented by a qualitative class whose is similar by the usual operations of signatures and permutations, to

$$\begin{pmatrix} +/0 & - & + \\ + & -/0 & - \\ + & + & +/0 \end{pmatrix}. \tag{55}$$

One can easily construct a matrix in this class with a positive eigenvalue. □

Theorem 5.14. *For a three-dimensional system the following four loop equivalence classes cannot be stable.*

1. *The loop structure ⟨⟩ consists of two 1-loops $L^{\pm}_1$ of opposite signs, and two negative 2-loops L^-_2, no 3-loop.*
2. *The loop structure ⟨⟩ consists of one positive 1-loop L^-_x, and two negative 1-loops L^-_1, one negative 2-loop L^-_2, and one positive 2-loop L^+_2, but no 3-loop.*
3. *The loop structure ⟨⟩ consists of two 1-loops $L^{\pm}_1$ of opposite signs, and one negative 2-loop L^-_2, and a positive 3-loop L^+_{xyz}.*
4. *The loop structure ⟨⟩ consists of two 1-loops $L^{\pm}_1$ of opposite signs, one negative 2-loop L^-_2, one positive 2-loop L^+_2, and a negative 3-loop L^-_{xyz}.*
5. *These loop classes cannot be unstable, and are the only ones that cannot be stable and unstable.*
6. *These equivalence classes are the ones necessary for any Hopf bifurcation ensuring the existence of limit cycles.*

Before presenting the proof we state the following conjecture to be illustrated in the next section [12].

Theorem 5.15 (Conjecture). *The previous loop equivalences are some of the necessary loop structures for the onset of any chaotic behavior in a three-dimensional system.*

Proof (Proof of Theorem 5.14). We consider in each case a representative qualitative equivalence class, that is, a sign-pattern with such a loop pattern, and then use some results for advanced matrix theory on the so-called arbitrary spectrally sign pattern. By signatures and permutations, a representative sign-pattern of the classes above is similar to respectively the following sign-patterns well-known to be spectrally arbitrary, that is, they contains a matrix whose spectrum has a positive eigenvalue, root of an arbitrary monic cubic polynomial.

1. Case one

$$\begin{pmatrix} + & - & 0 \\ + & 0 & - \\ 0 & + & - \end{pmatrix} \tag{56}$$

2. Case Two

$$\begin{pmatrix} + & - & + \\ + & - & 0 \\ + & 0 & - \end{pmatrix} \tag{57}$$

3. Case three

$$\begin{pmatrix} + & - & 0 \\ + & 0 & - \\ + & 0 & - \end{pmatrix} \tag{58}$$

4. Case four

$$\begin{pmatrix} + & + & - \\ + & 0 & - \\ + & 0 & - \end{pmatrix} \tag{59}$$

□

We now will present a Jacobian loop analysis of the well-known Lorenz and Rossler systems, paradigms of chaotic dynamics, as a step toward the necessary and/or sufficient conditions in terms of Feedback Loops for the onset of chaos.

5.4.2 A Loop Analysis of the Lorenz System

The Lorenz system [16, 33] is described by

$$\begin{aligned} \dot{x} &= \sigma(y - x) \\ \dot{y} &= \rho x - y - xz \\ \dot{z} &= -\beta z + xy, \end{aligned} \tag{60}$$

where $(x, y, z) \in \mathbb{R}^3$, and $\sigma, \rho, \beta > 0$. The Lorenz equations can also be found in simplified models for lasers, dynamos, electric circuits, chemical reactions. The variable x is proportional to the intensity of convective motion, y is proportional to the temperature difference between ascending and descending currents and z is proportional to the distortion from linearity of the vertical temperature profile. The Lorenz flow is dissipative, contracting volume with a negative divergence $-(1+\beta+\sigma)$. The Jacobian matrix at a steady state $p* = (x*, y*, z*)$ is

$$A = \begin{pmatrix} -\sigma & \sigma & 0 \\ \rho - z* & -1 & -x* \\ y* & x* & -\beta \end{pmatrix}, \tag{61}$$

with the corresponding qualitative Jacobian

$$A_q = \begin{pmatrix} - & + & 0 \\ sgn(\rho - z*) & - & -sgn(x*) \\ sgn(y*) & sgn(x*) & - \end{pmatrix}, \tag{62}$$

The main characteristics of the Lorenz system are

1. The system is invariant under reflection in the z-axis.
2. Its equilibria are the origin $O = (0, 0, 0)$ for all values of the parameter ρ, and $E_{\pm} = (\pm\sqrt{\beta(\rho-1)}, \pm\sqrt{\beta(\rho-1)}, \rho-1)$ appearing at $\rho > 1$ and symmetric with respect to the z-axis.

We give here a complete loop interpretation of this most celebrated system. First the stability parameters given by the k-order feedback loops lead to

$$\begin{aligned} \mathscr{F}_1 &= -(\sigma+\beta+1) < 0 \\ \mathscr{F}_2 &= \beta + \sigma(1+\beta) - \sigma\rho + \sigma z * + x* \\ \mathscr{F}_3 &= -\sigma(\beta(1-\rho) + \beta z * + x * y * + x*) \end{aligned} \tag{63}$$

Hence we have

1. $\mathscr{F}_1$ is always negative, ensuring the constant presence of at least one negative 1-loop $L^-_{x/y/z}$.
2. $\mathscr{F}_2 = 0$ for $\rho = \rho_c = \frac{\beta+\sigma(1+\beta)-\sigma\rho+\sigma z*+x*}{\sigma}$. This entails $\mathscr{F}_2 > 0$ for $\rho < \rho_c$ to promote stability and $\mathscr{F}_2 > 0$ for $\rho < \rho_c$ to promote instability.
3. $\mathscr{F}_3 = 0$ for $\rho = \rho_1 = \frac{\sigma(\beta(1-\rho)+\beta z*+x*y*+x*}{\sigma}\beta$

From the qualitative Jacobian A_q, there are three persistent negative 1-loops L^-_x, L^-_x, L^-_z. Then under the plane $z = \rho$ appears an additional positive 2-loop L^+_{xy} in the loop structure corresponding to the qualitative matrix

$$A^1_q = \begin{pmatrix} - & + & 0 \\ + & - & -sgn(x*) \\ sgn(y*) & sgn(x*) & - \end{pmatrix}. \tag{64}$$

Whereas in the plane $z = \rho$ itself, the previous positive disappears, as indicated by

$$A_q^2 = \begin{pmatrix} - & + & 0 \\ 0 & - & -sgn(x*) \\ sgn(y*) & sgn(x*) & - \end{pmatrix}. \quad (65)$$

Now above the same plane, that is, for $\rho - z > 0$, a negative 2-loop L_{xy}^- emerges as seen in

$$A_q^3 = \begin{pmatrix} - & + & 0 \\ - & - & -sgn(x*) \\ sgn(y*) & sgn(x*) & - \end{pmatrix}. \quad (66)$$

The symmetry of the Lorenz system with respect to the z-axis imposes the analysis along that axis and the region $\mathscr{R}^{\pm}$ with x and y respectively of the same and opposite sign. On the z-axis, we have

$$A_q^z = \begin{pmatrix} - & + & 0 \\ +/0/- & - & 0 \\ 0 & 0 & - \end{pmatrix}, \quad (67)$$

that is, s_{21} is respectively (+), (0), (−) under, in, and above the plane $z = \rho$. Therefore the loop structure has only the previous persistent ones. In the region $\mathscr{R}^+$ the loop structure is given by

$$A_q^+ = \begin{pmatrix} - & + & 0 \\ +/0/- & - & \mp \\ \pm & \pm & - \end{pmatrix}. \quad (68)$$

with again s_{21} taking respectively (+), (0), (−) as we move vertically. Now a negative 3-loop L_{xyz}^- persists throughout the region, coupled successively under the plane $z = \rho$ with a positive 2-loop L_{xy}^- and a negative 2-loop L_{yz}^-, in the plane, only with the negative 2-loop L_{yz}^-, and above with a negative 2-loop L_{xy}^- and a negative 2-loop L_{yz}^-.

In the negative region $\mathscr{R}^-$ the loop structure corresponds to

$$A_q^- = \begin{pmatrix} - & + & 0 \\ +/0/- & - & \mp \\ \mp & \pm & - \end{pmatrix}. \quad (69)$$

The previous 2-loops are now coupled with a positive 3-loop L_{xyz}^+.

We now consider the loop structures around the steady states. First around the origin, we obtain

$$A_q^O = \begin{pmatrix} - & + & 0 \\ + & - & 0 \\ 0 & 0 & - \end{pmatrix}. \quad (70)$$

Therefore in addition to the persistent negative 1-loops, a positive 2-loop L^+_{xy} appears. And the feedback parameters $\mathscr{F}_k$ adjust to

$$\begin{aligned}\mathscr{F}_1 &= -(\sigma + \beta + 1) < 0 \\ \mathscr{F}_2 &= \beta(1 + \sigma) + \sigma(1 - \rho) \\ \mathscr{F}_3 &= -\sigma\beta(1 - \rho))\end{aligned} \tag{71}$$

Thus the stability breaks down at $\rho = 1$ and at $\rho = 1 + \frac{\beta(1+\rho)}{\sigma}$.

Around the steady states $E_\pm = (\pm\sqrt{\beta(\rho - 1)}, \pm\sqrt{\beta(\rho - 1)}, \rho - 1)$ for $\rho > 1$ the qualitative Jacobian becomes

$$A_q^O = \begin{pmatrix} - & + & 0 \\ + & - & \mp \\ \pm & \pm & - \end{pmatrix}. \tag{72}$$

Therefore a negative 3-loop L^-_{xyz} emerges coupled with a positive 2-loop L^+_{xy} and a negative 2-loop L^-_{yz}. And the feedback parameters $\mathscr{F}_k$ become

$$\begin{aligned}\mathscr{F}_1 &= -(\sigma + \beta + 1) < 0 \\ \mathscr{F}_2 &= \beta(\sigma + \rho)) > 0 \\ \mathscr{F}_3 &= -2\sigma\beta(\rho - 1) < 0.\end{aligned} \tag{73}$$

meeting the above stability criteria. However as ρ approaches 1 from above, stability is lost. Recall from nonlinear theory on the Lorenz system that there is a subcritical Hopf bifurcation at the equilibria $E_\pm$ when $1 < \rho_0 = \frac{\sigma(\sigma+\beta+3)}{\sigma-\beta-1}$, for a range of values of σ and β, with $E_\pm$ stable for $\sigma < \beta + 1$ or $1 < \rho < \rho_0$ and $\sigma > \beta + 1$.

The overall loop interpretation for the Lorenz systems goes as follows:

1. For any value of $\rho > 0$ the positive 2-loop L^+_{xy} promotes multistationarity under the plane $z = \rho$ to become a generator of periodicity L^-_{xy}, with the corresponding oscillations normal to the xy plane due to L^-_x and L^-_y.
2. Around the origin, indeed stable for $\rho < 1$, the previous positive 2-loop L^+_{xy}, actually *functional* ensures multistationarity realized with the emergence of the two steady states $E_\pm$ and a loss of stability at $\rho = 1$. These steady states $E_\pm$ could be seen as symmetrical foci generated by the negative 2-loop L^-_{yz}.
3. In the region $\mathscr{R}^+$ the periodicity generated by the negative 2-loop L^-_{yz} could be damped by the two negative 1-loops L^-_z and L^-_z, made into a stable attractor along the x-axis due to the negative 1-loop L^-_x, and transient due to the negative 3-loop L^-_{xyz}. Whereas under the plane this periodicity could be destabilized by the positive 2-loop L^+_{xy} into limit cycles.

4. However all these trajectories are forced to switch from one attractor to another, crossing the z-axis periodically.
5. Crossing the plane $z = \rho$ the positive 2-loop L^{+}_{xy} disappears to reappear negative in the plane; thus the existence of the two generators of periodicity, distinct yet coupled at y. The oscillations are damped due to the negative 1-loop L^{-}_{y}.
6. The negative 3-loop L^{-}_{xyz} is functional and contributes to destabilize the periodic behavior, and sooner or later, forcing the trajectories to cross the z-axis, periodically due to the appearance of the negative 2-loop L^{-}_{xy} on the z-axis.
7. The combination of these local periodic movements, together with the global destabilizing influence generated by the negative 3-loop which may indicate the presence of the so-called *strange attractor*.
8. In addition a positive 3-loop L^{+}_{xyz} (of the dimension of the system) is present in the other region $\mathscr{R}^{-}$ (x and y of opposite signs), with a destabilizing effect in the whole region, making it the appearance of the attractors impossible under the plane, as well as that of the *strange attractor* above the plane, with the additional effect of directing the trajectories toward the attractors or the *strange attractor* whenever appropriate.

Remark 5.16. In conclusion, considering a qualitative Lorenz system, that is, in terms only of the sign of Jacobian entries and the corresponding loop structure, we obtain the presence of a destabilizing positive 2-loop L^{+}_{xy}, a negative 2-loop L^{-}_{yz} unique generator of periodicity promoting the two symmetrical attractors. And in addition we have the presence of a negative 3-loop, of the dimension of the system, which contributes to confining the system to bounded regions of the variables, forces the switching from an attractor to the other while ensuring the periodic recurrence of the entire process.

5.4.3 A Loop Analysis of the Rossler System

The Rossler system [23] is given by

$$\begin{aligned} \dot{x} &= -y - z \\ \dot{y} &= x + ay \\ \dot{z} &= b + xz - cz \end{aligned} \tag{74}$$

The system has two steady states or equilibria $E\pm$ located at

$$(x_{\pm}, y_{\pm}, z_{\pm}) = (\frac{c \pm \sqrt{c^2 - 4ab}}{2}, -\frac{c \pm \sqrt{c^2 - 4ab}}{2a}, \frac{c \pm \sqrt{c^2 - 4ab}}{2a}) \tag{75}$$

associated with the Jacobian matrix

$$\begin{pmatrix} 0 & -1 & -1 \\ 1 & a & 0 \\ z & 0 & -c \end{pmatrix}, \tag{76}$$

and its qualitative equivalence class

$$\begin{pmatrix} 0 & - & - \\ + & + & 0 \\ + & 0 & - \end{pmatrix}. \tag{77}$$

For a wide range of the parameters a, b, c the system exhibits two unstable equilibria of type *saddle-focus* periodically repulsive (resp. attractive) in a plane while attractive (resp. repulsive) along a normal direction.

The loop structure contains two negative 2-loops L_{xy}^- and L_{xz}^- along with two 1-loops of opposite signs L_y^+ and L_z^-. Note, as conjectured, the presence of two proper three-dimensional composite loops $\mathscr{L}_3^2$ given by the negative (L_y^+, L_{xz}^-) and the positive (L_z^-, L_{xy}^-).

A complete loop analysis as in the Lorenz case above leads to predict the chaotic behavior of the system.

6 Concluding Remarks

Given its fundamental qualitative nature, the methodology of Jacobian feedback loops allows only necessary conditions, not sufficient ones. However such necessary conditions are powerful enough to predict the effect to different structural or parameter changes. Indeed changes in the equations leading to a violation of any necessary conditions in terms of loops should yield the change as well in the dynamic behavior. And changes in the equations which do not affect the feedback loop structure of the Jacobian should preserve the dynamic behavior.

The above definitions, properties and theorems are the fundamentals of the theory of the Jacobian Loops analysis. They show that Jacobian loops and their combinations play an important dynamical role in a system, even when only the signs, not the magnitudes of the Jacobian terms, are known. Using stability analysis one can attain only the local dynamics of the system, and hence the need to use, for instance, numerical integration in conjunction to obtain global dynamics which can be predicted by the feedback loop methodology, as for the Lorenz and Rossler systems. This is a great advantage over the classical approach. Indeed it allows one to assert whether sustained oscillations, multistationarity, or chaotic dynamics are possible. As such this analysis is certainly an efficient tool in the qualitative modeling of complex systems. It allows to:

1. Stress qualitative understanding as the primary goal rather than numerical prediction.
2. Supplement the more familiar large scale quantitative methods made possible by improved computer technology.
3. Include variables difficult or even impossible to measure, e.g., a diabetes model should include measurable variables such as glucose, insulin and other chemicals but also real variables such as anxiety or stress but any attempt to measure stress is itself stress inducing.

In economics [24], behavioral and social sciences, as well as in complex physical sciences relevant informations about the underlying dynamics reside in the rules of construct of the system and not in the absolute values.

The Jacobian loops technique, easy to implement, intends to quickly demarcate both parameter and phase spaces into exciting regions (limit cycles, multiple equilibria, chaotic behavior), non-exciting regions (single stable fixed point), and hard-instance regions (ergodic behavior). Hence it could prove useful in surveying dynamical response of models simulating physico-chemical, biological and bio-chemical, and economical systems, as well as in game theory.

6.1 Open Problem and Future Research

We present here some directions to improve the effectiveness of the qualitative analysis of systems based on the Jacobian feedback loops. Recall a loop is determined by a set of nonzero terms $a_{ij} = \frac{\partial f_i}{\partial x_j}$ of the Jacobian matrix whose i (row) and j (column) indices are in cyclic permutation. Its oriented edges (arrows) are the a_{ij} elements considered with their signs to indicate positive, negative or no interaction. A loop is usually symbolized by the product of its elements: for example, a 3-loop L_{xyz} is given by $a_{12}a_{23}a_{31}$. A loop is positive or negative depending on the sign of this product, this is, depending on whether it comprises an even or an odd number of negative elements. A positive feedback loop is destabilizing, whereas a negative feedback loop is stabilizing. For instance a minimal requirement for oscillations is the existence of at least one positive and a negative feedback loop, e.g. in chemical reactions systems [25, 37].

1. We want to emphasize that qualitative modeling should consist of strictly qualitative relations and assumptions, as opposed to studying the qualitative structure of models consisting of some quantitative features which actually describes the classic qualitative analysis.
2. In qualitative modeling the rate of change $\dot{x}$ should be defined as a qualitative rate of change $\frac{dx}{dt}$, that is, rather than just taking the sign of the entries of the Jacobian, one should consider directly the qualitative interaction *per se* of the variables $x = (x_i, i = 1, \cdots, n)$.
3. A qualitative algebraic structure should be developed independently, not just by "signing" the current algebras. The study of the qualitative and loop equivalence classes is a first step.
4. There is a need of strong qualitative or loop stability theorem similar to the Lyapunov's theorem; this could be done only through a direct qualitative study, rather than translating from the signs of quantitative values. That is, proving the existence of a Lyapunov function using only qualitative properties and relations of the variables and/or parameters interactions.

5. Complexity is the keyword in the evolution of systems. And its main tool of analysis is a qualitative one, which could achieve a greater generality and realism than does the usual quantitative idealization of most mathematical models.

References

1. Bellman, R. (1997). *Introduction to matrix analysis*. Classics in Applied Matematics, vol. 19, 2nd ed. SIAM Philadelphia
2. Cinquin, P., & Demongeot, J. (2002). Positive and negative feedback: striking a balance between necessary antagonists. *Journal of Theoretical Biology, 216*, 229–241.
3. Delbruck, M. (1949). Discussion in "Unités biologiques douées de continuité génétique". *Colloq. Int. CNRS, 8*, 33–35.
4. Demongeot, J., Kaufman, M., & Thomas, R. (2000). Positive feedback circuits and memory. *Comptes rendus de lÁcadémie des sciences Paris Life Sciences, 323*, 69–79.
5. Eisenfeld, J., & De Lisi, C. (1994). *On conditions for qualitative instability of regulatory loops with applications to immunoqualitative control loops*. Mathematics and Computers in Biomathematical Applications, pp. 39–53. New York: Elsevier.
6. Gantmacher, F. R. (1966). *Théorie des Matrices*. Collection Universitaire de Mathématiques, vol. 1,2. Paris: Dunod.
7. Gouzé, J. L. (1998). Positive and negative circuits in dynamical systems. *Journal of Biological Systems, 6*, 11–15.
8. Hirsch, M. W. (1985). Systems of differential equations that are competitive or cooperative. *SIAM Journal on Mathematical Analysis, 16*, 432–439.
9. Justus, J. (2006). Loop analysis and qualitative modeling: limitations and merits. *Biology and Philosophy, 21*, 647–666.
10. Justus, J. (2005). Qualitative scientific modeling and loop analysis. *Philosophy of Science, 72*, 1272–1286.
11. Kaufman, M., Soulé, C., & Thomas, R. (2007). A new necessary condition on interaction graphs for multistationarity *Journal of Theoretical Biology, 248*, 675–685.
12. Kirkland, S. J., McDonald, J. J., & Tsatsomeros, M. J. (1996). Sign-patterns which require a positive eigenvalue. *Linear and Multilinear Algebra, 41*(3), 199–210.
13. Lancaster, P., & Tismenetsky, M. (1985). *The theory of matrices with applications*. Academic Press, Inc., Harcourt Brace Jovanovich, Publishers. San Diego, New York, Berkeley, Boston.
14. Levins, R. (1974). The qualitative analysis of partially specified systems. *Annals of the New York Academy of Sciences, 231*, 123–138.
15. Liénard, A., & Chipart, H. (1914). Sur le signe de la partie réelle des racines d'une équation algébrique. *Journal de Mathématiques Pures et Appliquées, 10*, 291–346.
16. Lorenz, E. N. (1963). Deterministic nonperiodic flow. *Journal of the Atmospheric Sciences, 20*, 130–141.
17. Lyapunov, A. ([1892] 1992). *The general problem of the stability of motion*. London: Taylor and Francis.
18. May, R. M. (1974). *Stability and complexity in model ecosystems*. Princeton, NJ: Princeton University Press.
19. Plahte, E., Mestl, T., & Omholt, W. (1995). Feedback loop, stability and multistationarity in dynamical systems. *Journal of Biological Systems, 3*(2), 409–413.
20. Puccia, C., & Richard, L. (1985). *Qualitative modeling of complex systems*. Cambridge, MA: Harvard University Press.
21. Quirk, R., & Ruppert, R. (1965). Qualitative economics and the stability of equilibrium. *Review of Economic Studies, 32*, 311–326.

22. Rosen, R. (1968). Recent developments in theory of control and regulation of cellular processes. *International Review of Cytology, 23*, 25–88.
23. Rossler, O. E. (1979). Continuous chaos—four prototype equations. *Annals of the New York Academy of Sciences, 316*, 376–392.
24. Samuelson, P. (1947). *Foundations of economic analysis.* Cambridge: Harvard University Press.
25. Sensse, A., Hauser, M., & Eiswirth, M. (2006). Feedback loops for Shil'nikov chaos: The peroxidase-oxidase reaction. *Journal of Chemical Physics, 125*, 014901-12.
26. Snoussi, E. H. (1998). Necessary condition for Multistationarity and stable periodicity. *Journal of Biological Systems, 6*, 3–9.
27. Soulé, C. (2003). Graphic requirements for multistationarity. *ComplexUs, 1*, 123–133.
28. Soulé, C. (2006). Mathematical approaches to differentiation and gene regulation. *Comptes Rendus Biologies, 329*, 13–20.
29. Thomas, R. (1994). The role of Feedback Circuits: positive feedback circuits are a necessary condition for positive eigenvalues in the feedback matrix. *Berichte der Bunsengesellschaft für physikalische Chemie, 98*, 1148–1151.
30. Thomas, R. (1996). Analyse et synthèse de systèmes à dynamique chaotique en terme de loops de rétroaction. *Académie Royale de Belgique* 6*e* *série*, Tome VII, 101–124.
31. Thomas, R., & D'Ari, R. (1990). Biological feedback. Boca Raton: CRC Press.
32. Thomas, R., Thieffry, D., & Kaufman, M. (1995). Dynamical behaviour of biological regulatory networks. I. Biological role of feedback loops and practical use of the concept of the loop-characteristic state. *Bulletin of Mathematical Biology, 57*, 247–276.
33. Toni, B., Thieffry, D., & Bulajich, R. (1999). Feedback Loops analysis for chaotic dynamics with an application to Lorenz system. *Fields Institute Communications, 21*, 473–483.
34. Toni, B. (2005). Jacobian feedback loops analysis I. *International Journal of Evolution Equations, 1*(4), 415–428.
35. Toni, B. (2008). Jacobian feedback loops analysis II: stabilty and instability. *International Journal of Evolution Equations, 2*(4), 355–366.
36. Toni, B., Parmananda, P., Bulajich, R., & Thieffry, D. (1998). Dynamics of a two-dimensional modelfor electrochemical corrosion using feedback circuit and nullcline analysis. *Journal of Physical Chemistry B, 102*, 4118–4122.
37. Tyson, J. (1975). Classification of instabilities in chemical reaction systems. *Journal of Physical Chemistry, 62*, 1010–1015.
38. Verhulst, F. (1990). *Nonlinear differential equations and dynamical systems.* Springer-Verlag Universitext. Berlin, Heidelberg, New York.

Forecasting of Time Series Data Using Multiple Break Points and Mixture Distributions

Rajan Lamichhane, Norou Diawara, and Cynthia M. Jones

Abstract Stochastic processes have applications in many areas such as oceanography and engineering. Special classes of such processes deal with time series of sparse data. Studies in such cases focus in the analysis, construction and prediction in parametric models. Here, we assume several non-linear time series with additive noise components, and the model fitting is proposed in two stages. The first stage identifies the density using all the clusters information, without specifying any prior knowledge of the underlying distribution function of the time series. In the second stage, we partition the time series into consecutive non-overlapping intervals of quasi stationary increments where the coefficients shift from one stable regression relationship to a different one using breakpoint detection algorithm. These breakpoints are estimated by minimizing the likelihood from the residuals. We approach time series prediction through the mixture distribution of combined error components. Parameter estimation of mixture distribution is done by using the EM algorithm. We apply the method to a simulated data.

R. Lamichhane (✉)
Department of Mathematics, Texas A&M University-Kingsville, Kingsville, TX 78363, USA
e-mail: rajan.lamichhane@tamuk.edu

N. Diawara
Department of Mathematics and Statistics, Old Dominion University, Norfolk, VA 23529, USA
e-mail: ndiawara@odu.edu

C.M. Jones
Department of Ocean, Earth and Atmospheric Sciences, Old Dominion University, Norfolk, VA 23523
e-mail: cjones@odu.edu

B. Toni (ed.), *New Frontiers of Multidisciplinary Research in STEAM-H (Science, Technology, Engineering, Agriculture, Mathematics, and Health)*, Springer Proceedings in Mathematics & Statistics 90, DOI 10.1007/978-3-319-07755-0_11,

1 Introduction

Stochastic processes for longitudinal data are fundamental in probability and statistics and have applications in many areas such as oceanography and engineering. Special classes of such processes deal with time series of sparse data. Studies in such cases focus on the analysis, construction and prediction in parametric models.

In this work, the prediction of time series is revisited and a different approach of prediction based on mixture distribution is explained with a simulated example.The density uses all the clusters information, without specifying any prior knowledge of the underlying distribution function of time series. The change in stability of regression coefficients during the time course can be accounted by creating different breakpoints. The time course is partitioned into consecutive non-overlapping intervals where the coefficients shift from one stable regression relationship to a different one. These breakpoints are estimated by minimizing the residual sum of squares (RSS) using the algorithm described by Bai and Peron [15]. The foundation for estimating breaks in time series regression models was given by Bai [11] and was extended to multiple breaks by Bai [12] and [13] and Bai and Perron [14] and [15]. The algorithm in selecting the number of change points is based on a simple iterative step in which the maximum difference is less than a critical value of the difference of two consecutive values and is less than an optimal threshold chosen in a Bayesian framework. The partition algorithm fits a different probability model maximizing likelihood within each block interval.

Since different parts of data fit different models, forecasting depends not just on one model, but on all the relevant models. Method based on mixture of different distributions to forecast this type of model is explained. The Expectation-Maximization (EM) algorithm, with initial values obtained from the empirical estimates, gives the estimates of the mixture distribution. Further improvement in the parameter estimation has been observed by using bootstrap re-sampling combined with EM algorithm. For simplicity, we name this method as Break Point Bootstrap Filtering (BPBF) method.

This work is an extension of the ideas developed akin to the cited references and related work. It presents a novel concept in time series prediction and some supporting empirical evidence in terms of real data. The concept of using multiple break points based on minimum RSS or Bayesian Information Criteria (BIC) does not always create desirable partitioning of intervals. There could be very few observations in some intervals and the estimates based on those observations may be suspicious. In such cases, the estimation of parameters are improved by using block bootstrap. The block bootstrap as described in Bai [11–13] is the most general method to improve the accuracy of bootstrap for time series data. By dividing the data into different blocks, it can preserve the original time series structure within a block. However, the accuracy of the block bootstrap is sensitive to the choice of block length, and the optimal block length depends on the sample size, the data generating process and the statistic considered. In our examples, we are using the approach proposed by Patton et al. [26] to identify the optimal block size. Varying block lengths that follow the geometric distribution are considered, and thus we avoid the problem of non-stationary by its construction [27].

This chapter is organized as follows. Section 2 presents the guidelines and theory of the different procedures in model fitting. The distributions of the models are specified, and our new method is provided. In Sect. 3, we apply the method to simulated data and get estimation of parameters as well as model forecasting. Conclusion is presented in Sect. 4.

2 Model Building

Partially observed time series models are studied under various conditions, e.g. state space models [19], dynamic models [31], and hidden markov models [17]. All of these methods work if we have regular time series data where the model structure does not change locally. In other words, if the variance changes locally, then it is hard to build the model based on regular time series approach. However, there are cases where structural changes or breaks appear to affect models, for example in the evolution in key economic and financial time series such as output growth, inflation, exchange rates, interest rates and stock returns.[1] If data are collected over a long period of time, we are more likely to observe the structural change. This change could be the result of many possible factors such as institutional or technological changes, environmental changes, shifts in economic policy, or could even be due to large macroeconomic shocks such as the doubling or quadrupling of commodity prices experienced over the past decades.

One main goal that arises in the context of time-series forecasting of such models is to incorporate these different model structures to estimate the overall model parameters. Sometimes it is reasonable to assume that if breaks have occurred in the past, surely they are also likely to happen in the future. Approaches that view breaks as being generated deterministically are not applicable when forecasting future events unless, of course, future break dates as well as the size of such breaks are known in advance. In most applications, this is not a plausible assumption and modelling of the stochastic process underlying the breaks is needed. In this section, we provide a general framework for forecasting time series under structural breaks that is capable of handling the different above scenarios.

Regular time series linear model of responses $\boldsymbol{Y}$ based on predictors X can be defined as:

$$\boldsymbol{Y} = X\boldsymbol{\beta} + \boldsymbol{\zeta},$$

where the errors $\boldsymbol{\zeta}$'s are not independent and assume stationarity process. These errors are assumed to be unobserved, and our goal is to formulate their distributional form.

Also, the lag h autocovariance for the $\boldsymbol{\zeta}$ is given by:

$$Cov(\zeta_t, \zeta_{t-h}) = Cov(\zeta_t, \zeta_{t+h}) = \gamma(h) = \sigma^2 \rho_h,$$

[1] A small subset of the many papers that have reported evidence of breaks in economic and financial time series includes Garcia and Perron [21], Koop and Potter [24], and Pastor and Stambaugh [25].

and the ζ follows an autoregressive moving average process of order (p, q), we denote as ARMA(p, q) which is:

$$\zeta_t - \phi_1\zeta_{t-1} - \phi_2\zeta_{t-2} - \ldots - \phi_p\zeta_{t-p} = Z_t + \theta_1 Z_{t-1} + \ldots + \theta q Z_{t-q},$$

with $\{Z_t\}$ being the white noise of the ζ process and $\phi_1, \ldots, \phi_p$ and $\theta_1, \ldots, \theta_q$ are AR and MA components, respectively.

Also, we can further extend the model to autoregressive integrated moving average, ARIMA (p, d, q), where $\{\zeta_t\}$ satisfies a difference equation of the form

$$\phi(B)(1 - B)^d \zeta_t = \theta(B)Z_t, \ \{Z_t\} \sim WN(0, \sigma^2),$$

with $\phi(z)$ and $\theta(z)$ are polynomials of degrees p and q, respectively,

$\phi(z) \neq 0$ for $|z| \leq 1$, d is the difference indicator and B is the backshift operator such that $B^j\zeta_t = \zeta_{t-j}$ and $B^j Z_t = Z_{t-j}$. Notice that in ARIMA models, the nonstationary time series is converted into stationary by differencing. For $d = 0$, an ARIMA(p, d, q) reduces to an ARMA(p, q) process.

If the structure of data is such that there is heterogeneous variance structure among different intervals, then parameter estimates based on a regular time series model is very unrealistic. So the data is divided into different parts using multiple breakpoints. The distribution function used for the confidence intervals for the breakpoints is given in Bai [13]. The ideas behind this implementation are described in Zeileis et al. [32]. The break points are obtained by testing or assessing deviations from stability in the classical linear regression model

$$y_j = x_j^T \boldsymbol{\beta} + u_j,$$

where at time j, y_j is the observation of the dependent variable, $x_j = (1, x_{j1}, \ldots, x_{jk})^T$ is a $(k + 1) \times 1$ vector of observations of the independent variables, and u_j are *iid* with 0 mean and variance σ^2, and $\boldsymbol{\beta}$ is the $(k + 1) \times 1$ vector of regression coefficients.

In many applications, it is reasonable to assume that there are m breakpoints, where the coefficients shift from one stable regression relationship to a different one. Thus, there are $m + 1$ segments, $I_1, \cdots, I_{m+1}$ in which the regression coefficients are constant, and the model can be rewritten as:

$$y_j = x_j^T \boldsymbol{\beta}_i + u_j, \tag{1}$$

where $\boldsymbol{\beta}_i, i = 1, 2, \cdots, m + 1$ is the vector of regression coefficients within each segment, i denotes the segment index and $j = j_{i-1} + 1, \ldots, j_i$. In practice, the breakpoints are rarely given exogenously, but have to be estimated. They are estimated by minimizing the residual sum of squares (RSS) from Eq. (1). The algorithm for computing the optimal breakpoints given the number of breaks is based on a dynamic programming approach based on the Bellman principle [16]. The main computational effort is to compute a triangular RSS matrix, which gives the RSS for a segment starting at observation indexed j and ending at indexed j' with $j < j'$. Also, the adjacent intervals separated by break points are significantly different.

Let I_i denote the ith interval with density function $f_{ij}(y_{ij}, \boldsymbol{\theta}_i)$ where $i = 1, 2, \ldots, m+1$ represents the number of intervals and $j = 1, \ldots, n_i$ represents the number of values within that interval and θ_i is the vector of time series parameters within each interval. Thus, we have $m + 1$ time series models and each model is based only on the data of corresponding interval. So our main challenge is to combine all this model information to create a common model that can be used for forecasting. Several studies have been done in the past to combine the multiple time series regression models. Qin [28], Qin and Lawless [29], Qin and Zhang [30], Gilbert [22], Zhang [33] and Fokianos et al. [20] worked on some semi-parametric methods. Kedem and Gagnon [23] further extended those ideas by showing the estimation of the probability distribution of a "reference" time series and using them in conditional prediction. All these aforementioned ideas use multiple time series regressions where different time series structures are related to different covariates but the ideas do not extend into the different time intervals.

2.1 Parameter Estimation

Let's assume that there are m breakpoints, so there are m+1 time series intervals. We assume that for each interval, different models fit the data so there are m+1 distinct models. Let $y_{ij}, i = 1, 2, \ldots, m, (m+1); j = 1, 2, \ldots, n_i$ be the jth observation in ith interval.

Let $f_i(\boldsymbol{y}_i, \boldsymbol{\theta}_i)$ be the density function at ith interval. Notice that this density function is the function of past values and time series parameters $\boldsymbol{\theta}_i$.

Let $\boldsymbol{t}_i, i = 1, 2, \ldots., m+1$ be the vector of discrete time components.

Then,

$$\begin{aligned} y_{1,t_1} &= f_1(z_{1,t_1-1}) + \zeta_{t_1}, t_1 = 1, 2, \ldots, n_1\,, \\ &\vdots \\ y_{(m+1),t_{m+1}} &= f_{m+1}(z_{m+1,t_{m+1}-1}) + \zeta_{t_{m+1}}\,, \end{aligned} \tag{2}$$

where $t_{m+1} = t_m + 1, t_m + 2, \ldots, t_m + n_{m+1}$ and $\mathbf{z}_{i,t_i-1}$ contains past values of covariate time series possibly including even past values of $y_{1,t_1}, \ldots, y_{m,t_m}, y_{m+1,t_{m+1}}$. Also, n_i is the number of observations in the ith interval. Throughout our discussion it will be assumed that data have been "mean corrected" by subtraction of the sample mean, so that it is appropriate to fit a zero-mean ARMA model to the adjusted data.

Since any ARMA model can be expressed in the linear form of $Y_t = \sum_{j=0}^{\infty} \psi_j Z_{t-j}$ where

$$\boldsymbol{Z} \sim WN(0, \sigma^2).$$

We have:

$$f(z_{i,t_i-1}) = \sum_{j=1}^{\infty} \psi_{ij} \zeta_{t_i-j}\ ; \boldsymbol{\zeta}_i \sim WN(0, \sigma_i^2)\,, i = 1, 2, \ldots, m+1.$$

If the ARMA process is driven by Gaussian white noise, we can take $\{\zeta_{t_i}\} \overset{iid}{\sim} N(0, \sigma_i^2)$. So, the predicted values are:

$$\hat{Y}_{i,t_i} = \sum_{j=0}^{\infty} \hat{\psi}_{ij} \zeta_{t_i - j}.$$

Let

$$\gamma_i(h) = E(Y_{i,t_i} Y_{i,t_i+h}). \qquad \text{[Since } E(Y_{i,t_i}) = 0]$$

Then estimate of h lag covariance can be written as:

$$\begin{aligned}
\hat{\gamma}_i(h) &= \widehat{COV}(Y_{i,t_i}, Y_{i,t_i+h}), \\
&= E(\hat{Y}_{i,t_i} \hat{Y}_{i,t_i+h}), \qquad \left[\text{Since } E(\hat{Y}_{i,t_i}) = 0\right] \\
&= \sum_{j=0}^{\infty} \sum_{k=0}^{\infty} \hat{\psi}_{ij} \hat{\psi}_{ik} \gamma_i(j - h - k).
\end{aligned}$$

Taking $k = j - h$, we get

$$\hat{\gamma}_i(h) = \sum_{j=0}^{\infty} \hat{\psi}_{ij} \hat{\psi}_{i(j-h)} \gamma_i(0).$$

Since, the covariance structure is symmetric so the lag h coefficients $\hat{\psi}_{i(j-h)}$ and $\hat{\psi}_{i(h+j)}$ are equal. Also, $\hat{\gamma}_i(0) = \widehat{Var}(\boldsymbol{\zeta}_i) = \hat{\sigma}_i^2$.

Hence,

$$\hat{\gamma}_i(h) = \hat{\sigma}_i^2 \sum_{j=0}^{\infty} \hat{\psi}_{ij} \hat{\psi}_{i(j+h)}.$$

$$\Longrightarrow \widehat{VAR}(Y_{i,t_i}) = \hat{\gamma}_i(0) = \hat{\sigma}_i^2 \sum_{j=0}^{\infty} \hat{\psi}_{ij}^2.$$

The parameters $\boldsymbol{\psi}_i$ are composed of both autoregressive components, $\boldsymbol{\phi}_i$ and moving average components, $\boldsymbol{\theta}_i$. The preliminary estimates of parameters $\boldsymbol{\phi}_i$ and $\boldsymbol{\theta}_i$ are obtained by several methods such as Yule-Walker method, Burg procedure, innovations algorithm, Hannan-Rissanen algorithm and maximum likelihood method. Each method has its own advantages and limitations. Apart from the theoretical properties of the estimators such as consistency, efficiency etc., practical issues like the speed of computation and size of the data must also be taken into account in choosing an appropriate method for a given problem.

Yule-Walker and Burg procedures apply to the fitting of pure autoregressive models but innovations algorithm and Hannan-Rissanen algorithm are used for mixed models. Innovations algorithm is applicable to all series with finite second moments, regardless of whether they are stationary or not (Brockwell and Davis 2002). We also prefer innovation algorithm for the preliminary estimation of the parameters. Parameter estimation is improved by using innovations algorithm in conjunction with maximum likelihood method. The maximum likelihood method of estimating model parameters is often favored because it has the advantage among others that its estimators are more efficient (have smaller variance) and many large sample properties are known under rather general conditions. In our case, we do the parameter estimation as follows:

(I) We first identify the order (p, q) of ARMA model based on minimum value of corrected version of Akaike Information Criterion (AICc).
(II) Based on order (p, q) from previous step, we use one-step innovations algorithms to get preliminary estimates of $\boldsymbol{\phi}_i$ and $\boldsymbol{\theta}_i$.
(III) One step prediction errors obtained from innovations algorithm by using different values of $\boldsymbol{\phi}_i$ and $\boldsymbol{\theta}_i$ are then used to numerically maximize the likelihood function based on Gaussian noise.

2.2 *Maximum Likelihood Estimation of Time Series Parameters*

We fit different time series models for different intervals. Parameters are estimated based on maximum likelihood method in which preliminary estimates are obtained through innovations algorithm. Even though the maximum likelihood method is based on the assumption of Gaussian noise, it still makes sense to use this method as a measure of goodness of fit of the model to the data and it has well defined asymptotic properties. A justification for using maximum Gaussian likelihood estimators of ARMA coefficients is that the large sample distribution of the estimators is the same for white noise $\{\zeta_t\} \sim IID(0, \sigma^2)$, regardless of whether or not Z_t is Gaussian (Brockwell and Davis 2002). For convenience, in our discussion we use a general case to derive the expressions for maximum likelihood rather than defining it for different intervals.

Let $\{Y_t\}$ be causal $ARMA(p, q)$ process then

$$Y_t = \Psi(B)\zeta_t = \sum_{j=0}^{\infty} \psi_j B^j \zeta_t \quad ; \zeta_t \sim WN(0, \sigma^2); \sum_{j=0}^{\infty} |\psi_j| < \infty.$$

Also for ARMA model,

$$Y_t = \frac{\theta(B)}{\phi(B)} \zeta_t.$$

So,

$$\Psi(B) = \frac{\theta(B)}{\phi(B)}$$

$$\Longrightarrow (1 - \phi_1 B - \phi_2 B^2 - \ldots - \phi_p B^p)(\psi_0 + \psi_1 B + \ldots) = 1 + \theta_1 B + \theta_2 B^2 + \ldots + \theta_q B^q.$$

Equating the like coefficients of B's, we get

$$\begin{aligned}
1 &= \psi_0 \, , \\
\theta_1 &= \psi_1 - \psi_0 \phi_1 \Longrightarrow \psi_1 = \theta_1 + \psi_0 \phi_1 \, , \\
\theta_2 &= \psi_2 - \psi_1 \phi_1 - \psi_0 \phi_2 \Longrightarrow \psi_2 = \theta_2 + \psi_1 \phi_1 + \psi_0 \phi_2 \, , \\
&\vdots
\end{aligned}$$

$$\psi_j = \theta_j + \sum_{i=1}^{min.(j,p)} \phi_i \psi_{j-i} \, , \; j = 0, 1, \ldots, \tag{3}$$

and we define $\theta_0 = 1$ and $\theta_j = 0$ for $j > q$. The innovations estimates $\hat{\theta}_{n1}, \hat{\theta}_{n2}, \ldots, \hat{\theta}_{n,(p+q)}$ are used to estimate $\psi_1, \psi_2, \ldots, \psi_p + q$. Replacing ψ_j by $\hat{\theta}_{nj}$ in Eq. (3) we get

$$\hat{\theta}_{nj} = \theta_j + \sum_{i=1}^{min.(j,p)} \phi_i \hat{\theta}_{n,(j-i)} \, , \; j = 1, 2 \ldots, p + q, \tag{4}$$

From last q equations we first estimate $\hat{\boldsymbol{\phi}}$ as:

$$\begin{pmatrix} \hat{\theta}_{n,q+1} \\ \hat{\theta}_{n,q+2} \\ \vdots \\ \hat{\theta}_{n,q+p} \end{pmatrix} = \begin{pmatrix} \hat{\theta}_{nq} & \hat{\theta}_{n,q-1} & \ldots & \hat{\theta}_{n,q+1-p} \\ \hat{\theta}_{n,q+1} & \hat{\theta}_{n,q} & \ldots & \hat{\theta}_{n,q+2-p} \\ \vdots & \vdots & . & \vdots \\ \hat{\theta}_{n,q+p-1} & \hat{\theta}_{n,q+p-2} & \ldots & \hat{\theta}_{n,q} \end{pmatrix} \begin{pmatrix} \phi_1 \\ \phi_2 \\ \vdots \\ \phi_p \end{pmatrix}.$$

Then $\boldsymbol{\theta}$ can be estimated from Eq. (3) as:

$$\hat{\theta}_j = \hat{\theta}_{nj} - \sum_{i=1}^{min.(j,p)} \hat{\phi}_i \hat{\theta}_{n,j-i} \, , \; j = 1, 2, \ldots, q.$$

After these preliminary estimation of $\boldsymbol{\phi}$ and $\boldsymbol{\theta}$, we use these values as the initial values to get the maximum likelihood estimates. The maximization is nonlinear in the sense that the function to be maximized is not a quadratic function of the unknown parameters, so the estimators cannot be found by solving a system of linear equations. They are found instead by searching numerically for the maximum of the likelihood surface. When the order p and q of ARMA model is known, good estimators of $\boldsymbol{\phi}$ and $\boldsymbol{\theta}$ can be found by imagining the data to be observations of a stationary Gaussian time series and maximizing the likelihood with respect to the $p+q+1$ parameters $\phi_1, \ldots, \phi_p, \theta_1, \ldots, \theta_q$ and σ^2.

Suppose that $\{Y_t\}$ is a Gaussian time series with mean zero and autocovariance function $\kappa(i, j) = E(Y_i Y_j)$. Let $\boldsymbol{Y}_n = (Y_1, Y_2, \ldots, Y_n)'$ and let the one step predictors $\hat{\boldsymbol{Y}}_n = (\hat{Y}_1, \hat{Y}_2, \ldots, \hat{Y}_n)'$ where $\hat{Y}_1 = 0$ and $\hat{Y}_j = P_{j-1}Y_j, j \geq 2$. Let Γ_n denote the covariance matrix $\Gamma_n = E(\boldsymbol{Y}_n \boldsymbol{Y}_n')$, and it is non-singular.

Then the likelihood of $\boldsymbol{Y}_n$ is

$$L(\Gamma_n; \boldsymbol{Y}_n) = (2\pi)^{-\frac{n}{2}} |\Gamma_n|^{-\frac{1}{2}} exp\left(-\frac{1}{2}\boldsymbol{Y}_n' \Gamma_n^{-1} \boldsymbol{Y}_n\right). \quad (5)$$

The direct calculation of Γ_n is cumbersome and in many situation not possible and it is avoided by using the one step prediction errors $Y_n - \hat{Y}_n$ and mean squared error, $E(Y_n - \hat{Y}_n)^2$ instead of Y_n and Γ_n. Both of prediction error and mean squared errors are calculated recursively from the innovations algorithm.

Also,

$$\begin{aligned}
\hat{\boldsymbol{Y}}_n &= \boldsymbol{\Theta}_n(\boldsymbol{Y}_n - \hat{\boldsymbol{Y}}_n) \\
&= C_n(\boldsymbol{Y}_n - \hat{\boldsymbol{Y}}_n) - I_n(\boldsymbol{Y}_n - \hat{\boldsymbol{Y}}_n) \\
&= C_n(\boldsymbol{Y}_n - \hat{\boldsymbol{Y}}_n) - \boldsymbol{Y}_n + \hat{\boldsymbol{Y}}_n \\
\Longrightarrow \boldsymbol{Y}_n &= C_n(\boldsymbol{Y}_n - \hat{\boldsymbol{Y}}_n). \quad (6)
\end{aligned}$$

Since the components $Y_n - \hat{Y}_n$ are uncorrelated, the covariance matrix of $Y_n - \hat{Y}_n$ is

$$\Sigma_n = \begin{bmatrix} \upsilon_0 & 0 & \ldots & 0 \\ 0 & \upsilon_1 & \ldots & 0 \\ \vdots & \vdots & \ddots & \vdots \\ 0 & 0 & \ldots & \upsilon_{n-1} \end{bmatrix}.$$

From Eq. (6),

$$\begin{aligned}
Var(\boldsymbol{Y}_n) &= Var[C_n(\boldsymbol{Y}_n - \hat{\boldsymbol{Y}}_n)] \\
\Longrightarrow \Gamma_n &= C_n \Sigma_n C_n'.
\end{aligned}$$

So,

$$|\Gamma_n| = |C_n|^2 |\Sigma_n| = \upsilon_0 \upsilon_1 \ldots \upsilon_{n-1},$$

and

$$\boldsymbol{Y}_n'\Gamma_n^{-1}\boldsymbol{Y}_n = \left[C_n(\boldsymbol{Y}_n - \hat{\boldsymbol{Y}}_n)\right]'\Gamma_n^{-1}C_n(\boldsymbol{Y}_n - \hat{\boldsymbol{Y}}_n) \qquad \text{[From Eq. (6)]}$$
$$= (\boldsymbol{Y}_n - \hat{\boldsymbol{Y}}_n)'C_n'\Gamma_n^{-1}C_n(\boldsymbol{Y}_n - \hat{\boldsymbol{Y}}_n)$$
$$= (\boldsymbol{Y}_n - \hat{\boldsymbol{Y}}_n)'\Sigma_n^{-1}(\boldsymbol{Y}_n - \hat{\boldsymbol{Y}}_n)$$
$$= \frac{\sum_{j=1}^{n}(Y_j - \hat{Y}_j)^2}{\upsilon_{j-1}}.$$

Hence, from Eq. (5) likelihood of vector $\boldsymbol{Y}_n$ reduces to:

$$L(\Gamma_n; \boldsymbol{Y}_n) = \frac{1}{\sqrt{(2\pi)^n \upsilon_0\upsilon_1 \dots \upsilon_{n-1}}} exp\left\{\frac{-1}{2}\frac{\sum_{j=1}^{n}(Y_j - \hat{Y}_j)^2}{\upsilon_{j-1}}\right\}.$$

The likelihood for data from $ARMA(p,q)$ process is easily computed from the innovations form by replacing $\hat{Y}_j$ by one-step predictor and υ_j by $\sigma^2 r_n$.

Hence, the Gaussian likelihood for an ARMA process can be written as:

$$L(\Gamma_n; \boldsymbol{Y}_n) = \frac{1}{\sqrt{(2\pi\sigma^2)^n r_0 r_1 \dots r_{n-1}}} exp\left\{\frac{-1}{2\sigma^2}\frac{\sum_{j=1}^{n}(Y_j - \hat{Y}_j)^2}{r_{j-1}}\right\}. \qquad (7)$$

So, maximum likelihood estimator of σ^2 is:

$$\hat{\sigma}^2 = \frac{1}{n}\frac{\sum_{j=1}^{n}(Y_j - \hat{Y}_j)^2}{r_{j-1}},$$

and $\hat{\boldsymbol{\phi}}$ and $\hat{\boldsymbol{\theta}}$ are the values of $\boldsymbol{\phi}$ and $\boldsymbol{\theta}$ that maximize the likelihood in Eq. (7).

Also, we used minimum AICc (Akaike Information Criteria Corrected) value as a major criterion for the selection of the orders p and q. AICc criterion can be defined as:

$$AICc = -2lnL(\boldsymbol{\phi}_p, \boldsymbol{\theta}_q; Y_n) + \frac{2(p+q+1)n}{n-p-q-2},$$

where $lnL(\boldsymbol{\phi}_p, \boldsymbol{\theta}_q; Y_n)$ is the log of likelihood function defined in Eq. (7) using maximum likelihood estimators $\hat{\boldsymbol{\phi}}_p$ and $\hat{\boldsymbol{\theta}}_q$. For any fixed p and q, it is clear that the AICc is minimized when $\boldsymbol{\phi}_p$ and $\boldsymbol{\theta}_q$ are the maximum likelihood estimators. Final decisions with respect to order selection should therefore be made on the basis of maximum likelihood estimators.

2.3 Forecasting

As we have seen, autoregressive moving average time series models can be regarded as means of transforming the data to white noise, that is, to an uncorrelated sequence of errors. If the appropriate model has been chosen, there will be zero autocorrelation in the errors. For large samples the residuals from a correctly fitted model resemble very closely the true errors of the process (Box and Pierce 1970). Since there are differences in trends, forecasting of multiple time series data based on well behaved residuals and certain joint relationship between their probability density functions are explored by Kedem and Gagnon [23]. We are also exploiting the similar idea but by using the mixture distribution of residual densities as the reference distribution. The mixture parameters are estimated through the distributions of combined noise.

Since, each part of the interval is fitted with different models, the residuals for each part are independent to each other and to the errors from other intervals. So,the error sequence $\{\zeta_{t_i}\}$ is the sequence of iid random variables.

Define,

$$\zeta_{t_i} \overset{iid}{\sim} g_i(\varsigma), i = 1, \ldots, m, m+1 \ ,$$

where $g_i(\varsigma)$ is the density function of ζ_{t_i} for ith interval. We approach time series prediction through the mixture distribution of these error components. Noises from different intervals are combined to form combined noise.

Let

$$\begin{aligned} \boldsymbol{\zeta} &= (\boldsymbol{\zeta}_1, \boldsymbol{\zeta}_2, \ldots, \boldsymbol{\zeta}_{m+1}) \\ &= \{(\zeta_1, \ldots, \zeta_{n_1}), \ldots, (\zeta_{t_i+1}, \ldots, \zeta_{t_i+n_i}), \ldots, (\zeta_{t_m+1}, \ldots, \zeta_{t_m+n_{m+1}})\}. \end{aligned}$$

The joint density of combined noise is the mixture of '$m+1$' noise distributions. So, the joint density of finite mixture of combined noise is:

$$g(\varsigma) = \sum_{i=1}^{m+1} p_i g_i(\varsigma),$$

where p_i be the mixing proportion with the constraints $p_i \geq 0$, $i = 1, \ldots, m$, $m+1$, and $\sum_{i=1}^{m+1} p_i = 1$.

Hence, the cumulative distribution function of combined error is:

$$P(\zeta \leq \varsigma) = G(\varsigma) = \sum_{i=1}^{m+1} p_i G_i(\varsigma), \tag{8}$$

where $G_i(\varsigma)$ is the cumulative distribution function of $\boldsymbol{\zeta}_i$, $i = 1, \ldots, m, m+1$.

Our main objective is to predict the future reference values $y_{m+2,t_{(m+1)}+1}$ conditional on past values $z_{m+1,t_{m+1}}$. Future probability of events conditional in past values can be written as:

$$\begin{aligned} P(y_{m+2,t_{(m+1)}+1} \leq y | \mathbf{z}_{(m+2),t_{m+1}}) &= P(f_{m+2}(z_{(m+2),t_{m+1}}) + \zeta_{t_{m+1}+1} \leq y) \quad \text{[From Eq. (2)]} \\ &= P(\zeta_{t_{m+1}+1} \leq y - f_{m+2}(z_{(m+2),t_{m+1}}) \\ &= G(y - f_{m+2}(z_{(m+2),t_{m+1}})) \\ &= \sum_{i=1}^{m+1} p_i G_i(y - f_{m+2}(z_{(m+2),t_{m+1}})) \quad \text{[From Eq. (8)]} \\ &= \sum_{i=1}^{m+1} \left[p_i G_i(y - \sum_{j=1}^{p} \phi_{ij} y_{m+2,t_{(m+1)}-j} \right. \\ &\quad \left. + \sum_{k=1}^{q} \theta_{ik} \zeta_{m+2,t_{(m+1)}-k}) \right]. \end{aligned} \tag{9}$$

Since we are using a sample of observed values, the cumulative distribution function can be approximated by empirical distribution function.

Let $\hat{G}_i(\varsigma)$ be the empirical distribution function of error components in ith interval. Then,

$$\hat{G}_i(\varsigma) = \frac{\sum_{j=1}^{n_i} I(\zeta_{ij} \leq \varsigma)}{n_i},$$

where

$$I(\zeta_{ij} \leq \varsigma) = \begin{cases} 1, & \text{if } \zeta_{ij} \leq \varsigma; \qquad j = 1, 2, \ldots, n_i, \\ 0, & \textit{otherwise}, \end{cases}$$

and ζ_{ij} be the jth error component in ith interval. Also, *strong law of large numbers* indicates that empirical cumulative distribution function converges almost surely to cumulative distribution function.

Here, for a fixed point ς the quantity $n_i \hat{G}_i(\varsigma) \sim Bin(n_i, G_i(\varsigma))$. Therefore

$$E(\hat{G}_i(\varsigma)) = G_i(\varsigma) \text{and} Var(\hat{G}_i(\varsigma)) = \frac{G_i(\varsigma)(1 - G_i(\varsigma))}{n_i}.$$

By using Chebyshev's inequality,

$$P(|\hat{G}_i(\varsigma) - G_i(\varsigma)| \geq \epsilon) \leq \frac{G_i(\varsigma)(1 - G_i(\varsigma))}{n_i \epsilon^2} \qquad \text{for any } \epsilon > 0.$$
$$\longrightarrow 0 \qquad \text{as } n_i \to \infty.$$

Hence,

$$\hat{G}_i(\varsigma) \xrightarrow{a.s} G_i(\varsigma).$$

So the future probability of events conditional in past values from Eq. (9) can be approximated as

$$P(y_{m+2,t_{(m+1)}+1} \leq y|\mathbf{z}_{(m+2),t_{m+1}})$$
$$= \sum_{i=1}^{m+1} \left[\hat{p}_i \hat{G}_i \left(y - \sum_{j=1}^{p} \hat{\phi}_{ij} y_{m+2,t_{(m+1)}-j} + \sum_{k=1}^{q} \hat{\theta}_{ik} \zeta_{m+2,t_{(m+1)}-k} \right) \right]. \tag{10}$$

The parameters $\boldsymbol{\phi}$ and $\boldsymbol{\theta}$ for each interval are estimated by using the method of maximum likelihood in conjunction with innovations algorithm as discussed in Sect. 2.2. Estimation of future values in Eq. (10) also requires the estimation of p_i which is discussed in the next section.

2.4 Estimation of Mixture Proportions

In general time series, noises are either uncorrelated white noises or Gaussian noises. White noises are assumed to be a sequence of uncorrelated random variables generated from uniform probability distribution while Gaussian noises are generated from Gaussian distribution. The parameter estimation of mixture of Gaussian or other exponential family distribution can be done by using EM algorithm since the likelihood function of these kind of distribution is well defined [18]. A general method of parameter estimation for mixture of exponential family distribution is already discussed in Sect. 2. But when dealing with the mixture of any location family distribution or particularly white noises EM algorithm may not be the appropriate method to estimate the parameters. The problem of identifiability should also be handled.

In our discussion we will focus more on Gaussian noise since it has some well defined properties. The time series parameter estimation using maximum likelihood method we proposed is based on the assumption of Gaussian noise. Another justification for using Gaussian noise is that, the large sample distribution of estimators is the same whether or not we use Gaussian (Brockwell and Davis 1991). Even though our primary focus is on Gaussian noise, we will also discuss the alternative way of parameter estimation for mixture distribution of white noises.

2.4.1 Gaussian Noise and EM Algorithm

For each interval, without loss of generality, assume that $\boldsymbol{\zeta}_i \sim N(0, \sigma_i^2)$. Gaussian mixture model is a simple linear superposition of Gaussian components. Gaussian mixture distribution can be written as linear combination of Gaussians in the form

$$g(\varsigma|\boldsymbol{\theta}) = \sum_{i=1}^{m+1} p_i g_i(\varsigma),$$

where $\boldsymbol{\theta} = (p_i, \mu_i, \sigma_i^2)$. So,

$$g(\varsigma|\boldsymbol{\theta}) = \frac{1}{\sqrt{2\pi}} \sum_{i=1}^{m+1} p_i \frac{1}{\sigma_i} \exp\left[\frac{-1}{2\sigma_i^2}(\varsigma - \mu_i)^2\right], \tag{11}$$

This gives us the incomplete likelihood as:

$$L(\boldsymbol{\theta}|\varsigma) = \left(\frac{1}{\sqrt{2\pi}}\right)^n \prod_{j=1}^{n} \left(\sum_{i=1}^{m+1} p_i \frac{1}{\sigma_i} \exp\left[\frac{-1}{2\sigma_i^2}(\varsigma - \mu_i)^2\right]\right); n = n_1, \ldots, n_{m+1},$$

and the log likelihood is:

$$l(\boldsymbol{\theta}|\varsigma) \propto \sum_{j=1}^{n} log \left(\sum_{i=1}^{m+1} p_i \frac{1}{\sigma_i} \exp\left[\frac{-1}{2\sigma_i^2}(\varsigma_j - \mu_i)^2\right]\right). \tag{12}$$

Maximizing the log likelihood of Eq. (12) turns out to be a more complex problem than for the case of a single Gaussian. The difficulty arises from the presence of summation over i that appears inside the logarithm, so that the logarithm function no longer acts directly on the Gaussian. If we set the derivatives of the log likelihood to zero, we will no longer obtain a closed form solution. Also, the maximum likelihood framework applied to the Gaussian mixture model has significant problem due to the presence of singularities. Whenever one of the Gaussian components collapses onto a specific data point, the log likelihood function will go to infinity as $\sigma_i \to 0$. This creates a singularity problem and inverse covariance matrix, which is often required in maximum likelihood framework, is unattainable. So we consider an alternative approach known as EM algorithm which is an elegant and powerful method for finding maximum likelihood solutions for models with latent variables [18].

Let us introduce a $m+1$ dimensional binary random variable $\boldsymbol{w} = (w_1, w_2, \ldots, w_{m+1})'$ in which a particular element w_i is equal to 1 and all other elements are equal to 0. The value of the latent indicator w_i therefore satisfies $w_i \epsilon \{0, 1\}$ and $\sum_{i=1}^{m+1} w_i = 1$.

Also, the probability

$$p(w_i = 1) = p_i,$$

where the parameters $\{p_i\}$ must satisfy

$$0 \leq p_i \leq 1 \qquad \text{and} \qquad \sum_{i=1}^{m+1} p_i = 1.$$

Hence, the marginal density

$$p(\boldsymbol{w}) = \prod_{i=1}^{m+1} p_i^{w_i}.$$

Similarly, the conditional distribution of ς given a particular value of $\boldsymbol{w}$ is:

$$p(\varsigma | w_i = 1, \boldsymbol{\theta}) = \frac{1}{\sqrt{2\pi}\sigma_i} \exp\left[\frac{-1}{2\sigma_i^2}(\varsigma - \mu_i)^2\right].$$

So

$$p(\varsigma | \boldsymbol{w}, \boldsymbol{\theta}) = \prod_{i=1}^{m+1} \left(\frac{1}{\sqrt{2\pi}\sigma_i}\right)^{w_i} \left[\exp\left\{\frac{-1}{2\sigma_i^2}(\varsigma - \mu_i)^2\right\}\right]^{w_i}.$$

Using conditional probability, the joint density of ς and $\boldsymbol{w}$ is:

$$\begin{aligned} p(\varsigma, w|\boldsymbol{\theta}) &= p(\varsigma|w, \boldsymbol{\theta})p(w|\boldsymbol{\theta}) \\ &= \prod_{i=1}^{m+1} \left(\frac{p_i}{\sqrt{2\pi}\sigma_i}\right)^{w_i} \left[\exp\left\{\frac{-1}{2\sigma_i^2}(\varsigma - \mu_i)^2\right\}\right]^{w_i}. \end{aligned}$$

Hence, the complete likelihood is:

$$L(\boldsymbol{\theta}; \varsigma, \boldsymbol{w}) = \prod_{j=1}^{n} \prod_{i=1}^{m+1} \left(\frac{p_i}{\sqrt{2\pi}\sigma_i}\right)^{w_i} \left[\exp\left\{\frac{-1}{2\sigma_i^2}(\varsigma_j - \mu_i)^2\right\}\right]^{w_i} \text{ where } n = n_1 + \ldots + n_{m+1}.$$

And the complete log likelihood becomes:

$$\begin{aligned} l(\boldsymbol{\theta}; \varsigma, \boldsymbol{w}) &= \sum_{j=1}^{n} \left(\sum_{i=1}^{m+1} w_i log p_i\right) - \frac{1}{2} \sum_{j=1}^{n} \left(\sum_{i=1}^{m+1} w_i log(2\pi\sigma_i^2)\right) \\ &+ \sum_{i=1}^{m+1} \sum_{j=1}^{n} w_i \left[\frac{-1}{2\sigma_i^2}(\varsigma_j - \mu_i)^2\right]. \end{aligned} \tag{13}$$

Notice that this log likelihood can be solved easily in the closed form once we identify the conditional distribution of $\boldsymbol{w}$ given ς.

The conditional probability of $\boldsymbol{w}$ given $\boldsymbol{\varsigma}$ when $w_i = 1$ plays an important role to define expectation function. We shall view p_i as the prior probability of $w_i = 1$, and the conditional distribution $p(w_i = 1|\boldsymbol{\varsigma}, \boldsymbol{\theta})$ as the corresponding posterior probability once we have observed $\boldsymbol{\varsigma}$ for some known parameter estimates $\boldsymbol{\theta}$. We can use Bayes rule to estimate the conditional probability of $\boldsymbol{w}$ given $\boldsymbol{\varsigma}$ and $\boldsymbol{\theta}$ as follows:

$$p(w_i = 1|\boldsymbol{\varsigma}, \boldsymbol{\theta}) = \frac{p(\boldsymbol{\varsigma}|w_i = 1, \boldsymbol{\theta}).p(w_i = 1|\boldsymbol{\theta})}{p(\boldsymbol{\varsigma}|\boldsymbol{\theta})}$$

$$\left[\because P(A|B) = \frac{P(B|A)P(A)}{P(B)}\right]$$

$$= \frac{\dfrac{p_i}{\sqrt{2\pi}\sigma_i} \exp\left\{\dfrac{-1}{2\sigma_i^2}(\varsigma_i - \mu_i)^2\right\}}{\dfrac{1}{\sqrt{2\pi}} \sum_{j=1}^{m+1} \dfrac{p_j}{\sigma_j} \exp\left[\dfrac{-1}{2\sigma_j^2}(\varsigma_j - \mu_j)^2\right]}.$$

Hence,

$$p(w_i = 1|\boldsymbol{\varsigma}, \boldsymbol{\theta}) = \frac{\dfrac{p_i}{\sigma_i} \exp\left\{\dfrac{-1}{2\sigma_i^2}(\varsigma_i - \mu_i)^2\right\}}{\sum_{j=1}^{m+1} \dfrac{p_j}{\sigma_j} \exp\left[\dfrac{-1}{2\sigma_j^2}(\varsigma_j - \mu_j)^2\right]}. \tag{14}$$

Thus, the conditional expectation of $\boldsymbol{w}$ given $\boldsymbol{\varsigma}$ and $\boldsymbol{\theta}$ is:

$$E(\boldsymbol{w}|\boldsymbol{\varsigma}, \boldsymbol{\theta}) = p(w_i = 1|\boldsymbol{\varsigma}, \boldsymbol{\theta})$$

Let's assume that we start with some initial values $\boldsymbol{\theta}^{(0)}$ and cycle up to kth step. Let $\boldsymbol{\theta}^{(k)} = (p_i^{(k}, \mu_i^{(k)}, \sigma_i^{2(k)})$ be the parameter values at kth step. Then, conditional expectation at kth step can be written as:

$$w_i^{(k)} = E(\boldsymbol{w}|\boldsymbol{\varsigma}, \boldsymbol{\theta}^{(k)})$$

$$= \frac{\dfrac{p_i^{(k)}}{\sigma_i^{(k)}} \exp\left\{\dfrac{-1}{2\sigma_i^{2(k)}}(\varsigma_i - \mu_i^{(k)})^2\right\}}{\sum_{j=1}^{m+1} \dfrac{p_j^{(k)}}{\sigma_j^{(k)}} \exp\left[\dfrac{-1}{2\sigma_j^{2(k)}}(\varsigma_j - \mu_j^{(k)})^2\right]}. \tag{15}$$

Now, in E step we replace w_i with the conditional expectation of $\boldsymbol{w}$ at kth step from Eq. (15) into the complete log likelihood obtained in Eq. (13). Hence, the expectation function, $Q(\boldsymbol{\theta}|\boldsymbol{\theta}^{(k)})$, becomes:

$$Q(\boldsymbol{\theta}|\boldsymbol{\theta}^{(k)}) = \sum_{j=1}^{n}\left(\sum_{i=1}^{m+1} w_i^{(k)} log p_i\right) - \frac{1}{2}\sum_{j=1}^{n}\left(\sum_{i=1}^{m+1} w_i^{(k)} log(2\pi\sigma_i^2)\right)$$
$$+ \sum_{i=1}^{m+1}\sum_{j=1}^{n} w_i^{(k)}\left[\frac{-1}{2\sigma_i^2}(\varsigma_j - \mu_i)^2\right]. \tag{16}$$

In the M step, we determine the revised parameter estimate $\boldsymbol{\theta}^{(k+1)}$ by maximizing Eq. (16) with respect to relative parameters, p_i, μ_i and σ_i^2. Equation (16) can be maximized with respect to p_i under the condition that $\sum_{i=1}^{m+1} p_i = 1$. So we need to maximize the Lagrange function. Lagrange function is

$$\Lambda(\boldsymbol{p}, \lambda) = \sum_{j=1}^{n}\sum_{i=1}^{m+1} w_i^{(k)} log p_i + Constant + \lambda\left(\sum_{i=1}^{m+1} p_i - 1\right).$$

Maximizing with respect to p_i and λ and substituting the value of λ, we get the estimate of p_i at $(k+1)$th step as:

$$p_i^{(k+1)} = \frac{\sum_{j=1}^{n} w_i^{(k)}}{n},$$

where $w_i^{(k)}$ is the conditional expectation as discussed in Eq. (15). Also,

$$\frac{\partial Q(\boldsymbol{\theta}|\boldsymbol{\theta}^{(k)})}{\partial \mu_i} = 0$$

$$\Longrightarrow \quad \sum_{j=1}^{n} w_i^{(k)}(\varsigma_j - \mu_i) = 0$$

$$\Longrightarrow \quad \mu_i = \frac{\sum_{j=1}^{n} w_i^{(k)}\varsigma_j}{\sum_{j=1}^{n} w_i^{(k)}}$$

$$\Longrightarrow \quad \mu_i^{(k+1)} = \frac{\sum_{j=1}^{n} w_i^{(k)}\varsigma_j}{n p_i^{(k+1)}}.$$

Similarly,

$$\frac{\partial Q(\boldsymbol{\theta}|\boldsymbol{\theta}^{(k)})}{\partial \sigma_i^2} = 0$$

$$\Longrightarrow \qquad \sigma_i^{2(k+1)} = \frac{\sum_{j=1}^{n} \left(\varsigma_j - \mu_i^{(k)}\right)^2 w_i^{(k)}}{np_i^{(k+1)}}.$$

It can be shown that the sequence $\{\boldsymbol{\theta}^{(1)}, \boldsymbol{\theta}^{(2)}, \ldots, \boldsymbol{\theta}^{(k)}\}$ converges to the maximum likelihood estimator of $\boldsymbol{\theta}$, i,e. $\hat{\boldsymbol{\theta}}$ as $k \to \infty$ [18].

In our applications, we are using the mixture of two Gaussian distributions in Sect. 3. The mixture of two Gaussian population is given as:

$$g(\varsigma|\boldsymbol{\theta}) = p\frac{1}{\sigma_1}\varphi\left(\frac{\varsigma - \mu_1}{\sigma_1}\right) + (1-p)\frac{1}{\sigma_2}\varphi\left(\frac{\varsigma - \mu_2}{\sigma_2}\right),$$

where φ is the cumulative distribution function of the standard normal distribution and $\boldsymbol{\theta} = (p, \mu_1, \mu_2, \sigma_1^2, \sigma_2^2)\ ; \quad 0 < p < 1$.

Then, the indicator variable W be treated as missing data information such that:

$$W = \begin{cases} 1, \text{ if} & \zeta_{\mathrm{j}} \text{ belongs to first interval} \\ 0, \text{ if} & \zeta_{\mathrm{j}} \text{ belongs to second interval,} \end{cases}$$

where W_i is Bernoulli distributed with parameter p.

Therefore, the likelihood expression for complete data becomes:

$$L_n(\theta|\varsigma, w) = \prod_{j=1}^{n} p^w(1-p)^{1-w}\frac{1}{\sigma_1^w}\varphi\left(\frac{\varsigma_j - \mu_1}{\sigma_1}\right)^w \frac{1}{\sigma_2^{1-w}}\varphi\left(\frac{\varsigma_j - \mu_2}{\sigma_2}\right)^{1-w}.$$

And the corresponding log-likelihood function for the density becomes:

$$\begin{aligned} l_n(\theta|\varsigma, w) = {} & \sum_{j=1}^{n} wlog(p) + \sum_{j=1}^{n}(1-w)log(1-p) - \frac{1}{2}\sum_{j=1}^{n} wlog(2\pi\sigma_1^2) \\ & -\frac{1}{2\sigma_1^2}\sum_{j=1}^{n} w(\varsigma_j - \mu_1)^2 - \frac{1}{2}\sum_{j=1}^{n}(1-w)log(2\pi\sigma_2^2) \\ & -\frac{1}{2\sigma_2^2}\sum_{j=1}^{n}(1-w)(\varsigma_j - \mu_2)^2. \end{aligned}$$

The conditional distribution of W given $\boldsymbol{\zeta}$ is:

$$W|\varsigma_j, \theta^{(k)} \sim Bin(1, w^{(k)}),$$

with

$$w^{(k)} = \frac{p^{(k)}\frac{1}{\sigma_1^{(k)}}\varphi(\frac{\varsigma_j-\mu_1^{(k)}}{\sigma_1^{(k)}})}{p^{(k)}\frac{1}{\sigma_1^{(k)}}\varphi(\frac{\varsigma_j-\mu_1^{(k)}}{\sigma_1^{(k)}}) + (1-p^{(k)})\frac{1}{\sigma_2^{(k)}}\varphi(\frac{\varsigma_j-\mu_2^{(k)}}{\sigma_2^{(k)}})},$$

where $p^{(k)}$ is a set of known or estimated parameters at kth step. The initial value $p^{(0)}$ can be obtained from the empirical distribution.

Hence, the conditional mean at kth step is:

$$E(w|\varsigma_j, \theta^{(k)}) = w^{(k)}.$$

The expectation function becomes:

$$Q(\theta|\theta^{(k)}) = \sum_{j=1}^{n} w^{(k)} log(p) + \sum_{j=1}^{n}(1-w^{(k)}) log(1-p) - \frac{1}{2}\sum_{j=1}^{n} w^{(k)} log(2\pi\sigma_1^2)$$
$$-\frac{1}{2\sigma_1^2}\sum_{j=1}^{n} w^{(k)}(\varsigma_j-\mu_1)^2 - \frac{1}{2}\sum_{j=1}^{n}(1-w^{(k)}) log(2\pi\sigma_2^2)$$
$$-\frac{1}{2\sigma_2^2}\sum_{j=1}^{n}(1-w^{(k)})(\varsigma_j-\mu_2)^2.$$

Now, we maximize the expectation function as discussed above.

Hence, the parameter estimates at the $(k+1)$th step are:

$$p^{(k+1)} = \frac{1}{n}\sum_{j=1}^{n} w^{(k)},$$

$$\mu_1^{(k+1)} = \frac{\sum_{j=1}^{n} w^{(k)}\varsigma_j}{\sum_{j=1}^{n} w^{(k)}}, \quad \mu_2^{(k+1)} = \frac{\sum_{j=1}^{n}(1-w^{(k)})\varsigma_j}{\sum_{j=1}^{n}(1-w^{(k)})},$$

$$\sigma_1^{(k+1)2} = \frac{\sum_{j=1}^{n} w^{(k)}(\varsigma-\mu_1^{(k+1)})^2}{\sum_{j=1}^{n} w^{(k)}}, \quad \text{and} \quad \sigma_2^{(k+1)2} = \frac{\sum_{j=1}^{n}(1-w^{(k)})(\varsigma_j-\mu_2^{(k+1)})^2}{\sum_{j=1}^{n}(1-w^{(k)})}.$$

The initial values of $\boldsymbol{\theta} = (p^{(0)}, \mu_1^{(0)}, \mu_2^{(0)}, \sigma_1^{2\,(0)}, \sigma_2^{2\,(0)})$ are obtained from the empirical distribution.

Each update to the parameters resulting from an E step followed by M step is guaranteed to increase the log likelihood function. In practice, the algorithm is deemed to have converged when the change in the log likelihood function, or alternatively in the parameters, falls below some threshold.

2.4.2 Mixture of White Noises

In the previous subsection, we have discussed more general case of mixture of Gaussian noises. As mentioned earlier, Gaussian noise have some well defined properties and they are easy to deal with. But it's not uncommon to assume time series as a linear combination of white noises. White noises are the random variables from some uniform distribution defined in a particular interval. Teicher (1963) showed that univariate normal mixtures are identifiable, while in general mixture of uniform distributions are not. Identifiability is a necessary condition for the possibility to estimate the parameters of a mixture model consistently. It makes sure that no two essentially different mixture parameter vectors parameterize the same distribution. According to Casella and Berger (2002), "A parameter θ for a family of distributions $\{f(x|\theta) : \theta \in \Theta\}$ is *identifiable* if distinct values of θ correspond to distinct probability density or mass functions. That is, if $\theta \neq \theta^{'}$, then $f(x|\theta)$ is not the same function of x as $f(x|\theta^{'})$."

Maximum likelihood method of parameter estimation for exponential family distribution give robust estimates. EM algorithm can also be thought as an adjusted maximum likelihood method. So the parameter estimates of mixture of Gaussian are robust and identifiable. Maximum likelihood estimates for a wide class of location-scale mixtures are not robust (Hennig 2004). So the parameter estimation based on EM algorithm may not be the appropriate choice when we deal with mixture of white noises.

Parameter estimation of mixture of uniform distributions using the method of moments and method of maximum likelihood is discussed in Craigmile and Titterington (1997) [5]. In this subsection, we briefly discuss the alternative way of estimating the parameters for mixture of white noises. Since our main interest lies in the Gaussian noise and parameter estimation through EM algorithm, we just outline some of the parameter estimates based on method of moments for mixture of uniform distribution without giving detailed explanation.

So the apparent simplicity of uniform mixtures conceals a hidden danger of non-identifiability.

For example, let us assume a two component mixture of uniform distribution:

$$f(\varsigma|p,\theta) = pU(\varsigma;0,\theta) + (1-p)U(\varsigma;\theta,1),$$

where $U(\varsigma;a,b)$ denotes the uniform density on the interval $[a,b)$ and $0 \leq p \leq 1$, and $0 \leq \theta \leq 1$.

Let us take $p = \theta$. Then,

$$f(\varsigma|p,p) = U(0,1).$$

So, for any values of p, we get the same distribution function. The problem of non-identifiability arises and the estimate is not consistent. But if p or θ is known, the mixture is identifiable even if the true values of p and θ are equal. So care should be taken when dealing with mixture of uniform distribution and we should avoid the condition of non-identifiability.

For convenience, let us assume that we have the mixture of two white noises from each of the interval, so $\zeta_i \overset{iid}{\sim} WN(0, \sigma_i^2); i = 1, 2$. Notice that σ_i^2 is the variance of the uniform distribution in the ith interval. Since, uniform distribution is a location-scale family, we can consider two disjoint intervals for uniform mixture and can write the density function as:

$$f(\varsigma | p, \theta) = pU(\varsigma; 0, \theta) + (1 - p)U(\varsigma; \theta, 1), \tag{17}$$

As discussed earlier, this is non-identifiable when $p = \theta$, so we take the cases when $p \neq q$. Equation (17) can be written as

$$f(\varsigma | p, \theta) = \frac{p}{\theta} I(0 \leq \varsigma < \theta) + \frac{(1 - p)}{(1 - \theta)} I(\theta \leq \varsigma < 1),$$

where I be the indicator function. Since the kth raw moment of uniform distribution $U(\varsigma; a, b)$ is

$$\frac{1}{k+1} \frac{b^{k+1} - a^{k+1}}{b - 1},$$

we can write the kth raw moment of the mixture density as:

$$\begin{aligned} m_k = E(\zeta^k) &= \frac{p\theta^k}{k+1} + \frac{(1-p)(1-\theta^{k+1})}{(1-\theta)(k+1)} \\ &= \frac{1 - \theta^{k+1} - p(1 - \theta^k)}{(1-\theta)(k+1)} . \end{aligned} \tag{18}$$

Also, kth sample moment can be represented as:

$$M_k = \frac{1}{n} \sum_{j=1}^{n} x_j^k. \tag{19}$$

By equating kth raw moment with kth sample moment, we estimate the parameters of mixture distribution. We choose the cases when $p \neq \theta$ to avoid non-identifiability. There are three cases of parameter estimation:

(i) θ known, p unknown
(ii) p known, θ unknown and
(iii) both p and θ are unknown

But in our situation, we don't have known p, so second case is irrelevant to our discussion. Here we'll discuss case (i) and case (iii) briefly.

Case I: θ known, p unknown

When θ is known, p is estimated by equating kth raw moment with kth sample moment as

$$M_k = m_k,$$

$$M_k = \frac{p\theta^k}{k+1} + \frac{(1-p)(1-\theta^{k+1})}{(1-\theta)(k+1)}, \qquad \text{[From Eq. (18)]}$$

$$\Longrightarrow \quad \tilde{p} = \frac{-(1-\theta)(k+1)}{(1-\theta^k)} M_k + \frac{(1-\theta^{k+1})}{(1-\theta^k)},$$

where M_k is the kth sample moment defined in Eq. (19). For simplicity we can write

$$\tilde{p} = c_k M_k + d_k.$$

The order k is determined based on the optimal variance of $\tilde{p}$.

$$\begin{aligned} var(\tilde{p}) &= c_k^2 var\left(\frac{1}{n}\sum_{j=1}^{n} x_j^k\right) \\ &= \frac{c_k^2}{n^2}\left[E\left(\left\{\sum_{i=1}^{n} x_i^k\right\}^2\right) - E\left(\sum_{i=1}^{n} x_i^k\right)^2\right] \\ &= \frac{c_k^2(m_{2k} - m_k^2)}{n}. \end{aligned}$$

We choose k and estimate of p in such a way that variance of $\tilde{p}$ will be minimum. Gupta and Miyawaki (1978) suggested $k = 1$ for estimation of p.

Case III: Both p and θ unknown

Gupta and Miyawaki (1978) has suggested using first and second order moments to estimate the parameters of θ and p. Here, we will derive the expression based on first and second order moments. As suggested by method of moments,

$$\begin{aligned} M_1 &= m_1 \\ M_1 &= \frac{p\theta}{2} + \frac{(1-p)(1-\theta^2)}{2(1-\theta)} \\ 2M_1 &= 1 + \theta - p. \end{aligned} \tag{20}$$

Again,

$$\begin{aligned} M_2 &= m_2 \\ M_2 &= \frac{p\theta^2}{3} + \frac{(1-p)(1-\theta^3)}{3(1-\theta)} \\ 3M_2 &= 1 + \theta - p + \theta^2 - p\theta. \end{aligned} \tag{21}$$

From Eqs. (20) and (21), we have:

$$\tilde{\theta} = \frac{3M_2 - 2M_1}{2M_1 - 1},$$

and

$$\tilde{\mathrm{p}} = 1 - \frac{4M_1^2 - 3M_2}{2M_1 - 1}.$$

2.4.3 Caveat: Mixture of White Noises

The extension of more than 2 component mixture of uniform distribution and their parameter estimation is discussed briefly by Craigmile and Titterington (1997) [5] giving an example for mixture of 3 component uniform distributions. Even for three component mixture distribution, there are several cases of non-identifiability and if we have higher component mixture distributions we will encounter multiple cases of non-identifiability. So the parameter estimation is restricted by several conditions. It is not possible to track all the restricted conditions so higher component mixture of uniform distribution is not suggested. The parameter estimation could be very inconsistent and in many cases not possible. Also, one should be very careful when assuming the mixture of uniform distribution, since mixture is defined as the combination of non-overlapping uniform distribution. In the cases with large number of breakpoints, the number intervals $m + 1$ may not be equal to the number of clusters for the mixture of uniform distributions. If this situation arises, the method of forecasting based on $m + 1$ mixtures that we have proposed will not be appropriate and some other methods with reduced dimension should be considered. But this is not the case for mixture of Gaussian noises. So, we consider the case of mixture of white noises as just an alternative approach and preference is given to the framework based on mixture of Gaussian noises.

2.5 Confidence Interval Estimation and Large Sample Properties

Time series parameters are estimated using innovations algorithm together with the maximum likelihood method. We can use the asymptotic distribution of ARMA parameters $(\hat{\boldsymbol{\phi}}, \hat{\boldsymbol{\theta}})$ to derive approximate large sample confidence regions for the true coefficient vectors $(\boldsymbol{\phi}, \boldsymbol{\theta})$.

Let Y_t be the stationary and invertible time series process. An ARMA model can be written as:

$$Y_t - \sum_{i=1}^{p} \phi_i Y_{t-i} = \zeta_t - \sum_{j=1}^{q} \theta_j \zeta_{t-j}; \qquad \zeta \overset{iid}{\sim} N(0, \sigma^2).$$

This is equivalent to

$$\prod_{i=1}^{p}(1-A_i)\,Y_t=\prod_{j=1}^{q}\left(1-M_j\right)\zeta_t, \tag{22}$$

$$\Longrightarrow \qquad \zeta_t=\prod_{i=1}^{p}(1-A_i)\prod_{j=1}^{q}\left(1-M_j\right)^{-1}Y_t\,.$$

For example, if we have $ARMA(2,2)$ model

$$Y_t-\phi_1 Y_{t-1}-\phi_2 Y_{t-2}=\zeta_t-\theta_1\zeta_{t-1}-\theta_1\zeta_{t-1},$$

Then we can easily derive

$$\phi_1=A_1+A_2 \qquad\qquad \phi_2=-A_1A_2$$
$$\theta_1=M_1+M_2 \qquad\qquad \theta_2=-M_1M_2.$$

ARMA model are the superposition of both AR and MA models, so we can write AR and MA components of Eq. (22) in terms of past errors as:

$$\begin{aligned} u_{t,i}&=-\frac{\partial\zeta_t}{\partial A_i}=(1-A_iB)^{-1}\zeta_{t-1},\\ v_{t,i}&=-\frac{\partial\zeta_t}{\partial M_i}=-(1-M_jB)^{-1}\zeta_{t-1}. \end{aligned} \tag{23}$$

For the mixed $ARMA$ models, the information matrix can be written as:

$$I(\boldsymbol{\phi},\boldsymbol{\theta})=$$

$$\frac{n}{\sigma^2}\left[\begin{array}{cccc|cccc} \gamma_{uu}(0) & \gamma_{uu}(1) & \dots & \gamma_{uu}(p-1) & \gamma_{uv}(0) & \gamma_{uv}(-1) & \dots & \gamma_{uv}(1-q)\\ \gamma_{uu}(1) & \gamma_{uu}(0) & \dots & \gamma_{uu}(p-2) & \gamma_{uv}(1) & \gamma_{uv}(0) & \dots & \gamma_{uv}(2-q)\\ \vdots & & & \vdots & & \vdots & & \vdots\\ \gamma_{uu}(p-1) & \gamma_{uu}(p-2) & \dots & \gamma_{uu}(0) & \gamma_{uv}(p-1) & \gamma_{uv}(p-2) & \dots & \gamma_{uv}(p-q)\\ \hline \gamma_{uv}(0) & \gamma_{uv}(-1) & \dots & \gamma_{uv}(1-q) & \gamma_{uu}(0) & \gamma_{uu}(1) & \dots & \gamma_{uu}(p-1)\\ \gamma_{uv}(1) & \gamma_{uv}(0) & \dots & \gamma_{uv}(2-q) & \gamma_{uu}(1) & \gamma_{uu}(0) & \dots & \gamma_{uu}(p-2)\\ \vdots & & & \vdots & & \vdots & & \vdots\\ \gamma_{uu}(1-q) & \gamma_{uu}(2-q) & \dots & \gamma_{uu}(p-q) & \gamma_{uv}(q-1) & \gamma_{uv}(q-2) & \dots & \gamma_{uv}(0) \end{array}\right] \tag{24}$$

where $\gamma_{uu}(h)=E(u_t,u_{t+h})$,$\gamma_{uv}(h)=E(u_t,v_{t+h})$ and so on. The components u and v are related to autoregressive and moving average components of Eq. (23). Using Eqs. (23) and (24) for the large sample, we get the information matrix in terms of A_i and M_j as

$I(\boldsymbol{\phi}, \boldsymbol{\theta}) =$

$$n \left[\begin{array}{cccc|ccc} (1-A_1^2)^{-1} & (1-A_1A_2)^{-1} & \ldots & (1-A_1A_p)^{-1} & -(1-A_1M_1)^{-1} & \ldots & -(1-A_1M_q)^{-1} \\ \vdots & \vdots & & \vdots & \vdots & & \vdots \\ (1-A_1A_p)^{-1} & (1-A_2A_p)^{-1} & \ldots & (1-A_p^2)^{-1} & -(1-A_pM_1)^{-1} & \ldots & -(1-A_pM_q)^{-1} \\ \hline -(1-A_1M_1)^{-1} & -(1-A_2M_1)^{-1} & \ldots & -(1-A_pM_1)^{-1} & (1-M_1^2) & \ldots & -(1-M_1M_q)^{-1} \\ \vdots & \vdots & & \vdots & \vdots & & \vdots \\ -(1-A_1M_q)^{-1} & -(1-A_2M_q)^{-1} & \ldots & -(1-A_pM_q)^{-1} & (1-M_1M_q)^{-1} & \ldots & (1-M_q^2)^{-1} \end{array}\right].$$

This matrix can be partitioned after pth row and qth column. So,

$$I(\boldsymbol{\phi}, \boldsymbol{\theta}) = n \left[\begin{array}{c|c} I_{AA} & I_{AM} \\ \hline I'_{AM} & I_{MM} \end{array}\right].$$

Notice that matrix $I^{-1}(\boldsymbol{\phi}, \boldsymbol{\theta})$ is non-singular, so the covariance matrix of estimators of $\boldsymbol{\phi}$ and $\boldsymbol{\theta}$ for ARMA model is

$$\boldsymbol{\Sigma}(\hat{\boldsymbol{\phi}}, \hat{\boldsymbol{\theta}}) = I^{-1}(\boldsymbol{\phi}, \boldsymbol{\theta}).$$

In case of pure AR and MA models, this covariance matrix can further split by removing cross covariance matrices of AR and MA components (i.e. $I_{AM} = \mathbf{0}$). This gives

$$\boldsymbol{\Sigma}(\hat{\boldsymbol{\phi}}) = I^{-1}(\boldsymbol{\phi}) = \frac{1}{n} \quad I_{AA}^{-1},$$

$$\boldsymbol{\Sigma}(\hat{\boldsymbol{\theta}}) = I^{-1}(\boldsymbol{\theta}) = \frac{1}{n} \quad I_{MM}^{-1}.$$

Let $\boldsymbol{\beta} = (\boldsymbol{\phi}, \boldsymbol{\theta})$ be the vector of $ARMA$ model parameters.

The large sample distribution of maximum likelihood estimators of $ARMA(p, q)$ can be written as

$$\hat{\boldsymbol{\beta}} \sim N_{p+q}(\boldsymbol{\beta}, n^{-1}\Sigma(\hat{\boldsymbol{\phi}}, \hat{\boldsymbol{\theta}}),$$

and for pure autoregressive (AR) and moving average (MA) models,

$$\hat{\boldsymbol{\beta}}_p \sim N_p(\boldsymbol{\beta}_p, n^{-1}\Sigma(\hat{\boldsymbol{\phi}}))$$

and

$$\hat{\boldsymbol{\beta}}_q \sim N_q(\boldsymbol{\beta}_q, n^{-1}\Sigma(\hat{\boldsymbol{\theta}})).$$

As we have seen that any ARMA(p, q) model can be obtained using the linear filter ψ from white noise or Gaussian noise, $\zeta \overset{iid}{\sim} N(0, \sigma^2)$, let's write:

$$Y_t = \psi(B)\zeta_t = \sum_{j=0}^{\infty} \psi_j \zeta_{t-j}.$$

Let $\hat{Y}_{t+h}$ be the h step predictor, then

$$\hat{Y}_{t+h} = \sum_{j=0}^{h-1} \hat{\psi}_j \zeta_{t+h-j}.$$

Hence, the mean squared error is:

$$S^2 = E(Y_t - \hat{Y}_{t+h})^2 = \sum_{j=0}^{h-1} \hat{\psi}_j^2 \operatorname{Var}(\zeta_{t+h-j}).$$

$$= \hat{\sigma}^2 \sum_{j=0}^{h-1} \hat{\psi}_j^2 . \tag{25}$$

Here, $\hat{\psi}_j^2$ is the function of estimators $\boldsymbol{\phi}$ and $\boldsymbol{\theta}$ such that:

$$\hat{\psi}_j = \sum_{k=1}^{p} \hat{\phi}_k \hat{\psi}_{j-i} - \hat{\theta}_j, \quad j = 0, 1, 2, \ldots,$$

where $\hat{\theta}_0 = 1$, $\hat{\theta}_j = 0$ if $j > q$, $\hat{\psi}_0 = 1$, and $\hat{\psi}_j = 0$ if $j < 0$.

Other estimators $\hat{\boldsymbol{\phi}}_k, k = 1, 2, \ldots, p$ and $\hat{\boldsymbol{\theta}}_j, j = 1, 2, \ldots, q$, are estimated using innovations algorithm together with maximum likelihood method. And the estimator $\hat{\sigma}^2$ is obtained from the empirical data.

If the ARMA(p, q) process $\{Y_t\}$ (for each interval separated by breakpoints) is driven by Gaussian white noise, then the prediction error $Y_{t+k} - \hat{Y}_{t+k}$ is normally distributed with mean 0 and variance S^2 given by the Eq. (25). In our case we are using one-step prediction so $k = 1$. Hence, the prediction interval of Y_{t+k} is

$$Y_{t+k} = \hat{Y}_{t+k} \pm \Phi_{1-\alpha/2} S$$

Let us assume that there are $m + 1$ mixture components, then for the forecasting based on mixture distribution we can rewrite the h step prediction as:

$$\hat{Y}_{n+h} = \sum_{i=1}^{m+1} p_i \sum_{j=0}^{h-1} \hat{\boldsymbol{\psi}}_{ij} \zeta_{n+h-j}.$$

where $\sum_{i=1}^{m+1} p_i = 1$ and $0 \leq p_i \leq 1$, $n = n_1 + n_2 + \ldots + n_{m+1}$, n_i be the number of observations in each component of mixtures and $\hat{\boldsymbol{\psi}}_i$ is composed of ARMA parameters $(\hat{\boldsymbol{\phi}}_i, \hat{\boldsymbol{\theta}}_i)$ from each breakpoint groups. Hence, the mean squared error is

$$S_m^2 = \hat{\sigma}_{m+1}^2 \sum_{i=1}^{m+1} \sum_{j=1}^{h-1} p_i^2 \hat{\boldsymbol{\psi}}_{ij}^2 ,$$

where $\hat{\sigma}_{m+1}^2$ be the white noise variance estimator of $(m+1)$th component.

Assuming that the ARMA process $\{Y_n\}$ is driven by Gaussian white noise so if $\zeta_t \overset{iid}{\sim} N(0, \sigma^2)$, then for each $h \geq 1$ the prediction error is normally distributed with mean 0 and variance S_m^2. It follows that Y_{n+h} lies between the bounds

$$\hat{Y}_{n+h} \pm \Phi_{1-\alpha/2} S_m. \tag{26}$$

with probability $(1 - \alpha)$. In the above equation, $\Phi_{1-\alpha/2}$ is the $(1 - \alpha/2)$ quintile of standard normal distribution. We can call this bound as prediction bound for Y_{n+h}.

2.6 Block Bootstrap

We use block bootstrap to generate bootstrap replicates of a statistic applied to time series. By dividing the data into several blocks, The original time series structure as well as the properties of original data generating process are preserved within a block.

Let $\{Y_t : t = 1, \ldots, n\}$ be time series data then we construct bootstrap sample in the following steps:

1. Pick the optimal block size, l. The block size is chosen according to Patton et al. [26].
2. Consider the overlapping blocks with varying block lengths. The optimal block size l is the mean of geometric distribution used to generate the block length. This avoids the problem of non-stationarity by construction [27]. For the overlapping method, we divide the data into $n-l+1$ blocks, which block 1 being $\{Y_1, \cdots, Y_l\}$, block 2 being $\{Y_2, \cdots, Y_{l+1}\}, \cdots, etc.$
3. Resample the blocks randomly with replacement and generate bootstrap sample $\{Y_t^* : t = 1, \cdots, n\}$ by gluing blocks together in the order that they were sampled.
4. Calculate the estimator.

For simplicity, this combination of identification of breakpoints together with bootstrap is named as Breakpoints Bootstrap Filtering (BPBF) method.

3 Application

In this section, we test our proposed methodology on a simulation data. In order to justify the methodology we simulate an ARMA model with mixture of Gaussian noise. We implement the proposed method for forecasting and prediction and compare the result with the classical time series approach.

3.1 ARMA Model with Mixture of Gaussian Noise

We simulate an ARMA model with mixture of Gaussian noise. An AR model with zero mean, AR component $(\phi) = 0.4$ with mixture of two component Gaussian is simulated. The simulated mixture of Gaussian noise has the following parameters

$$p = 0.3, \mu_1 = 1, \sigma_1 = 0.6, \mu_2 = 3, \sigma_2 = 2.$$

First, we fit the model based on classical approach, which fits an ARMA model for the entire data. Looking at Fig. 1, we can see that the data structure does not look

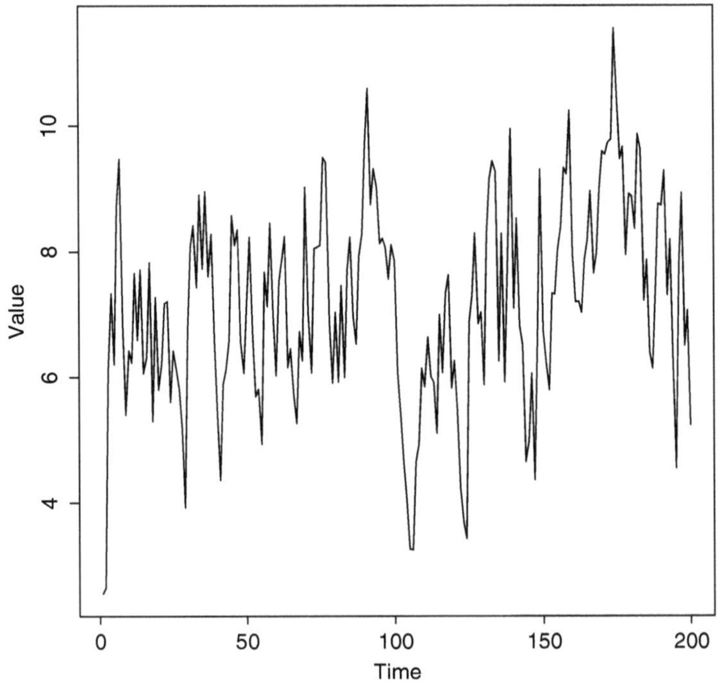

Fig. 1 AR(0.4) model with noises from mixture of two Gaussians

Table 1 Parameter estimates and se() of MA(4) model for simulated data

$\hat{\mu}$	$\hat{\theta}_1$	$\hat{\theta}_2$	$\hat{\theta}_3$	$\hat{\theta}_4$	$\hat{\sigma}^2$	AIC	BIC	Log-lik
3.97	0.28	0.13	0.14	−0.12	3.05	802.90	822.69	−395.45
(0.18)	(0.07)	(0.08)	(0.08)	(0.07)				

Table 2 Summary of AICs using different combinations of p and q for best model selection of simulated data

$q \rightarrow$ $p \downarrow$	0	1	2	3	4	5
0	0.0	806.8	807.6	804.0	**802.9**	803.9
1	804.9	806.6	808.5	804.1	803.5	805.7
2	806.6	808.6	808.6	805.5	805.5	805.6
3	808.6	807.4	806.1	805.1	806.8	801.5
4	803.7	803.9	805.8	806.6	807.9	806.4
5	803.7	805.4	806.4	802.5	802.5	804.5

same and it changes over two different time intervals. Also, there is no seasonal component associated with these data. So, we don't need to worry about fitting an ARIMA model. Also, both augmented Dickey-Fuller test and Philips-Perron unit root tests suggest that data is stationary (p-value=0.01). Once stationarity is established, now we want to see which model best fit the data. Based on maximum likelihood method and minimum Akaike Information Criterion (AICc), we choose MA(4), model. The estimated parameters for this model are given in the Table 1. The values in the parenthesis are the standard error estimates.

Also, Table 2 shows the AICs using maximum likelihood method for different combinations or AR(p) and MA(q) components. Notice that for MA(4) we achieve minimum AIC.

Also, predicted values and twenty future forecast values using MA(4) model together with the original data are plotted in Fig. 2. In the figure, the yellow band after time 200 represents the prediction bound for forecasting.We can clearly see that model fitting is not very good and many cyclic variations are not captured. Forecasting is even worse, it has large prediction bounds and forecast looks constant. Here, the model based on classical approach fails to incorporate the cyclic variation in the forecast and it's not capturing the change of data structure over different intervals of time. We overcome these problems by using breakpoints and mixture distribution based forecasting discussed in previous sections.

In the next step, we use our proposed method to the data. First, we identify the breakpoints in the data set and divide it into different intervals. Then, we fit separate models for the data in each interval. Breakpoints are identified according to the method discussed in the Sect. 2. Using R package *strucchange* it is reasonable to use one breakpoint in the data set. Figure 3 shows different values of Bayesian Information Criterion (BIC) and Residual Sums of Squares (RSS) for different breakpoints. Our goal is to take the optimal solution and it is reasonable to consider one breakpoint. Also, if we choose more than one breakpoint, we may encounter the problem of overfitting.

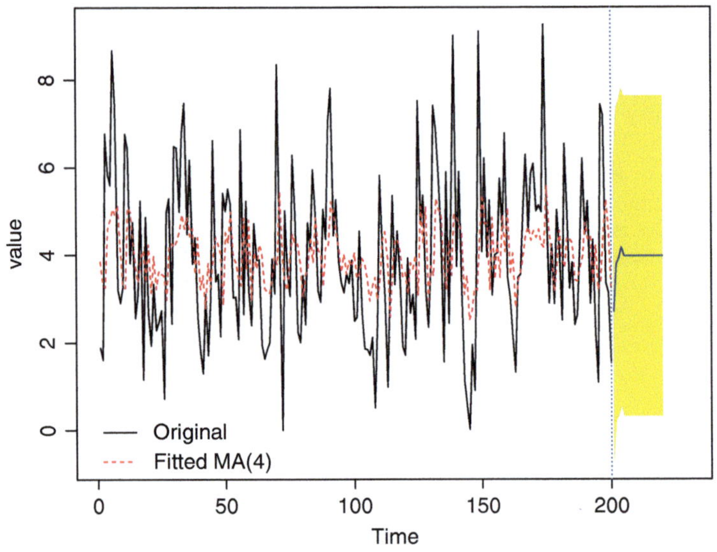

Fig. 2 Forecasting and model fitting of simulated data

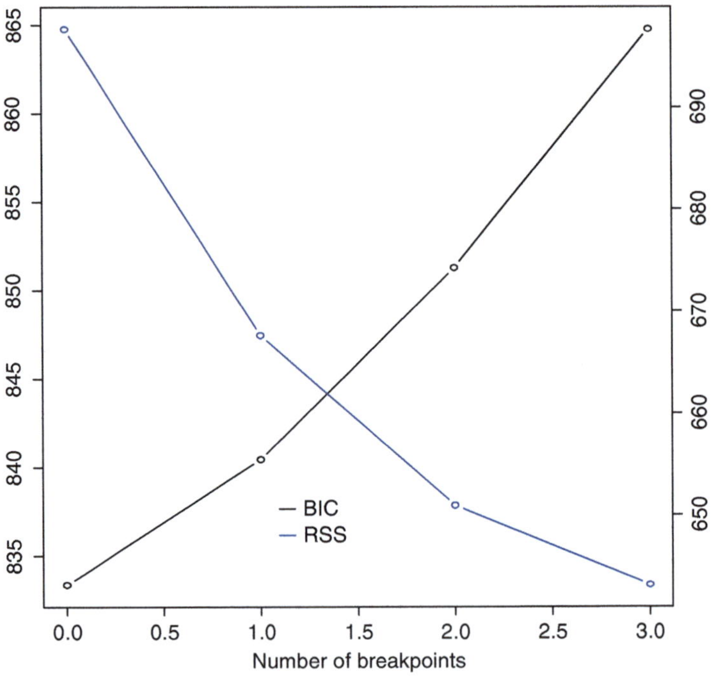

Fig. 3 Breakpoint identification of simulated data

Table 3 Summary of AICs using different combinations of p and q for first part (1–124 observations) of simulated data

q→ p↓	0	1	2	3	4	5
0	0.0	487.8	489.6	**486.1**	486.7	488.5
1	486.8	488.3	490.2	487.0	488.6	490.5
2	488.5	490.3	488.9	487.7	489.1	489.8
3	489.2	488.0	490.0	488.1	489.6	491.2
4	488.3	489.8	490.3	489.4	490.9	492.9
5	489.3	491.0	486.5	487.0	492.9	494.9

Table 4 Summary of AICs using different combinations of p and q for second part (125–200 observations) of simulated data

q→ p↓	0	1	2	3	4	5
0	0.0	319.5	320.7	320.9	319.4	320.1
1	318.9	320.9	322.5	321.6	316.3	318.3
2	320.9	322.9	**315.2**	321.3	318.3	320.3
3	321.9	318.3	317.2	317.9	320.2	322.3
4	316.2	317.9	319.7	319.5	315.8	317.5
5	318.0	319.8	321.7	320.9	317.4	316.7

Table 5 Parameter estimates and se() of MA(3) model for part 1 of simulated data

$\hat{\mu}$	$\hat{\theta}_1$	$\hat{\theta}_2$	$\hat{\theta}_3$	$\hat{\sigma}^2$	AIC	BIC	Log-lik
3.76	0.35	0.14	0.24	2.72	486.06	500.16	−238.03
(0.25)	(0.09)	(0.11)	(0.10)				

Table 6 Parameter estimates and se() of ARMA(2,2) model for part 2 of simulated data

$\hat{\mu}$	$\hat{\phi}_1$	$\hat{\phi}_2$	$\hat{\theta}_1$	$\hat{\theta}_2$	$\hat{\sigma}^2$	AIC	BIC	Log-lik
4.30	1.34	−0.94	−1.13	0.87	3.08	315.23	329.21	−151.61
(0.25)	(0.07)	(0.06)	(0.12)	(0.15)				

From Fig. 3 it is clear that at one breakpoint we get BIC$=$ 840.4 and RSS$=$ 667.6. These values are close to the possible minimum values of BIC ($=$ 833.4) and RSS ($=$ 697.7). In the data set, this breakpoint lies in the 124th observation, so we divide the data into two parts, 1–124 and 125–200. Now, we fit different ARMA models to these two parts and combine the error distributions.

In the data of both parts no non-stationarity is evident. Phillips-Perron Unit root tests suggest the stationarity of the data in both intervals. Also, there are no seasonal or periodic components in the data set, so we use ARMA based models on both parts. Based on the method of maximum likelihood and minimum AIC, We choose MA(3) and ARMA$(2,2)$ models for first and second parts respectively. Tables 3 and 4 show the AIC values for different combinations of AR (p) and MA(q) components.

Parameter estimates of best models are given in Tables 5 and 6.

Notice that all comparative measures such as AIC, BIC and Log-likelihood of the models obtained by using breakpoints (Tables 5 and 6) are significantly improved compared to the model obtained by using classical approach (Table 1).

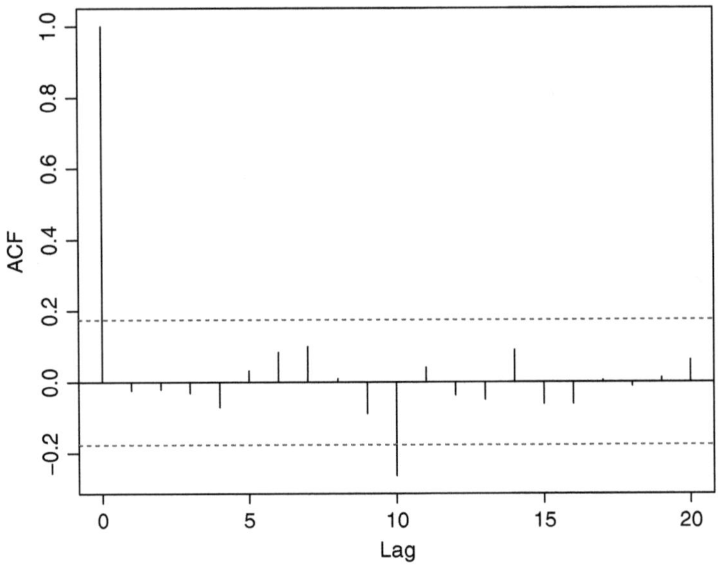

Fig. 4 Autocorrelation function of residuals from part 1 (1–124 observations)

We also check the residuals from each of these fits to see whether or not they meet the autocorrelation test and normality test with homoscedastic variance. Both Box-Pierce and Ljung-Box portmanteau tests suggest independence and white noise property of the data. P-values for Box-Pierce and Ljung-Box tests are 0.79 and 0.78, so we fail to reject the null hypothesis that "data are independently distributed". Figures 4 and 5 show there is no serious autocorrelation between the residuals. Also, residuals from both intervals meet the criterion of normality separately. Also, for the model based on classical time series approach, the residuals are not normal. Several model selection methods based on AIC, BIC and minimum variance were tried and in all cases residuals were not normally distributed. This is reasonable, because we intentionally simulated the model with mixture Gaussian noise and classical time series approach fails to handle this situation.

The process is invertible, so we can get the actual data from the errors. So, for forecasting we combine the errors from both parts and estimate the mixture parameters of mixture of two component Gaussian distributions using the EM algorithm. Figure 6 shows the histogram of combined noise which seems right skewed from normal distribution, infact it is the mixture of normal distribution. Also, for combined noise we don't see significant autocorrelation (Fig. 7), we fail to reject both Box-Pierce (p-value$=0.74$) and Box-Ljung (p-value$=0.74$).

It's a reasonable assumption to consider that the joint distribution is the mixture of Gaussian distributions since the source of residuals are different. We use EM algorithm to estimate the model parameters of this mixture distribution. Parameter

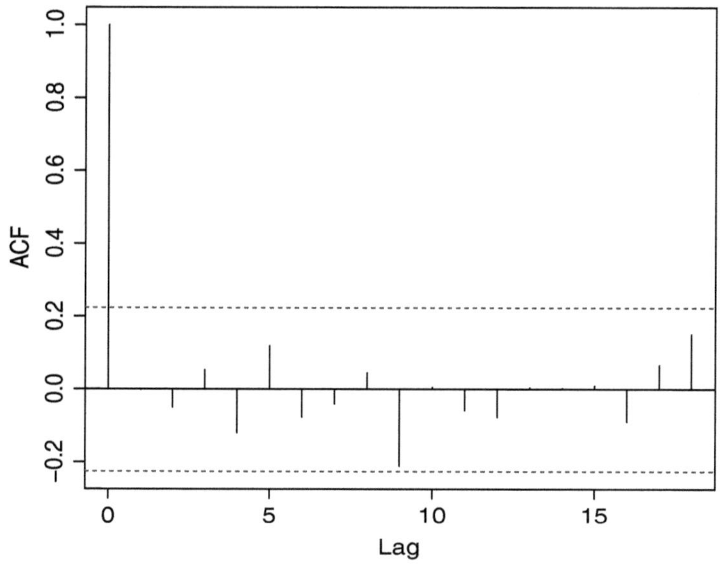

Fig. 5 Autocorrelation function of residuals from part 2 (125–200 observations)

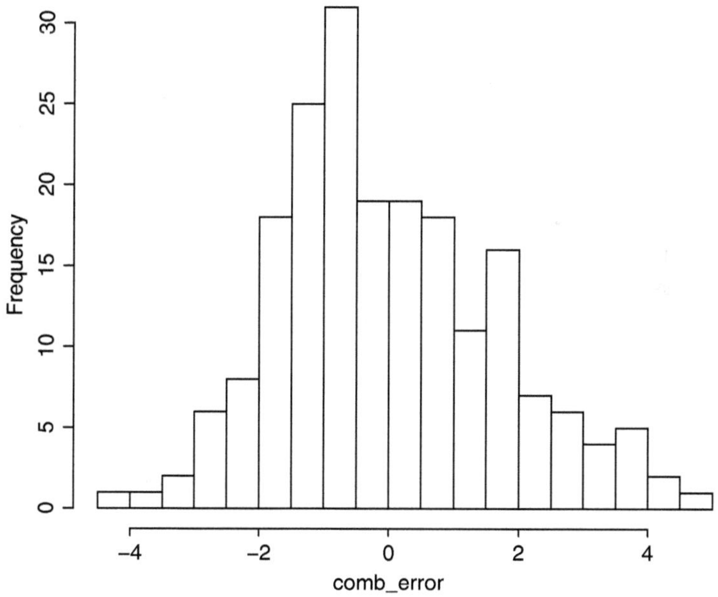

Fig. 6 Histogram of combined residuals of simulated data

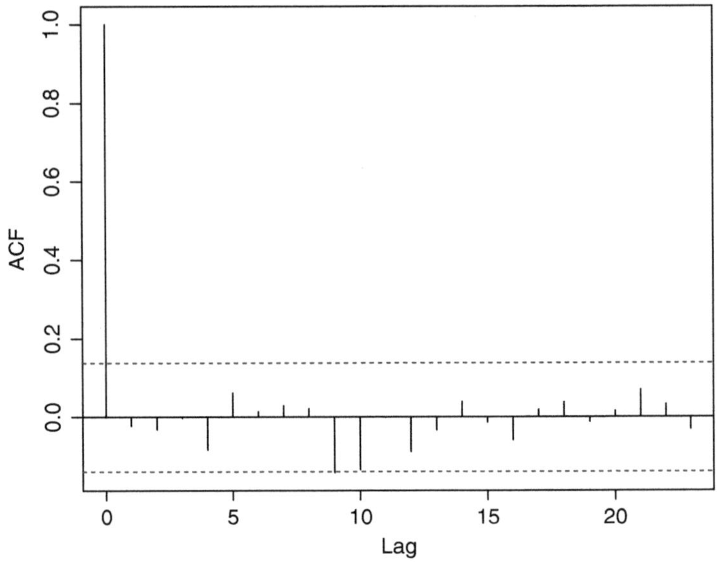

Fig. 7 Autocorrelation of combined residuals of simulated data

Table 7 Parameter estimates of combined residuals using EM algorithm

p	$\hat{\mu}_1$	$\hat{\mu}_2$	$\hat{\sigma}_1$	$\hat{\sigma}_1$
0.32	−1.01	0.48	0.76	1.80

estimation of mixture of two component Gaussian using EM algorithm is presented in Table 7. Also, Fig. 8 shows the estimated mixture density together with individual Gaussian component density for combined data.

Now we forecast the next 20 future values using the theory already discussed in Sect. 2.4.1.

In Fig. 9, we can see that the model fitting has improved significantly by using the mixture model. Most importantly, forecasting of the data has significantly improved. The mixture model forecasting also incorporates the cyclic factor of the data and the prediction intervals are narrower than those of classical approach.

4 Conclusion

We have introduced a non-linear dynamical probability time series model which exploits the idea of breakpoints together with bootstrapping and mixture distribution. Breakpoints partition the time course into consecutive non-overlapping intervals where the coefficients shift from one stable regression relationship to a different one. Also, because there are limited observations in some intervals, we

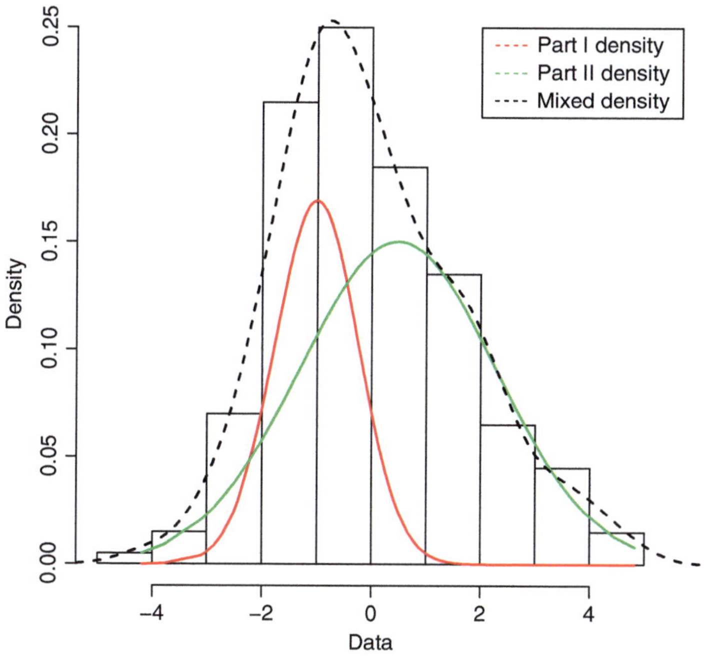

Fig. 8 Density of the mixture distribution together with individual component distribution for combined residuals

use block bootstrapping to improve the parameter estimates. The optimal size of the blocks needed is chosen such that the RSS will be minimum. Once we fit the model for different intervals, such information is combined and used in forecasting.

Forecasting partitioned data which has different model structures at different partitions is a challenging task. Over the last decade, there has been much interest in developing breakpoints to time series data in a small sample scale. To our knowledge, there are no existing methods that discuss the prediction of this type of data. We have shown numerically that the model accommodates data with different variance structures with the introduction of the breakpoints. The regression and the dependency of the parameters in the model have been included in a consistent and efficient manner. Regression models with unequal mixed sample frequencies and their advantages is still relatively the unexplored area (Andreou et al. 2002). Consistency is guaranteed since we are using the maximum likelihood method, and efficiency is guaranteed since the bootstrap method is used to resample the data once the blocks have been identified and the predictions lie within the smaller range intervals than the classical time series modelling.

The method discussed in this work is different from other existing methods that are based on time series data in which different covariates have different covariance structures. Typically, the models that are built with the predictors without the breakpoint inclusion do not provide substantial forecasting (Stock 2008). We have

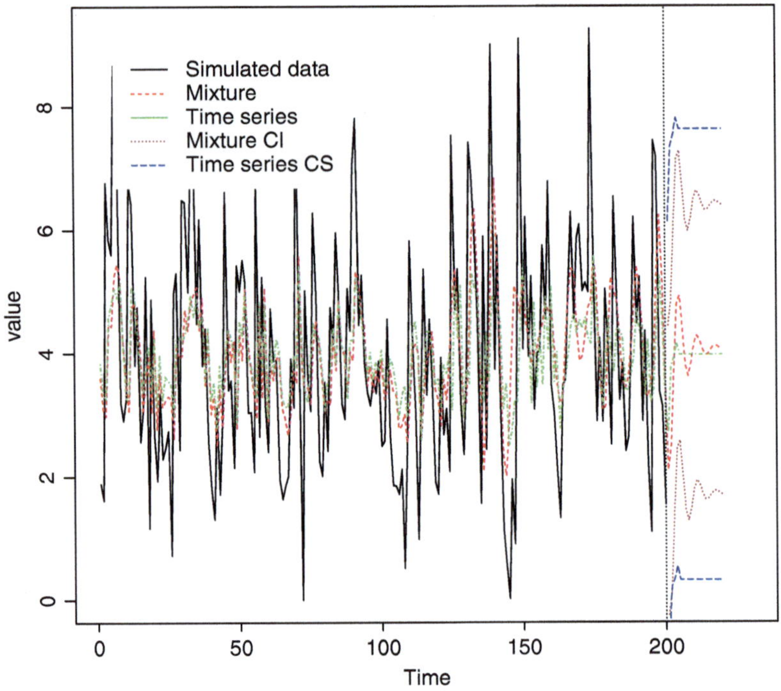

Fig. 9 Model fitting and forecasting of simulated data by using classical time series and mixture model approaches

developed a new approach which advances previous concepts with new ideas for forecasting time series data that are subject to the structural breaks and non-equidistant time. Our approach is based on the mixture distribution where the parameters are estimated by using EM algorithm combined with bootstrapping. Our approach together with block bootstrapping performs very well when faced with small and sparse data sets. Our approach is quite general and can be implemented in different ways other than those documented. Pesaran et al. (2006) discussed similar type of data by using Bayesian approach and by allowing the possibility of new breaks occurring over the forecast horizon. We assume that the existence of breakpoints in the forecast horizon is somehow unrealistic. Our approach is based on past data within the intervals and we do not use the information of systematic breakpoints in the forecast horizon.

Further questions are being explored. One of the questions is related to the identification of optimal block size for block bootstrapping as discussed in Patton et al. [26]. Another concern is related to finding a procedure of choosing initial value in EM algorithm for faster convergence.

Acknowledgements The authors would like to express sincere thanks to the late Professor Dayanand Naik for his insightful comments and suggestions during the research which resulted in an improvement in the presentation of this work. We dedicate this work to him.

References

1. Andreou, E., & Ghysels, E. (2002). Detecting multiple breaks in financial market volatility dynamics. Journal of Applied Econometric, 17, 579-600.
2. Brockwell, P.J., & Davis, R.A. (1991). Time Series: Theory and Methods, 2nd Ed., New York: Springer.
3. Brockwell, P.J., & Davis, R.A. (2002). Introduction to Time Series and Forecasting, 2nd Ed., New York: Springer-Verlag.
4. Casella, G., & Berger, R.L. (2002). Statistical Inference, 2nd Ed., Pacific Grove: Duxbury.
5. Craigmile, P.F., & Titterington, D.M. (1997) Parameter estimation for finite mixtures of uniform distributions. *Communications in Statistics: Theory and Methods, 26*(8), pp. 1981–1995.
6. Gupta, A.K., & Miyawaki, T. (1978). On a uniform mixture model. Biometrical Journal, 20, 631–637.
7. Henning, E. (2004). Finding Your Way in Qualitative Research. Pretoria: Van Schaik Publishers.
8. Pesaran, M.H.,, Pettenuzzo, D., & Timmermann, A. (2006). Forecasting time series subject to multiple structural breaks. Review of Economic Studies, 73, 1057–1084.
9. Stock, J.H. (2008). Introduction to Econometrics, Pearson Education.
10. Teicher, H. (1963). Identifiability of Finite Mixtures, The Annals of Mathematical Statistics, 34, 1265–1269
11. Bai, J. (1994). Least squares estimation of a shift in linear processes. *Journal of Time Series Analysis, 15*, 453–472.
12. Bai, J. (1997a). Estimating multiple breaks one at a time. *Econometric Theory, 13*, 315–352.
13. Bai, J. (1997b). Estimation of a change point in multiple regression models. *Review of Economics and Statistics, 79*, 551–563.
14. Bai, J., & Perron, P. (1998). Estimating and testing linear models with multiple structural changes. *Econometrica, 66*, 47–78.
15. Bai, J., & Perron, P. (2003). Computation and analysis of multiple structural change models. *Journal of Applied Econometrics, 18*, 1–22.
16. Bellman, R. (1952). On the theory of dynamic programming. *Proceedings of the National Academy of Sciences*, 1952.
17. Cappè, O., Moulines, E., & Rydèn, T. (2005). *Inference in hidden markov models*. New York: Springer.
18. Dempster, A. P., Laird, N. M., & Rubin, D. B. (1977). Maximum likelihood from incomplete data via the EM algorithm. *Journal of the Royal Statistical Society, B, 39*, 1–38.
19. Durbin, J., & Koopman, S. J. (2001). *Time series analysis by state space methods*. Oxford: Oxford University Press.
20. Fokianos, K., Kedem, B., Qin, J. , & Short, D. A. (2001). A semiparametric approach to the one-way layout. *Technometrics, 43*, 56–65.
21. Garcia, R., & Perron, P. (1996). An analysis of the real interest rate under regime shifts. *Review of Economics and Statistics, 78*, 111–125.
22. Gilbert, P. B. (2000). Large sample theory of maximum likelihood estimation in semiparametric biased sampling models. *Annals of Statistics, 28*, 151–194.
23. Kedem, B., & Gagnon, R. (2010). Semiparametric distribution forecasting. *Journal of Statistical Planning and Inference, 140*, 3734–3741.
24. Koop, G., & Potter, S. (2001). Are apparent findings of nonlinearity due to structural instability in economic time series? *Econometric Journal, 4*, 37–55.

25. Pastor, L., & Stambaugh, R. F. (2001). The equity premium and structural breaks. *Journal of Finance, 56*, 1207–1239.
26. Patton, A., Politis D. N., & White H. (2009). CORRECTION TO "Automatic block-length selection for the dependent bootstrap by D. Politis and H. White". *Econometric Reviews 28*(4), 372–375.
27. Politis, D. N., & Romano, J. P. (1994). The stationary bootstrap. *Journal of American Statistical Association, 89*, 1303–1313.
28. Qin, J. (1993). Empirical likelihood in biased sampling problems. *Annals of Statistics, 21*, 1182–1186.
29. Qin, J., & Lawless, J. F. (1994). Empirical likelihood and general estimating equations. *Annals of Statistics, 22*, 300–325.
30. Qin, J., & Zhang, B. (1997). A goodness of fit test for logistic regression models based on case-control data. *Biometrica, 84*, 609–618.
31. West, M., & Harrison, J. (1997). *Bayesian forecasting and dynamic models* (2nd Ed.). New York: Springer.
32. Zeileis, A., Leisch, F., Hornik, K., & Kleiber, C. (2003). Strucchange: An R package for testing for structural change in linear regression models. *Journal of Statistical Software, 7*(2), 1–38.
33. Zhang, B. (2000). M-estimation under a two sample semiparametric model. *Scandinavian Journal of Statistics, 27*, 263–280.

Direct Differentiation of Human Pluripotent Stem Cells into Advanced Spermatogenic Cells: In Search of an In Vitro System to Model Male Factor Infertility

Charles A. Easley IV, Calvin R. Simerly, and Gerald Schatten

Abstract Differentiation of stem cells into spermatogenic lineages in vitro provides a unique window into the biological mechanisms responsible for driving pluripotent stem cells into essential progeny—haploid spermatids and viable sperm—as well as provides an innovative approach for determining novel root causes for male infertility. Our recent work outlined a novel approach for differentiating human embryonic stem cells (hESCs) and induced pluripotent stem cells (hiPSCs) into advanced spermatogenic lineages including haploid spermatids with correct parent-of-origin genomic imprints on two loci. The work described here in this chapter provides a foundation for building a true in vitro model for human spermatogenesis with which to model, diagnose and potentially treat male factor infertility.

1 Introduction

As previously discussed [13], the sharp biological distinction between mortal somatic cells and potentially immortal germ cells has been held as a central tenet in developmental biology for well over a century dating back to August

C.A. Easley IV (✉)
Department of Cell Biology, Laboratory of Translational Cell Biology, Emory University School of Medicine, 615 Michael St, Whitehead Biomedical Research Building RM405H, Atlanta, GA 30322, USA
e-mail: caeasle@emory.edu

C.R. Simerly • G. Schatten
Department of OB/GYN and Reproductive Sciences, University of Pittsburgh School of Medicine, Pittsburgh, PA, USA

Magee Womens Research Institute, Pittsburgh Development Center, 204 Craft Ave Pittsburgh, Pittsburgh, PA 15213, USA
e-mail: simerlycr@upmc.edu; schattengp@upmc.edu

B. Toni (ed.), *New Frontiers of Multidisciplinary Research in STEAM-H (Science, Technology, Engineering, Agriculture, Mathematics, and Health)*, Springer Proceedings in Mathematics & Statistics 90, DOI 10.1007/978-3-319-07755-0_12,

Weismann's Germ-Plasm Theory (for review, [53]). This theory holds that whereas the germ line lineage can both maintain itself and also differentiate into somatic progeny, it is a rectified pathway in which somatic cells cannot themselves generate gametes. Cracks in this seemingly impregnable wall separating somatic and germ cells first appeared when Dolly and other cloned animals had offspring and were therefore reproductively fertile [7], since the transferred somatic cell nucleus was reprogrammed within the oocyte into a germ line lineage; explanations incorporated the idea that the oocyte's germplasm or ooplasm was vital in this process as in other systems [54]. Breakthroughs in induced pluripotency and the generation of fertile mice using tetraploid complementation embryo transfers (for review, [64]) opened the floodgate by demonstrating that exposure to a just a few transcription factors could reprogram somatic cells which were rigidly committed differentiated cells into most every other cell, including cells in the germ line. Derivations of cells in the spermatogenic lineage show the promiscuity of pluripotent stem cells, and now findings of oocyte stem cells in mice capable of generating pups [65] and recently similar oocyte-like stem cells from women [58], might be another example of this cellular promiscuity in vitro. Whether these in vitro generated gamete precursors have reproductive capabilities in vivo, helpful for infertility patients, will be important to evaluate pre-clinically, though they will be of keen biological importance regardless.

The quest to generate viable sperm and spermatids in vitro from pluripotent stem cells and even somatic cells in humans and other primates has many biomedical justifications even though the endeavor is fraught with experimental and bioethical challenges [10, 31, 35]. Furthermore the stringencies which with these 'artificial sperm' are evaluated vary according the necessary endpoint. The greatest stringency is for the generation of fully functional sperm or spermatids useful and safe for reproduction in ART clinics. This objective is well justified by the Oncofertility Consortium, which seeks the benevolent objective of preserving fertility in male cancer survivors who were rendered infertile during their therapies but were also too young or fragile to produce a sperm specimen for cryobanking [25, 26, 34, 52, 57, 59, 61]. It is also justified for the treatment of infertile men suffering from either diagnosed [24] or idiopathic male infertility in cases in which neither sperm nor elongated spermatids useful for either ICSI or ELSI can be obtained [25, 26, 34, 52, 57, 59, 61]. Discovering of the stages during spermatogenesis at which various forms of idiopathic male infertility arrest would greatly aid in the diagnoses, and perhaps eventual treatments, of these still mysterious processes. Learning of these spermatogenic arrest sites might also contribute to the design of novel contraceptives. Additionally the epigenetic modifications enabling the properly imprinted sperm chromatin and the replacement of nuclear proteins to form the sperm nucleus could be better investigated in these types of cell cultures versus in intact tissues. Anticipated improvements in the efficiency of in vitro spermatogenesis may also help understanding how mitochondria are modified to create the sperm mitochondria as well as how the somatic centrosome is reduced during male meiosis to form the sperm tail's basal body and the sperm centrosome [49].

Recent studies suggest that human pluripotent stem cells (PSCs) can enter meiosis, and in some cases produce haploid products, in vitro [15, 29, 44]. In this chapter, we examine the article recently published entitled ***Direct Differentiation of Human Pluripotent Stem Cells into Haploid Spermatogenic Cells*** [12], in which we developed an in vitro method which achieves two significant endpoints. First, male hESCs and hiPSCs are directly differentiated into adult-type spermatogonia. Secondly, differentiating stem cells give rise to cells which are phenotypically consistent with post-meiotic round spermatids. These results highlight the full plasticity of human PSCs by showing the ability to undergo spermatogenesis in vitro culminating in the production of round spermatid-like, haploid cells with correct parent-of-origin genomic imprints on at least two loci. These results also contribute to the overall goal of ultimately generating gametes that may prove invaluable for understanding infertility mechanisms.

2 Differentiation of Human PSCs in Mouse Spermatogonial Stem Cell Conditions Significant Increases VASA Expression

Because human testis cells have been shown to directly de-differentiate into PSCs by culturing cells in PSC conditions [8, 32, 33], we examined whether ESCs could directly differentiate into germline stem cells. Our goal was to differentiate PSCs into spermatogonial stem cell (SSC)-like cells because this spermatogenic lineage has shown an exceptional ability to re-colonize sterilized testes and thus restore fertility in certain species including mice and non-human primates (NHPs) [1, 22, 27]. One advantage of this strategy is that there are established protocols for culturing and expanding rodent SSCs in vitro [28]. Using these established protocols, we cultured male hESCs and hiPSCs into mouse SSC medium on specialty STO feeders for 10 days (Fig. 1a). After 10 days, we observed a significant increase in VASA-expressing cells (Fig. 1b, c) with VASA expression showing similar perinuclear localization to that of germ cells found in human testis histological sections (Fig. 1d).

3 SSC Conditions Elevate Additional Germ Cell Markers in Human PSCs

We further analyzed whether hESCs and hiPSCs cultured in mouse SSC conditions express additional germ cell markers. Deleted-in-Azoospermia-like (DAZL) and VASA are two germline specific, RNA binding proteins that are important in germ cell development and normal spermatogenesis [5, 29]. Here, both male hESC and

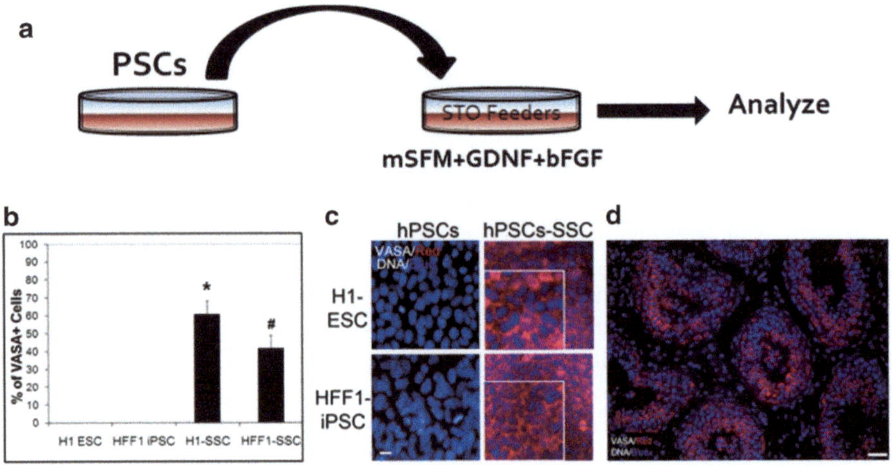

Fig. 1 Differentiation of human pluripotent stem cells (PSCs) in mouse spermatogonial stem cell (SSC) conditions significantly increases VASA expression. Adapted from [12]. (**a**) Schematic depicting our differentiation methodology by culturing hPSCs onto specialty STOs in mouse SSC medium containing bFGF and GDNF. Mouse serum-free medium (mSFM). (**b**) hESCs and hiPSCs cultured in mouse SSC conditions for 10 days and then stained for VASA. Percentage of VASA expression was quantified in the parent PSC lines and the differentiated lines. Representative graphical analysis from five separate trials, >5,000 cells counted for each condition, is shown. *Asterisk* signifies $p < 0.01$ comparing H1 ESC to H1 SSC. *Hash* signifies $p < 0.01$ comparing HFF1 iPSC to HFF1 SSC. (**c**) Representative images of PSCs and PSCs differentiated in SSC culture conditions for 10 days and stained for VASA. DNA labeled with Hoechst. *Scale* 50 μm. Enlarged insets show typical, perinuclear localization of VASA. (**d**) Human testis tissue was processed for immunohistochemistry and stained for VASA expression. In VASA-expressing cells within the seminiferous tubules, VASA localizes to the perinuclear region. DNA is counterstained with Hoechst. *Scale bar* 500 μm

hiPSC lines do not exhibit expression of germ cell marker mRNAs (Fig. 2b). Differentiated cells show an increase in all germ cell markers tested, including *CXCR4* and *PIWIL1*, by RT-PCR, suggesting that this is an efficient way to generate germ cell lineages (Fig. 2b). VASA and DAZL protein expression was also elevated in differentiated human PSCs compared to the undifferentiated, parent PSC lines (Fig. 2c). We also observed that germ cell differentiation was dependent on the growth factor GDNF (glial-derived neurotrophic factor, hPSCs + Complete). Cells differentiated without GDNF (hPSCs + FGF only) demonstrated no increase in VASA or DAZL protein expression but did show a loss of the pluripotent marker Nanog, suggesting that both lines differentiated (Fig. 2c). These results suggest that GDNF containing SSC medium efficiently and rapidly differentiates hPSCs into germ cell lineages.

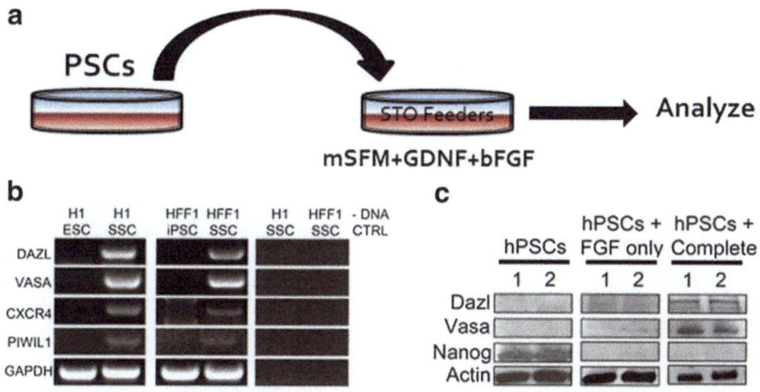

Fig. 2 SSC conditions elevate additional germ cell markers in human pluripotent stem cells. Adapted from [12]. (**a**) Schematic depicting our differentiation methodology by culturing hPSCs onto specialty STOs in mouse SSC medium containing bFGF and GDNF. Mouse serum-free medium (mSFM). (**b**) Reverse transcription (RT) PCR for germ cell markers DAZL, VASA, CXCR4 and PIWIL1 in PSCs and their differentiated counterparts. GADPH is shown as a loading control. No DNA (−DNA) is also shown as a negative control. (**c**) Representative western blot analyses showing upregulation of germ cell marker expression and a concomitant loss of the pluripotent marker Nanog in complete SSC culture conditions (with GDNF and FGF). Despite loss of Nanog in FGF only SSC medium (i.e. without GDNF), germ cell markers were not expressed. Actin is a loading control

4 Human Pluripotent Stem Cells Cultured in Mouse SSC Conditions Express PLZF, a marker of Stem and Progenitor Spermatogonia

We next evaluated whether hPSCs differentiated in mouse SSC culture conditions expressed PLZF, a zinc-finger transcription factor that is a consensus marker of stem and progenitor spermatogonia. PLZF, or ZBTB16, plays a critical role in SSC self-renewal and growth [2, 9, 23]. Whereas hPSCs do not express PLZF, 10 day culture in mouse SSC conditions induced expression of PLZF, localized to the nucleus, in both differentiating hESCs and hiPSCs (Fig. 3). This nuclear expression of PLZF mirrors that observed in human testes (Fig. 3, fourth row for each cell type). Furthermore our protocol generates a high percentage of PLZF-positive cells within differentiating colonies (Fig. 3, low magnification views, third row for each column). Unlike other germ cell differentiation methods, our protocol induces PLZF expression. This suggests that we are more closely mirroring the early events of in vivo spermatogenesis.

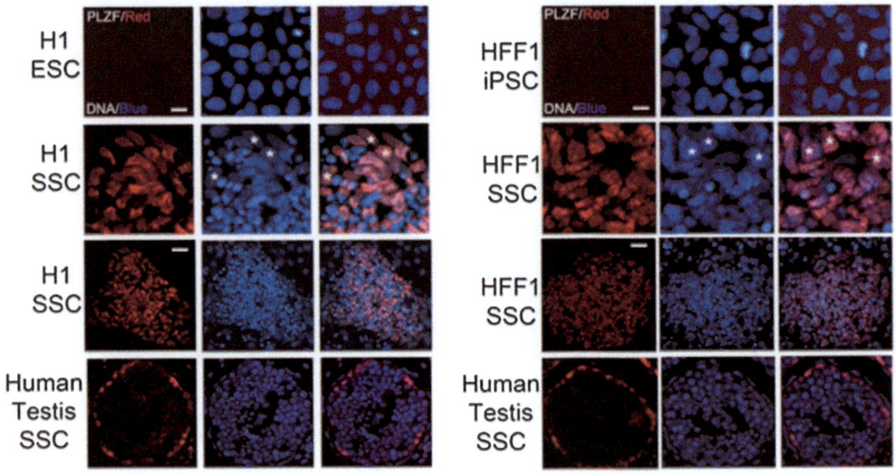

Fig. 3 Human pluripotent stem cells cultured in mouse SSC conditions express PLZF, a marker of stem and progenitor spermatogonia. Adapted from [12]. While the parent PSC lines do not express detectable levels of PLZF, 10 day culture in SSC conditions upregulates PLZF (*red*) expression in both lines (hESC *left column*, hiPSC *right column*). Hoechst (*blue*): DNA. *Scale* 40 μm. Global view (third row of each column) of differentiated colonies shows a large portion of cells expressing PLZF. *Scale* 100 μm. Fourth row panels in each column depict PLZF staining in human testis sections

5 Haploid Cells are Generated from hPSCs Culturedin SSC Conditions

SSCs are defined in part by their ability to produce gametes through a complex combination of division and differentiation. Mouse SSCs can differentiate into haploid cells in vitro [16, 18, 41], so we next quantified whether haploid cells were produced in differentiating hESCs and hiPSCs in mouse SSC conditions. Flow cytometry analyses indicated that a haploid population exists in hESCs (4.5 %) and iPSCs (3.9 %) differentiated in mouse SSC conditions corresponding to haploid peaks observed with human sperm (Fig. 4). Parent hESC and hiPSC lines showed no presence of haploid cells in their respective cultures (Fig. 4). We further confirmed haploidy of isolated cells by fluorescence in situ hybridization (FISH) with an LNA probe to satellite DNA found on chromosomes 1, 9, 16 and Y (Fig. 4, lower right inset). These results suggest that we are able to generate a small percentage of haploid cells in vitro from hPSCs within 10 days of SSC culture.

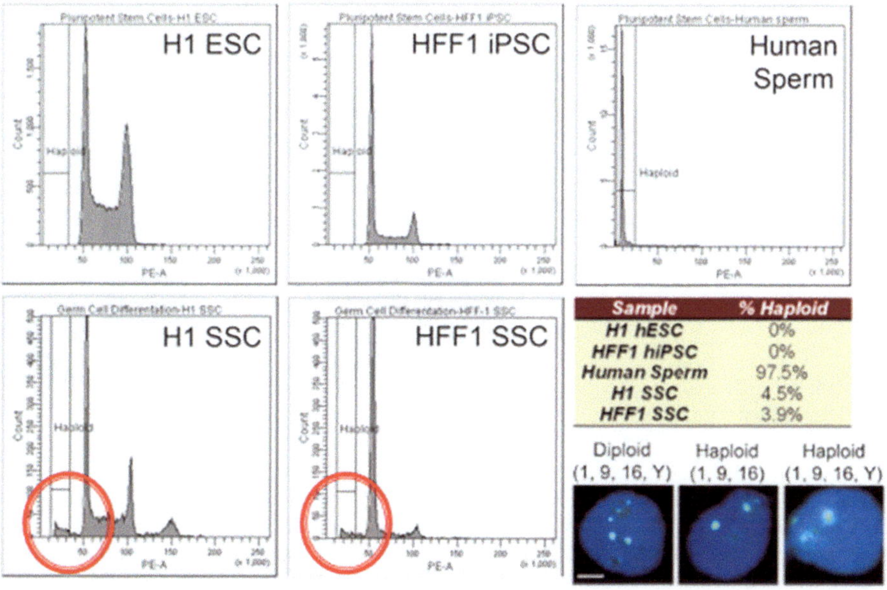

Fig. 4 Haploid cells are generated from hPSCs cultured in SSC conditions. Adapted from [12]. FACS ploidy analysis reveals a small haploid peak in hPSCs cultured in SSC culture conditions for 10 days. This peak corresponds to the haploid peak observed in human sperm. Chart below represents % of haploid cells in undifferentiated and SSC-mediated differentiated hPSCs. Data is representative of five cell sorts with 500,000 cells sorted per experiment. Included table shows a summary of the average % of haploid cells produced. *Lower right inset*, cells by FACS from the haploid peak are confirmed as haploid by FISH, using an LNA probe directed at satellite DNA found on chromosomes 1, 9, 16 and Y. *Left*, undifferentiated H1 hESCs show diploid probe expression with seven "dots" present. Haploid cells isolated by FACS show both appearance of three "dots" and four "dots" signifying the generation of X-haploid cells and Y-haploid cells, respectively. *Scale bar* 2 μm

6 Haploid Cells Isolated from hPSCs Cultured in SSC Conditions Resemble Round Spermatids

Because differentiation in SSC conditions yielded a small percentage of haploid cells in addition to a large population of PLZF-positive spermatogonia, we next evaluated whether hESCs and hiPSCs differentiated into intermediate cell types observed in in vivo spermatogenesis during culture in mouse SSC conditions. In addition to PLZF, we observed expression of UTF1 and CDH1 (Fig. 5, left column), proteins expressed both in spermatogonia and PSCs. Unlike PSCs, we observed an increase in protein expression of RET and GFRα1 (Fig. 5, western blots), receptors for GDNF found on spermatogonia.

Differentiation of hPSCs in SSC conditions showed an increase in *PIWIL1* RNA expression (Fig. 2b). *PIWIL1*, also known as *HIWI,* is essential in spermatogenic progression from SSCs to round spermatids [11]. We examined expression of three

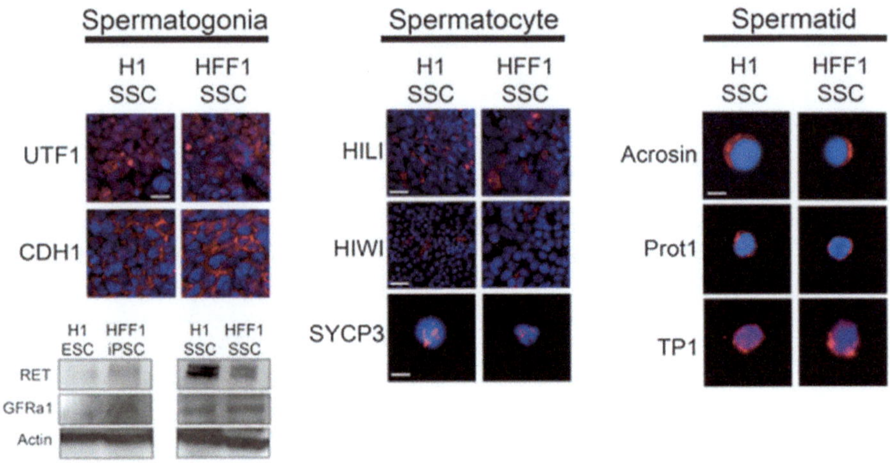

Fig. 5 Haploid cells isolated from hPSCs cultured in SSC conditions resemble round spermatids. Adapted from [12]. *Left*: 10 days post differentiation cultures of hESCs and hiPSCs express pre-meiotic spermatogonial markers UTF1 and CDH1. *Scale* 50 μm. Differentiation also induces expression of two membrane receptors: RET and GFRa1. Actin is a loading control. *Center*: expression of spermatogonia-to-spermatocyte marker HILI, spermatocyte-to-spermatid marker HIWI and meiotic marker SYCP3. Scale for HILI 200 μm, scale for HIWI, 500 μm and scale for SYCP3, 10 μm. *Right*: expression of post-meiotic spermatid markers Acrosin, Protamine 1 (Prot1) and Transition Protein 1 (TP1). Haploid cells were sorted by FACS and immunostained with antibodies directed at the indicated protein. *Scale* 10 μm

spermatocyte markers for pre-meiotic spermatocytes/differentiating spermatogonia, meiotic spermatocytes and post-meiotic spermatocytes. We identified cells in both differentiating hESCs and hiPSCs expressing pre-meiotic HILI protein, meiotic marker SYCP3 (synaptonemal complex 3), involved in recombination and segregation of meiotic chromosomes; and post-meiotic HIWI (Fig. 5, center column). While there were a large number of HILI-positive cells, very few cells expressed SYCP3 or HIWI, suggesting that there is bottleneck prior to meiosis.

We next isolated the haploid peaks from FACS and immunostained isolated cells for spermatid markers. During spermiogenesis, acrosin expression is turned on and histones are replaced by protamines via transition proteins [4]. Haploid cells isolated from differentiated hESC and hiPSC cultures express post-meiotic, sperm markers: acrosin, protamine 1 and transition protein 1 (Fig. 5, right column). In particular, acrosin staining exhibits polar localization in both cell lines (Fig. 5, first row). These haploid cells resemble round spermatids by acrosin localization, the nuclear/perinuclear localization of transition protein 1 (TP1) and the perinuclear localization of protamine 1 (Prot1) (Fig. 5, right column), which localizes to the perinuclear region of haploid cells and enters the nucleus at the elongated spermatid stage [4]. These haploid cells also resemble round spermatids observed in human and NHPs [4, 40, 47]. These results coupled with the preceding *PIWIL1* expression data suggest that PSCs are able to directly differentiate into post-meiotic, round-spermatid like cells in vitro.

7 Haploid Spermatids from Pluripotent Stem Cells Show Similar Imprint Patterns to Human Sperm

During in vivo germ cell specification, genomic imprints are removed at the primordial germ cell stage and then re-established during spermatogenesis [36]. In mice, differentiating PSCs into functional germ cells results in progeny that exhibit epigenetic disease phenotypes [41, 42]. One explanation was improper imprinting during gametogenesis [37]. To evaluate imprinting statuses on haploid spermatids differentiated here, we isolated haploid cells by FACS and examined the methylation status of the imprinting control region (ICR) for paternally imprinted (H19) and maternally imprinted genes (IGF2). As previously reported, hiPSCs showed aberrant imprinting [46], but hESCs showed typical somatic cell imprinting on ICRs for H19 and IGF2 (Fig. 6). Isolation of haploid cells from differentiated hESC cultures showed imprinting patterns similar to those observed in human sperm with H19 ICR methylation around 90 % and IGF2 ICR methylation around 5 % (Fig. 6). Haploid cells from differentiated hiPSC cultures showed similar levels of H19 ICR methylation to human sperm (90 %), but IGF2 methylation (14 %) was slightly elevated above methylation observed in human sperm (Fig. 6). These results suggest that haploid products obtained show similar DNA methylation patterns on at least two parent-of-origin genomic imprints.

8 Discussion

Infertility affects perhaps 15 % of couples worldwide, with male factors responsible for 40–60 % of all cases [51]. In men without a genetic root cause for infertility, stem cell transplantation represents a possible treatment option to restore fertility

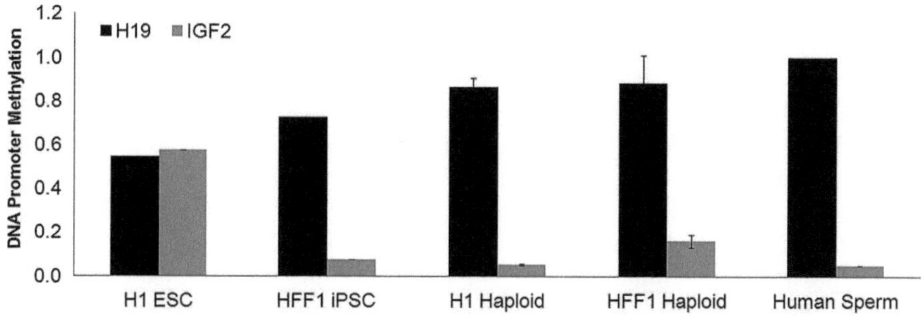

Fig. 6 Haploid spermatids from pluripotent stem cells show similar imprint patterns to human sperm. Adapted from [12]. hESCs, hiPSCs, fertile human sperm, and haploid cells obtained by FACS from differentiated hESC and hiPSC cultures were examined for methylation on imprinting control regions (ICRs) for H19 (paternally imprinted) and IGF2 (maternally imprinted). Methylation statuses were examined using Qiagen Epitect Methyl II PCR Array. Graph shows average % methylation with error bars

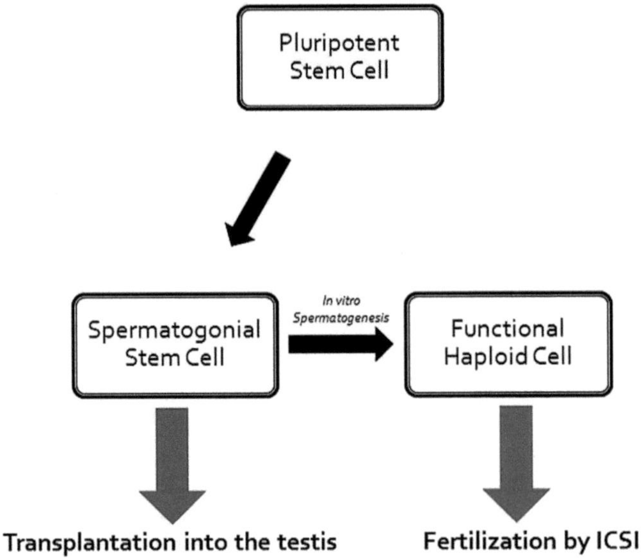

Fig. 7 Proposed model for treating male infertility with stem cells. Adapted from [13]. Diagram depicting how our differentiation strategy yields SSCs, which can be useful for restoring fertility by transplantation into the testis, and haploid spermatids which can potentially be used to fertilize an oocyte by IVF

[38, 39, 43, 63]. Clinical interventions such as chemotherapy and immune suppressant treatments often render male patients sterile. Protocols to preserve future fertility in boys undergoing cancer therapies who cannot yet bank their own sperm are under development [19, 21, 30, 48, 50, 60]. However for adult and prepubescent patients rendered sterile prior to sperm collection, there are no current treatments to restore fertility.

Our differentiation protocol generated two endpoints critical for driving in vitro spermatogenesis to the clinic to treat infertility in patients without a known genetic etiology (Fig. 7). First, human PSCs were differentiated into SSC-like cells, cells that reside at the foundation of spermatogenesis. Several previous studies have shown the ability of human and non-human primates PSCs to differentiate into PGCs [3, 17, 29, 44, 45, 55, 56, 62]. Although this cell lineage has the capability of restoring fertility in rodents, including primordial germ cells (PGCs) derived from mouse PSCs [6, 20], SSCs remain the gold standard for colonizing cells which recapitulate spermatogenesis following transplantation [1, 27]. Thus differentiating hPSCs into SSCs is an important step in the future ability for using patient-specific PSCs to restore fertility, as SSCs derived from PSCs can be transplanted into the sterilized testes to restore spermatogenesis (Fig. 2). Furthermore, the sperm generated following transplant would, in theory, be the patient's own genetic material.

However, transplantation of SSCs derived from PSCs supposes that the somatic environment of the testis remains intact. Prolonged clinical interventions, injury, exposure to environmental toxins, etc. can cause sterility and render the somatic environment useless for SSC transplantation. For these patients, complete spermatogenesis in vitro is critical for generating haploid products useful for ART procedures to fertilize a partner's oocyte and pass along their own genetic material (Fig. 7). Our differentiation protocol generates haploid products consistent with round spermatids. While techniques for utilizing round spermatids to fertilize oocytes have not been proven in human and non-human primates, our differentiation protocol at least shows the feasibility of generating haploid products that could be useful in IVF. This would suggest that functional haploid cells may be obtained from no greater starting material than a skin biopsy needed for iPSC derivations.

In vitro spermatogenesis also holds great promise to diagnose male infertility and provides a novel tool for exploring root causes for male infertility [14]. By deriving hiPSCs from infertile men, such as from patients with Sertoli-cell-only (SCO) syndrome, followed by direct differentiation with our protocol, we can examine where spermatogenesis arrests, and in the case of SCO patients, identify whether hiPSCs can differentiate into SSCs and whether viability of SSCs is a major concern. A similar strategy can be implored for men with defects in Leydig Cell function, DAZ-family deletions and even Klinefelter Syndrome. In cases where spermatogenesis arrests in vitro, chemical screens can be employed with a read-out for haploid cell production to identify novel compounds that could treat known causes for male infertility. In this same light, chemical screens can be utilized to discover male forms of birth control that temporarily arrest spermatogenesis but do not endanger SSC survival. Thus the clinical uses for in vitro spermatogenesis are substantial and could lead to the first cures for male sterility.

9 Conclusion

While the risks and ethical considerations for moving in vitro spermatogenesis to the clinic are great, the potential rewards are sufficient to continue to explore this option to treat male infertility. To date, our methodology needs to be refined to use xeno-free conditions to generate haploid spermatids for use in the clinic. As advances in in vitro spermatogenesis are made, this technique may become fundamental in diagnosing and treating a currently incurable disorder: male infertility.

References

1. Brinster, R. L., & Avarbock, M. R. (1994). Germline transmission of donor haplotype following spermatogonial transplantation. *Proceedings of the National Academy of Sciences of the United States of America, 91*(24), 11303–11307.

2. Buaas, F. W., Kirsh, A. L., Sharma, M., McLean, D. J., Morris, J. L., Griswold, M. D., et al. (2004). Plzf is required in adult male germ cells for stem cell self-renewal. *Nature Genetics, 36*(6), 647–652.
3. Bucay, N., Yebra, M., Cirulli, V., Afrikanova, I., Kaido, T., Hayek, A., et al. (2009). A novel approach for the derivation of putative primordial germ cells and sertoli cells from human embryonic stem cells. *Stem Cells, 27*(1), 68–77.
4. Carrell, D. T., Emery, B. R., & Hammoud, S. (2007). Altered protamine expression and diminished spermatogenesis: What is the link? *Human Reproduction Update, 13*(3), 313–327.
5. Castrillon, D. H., Quade, B. J., Wang, T. Y., Quigley, C., & Crum, C. P. (2000). The human VASA gene is specifically expressed in the germ cell lineage. *Proceedings of the National Academy of Sciences of the United States of America, 97*(17), 9585–9590.
6. Chuma, S., Kanatsu-Shinohara, M., Inoue, K., Ogonuki, N., Miki, H., Toyokuni, S., et al. (2005). Spermatogenesis from epiblast and primordial germ cells following transplantation into postnatal mouse testis. *Development, 132*(1), 117–122.
7. Cibelli, J. B., Campbell, K. H., Seidel, G. E., West, M. D., & Lanza, R. P. (2002). The health profile of cloned animals. *Nature Biotechnology, 20*(1), 13–14.
8. Conrad, S., Renninger, M., Hennenlotter, J., Wiesner, T., Just, L., Bonin, M., et al. (2008). Generation of pluripotent stem cells from adult human testis. *Nature, 456*(7220), 344–349.
9. Costoya, J. A., Hobbs, R. M., Barna, M., Cattoretti, G., Manova, K., Sukhwani, M., et al. (2004). Essential role of Plzf in maintenance of spermatogonial stem cells. *Nature Genetics, 36*(6), 653–659.
10. Daley, G. Q. (2007). Gametes from embryonic stem cells: A cup half empty or half full? *Science, 316*(5823), 409–410.
11. Deng, W., & Lin, H. (2002). Miwi, a murine homolog of piwi, encodes a cytoplasmic protein essential for spermatogenesis. *Developmental Cell, 2*(6), 819–830.
12. Easley, C. A., Phillips, B., McGuire, M., Barringer, J., Valli, H., Hermann, B. P., et al. (2012). Direct differentiation of human pluripotent stem cells into haploid spermatogenic cells. *Cell Reports, 2*(3), 440–446.
13. Easley, C. A., Phillips, B. T., Wu, G., Schatten, G. P., & Simerly, C. R. (2012). Clinical implications of human spermatogenesis initiation in vitro. *Journal of Medical Sciences, 32*(6), 257–263.
14. Easley, C. A., 4th, Simerly, C. R., & Schatten, G. (2013). Stem cell therapeutic possibilities: Future therapeutic options for male-factor and female-factor infertility? *Reproductive Biomedicine Online, 27*(1), 75–80.
15. Eguizabal, C., Montserrat, N., Vassena, R., Barragan, M., Garreta, E., Garcia-Quevedo, L., et al. (2011). Complete meiosis from human induced pluripotent stem cells. *Stem Cells, 29*(8), 1186–1195.
16. Feng, L. X., Chen, Y., Dettin, L., Pera, R. A., Herr, J. C., Goldberg, E., et al. (2002). Generation and in vitro differentiation of a spermatogonial cell line. *Science, 297*(5580), 392–395.
17. Fukunaga, N., Teramura, T., Onodera, Y., Takehara, T., Fukuda, K., & Hosoi, Y. (2010). Leukemia inhibitory factor (LIF) enhances germ cell differentiation from primate embryonic stem cells. *Cellular Reprogramming, 12*(4), 369–376.
18. Geijsen, N., Horoschak, M., Kim, K., Gribnau, J., Eggan, K., & Daley, G. Q. (2004). Derivation of embryonic germ cells and male gametes from embryonic stem cells. *Nature, 427*(6970), 148–154.
19. Ginsberg, J. P., Carlson, C. A., Lin, K., Hobbie, W. L., Wigo, E., Wu, X., et al. (2010). An experimental protocol for fertility preservation in prepubertal boys recently diagnosed with cancer: A report of acceptability and safety. *Human Reproduction, 25*(1), 37–41.
20. Hayashi, K., Ohta, H., Kurimoto, K., Aramaki, S., & Saitou, M. (2011). Reconstitution of the mouse germ cell specification pathway in culture by pluripotent stem cells. *Cell, 146*, 1–14.
21. Hermann, B. P., Sukhwani, M., Lin, C. C., Sheng, Y., Tomko, J., Rodriguez, M., et al. (2007). Characterization, cryopreservation, and ablation of spermatogonial stem cells in adult rhesus macaques. *Stem Cells, 25*(9), 2330–2338.

22. Hermann, B. P., Sukhwani, M., Winkler, F., Pascarella, J. N., Peters, K. A., Sheng, Y., et al. (2012). Spermatogonial stem cell transplantation into rhesus testes regenerates spermatogenesis producing functional sperm. *Cell Stem Cell, 11*(5), 715–726.
23. Hobbs, R. M., Seandel, M., Falciatori, I., Rafii, S., & Pandolfi, P. P. (2010). Plzf regulates germline progenitor self-renewal by opposing mTORC1. *Cell, 142*(3), 468–479.
24. Houk, C. P., Rogol, A., & Lee, P. A. (2010). Fertility in men with Klinefleter syndrome. *Pediatric Endocrinology Reviews, 8*(Suppl 1), 182–186.
25. Hwang, K., & Lamb, D. J. (2010). New advances on the expansion and storage of human spermatogonial stem cells. *Current Opinion in Urology, 20*(6), 510–514.
26. Jahnukainen, K., Ehmcke, J., Hou, M., & Schlatt, S. (2011). Testicular function and fertility preservation in male cancer patients. *Best Practice & Research. Clinical Endocrinology & Metabolism, 25*(2), 287–302.
27. Jahnukainen, K., Ehmcke, J., Quader, M. A., Saiful Huq, M., Epperly, M. W., Hergenrother, S., et al. (2011). Testicular recovery after irradiation differs in prepubertal and pubertal non-human primates, and can be enhanced by autologous germ cell transplantation. *Human Reproduction, 26*(8), 1945–1954.
28. Kanatsu-Shinohara, M., Ogonuki, N., Inoue, K., Miki, H., Ogura, A., Toyokuni, S., et al. (2003). Long-term proliferation in culture and germline transmission of mouse male germline stem cells. *Biology of Reproduction, 69*(2), 612–616.
29. Kee, K., Angeles, V. T., Flores, M., Nguyen, H. N., & Reijo Pera, R. A. (2009). Human DAZL, DAZ and BOULE genes modulate primordial germ-cell and haploid gamete formation. *Nature, 462*(7270), 222–225.
30. Keros, V., Hultenby, K., Borgstrom, B., Fridstrom, M., Jahnukainen, K., & Hovatta, O. (2007). Methods of cryopreservation of testicular tissue with viable spermatogonia in pre-pubertal boys undergoing gonadotoxic cancer treatment. *Human Reproduction, 22*(5), 1384–1395.
31. Ko, K., Huebner, K., Mueller-Keuker, J., & Schoeler, H. R. (2010). In vitro derivation of germ cells from embryonic stem cells. *Frontiers in Bioscience (Landmark Edition), 15*, 46–56.
32. Ko, K., Tapia, N., Wu, G., Kim, J. B., Bravo, M. J., Sasse, P., et al. (2009). Induction of pluripotency in adult unipotent germline stem cells. *Cell Stem Cell, 5*(1), 87–96.
33. Kossack, N., Meneses, J., Shefi, S., Nguyen, H. N., Chavez, S., Nicholas, C., et al. (2009). Isolation and characterization of pluripotent human spermatogonial stem cell-derived cells. *Stem Cells, 27*(1), 138–149.
34. Levine, J., Canada, A., & Stern, C. J. (2010). Fertility preservation in adolescents and young adults with cancer. *Journal of Clinical Oncology, 28*(32), 4831–4841.
35. Lokman, M., & Moore, H. (2010). An artificial sperm – next year or never? *Human Fertility (Cambridge, England), 13*(4), 272–276.
36. Lucifero, D., Mertineit, C., Clarke, H. J., Bestor, T. H., & Trasler, J. M. (2002). Methylation dynamics of imprinted genes in mouse germ cells. *Genomics, 79*(4), 530–538.
37. Lucifero, D., & Reik, W. (2006). Artificial sperm and epigenetic reprogramming. *Nature Biotechnology, 24*(9), 1097–1098.
38. Marques-Mari, A. I., Lacham-Kaplan, O., Medrano, J. V., Pellicer, A., & Simon, C. (2009). Differentiation of germ cells and gametes from stem cells. *Human Reproduction Update, 15*(3), 379–390.
39. Mathews, D. J., Donovan, P. J., Harris, J., Lovell-Badge, R., Savulescu, J., & Faden, R. (2009). Pluripotent stem cell-derived gametes: Truth and (potential) consequences. *Cell Stem Cell, 5*(1), 11–14.
40. Moreno, R. D., Palomino, J., & Schatten, G. (2006). Assembly of spermatid acrosome depends on microtubule organization during mammalian spermiogenesis. *Developmental Biology, 293*(1), 218–227.
41. Nayernia, K., Nolte, J., Michelmann, H. W., Lee, J. H., Rathsack, K., Drusenheimer, N., et al. (2006). In vitro-differentiated embryonic stem cells give rise to male gametes that can generate offspring mice. *Developmental Cell, 11*(1), 125–132.

42. Nolte, J., Michelmann, H. W., Wolf, M., Wulf, G., Nayernia, K., Meinhardt, A., et al. (2010). PSCDGs of mouse multipotent adult germline stem cells can enter and progress through meiosis to form haploid male germ cells in vitro. *Differentiation, 80*(4–5), 184–194.
43. Orwig, K. E., & Schlatt, S. (2005). Cryopreservation and transplantation of spermatogonia and testicular tissue for preservation of male fertility. *Journal of the National Cancer Institute. Monographs, 34*, 51–56.
44. Panula, S., Medrano, J. V., Kee, K., Bergstrom, R., Nguyen, H. N., Byers, B., et al. (2011). Human germ cell differentiation from fetal- and adult-derived induced pluripotent stem cells. *Human Molecular Genetics, 20*(4), 752–762.
45. Park, T. S., Galic, Z., Conway, A. E., Lindgren, A., van Handel, B. J., Magnusson, M., et al. (2009). Derivation of primordial germ cells from human embryonic and induced pluripotent stem cells is significantly improved by coculture with human fetal gonadal cells. *Stem Cells, 27*(4), 783–795.
46. Pick, M., Stelzer, Y., Bar-Nur, O., Mayshar, Y., Eden, A., & Benvenisty, N. (2009). Clone- and gene-specific aberrations of parental imprinting in human induced pluripotent stem cells. *Stem Cells, 27*(11), 2686–2690.
47. Ramalho-Santos, J., Schatten, G., & Moreno, R. D. (2002). Control of membrane fusion during spermiogenesis and the acrosome reaction. *Biology of Reproduction, 67*(4), 1043–1051.
48. Sadri-Ardekani, H., Akhondi, M. A., van der Veen, F., Repping, S., & van Pelt, A. M. (2011). In vitro propagation of human prepubertal spermatogonial stem cells. *JAMA, 305*(23), 2416–2418.
49. Schatten, G. (1994). The centrosome and its mode of inheritance: The reduction of the centrosome during gametogenesis and its restoration during fertilization. *Developmental Biology, 165*(2), 299–335.
50. Schlatt, S., Ehmcke, J., & Jahnukainen, K. (2009). Testicular stem cells for fertility preservation: Preclinical studies on male germ cell transplantation and testicular grafting. *Pediatric Blood & Cancer, 53*(2), 274–280.
51. Schlegel, P. N. (2009). Evaluation of male infertility. *Minerva Ginecologica, 61*(4), 261–283.
52. Silber, S. J. (2010). Sperm retrieval for azoospermia and intracytoplasmic sperm injection success rates–a personal overview. *Human Fertility (Cambridge, England), 13*(4), 247–256.
53. Stanford, P. K. (2005). August Weismann's theory of the germ-plasm and the problem of unconceived alternatives. *History and Philosophy of the Life Sciences, 27*(2), 163–199.
54. Strome, S., & Lehmann, R. (2007). Germ versus soma decisions: Lessons from flies and worms. *Science, 316*(5823), 392–393.
55. Teramura, T., Takehara, T., Kawata, N., Fujinami, N., Mitani, T., Takenoshita, M., et al. (2007). Primate embryonic stem cells proceed to early gametogenesis in vitro. *Cloning and Stem Cells, 9*(2), 144–156.
56. Tilgner, K., Atkinson, S. P., Golebiewska, A., Stojkovic, M., Lako, M., & Armstrong, L. (2008). Isolation of primordial germ cells from differentiating human embryonic stem cells. *Stem Cells, 26*(12), 3075–3085.
57. Wallace, W. H. (2011). Oncofertility and preservation of reproductive capacity in children and young adults. *Cancer, 117*(10 Suppl), 2301–2310.
58. White, Y. A., Woods, D. C., Takai, Y., Ishihara, O., Seki, H., & Tilly, J. L. (2012). Oocyte formation by mitotically active germ cells purified from ovaries of reproductive-age women. *Nature Medicine, 18*(3), 413–421.
59. Woodruff, T. K. (2010). The Oncofertility Consortium – addressing fertility in young people with cancer. *Nature Reviews. Clinical Oncology, 7*(8), 466–475.
60. Wyns, C., Curaba, M., Petit, S., Vanabelle, B., Laurent, P., Wese, J. F., et al. (2011). Management of fertility preservation in prepubertal patients: 5 years' experience at the Catholic University of Louvain. *Human Reproduction, 26*(4), 737–747.
61. Wyns, C., Curaba, M., Vanabelle, B., Van Langendonckt, A., & Donnez, J. (2010). Options for fertility preservation in prepubertal boys. *Human Reproduction Update, 16*(3), 312–328.
62. Yamauchi, K., Hasegawa, K., Chuma, S., Nakatsuji, N., & Suemori, H. (2009). In vitro germ cell differentiation from cynomolgus monkey embryonic stem cells. *PLoS One, 4*(4), e5338.

63. Yao, L., Yu, X., Hui, N., & Liu, S. (2011). Application of iPS in assisted reproductive technology: Sperm from somatic cells? *Stem Cell Reviews, 7*(3), 714–721.
64. Zhao, X. Y., Li, W., Lv, Z., Liu, L., Tong, M., Hai, T., et al. (2010). Viable fertile mice generated from fully pluripotent iPS cells derived from adult somatic cells. *Stem Cell Reviews, 6*(3), 390–397.
65. Zou, K., Yuan, Z., Yang, Z., Luo, H., Sun, K., Zhou, L., et al. (2009). Production of offspring from a germline stem cell line derived from neonatal ovaries. *Nature Cell Biology, 11*(5), 631–636.

Stepanov-Like Pseudo-Almost Periodic Functions in Lebesgue Spaces with Variable Exponents $L^{p(x)}$

Toka Diagana and Mohamed Zitane

Abstract In this paper we introduce and study a new class of functions called Stepanov-like pseudo-almost periodic spaces with variable exponents, which generalizes in a natural way the space of Stepanov-like pseudo-almost periodic spaces. Basic properties of these new spaces are established. The existence of pseudo-almost periodic solutions to some first-order differential equations with $S^{p,q(x)}$-pseudo-almost periodic coefficients will also be studied.

2000 *Mathematics Subject Classification.* 34C27; 35B15; 46E30

1 Introduction

The notion of pseudo-almost periodicity was introduced in the literature over a decade ago by Zhang [14]. Since then such a concept has been largely studied and extended in various directions. In particular, in Diagana [3], the notion of S^p-pseudo-almost periodicity (or Stepanov-like pseudo-almost periodicity), which generalizes the notion of pseudo-almost periodicity, was introduced and studied. The construction of S^p-pseudo-almost periodic spaces makes extensive use of the

T. Diagana (✉)
Department of Mathematics, Howard University, 2441 6th Street N.W., Washington DC 20059, USA
e-mail: tokadiag@gmail.com

M. Zitane
Department of Mathematics, Laboratory of An. Math and NCG, Faculty of Science, Ibn Tofaïl University, Kenitra, Morocco
e-mail: zitanem@gmail.com

B. Toni (ed.), *New Frontiers of Multidisciplinary Research in STEAM-H (Science, Technology, Engineering, Agriculture, Mathematics, and Health)*, Springer Proceedings in Mathematics & Statistics 90, DOI 10.1007/978-3-319-07755-0_13,

Lebesgue space L^p and its properties. Various important results on these spaces have recently been established including some composition theorems, e.g., [2, 4, 5, 7, 8, 10, 11], and [12].

The main objective of this paper is twofold. Our first goal consists of generalizing S^p-pseudo-almost periodic spaces to the case of variable exponents. These new spaces are called $S^{p,q(x)}$-pseudo-almost periodic functions with variable exponents. The construction of these new spaces makes extensive use of basic properties of Lebesgue spaces with variable exponents $L^{q(x)}$ (see [6, 9, 13]). Various properties of these new functions are investigated including some composition results (see Theorem 5.15).

The second goal of the paper consists of using the newly-introduced functions to study the existence of pseudo-almost periodic solutions to the first-order differential equation

$$u'(t) = Au(t) + F(t, u(t)), \quad t \in \mathbb{R}, \tag{1}$$

where the (possibly unbounded) linear operator $A : D(A) \subset \mathbb{X} \mapsto \mathbb{X}$ is the infinitesimal generator of a C_0-semigroup which is exponentially stable on a Banach space $\mathbb{X}$, and the forcing term $F : \mathbb{R} \times \mathbb{X} \mapsto \mathbb{X}$ is a $S^{p,q}$-pseudo-almost periodic function satisfying some additional conditions. Such a result generalizes most of the results encountered in the literature on pseudo-almost periodic solutions to Eq. (1).

In order to study the existence and uniqueness of pseudo-almost periodic solution to Eq.(1), we first study the existence of pseudo-almost periodic solutions to the linear differential equation

$$u'(t) = Au(t) + f(t), \quad t \in \mathbb{R}, \tag{2}$$

where the linear operator A satisfies the above-mentioned assumptions and the forcing term f belongs to $S_{pap}^{\theta,\vartheta(x)}(\mathbb{X}) \cap C(\mathbb{R}, \mathbb{X})$ for $\theta > 1$ and $\vartheta \in C_+(\mathbb{R})$.

This paper is organized as follows: Section 2 is devoted to some useful notations needed in the sequel. Section 3 collects the basic background on pseudo-almost periodic functions needed in the sequel. Section 4 gathers the basic results on the Lebesgue space with variable exponents $L^{p(x)}$. Section 5 introduces $S^{p,q(x)}$-pseudo-almost periodic functions and studies their properties. Section 6 studies the existence of pseudo-almost periodic solutions to Eqs. (1) and (2).

2 Preliminaries

Let $(\mathbb{X}, \|\cdot\|), (\mathbb{Y}, \|\cdot\|_{\mathbb{Y}})$ be two Banach spaces. Let $BC(\mathbb{R}, \mathbb{X})$ (respectively, $BC(\mathbb{R} \times \mathbb{Y}, \mathbb{X})$) denote the collection of all $\mathbb{X}$-valued bounded continuous functions (respectively, the class of jointly bounded continuous functions $F : \mathbb{R} \times \mathbb{Y} \to \mathbb{X}$). The space $BC(\mathbb{R}, \mathbb{X})$ equipped with the sup norm $\|\cdot\|_\infty$ is a Banach space.

Furthermore, $C(\mathbb{R}, \mathbb{Y})$ (respectively, $C(\mathbb{R} \times \mathbb{Y}, \mathbb{X})$) denotes the class of continuous functions from $\mathbb{R}$ into $\mathbb{Y}$ (respectively, the class of jointly continuous functions $F : \mathbb{R} \times \mathbb{Y} \to \mathbb{X}$). Let $B(\mathbb{X}, \mathbb{Y})$ stand for the Banach space of bounded linear operators from $\mathbb{X}$ into $\mathbb{Y}$ equipped with its natural operator topology; in particular, $B(\mathbb{X}, \mathbb{X})$ is denoted by $B(\mathbb{X})$.

3 Pseudo-Almost Periodic Functions

Definition 3.1 ([3]). A function $f \in C(\mathbb{R}, \mathbb{X})$ is called almost periodic if for each $\varepsilon > 0$ there exists $l(\varepsilon) > 0$ such that every interval of length $l(\varepsilon)$ contains a number τ such that $\|f(t+\tau) - f(t)\| < \varepsilon$ for each $t \in \mathbb{R}$. The collection of all almost periodic functions from $\mathbb{R}$ to $\mathbb{X}$ will be denoted by $AP(\mathbb{X})$.

Definition 3.2 ([3]). A jointly continuous function $F \in C(\mathbb{R} \times \mathbb{Y}, \mathbb{X})$ is called almost periodic in $t \in \mathbb{R}$ uniformly in $x \in \mathbb{Y}$ if for each $\varepsilon > 0$ and any $K \subset \mathbb{Y}$ a bounded subset, there exists $l(\varepsilon)$ such that every interval of length $l(\varepsilon)$ contains a number τ with the property that $\|F(t+\tau, y) - F(t, y)\| < \varepsilon$ for each $t \in \mathbb{R}$, $y \in K$. The collection of such functions will be denoted by $AP(\mathbb{R} \times \mathbb{X})$.

Define

$$PAP_0(\mathbb{X}) := \left\{ f \in BC(\mathbb{R}, \mathbb{X}) : \lim_{T \to \infty} \frac{1}{2T} \int_{-T}^{T} \|f(\sigma)\| d\sigma = 0 \right\}.$$

Similarly, define $PAP_0(\mathbb{R} \times \mathbb{X})$ as the collection of jointly continuous functions $F : \mathbb{R} \times \mathbb{Y} \to \mathbb{X}$ such that $F(\cdot, y)$ is bounded for each $y \in K$ (K being an arbitrary bounded subset of $\mathbb{Y}$) and

$$\lim_{T \to \infty} \frac{1}{2T} \int_{-T}^{T} \|F(s, y)\| ds = 0$$

uniformly in $y \in K$.

Definition 3.3 ([14]). A function $f \in BC(\mathbb{R}, \mathbb{X})$ is called pseudo-almost periodic if it can be expressed as $f = g + \phi$, where $g \in AP(\mathbb{X})$ and $\phi \in PAP_0(\mathbb{X})$. The collection of such functions will be denoted by $PAP(\mathbb{X})$.

Definition 3.4. A function $F \in C(\mathbb{R} \times \mathbb{Y}, \mathbb{X})$ is called pseudo-almost periodic if it can be expressed as $F = G + \Phi$, where $G \in AP(\mathbb{R} \times \mathbb{Y})$ and $\Phi \in PAP_0(\mathbb{R} \times \mathbb{X})$. The collection of such functions will be denoted by $PAP(\mathbb{R} \times \mathbb{X})$.

Theorem 3.5 ([14]). *The space $PAP(\mathbb{X})$ equipped with the sup norm $\|\cdot\|_\infty$ is a Banach space.*

Theorem 3.6 ([1]). *Assume* $f : \mathbb{R}\times\mathbb{Y} \mapsto \mathbb{X}$ *is pseudo-almost periodic and satisfies the Lipschitz condition, that is, there exists* $L > 0$ *such that*

$$\|f(t,u) - f(t,v)\| \le L \cdot \|u - v\|_{\mathbb{Y}} \quad \text{for all } u, v \in \mathbb{Y},\ t \in \mathbb{R}.$$

If $h \in PAP(\mathbb{Y})$, *then* $f(\cdot, h(\cdot)) \in PAP(\mathbb{X})$.

4 Lebesgue Spaces with Variable Exponents $L^{p(x)}$

This section is mainly devoted to the so-called Lebesgue spaces with variable exponents $L^{p(x)}(\mathbb{R}, \mathbb{X})$. Various basic properties of these functions are reviewed. For more on these spaces and related issues we refer to Diening et al. [6].

Let $(\mathbb{X}, \|\cdot\|)$ be a Banach space and let $\Omega \subseteq \mathbb{R}$ be a subset. Let $M(\Omega, \mathbb{X})$ denote the collection of all measurable functions $f : \Omega \mapsto \mathbb{X}$. Let us recall that two functions f and g of $M(\Omega, \mathbb{X})$ are equal whether they are equal almost everywhere. Set $m(\Omega) := M(\Omega, \mathbb{R})$ and fix $p \in m(\Omega)$. Let $\varphi(x,t) = t^{p(x)}$ for all $x \in \Omega$ and $t \ge 0$, and define

$$\rho(u) = \rho_{p(x)}(u) = \int_{\Omega} \varphi(x, \|u(x)\|)dx = \int_{\Omega} \|u(x)\|^{p(x)} dx,$$

$$L^{p(x)}(\Omega, \mathbb{X}) = \left\{ u \in M(\Omega, \mathbb{X}) : \lim_{\lambda \to 0+} \rho(\lambda u) = 0 \right\},$$

$$L^{p(x)}_{OC}(\Omega, \mathbb{X}) = \left\{ u \in L^{p(x)}(\Omega, \mathbb{X}) : \rho(u) < \infty \right\}, \quad \text{and}$$

$$E^{p(x)}(\Omega, \mathbb{X}) = \left\{ u \in L^{p(x)}(\Omega, \mathbb{X}) : \text{for all } \lambda > 0,\ \rho(\lambda u) < \infty \right\}.$$

Note that the space $L^{p(x)}(\Omega, \mathbb{X})$ defined above is a Musielak-Orliez type space while the space $L^{p(x)}_{OC}(\Omega, \mathbb{X})$ is a generalized Orliez type space. Further, the sets $E^{p(x)}(\Omega, \mathbb{X})$ and $L^{p(x)}(\Omega, \mathbb{X})$ are vector subspaces of $M(\Omega, \mathbb{X})$. In addition, $L^{p(x)}_{OC}(\Omega, \mathbb{X})$ is a convex subset of $L^{p(x)}(\Omega, \mathbb{X})$, and the following inclusions hold

$$E^{p(x)}(\Omega, \mathbb{X}) \subset L^{p(x)}_{OC}(\Omega, \mathbb{X}) \subset L^{p(x)}(\Omega, \mathbb{X}).$$

Definition 4.1 ([6]). A convex and left-continuous function $\psi : [0, \infty) \to [0, \infty]$ is called a Φ-function if it satisfies the following conditions,

(a) $\psi(0) = 0$;
(b) $\lim_{t \to 0+} \psi(t) = 0$; and
(c) $\lim_{t \to \infty} \psi(t) = \infty$.

Moreover, ψ is said to be positive whether $\psi(t) > 0$ for all $t > 0$.

Let us mention that if ψ is a Φ-function, then on the set $\{t > 0 : \psi(t) < \infty\}$, the function ψ is of the form

$$\psi(t) = \int_0^t k(t)dt,$$

where $k(\cdot)$ is the right-derivative of $\psi(t)$. Moreover, k is a non-increasing and right-continuous function. For more on these functions and related issues we refer to [6].

Example 4.2. (a) Consider the function $\varphi_p(t) = p^{-1}t^p$ for $1 \leq p < \infty$. It can be shown that φ_p is a Φ-function. Furthermore, the function φ_p is continuous and positive.

(b) It can be shown that the function φ defined above, that is, $\varphi(x,t) = t^{p(x)}$ for all $x \in \mathbb{R}$ and $t \geq 0$ is a Φ-function.

For any $p \in m(\Omega)$, we define

$$p^- := \text{ess}\inf_{x\in\Omega} p(x), \quad p^+ := \text{ess}\sup_{x\in\Omega} p(x).$$

Define

$$C_+(\Omega) := \left\{ p \in m(\Omega) : 1 < p^- \leq p(x) \leq p^+ < \infty, \text{ for each } x \in \Omega \right\}.$$

Let $p \in C_+(\Omega)$. Using similar argument as in [6, Theorem 3.4.1], it can be shown that

$$E^{p(x)}(\Omega, \mathbb{X}) = L_{OC}^{p(x)}(\Omega, \mathbb{X}) = L^{p(x)}(\Omega, \mathbb{X}).$$

In view of the above, we define the Lebesgue space with variable exponents $L^{p(x)}(\Omega, \mathbb{X})$ with $p \in C_+(\Omega)$, by

$$L^{p(x)}(\Omega, \mathbb{X}) := \left\{ u \in M(\Omega, \mathbb{X}) : \int_\Omega \|u(x)\|^{p(x)} dx < \infty \right\}.$$

Define, for each $u \in L^{p(x)}(\Omega, \mathbb{X})$,

$$\|u\|_{p(x)} := \inf \left\{ \lambda > 0 : \int_\Omega \left\| \frac{u(x)}{\lambda} \right\|^{p(x)} dx \leq 1 \right\}.$$

It can be shown that $\|\cdot\|_{p(x)}$ is a norm upon $L^{p(x)}(\Omega, \mathbb{X})$, which is referred to as the *Luxemburg norm*.

Remark 43. Let $p \in C_+(\Omega)$. If p is constant, then the space $L^{p(\cdot)}(\Omega, \mathbb{X})$, as defined above, coincides with the usual space $L^p(\Omega, \mathbb{X})$.

We now establish some of the basic properties of these spaces. For more on these functions and related issues we refer to [6].

Proposition 4.4 ([6]). *Let $p \in C_+(\Omega)$ and let $u, u_k, v \in M(\Omega, \mathbb{X})$ for $k = 1, 2, \ldots$. Then the following statements hold,*

(a) *If $u_k \to u$ a.e., then $\rho_p(u) \leq \lim_{k\to\infty} \inf(\rho_p(u_k))$.*
(b) *If $\|u_k\| \to \|u\|$ a.e., then $\rho_p(u) = \lim_{k\to\infty} \rho_p(u_k)$.*
(c) *If $u_k \to u$ a.e., $\|u_k\| \leq \|v\|$ and $v \in E^{p(x)}(\Omega, \mathbb{X})$, then $u_k \to u$ in $L^{p(x)}(\Omega, \mathbb{X})$.*

Proof. (a) Since $\varphi(x, \cdot)$ is a Φ-function, then by [6, Lemma 2.3.4], the function $\varphi(x, \cdot)$ is lower semicontinuous. Thus the usual Fatou's Lemma yields,

$$\begin{aligned}
\rho_p(u) &= \rho_p\left(\lim_{k\to\infty} \inf(u_k)\right) \\
&= \int_\Omega \left\| \lim_{k\to\infty} \inf(u_k) \right\|^{p(x)} dx \\
&\leq \lim_{k\to\infty} \inf \int_\Omega \|u_k\|^{p(x)} dx \\
&= \lim_{k\to\infty} \inf \rho_p(u_k).
\end{aligned}$$

(b) Let $\|u_k\| \to \|u\|$ *a.e.*. Then by the left-continuity and monotonicity of $\varphi(x, \cdot)$ we have

$$0 \leq \varphi(\cdot, \|u_k(\cdot)\|) \to \varphi(\cdot, \|u(\cdot)\|).$$

Thus the usual Theorem of Monotone Convergence yields,

$$\begin{aligned}
\rho_p(u) &= \int_\Omega \left\| \lim_{k\to\infty} (u_k) \right\|^{p(x)} dx \\
&= \int_\Omega \lim_{k\to\infty} \|u_k\|^{p(x)} dx \\
&= \lim_{k\to\infty} \int_\Omega \|u_k\|^{p(x)} dx \\
&= \lim_{k\to\infty} \rho_p(u_k).
\end{aligned}$$

(c) Let $u_k \to u$ a.e., $\|u_k\| \leq \|v\|$ and $v \in E^{p(x)}(\Omega, \mathbb{X})$. Then $\|u_k - u\| \to 0$ *a.e.*, $\|u\| \leq \|v\|$ and $\|u_k - u\| \leq 2\|v\|$. Since $\rho_p(2\lambda v) < \infty$, then using the usual Theorem of Dominated Convergence, we conclude that

$$\lim_{k\to\infty} \rho_p(\lambda\|u_k - u\|) = \int_\Omega \lim_{k\to\infty} \Big(\lambda\|u_k - u\|\Big)^{p(x)} dx = 0.$$

Now since $\lambda > 0$ was arbitrary, then [6, Lemma 2.1.9] yields $u_k \to u$ in $L^{p(x)}(\Omega, \mathbb{X})$.

□

Proposition 4.5 ([6,13]). *Let $p \in C_+(\Omega)$. If $u, v \in L^{p(x)}(\Omega, \mathbb{X})$, then the following properties hold,*

(a) $\|u\|_{p(x)} \geq 0$, *with equality if and only if* $u = 0$.
(b) $\rho_p(u) \leq \rho_p(v)$ *and* $\|u\|_{p(x)} \leq \|v\|_{p(x)}$ *if* $\|u\| \leq \|v\|$.
(c) $\rho_p(u\|u\|_{p(x)}^{-1}) = 1$ *if* $u \neq 0$.
(d) $\rho_p(u) \leq 1$ *if and only if* $\|u\|_{p(x)} \leq 1$.
(e) *If* $\|u\|_{p(x)} \leq 1$, *then*

$$\Big[\rho_p(u)\Big]^{\frac{1}{p^-}} \leq \|u\|_{p(x)} \leq \Big[\rho_p(u)\Big]^{\frac{1}{p^+}}.$$

(f) *If* $\|u\|_{p(x)} \geq 1$, *then*

$$\Big[\rho_p(u)\Big]^{\frac{1}{p^+}} \leq \|u\|_{p(x)} \leq \Big[\rho_p(u)\Big]^{\frac{1}{p^-}}.$$

Proposition 4.6 ([6]). *Let $p \in C_+(\Omega)$ and let $u, u_k, v \in M(\Omega, \mathbb{X})$ for $k = 1, 2, \ldots$. Then the following statements hold:*

(a) *If* $u \in L^{p(x)}(\Omega, \mathbb{X})$ *and* $0 \leq \|v\| \leq \|u\|$, *then* $v \in L^{p(x)}(\Omega, \mathbb{X})$ *and* $\|v\|_{p(x)} \leq \|u\|_{p(x)}$.
(b) *If* $u_k \to u$ a.e., *then* $\|u\|_{p(x)} \leq \lim_{k\to\infty} \inf(\|u_k\|_{p(x)})$.
(c) *If* $\|u_k\| \to \|u\|$ a.e. *with* $u_k \in L^{p(x)}(\Omega, \mathbb{X})$ *and* $\sup_k \|u_k\|_{p(x)} < \infty$, *then* $u \in L^{p(x)}(\mathbb{R}, \mathbb{X})$ *and* $\|u_k\|_{p(x)} \to \|u\|_{p(x)}$.

Using similar arguments as in Fan et al. [9], we obtain the following:

Proposition 4.7. *If $u, u_n \in L^{p(x)}(\Omega, \mathbb{X})$ for $k = 1, 2, \ldots$, then the following statements are equivalent:*

(a) $\lim_{k\to\infty} \|u_k - u\|_{p(x)} = 0$;
(b) $\lim_{k\to\infty} \rho_p(u_k - u) = 0$;
(c) $u_k \to u$ *and* $\lim_{k\to\infty} \rho_p(u_k) = \rho_p(u)$.

Theorem 4.8 ([6,9]). *The space $(L^{p(x)}(\Omega, \mathbb{X}), \|\cdot\|_{p(x)})$ is a Banach space that is separable and uniform convex. Its topological dual is $L^{q(x)}(\Omega, \mathbb{X})$, where $p^{-1}(x) + q^{-1}(x) = 1$. Moreover, for any $u \in L^{p(x)}(\Omega, \mathbb{X})$ and $v \in L^{q(x)}(\Omega, \mathbb{R})$, we have*

$$\left\| \int_{\Omega} uv dx \right\| \leq \left(\frac{1}{p^-} + \frac{1}{q^-} \right) \|u\|_{p(x)} \cdot |v|_{q(x)}.$$

Define

$$D_+(\Omega) := \left\{ p \in m(\Omega) : 1 \leq p^- \leq p(x) \leq p^+ < \infty, \text{ for each } x \in \Omega \right\}.$$

Corollary 4.9 ([13]). *Let $p, r \in D_+(\Omega)$. If the function q defined by the equation*

$$\frac{1}{q(x)} = \frac{1}{p(x)} + \frac{1}{r(x)}$$

is in $D_+(\Omega)$, then there exists a constant $C = C(p, r) \in [1, 5]$ such that

$$\|uv\|_{q(x)} \leq C \|u\|_{p(x)} \cdot |v|_{r(x)},$$

for every $u \in L^{p(x)}(\Omega, \mathbb{X})$ and $v \in L^{r(x)}(\Omega, \mathbb{R})$.

Corollary 4.10 ([6]). *Let $mes(\Omega) < \infty$ where mes stands for the Lebesgue measure and $p(x), q(x) \in D_+(\Omega)$. If $q(\cdot) \leq p(\cdot)$ almost everywhere in Ω, then the embedding $L^{p(x)}(\Omega, \mathbb{X}) \hookrightarrow L^{q(x)}(\Omega, \mathbb{X})$ is continuous whose norm does not exceed $2(mes(\Omega) + 1)$.*

5 Stepanov-Like Pseudo-Almost Periodic Functions with Variable Exponents

Definition 5.1 ([4]). The Bochner transform $f^b(t, s)$, $t \in \mathbb{R}$, $s \in [0, 1]$ of a function $f : \mathbb{R} \to \mathbb{X}$ is defined by $f^b(t, s) := f(t + s)$.

Remark 5.2. (i) A function $\varphi(t, s)$, $t \in \mathbb{R}$, $s \in [0, 1]$, is the Bochner transform of a certain function f, $\varphi(t, s) = f^b(t, s)$, if and only if $\varphi(t + \tau, s - \tau) = \varphi(s, t)$ for all $t \in \mathbb{R}$, $s \in [0, 1]$ and $\tau \in [s - 1, s]$.
(ii) Note that if $f = h + \varphi$, then $f^b = h^b + \varphi^b$. Moreover, $(\lambda f)^b = \lambda f^b$ for each scalar λ.

Definition 5.3 ([4]). The Bochner transform $F^b(t, s, u)$, $t \in \mathbb{R}$, $s \in [0, 1]$, $u \in \mathbb{X}$ of a function $F(t, u)$ on $\mathbb{R} \times \mathbb{X}$, with values in $\mathbb{X}$, is defined by $F^b(t, s, u) := F(t+s, u)$ for each $u \in \mathbb{X}$.

Definition 5.4. Let $p \in [1, \infty)$. The space $BS^p(\mathbb{X})$ of all Stepanov bounded functions, with the exponent p, consists of all measurable functions f on $\mathbb{R}$ with values in $\mathbb{X}$ such that $f^b \in L^\infty\big(\mathbb{R}, L^p((0, 1), \mathbb{X})\big)$. This is a Banach space with the norm

$$\|f\|_{S^p} = \|f^b\|_{L^\infty(\mathbb{R},L^p)} = \sup_{t\in\mathbb{R}}\left(\int_t^{t+1}\|f(\tau)\|^p\,d\tau\right)^{1/p}.$$

Note that for each $p \geq 1$, we have the following continuous inclusion:

$$(BC(\mathbb{X}), \|\cdot\|_\infty) \hookrightarrow (BS^p(\mathbb{X}), \|\cdot\|_{S^p}).$$

We introduce

Definition 5.5. Let $p \in C_+(\mathbb{R})$. The space $BS^{p(x)}(\mathbb{X})$ consists of all functions $f \in M(\mathbb{R},\mathbb{X})$ such that $\|f\|_{S^{p(x)}} < \infty$, where

$$\begin{aligned}\|f\|_{S^{p(x)}} &= \sup_{t\in\mathbb{R}}\left[\inf\left\{\lambda > 0 : \int_0^1 \left\|\frac{f(x+t)}{\lambda}\right\|^{p(x+t)} dx \leq 1\right\}\right]\\ &= \sup_{t\in\mathbb{R}}\left[\inf\left\{\lambda > 0 : \int_t^{t+1} \left\|\frac{f(x)}{\lambda}\right\|^{p(x)} dx \leq 1\right\}\right].\end{aligned}$$

Note that the space $\left(BS^{p(x)}(\mathbb{X}), \|\cdot\|_{S^{p(x)}}\right)$ is a Banach space, which, depending on $p(\cdot)$, may or may not be translation-invariant.

Definition 5.6. If $p, q \in C_+(\mathbb{R})$, we then define the space $BS^{p(x),q(x)}(\mathbb{X})$ as follows:

$$\begin{aligned}BS^{p(x),q(x)}(\mathbb{X}) &:= BS^{p(x)}(\mathbb{X}) + BS^{q(x)}(\mathbb{X})\\ &= \left\{f = h+\varphi \in M(\mathbb{R},\mathbb{X}) : h \in BS^{p(x)}(\mathbb{X}) \text{ and } \varphi \in BS^{q(x)}(\mathbb{X})\right\}.\end{aligned}$$

We equip $BS^{p(x),q(x)}(\mathbb{X})$ with the norm $\|\cdot\|_{S^{p(x),q(x)}}$ defined by

$$\|f\|_{S^{p(x),q(x)}} := \inf\left\{\|h\|_{S^{p(x)}} + \|\varphi\|_{S^{q(x)}} : \ f = h + \varphi\right\}.$$

Clearly, $\left(BS^{p(x),q(x)}(\mathbb{X}), \|\cdot\|_{S^{p(x),q(x)}}\right)$ is a Banach space, which, depending on both $p(\cdot)$ and $q(\cdot)$, may or may not be translation-invariant.

Lemma 5.7. *Let $p, q \in C_+(\mathbb{R})$. Then the following continuous inclusion holds,*

$$\left(BC(\mathbb{R},\mathbb{X}), \|\cdot\|_\infty\right) \hookrightarrow \left(BS^{p(x)}(\mathbb{X}), \|\cdot\|_{S^{p(x)}}\right) \hookrightarrow \left(BS^{p(x),q(x)}(\mathbb{X}), \|\cdot\|_{S^{p(x),q(x)}}\right).$$

Proof. The fact that $\left(BS^{p(x)}(\mathbb{X}), \|\cdot\|_{S^{p(x)}}\right) \hookrightarrow \left(BS^{p(x),q(x)}(\mathbb{X}), \|\cdot\|_{S^{p(x),q(x)}}\right)$ is obvious. Thus we will only show that $\left(BC(\mathbb{R},\mathbb{X}), \|\cdot\|_\infty\right) \hookrightarrow \Big(BS^{p(x)}(\mathbb{X}), \|\cdot$

$\|_{S^{p(x)}}\big)$. Indeed, let $f \in BC(\mathbb{R},\mathbb{X}) \subset M(\mathbb{R},\mathbb{X})$. If $\|f\|_\infty = 0$, which yields $f = 0$, then there is nothing to prove. Now suppose that $\|f\|_\infty \neq 0$. Using the facts that $0 < \left\|\frac{f(x)}{\|f\|_\infty}\right\| \leq 1$ and that $p \in C_+(\mathbb{R})$ it follows that for every $t \in \mathbb{R}$,

$$\int_t^{t+1} \left\|\frac{f(x)}{\|f\|_\infty}\right\|^{p(x)} dx \leq \int_t^{t+1} 1^{p(x)} dx = 1,$$

and hence $\|f\|_\infty \in \left\{\lambda > 0 : \int_t^{t+1} \left\|\frac{f(x)}{\lambda}\right\|^{p(x)} dx \leq 1\right\}$, which yields

$$\inf\left\{\lambda > 0 : \int_t^{t+1} \left\|\frac{f(x)}{\lambda}\right\|^{p(x)} dx \leq 1\right\} \leq \|f\|_\infty.$$

Therefore, $\|f\|_{S^{p(x)}} \leq \|f\|_\infty < \infty$. This shows that not only $f \in (BS^{p(x)}(\mathbb{X})), \|\cdot\|_{S^{p(x)}})$ but also the injection $(BC(\mathbb{R},\mathbb{X}), \|\cdot\|_\infty) \hookrightarrow (BS^{p(x)}(\mathbb{X}), \|\cdot\|_{S^{p(x)}})$ is continuous. □

Definition 5.8. Let $p \geq 1$ be a constant. A function $f \in BS^p(\mathbb{X})$ is said to be S^p-almost periodic (or Stepanov-like almost periodic) if $f^b \in AP\big(L^p((0,1),\mathbb{X})\big)$. That is, for each $\varepsilon > 0$ there exists $l(\varepsilon) > 0$ such that every interval of length $l(\varepsilon)$ contains a number τ with the property that

$$\sup_{t\in\mathbb{R}} \left(\int_0^1 \left\|f^b(t+\tau,s) - f^b(t,s)\right\|^p ds\right)^{1/p} = \sup_{t\in\mathbb{R}} \left(\int_t^{t+1} \left\|f(s+\tau) - f(s)\right\|^p ds\right)^{1/p} < \varepsilon.$$

The collection of such functions will be denoted by $S^p_{ap}(\mathbb{X})$.

Remark 5.9. There are some difficulties in defining $S^{p(x)}_{ap}(\mathbb{X})$ for a function $p \in C_+(\mathbb{R})$ that is not necessarily constant. This is mainly due to the fact that the space $BS^{p(x)}(\mathbb{X})$ is not always translation-invariant. In other words, the quantities $f^b(t+\tau,s)$ and $f^b(t,s)$ (for $t \in \mathbb{R}$, $s \in [0,1]$) that are used in the definition of S^p-almost periodicity, do not belong to the same space, unless p is constant.

We now introduce the concept of $S^{p,q(x)}$-pseudo-almost periodicity that obviously generalizes that of S^p-pseudo-almost periodicity.

Definition 5.10. Let $p \geq 1$ be a constant and let $q \in C_+(\mathbb{R})$. A function $f \in BS^{p,q(x)}(\mathbb{X})$ is said to be $S^{p,q(x)}$-pseudo-almost periodic (or Stepanov-like pseudo-almost periodic with variable exponents $p, q(x)$) if it can be decomposed as

$$f = h + \varphi,$$

where $h \in S^p_{ap}(\mathbb{X})$ and $\varphi \in S^{q(x)}_{pap_0}(\mathbb{X})$ with $S^{q(x)}_{pap_0}(\mathbb{X})$ being the space of all $\psi \in BS^{q(x)}(\mathbb{X})$ such that

$$\lim_{T\to\infty}\frac{1}{2T}\int_{-T}^{T}\inf\left\{\lambda>0:\int_t^{t+1}\left\|\frac{\psi(x)}{\lambda}\right\|^{q(x)}dx\le 1\right\}dt=0.$$

The collection of $S^{p,q(x)}$-pseudo-almost periodic functions will be denoted $S^{p,q(x)}_{pap}(\mathbb{X})$.

Proposition 5.11. *Let $p \geq 1$ be a constant and let $q \in C_+(\mathbb{R})$. If $f \in PAP(\mathbb{X})$, then f is $S^{p,q(x)}$-pseudo-almost periodic.*

Proof. Let $f \in PAP(\mathbb{X})$. Thus there exist two functions $h, \varphi : \mathbb{R} \to \mathbb{X}$ such that

$$f = h + \varphi,$$

where $h \in AP(\mathbb{X})$ and $\varphi \in PAP_0(\mathbb{X})$. We first show that $h \in S^p_{ap}(\mathbb{X})$. Indeed, since $h \in AP(\mathbb{X})$, for each $\varepsilon > 0$ there exists $l(\varepsilon) > 0$ such that every interval of length $l(\varepsilon)$ contains a number τ with the property that

$$\|h(t+\tau)-h(t)\| < \varepsilon$$

for each $t \in \mathbb{R}$.

Now

$$\int_t^{t+1}\left\|h(s+\tau)-h(s)\right\|^p ds \le \int_t^{t+1}\varepsilon^p dx = \varepsilon^p$$

for all $t \in \mathbb{R}$, which means that

$$\|h(t+\tau)-h(t)\|_{S^p} \le \varepsilon,$$

that is, $h^b \in AP\big(L^p((0,1),\mathbb{X})\big)$.

To complete the proof, we need to show that $\varphi^b \in PAP_0(L^{q^b(x)}((0,1),\mathbb{X}))$. Using (e)–(f) of Proposition 4.5 and the usual Hölder inequality, it follows that

$$\begin{aligned}
&\int_{-T}^{T}\inf\left\{\lambda>0:\int_0^1\left\|\frac{\varphi(x+t)}{\lambda}\right\|^{q(x+t)}dx\le 1\right\}dt\\
&\le \int_{-T}^{T}\left(\int_0^1\|\varphi(t+x)\|^{q(t+x)}\,dx\right)^{\gamma}dt\\
&\le (2T)^{1-\gamma}\left[\int_{-T}^{T}\left(\int_0^1\|\varphi(t+x)\|^{q(t+x)}\,dx\right)dt\right]^{\gamma}
\end{aligned}$$

$$
\begin{aligned}
&\leq (2T)^{1-\gamma}\Bigg[\int_{-T}^{T}\Bigg(\int_0^1 \|\varphi(t+x)\|.\,\|\varphi\|_\infty^{q(t+x)-1}\,dx\Bigg)\,dt\Bigg]^{\gamma}\\
&\leq (2T)^{1-\gamma}\Big(\|\varphi\|_\infty+1\Big)^{\frac{q^+-1}{\gamma}}\Bigg[\int_{-T}^{T}\Bigg(\int_0^1 \|\varphi(t+x)\|\,dx\Bigg)\,dt\Bigg]^{\gamma}\\
&= (2T)^{1-\gamma}\Big(\|\varphi\|_\infty+1\Big)^{\frac{q^+-1}{\gamma}}\Bigg[\int_0^1\Bigg(\int_{-T}^{T} \|\varphi(t+x)\|\,dt\Bigg)\,dx\Bigg]^{\gamma}\\
&= (2T)\Big(\|\varphi\|_\infty+1\Big)^{\frac{q^+-1}{\gamma}}\Bigg[\int_0^1\Bigg(\frac{1}{2T}\int_{-T}^{T} \|\varphi(t+x)\|\,dt\Bigg)\,dx\Bigg]^{\gamma},
\end{aligned}
$$

where

$$
\gamma = \begin{cases} \frac{1}{q^+} & \text{if } \|\varphi\| < 1,\\[2ex] \frac{1}{q^-} & \text{if } \|\varphi\| \geq 1. \end{cases}
$$

Using the fact that $PAP_0(\mathbb{X})$ is translation invariant and the (usual) Dominated Convergence Theorem, it follows that

$$
\begin{aligned}
&\lim_{T\to\infty}\frac{1}{2T}\int_{-T}^{T}\inf\Bigg\{\lambda>0:\int_0^1\Big\|\frac{\varphi(x+t)}{\lambda}\Big\|^{q(x+t)}dx\leq 1\Bigg\}\,dt\\
&\leq \Big(\|\varphi\|_\infty+1\Big)^{\frac{q^+-1}{\gamma}}\Bigg[\int_0^1\Bigg(\lim_{T\to\infty}\frac{1}{2T}\int_{-T}^{T}\|\varphi(t+x)\|\,dt\Bigg)\,dx\Bigg]^{\gamma}=0.
\end{aligned}
$$

□

Definition 5.12. Let $p \geq 1$ and $q \in C_+(\mathbb{R})$. A function $F : \mathbb{R}\times\mathbb{X} \to \mathbb{X}$ with $F(.,u) \in B^{p,q(x)}(\mathbb{X})$ for each $u \in \mathbb{X}$, is said to be $S^{p,q(x)}$-pseudo-almost periodic in $t \in \mathbb{R}$ uniformly in $u \in \mathbb{X}$ if $t \mapsto F(t,u)$ is $S^{p,q(x)}$-pseudo-almost periodic for each $u \in B$ where $B \subset \mathbb{X}$ is an arbitrary bounded set. This means, there exist two functions $G, H : \mathbb{R}\times\mathbb{X} \to \mathbb{X}$ such that $F = G + H$, where $G^b \in AP(\mathbb{R} \times L^p((0,1),\mathbb{X}))$ and $H^b \in PAP_0(\mathbb{R}\times L^{q^b(x)}((0,1),\mathbb{X}))$, that is,

$$
\lim_{T\to\infty}\frac{1}{2T}\int_{-T}^{T}\inf\Bigg\{\lambda>0:\int_0^1\Big\|\frac{H(x+t,u)}{\lambda}\Big\|^{q(x+t)}dx\leq 1\Bigg\}\,dt=0,
$$

uniformly in $u \in B$ where $B \subset \mathbb{X}$ is an arbitrary bounded set.

The collection of such functions will be denoted by $S_{pap}^{p,q(x)}(\mathbb{R}\times\mathbb{X})$.

Let $Lip^r(\mathbb{R},\mathbb{X})$ denote the set of functions $f : \mathbb{R}\times\mathbb{X}\to\mathbb{X}$ satisfying: there exists a nonnegative function $L_f \in L^r(\mathbb{R})$ such that

$$\|f(t,u)-f(t,v)\| \le L_f(t)\|u-v\| \text{ for all } u,v\in\mathbb{X},\ t\in\mathbb{R}. \tag{3}$$

Now, we recall the following composition theorem for S^p_{ap} functions.

Theorem 5.13 ([12]). *Let $p > 1$ be a constant. We suppose that the following conditions hold:*

(a) $f \in S^p_{ap}(\mathbb{R}\times\mathbb{X}) \cap Lip^r(\mathbb{R},\mathbb{X})$ *with* $r \ge \max\{p, \frac{p}{p-1}\}$.
(b) $\phi \in S^p_{ap}(\mathbb{X})$ *and there exists a set* $E \subset \mathbb{R}$ *with* $mes\,(E) = 0$ *such that*

$$K := \overline{\{\phi(t) : t\in\mathbb{R}\setminus E\}}$$

is compact in $\mathbb{X}$.

Then there exists $m \in [1,p)$ *such that* $f(\cdot,\phi(\cdot)) \in S^m_{ap}(\mathbb{R}\times\mathbb{X})$.

Lemma 5.14 ([12]). *Let $q > 1$ be a constant and let $K \subseteq \mathbb{X}$ be compact subset. If $f \in Lip^q(\mathbb{R},\mathbb{X})$ and $f^b \in PAP_0(L^q((0,1),\mathbb{X}))$, then $\tilde{f} \in PAP_0(\mathbb{R})$, where the function $\tilde{f}$ is defined by*

$$\tilde{f}(t) := \left\| \sup_{u\in K} \|f(t+\cdot,u)\| \right\|_q \tag{4}$$

for all $t\in\mathbb{R}$.

Theorem 5.15. *Let $p,q > 1$ be constants such that $p \le q$. Suppose that the following conditions hold:*

(a) $f = g + h \in S^{p,q}_{pap}(\mathbb{R}\times\mathbb{X})$ *with* $g^b \in AP(\mathbb{R}\times L^p((0,1),\mathbb{X}))$ *and* $h^b \in PAP_0(\mathbb{R}\times L^q((0,1),\mathbb{X}))$. *Further,* $f,g \in Lip^r(\mathbb{R},\mathbb{X})$ *with* $r \ge \max\{q, \frac{p}{p-1}\}$.
(b) $\phi = \alpha + \beta \in S^{p,q}_{pap}(\mathbb{X})$ *with* $\alpha \in S^p_{ap}(\mathbb{X})$ *and* $\beta \in S^q_{pap_0}(\mathbb{X})$, *and there exists a set* $E \subset \mathbb{R}$ *with* $mes\,(E) = 0$ *such that*

$$K := \overline{\{\alpha(t) : t\in\mathbb{R}\setminus E\}}$$

is compact in $\mathbb{X}$.

Then there exists $m \in [1,p)$ *such that* $f(\cdot,\phi(\cdot)) \in S^{m,m}_{pap}(\mathbb{R}\times\mathbb{X})$.

Proof. The proof is a sequel of Lemma 5.14 and Theorem 5.13. Indeed, decompose f^b as follows:

$$f^b(\cdot,\phi^b(\cdot)) = g^b(\cdot,\alpha^b(\cdot)) + f^b(\cdot,\phi^b(\cdot)) - f^b(\cdot,\alpha^b(\cdot)) + h^b(\cdot,\alpha^b(\cdot)).$$

Using Theorem 5.13, it easily follows that there exists $m \in [1, p)$ with $\frac{1}{m} = \frac{1}{p} + \frac{1}{r}$ such that $g^b(\cdot, \alpha^b(\cdot)) \in AP(\mathbb{R} \times L^m((0,1), \mathbb{X}))$.

Set

$$\varphi^b(\cdot) = f^b(\cdot, \phi^b(\cdot)) - f^b(\cdot, \alpha^b(\cdot)).$$

Clearly, $\varphi^b \in PAP_0(L^m((0,1), \mathbb{X}))$. Indeed, for $T > 0$,

$$\frac{1}{2T}\int_{-T}^{T}\left(\int_0^1 \|\varphi^b(t+s)\|^m ds\right)^{\frac{1}{m}} dt$$

$$= \frac{1}{2T}\int_{-T}^{T}\left(\int_0^1 \|f^b(t+s, \phi^b(t+s)) - f^b(t+s, \alpha^b(t+s))\|^m ds\right)^{\frac{1}{m}} dt$$

$$\leq \frac{1}{2T}\int_{-T}^{T}\left(\int_0^1 \left(L_f^b(t+s).\,\|\beta^b(t+s)\|\right)^m ds\right)^{\frac{1}{m}} dt$$

$$\leq \|L_f^b\|_{S^r}.\left[\frac{1}{2T}\int_{-T}^{T}\left(\int_0^1 \|\beta^b(t+s)\|^p ds\right)^{\frac{1}{p}} dt\right]$$

$$\leq \|L_f^b\|_{S^r}.\left[\frac{1}{2T}\int_{-T}^{T}\left(\int_0^1 \|\beta^b(t+s)\|^q ds\right)^{\frac{1}{q}} dt\right].$$

Using the fact that $\beta^b \in PAP_0(L^q((0,1), \mathbb{X}))$, it follows that $\varphi^b \in PAP_0(L^m((0,1), \mathbb{X}))$.

Now using the fact that $h = f - g \in Lip^r(\mathbb{R}, \mathbb{X}) \subset Lip^q(\mathbb{R}, \mathbb{X})$, it follows by Lemma 5.14 that

$$\lim_{T\to\infty} \frac{1}{2T}\int_{-T}^{T} \left\| \sup_{u\in K} \|h(t + \cdot, u)\| \right\|_q dt = 0,$$

which yields

$$\frac{1}{2T}\int_{-T}^{T}\left(\int_0^1 \|h^b(t+s, \alpha^b(t+s))\|^m ds\right)^{\frac{1}{m}} dt$$

$$\leq \frac{1}{2T}\int_{-T}^{T}\left(\int_0^1 \|h^b(t+s, \alpha^b(t+s))\|^q ds\right)^{\frac{1}{q}} dt$$

$$\leq \frac{1}{2T}\int_{-T}^{T}\left(\int_0^1 \left(\sup_{u\in K}\|h^b(t+s, u)\|\right)^q ds\right)^{\frac{1}{q}} dt \to 0 \quad \text{as} \quad T \to \infty,$$

which means that $h^b(\cdot, \alpha^b(\cdot)) \in PAP_0(L^m(0,1;\mathbb{X}))$. This completes the proof. $\square$

Remark 5.16. A general composition theorem in $S_{pap}^{p,q(x)}(\mathbb{R} \times \mathbb{X})$ is unlikely as compositions of elements of $S_{pap}^{p,q(x)}(\mathbb{R} \times \mathbb{X})$ may not be well-defined unless $q(\cdot)$ is the constant function.

6 Existence of Pseudo-Almost Periodic Solutions

Let $p, q > 1$ be constants such that $p \leq q$. This section is devoted to the search of a pseudo-almost periodic solution to the first-order differential equation (1). Throughout the rest of the paper we suppose that:

(A.1) The operator A is the infinitesimal generator of an exponentially stable C_0-semigroup $(T(t))_{t \geq 0}$, that is, there exist constants $M, \omega > 0$ such that

$$\|T(t)\| \leq M e^{-\omega t}$$

for each $t \geq 0$.

(A.2) $F = G + H \in S_{pap}^{p,q}(\mathbb{R} \times \mathbb{X}) \cap C(\mathbb{R} \times \mathbb{X})$ with $G^b \in AP(\mathbb{R} \times L^p((0,1),\mathbb{X}))$ and $H^b \in PAP_0(\mathbb{R} \times L^q((0,1),\mathbb{X}))$. Moreover; $F, G \in Lip^r(\mathbb{R}, \mathbb{X})$ with

$$r \geq \max\left\{q, \frac{p}{p-1}\right\}.$$

Definition 6.1. Under (A.1), if $f : \mathbb{R} \to \mathbb{X}$ is a bounded continuous function, then a mild solution to Eq. (2) is a continuous function $u : \mathbb{R} \to \mathbb{X}$ satisfying

$$u(t) = T(t-s)u(s) + \int_s^t T(t-\sigma)f(\sigma)d\sigma \tag{5}$$

for all $t, s \in \mathbb{R}$ and $t \geq s$.

Definition 6.2. Suppose (A.1) holds. If $F : \mathbb{R} \times \mathbb{X} \to \mathbb{X}$ is a bounded continuous function, then a mild solution to Eq. (1) is a continuous function $u : \mathbb{R} \to \mathbb{X}$ satisfying

$$u(t) = T(t-s)u(s) + \int_s^t T(t-\sigma)F(\sigma, u(\sigma))d\sigma \tag{6}$$

for all $t, s \in \mathbb{R}$ and $t \geq s$.

Theorem 6.3. *Let $\theta > 1$ be a constant and let $\vartheta \in C_+(\mathbb{R})$. Suppose that (A.1) holds. If $f \in S_{paa}^{\theta,\vartheta(x)}(\mathbb{X}) \cap C(\mathbb{R},\mathbb{X})$, then the Eq. (2) has a unique pseudo-almost periodic (mild) solution given by*

$$u(t) = \int_{-\infty}^{t} T(t-s)f(s)ds. \tag{7}$$

Proof. Define the function $u : \mathbb{R} \mapsto \mathbb{X}$ by

$$u(t) = \int_{-\infty}^{t} T(t-s)f(s)ds, \; t \in \mathbb{R}. \tag{8}$$

It is easy to check that u given in Eq. (8) satisfies Eq. (5) and hence it is a mild solution. Let us now show that $u \in PAP(\mathbb{X})$. Indeed, since $f \in S_{pap}^{p,q(x)}(\mathbb{X}) \cap C(\mathbb{R}, \mathbb{X})$, then $f = h + \varphi$, where $h^b \in AP\big(L^{\theta}((0,1), \mathbb{X})\big)$ and $\varphi^b \in PAP_0(L^{\vartheta^b(x)}((0,1), \mathbb{X})\big)$. Then u can be decomposed as $u(t) = X(t) + Y(t)$, where

$$X(t) = \int_{-\infty}^{t} T(t-s)h(s)ds, \text{ and } Y(t) = \int_{-\infty}^{t} T(t-s)\varphi(s)ds.$$

The proof that $X \in AP(\mathbb{X})$ is obvious and hence is omitted. See, e.g. [3, Theorem 3.2]. To prove that $Y \in PAP_0(\mathbb{X})$, we define for all $n = 1, 2, \ldots.$, the sequence of integral operators

$$Y_n(t) := \int_{n-1}^{n} T(s)\varphi(t-s)ds = \int_{t-n}^{t-n+1} T(t-s)\varphi(s)ds.$$

for each $t \in \mathbb{R}$.

Let $d \in m(\mathbb{R})$ such that $d^{-1}(x) + \vartheta^{-1}(x) = 1$. Using Hölder inequality (Theorem 4.8), it follows that

$$\begin{aligned}
\|Y_n(t)\| &\le M \int_{t-n}^{t-n+1} e^{-\omega(t-s)} \|\varphi(s)\| ds \\
&\le M\Big(\frac{1}{d^-} + \frac{1}{\vartheta^-}\Big)\Bigg[\inf\Bigg\{\lambda > 0 : \int_{t-n}^{t-n+1} \Big(\frac{e^{-\omega(t-s)}}{\lambda}\Big)^{d(s)} ds \le 1\Bigg\}\Bigg] \\
&\times \Bigg[\inf\Bigg\{\lambda > 0 : \int_{t-n}^{t-n+1} \Big\|\frac{\varphi(s)}{\lambda}\Big\|^{\vartheta(s)} ds \le 1\Bigg\}\Bigg].
\end{aligned}$$

Now since

$$\begin{aligned}
\int_{t-n}^{t-n+1} \Big[\frac{e^{-\omega(t-s)}}{e^{-\omega(n-1)}}\Big]^{d(s)} ds &= \int_{t-n}^{t-n+1} \Big[e^{\omega(s-t+n-1)}\Big]^{d(s)} ds \\
&\le \int_{t-n}^{t-n+1} \Big[1\Big]^{d(s)} ds
\end{aligned}$$

$$\leq 1$$

it follows that $e^{-\omega(n-1)} \in \left\{\lambda > 0 : \int_{t-n}^{t-n+1} \left(\frac{e^{-\omega(t-s)}}{\lambda}\right)^{d(s)} ds \leq 1\right\}$, which shows that

$$\left[\inf\left\{\lambda > 0 : \int_{t-n}^{t-n+1} \left(\frac{e^{-\omega(t-s)}}{\lambda}\right)^{d(s)} ds \leq 1\right\}\right] \leq e^{-\omega(n-1)}.$$

Consequently,

$$\|Y_n(t)\| \leq M\left(\frac{1}{d^-} + \frac{1}{q^-}\right) e^{-\omega(n-1)} \|\varphi\|_{S^{\vartheta(x)}}$$

Since the series

$$\sum_{n=1}^{\infty} (e^{-\omega(n-1)})$$

is convergent, we deduce from the well-known Weierstrass test that the series

$$\sum_{k=1}^{\infty} Y_n(t)$$

is uniformly convergent on $\mathbb{R}$. Furthermore,

$$Y(t) = \sum_{n=1}^{\infty} Y_n(t),$$

$Y \in C(\mathbb{R}, \mathbb{X})$, and

$$\|Y(t)\| \leq \sum_{n=1}^{\infty} \|Y_n(t)\| \leq K_1 \|\varphi\|_{S^{\vartheta(x)}},$$

where $K_1 = M\left(\frac{1}{d^-} + \frac{1}{\vartheta^-}\right) \sum_{n=1}^{\infty} (e^{-\omega(n-1)})$.

Next, we will show that

$$\lim_{T \to \infty} \frac{1}{2T} \int_{-T}^{T} \|Y(s)\| \, ds = 0.$$

Indeed,

$$\frac{1}{2T}\int_{-T}^{T}\|Y_n(t)\|\,dt \le M\Big(\frac{1}{d^-}+\frac{1}{\vartheta^-}\Big)e^{-\omega(n-1)}$$
$$\times\left[\frac{1}{2T}\int_{-T}^{T}\inf\left\{\lambda>0:\int_{t-n}^{t-n+1}\left\|\frac{\varphi(s)}{\lambda}\right\|^{\vartheta(s)}ds\le 1\right\}\right].$$

Since $\varphi^b \in PAP_0(L^{\vartheta^b(x)}((0,1),\mathbb{X}))$, the above inequality leads to $Y_n \in PAP_0(\mathbb{X})$. Using the following inequality

$$\frac{1}{2T}\int_{-T}^{T}\|Y(s)\|\,ds \le \frac{1}{2T}\int_{-T}^{T}\Big\|Y(s)-\sum_{n=1}^{\infty}Y_n(s)\Big\|\,dt+\sum_{n=1}^{\infty}\frac{1}{2T}\int_{-T}^{T}\|Y_n(s)\|\,ds,$$

we deduce that the uniform limit $Y(\cdot)=\sum_{n=1}^{\infty}Y_n(\cdot)\in PAP_0(\mathbb{X})$. Therefore $u \in PAP(\mathbb{X})$.

To complete the proof, we have to prove that the mild solution u is unique. Indeed, $u:\mathbb{R}\to\mathbb{X}$ is bounded and satisfies the homogeneous equation

$$u'(t)=Au(t),\quad t\in\mathbb{R}, \tag{9}$$

Then $u(t)=T(t-s)u(s)$, for any $t\ge s$. Thus $\|u(t)\|\le MKe^{-\omega(t-s)}$, where $\|u(s)\|\le K$. Take a sequence of real numbers (s_n) such that $s_n\to-\infty$ as $n\to\infty$. For any $t\in\mathbb{R}$ fixed, one can find a subsequence $(s_{n_k})\subset(s_n)$ such that $s_{n_k}<t$ for all $k=1,2,\ldots$. By letting $k\to\infty$, we get $u(t)=0$. Now if u,v are bounded solutions to Eq. (2), then $w=u-v$ is a bounded solution to Eq. (9). In view of the above, $w=u-v=0$ that is $u=v$. □

Using Theorem 6.3 one easily proves the following theorem

Theorem 6.4. *Let $p,q>1$ be constants such that $p\le q$. Under assumptions (A.1)–(A.2), the Eq. (1) has a unique pseudo-almost periodic (mild) solutions whenever $\|L_F\|_{S^r}$ is small enough.*

Proof. Define the function $u:\mathbb{R}\times\mathbb{X}\mapsto\mathbb{X}$ by

$$u(t)=\int_{-\infty}^{t}T(t-s)F(s,u(s))ds,\quad \text{for each } t\in\mathbb{R}. \tag{10}$$

It is easy to check that u given in Eq. (10) satisfies Eq. (6) and hence it is a mild solution.

Let $u=\alpha+\beta\in PAP(\mathbb{X})$, where $\alpha\in AP(\mathbb{X})$ and $\beta\in PAP_0(\mathbb{X})$. Then $u\in S^{p,q}_{pap}(\mathbb{X})$ and $K=\overline{\{\alpha(t):t\in\mathbb{R}\}}$ is compact in $\mathbb{X}$. Consequently, from Theorem 5.15, there exists $m\in[1,p)$ such that $F(\cdot,u(\cdot))\in S^{m,m}_{pap}(\mathbb{R}\times\mathbb{X})$.

Applying the proof of Theorem 6.3 to $f(.) = F(., u(.))$, one can easily check that u given in Eq. (10) is in $PAP(\mathbb{X})$.

To complete the proof, we make use of the Banach fixed-point theorem. Indeed, consider the nonlinear operator Υ defined by

$$(\Upsilon u)(t) := \int_{-\infty}^{t} T(t-s)F(s,u(s))ds, \quad \text{for each } t \in \mathbb{R}.$$

For all $u, v \in PAP(\mathbb{X})$, it is easy to see that

$$\begin{aligned}
\|(\Upsilon u)(t) - (\Upsilon v)(t)\| &\leq \int_{-\infty}^{t} \|T(t-s)\|.\,\|F(s,u(s)) - F(s,v(s))\|\,ds \\
&\leq \|u-v\|_{\infty}.\int_{-\infty}^{t} Me^{-\omega(t-s)}L_F(s)\,ds \\
&\leq \|u-v\|_{\infty}.\sum_{n=1}^{\infty}\int_{t-n}^{t-n+1} Me^{-\omega(t-s)}L_F(s)\,ds \\
&\leq M.\,\|u-v\|_{\infty}.\sum_{n=1}^{\infty}\left(\int_{t-n}^{t-n+1} e^{-r_0\omega(t-s)}\,ds\right)^{\frac{1}{r_0}}.\,\|L_F\|_{S^r} \\
&\leq M.\,\|u-v\|_{\infty}.\sum_{n=1}^{\infty}\left(\frac{e^{-r_0(n-1)\omega} - e^{-r_0 n\omega}}{r_0\omega}\right)^{\frac{1}{r_0}}.\,\|L_F\|_{S^r} \\
&\leq M.\,\|u-v\|_{\infty}.\sqrt[r_0]{\frac{1+e^{r_0\omega}}{r_0\omega}}.\sum_{n=1}^{\infty} e^{-n\omega}.\,\|L_F\|_{S^r},
\end{aligned}$$

for each $t \in \mathbb{R}$, where

$$\frac{1}{r} + \frac{1}{r_0} = 1.$$

Hence is whenever $\|L_F\|_{S^r}$ is small enough, that is,

$$M.\sqrt[r_0]{\frac{1+e^{r_0\omega}}{r_0\omega}}.\sum_{n=1}^{\infty} e^{-n\omega}.\,\|L_F\|_{S^r} < 1,$$

then Υ has a unique fixed point, which obviously is the unique pseudo-almost periodic solution to Eq. (1). □

References

1. Amir, B., & Maniar, L. (1999). Composition of pseudo-almost periodic functions and Cauchy problems with operator of nondense domain. *Annales Mathématiques Blaise Pascal, 6*(1), 1–11.
2. Cuevas, C., Sepúlveda, A., & Soto, H. (2011). Almost periodic and pseudo-almost periodic solutions to fractional differential and integro-differential equations. *Applied Mathematics and Computation, 218*(5), 1735–1745.
3. Diagana, T. (2008). Stepanov-like pseudo almost periodicity and its applications to some nonautonomous differential equations. *Nonlinear Analysis, 69*, 4277–2485.
4. Diagana, T. (2013). *Almost automorphic type and almost periodic type functions in abstract spaces*. New York: Springer.
5. Diagana, T. (2007). Stepanov-like pseudo-almost periodic functions and their applications to differential equations. *Communications in Mathematical Analysis, 3*(1), 9–18.
6. Diening, L., Harjulehto, P., Hästö, P., & Ruzicka, M. (2011). *Lebesgue and Sobolev spaces with variable exponents*. Lecture Notes in Mathematics, vol. 2011. Heidelberg: Springer.
7. Ding, H. S., Liang, J., & Xiao, T. J. (2009). Some properties of Stepanov-like almost automorphic functions and applications to abstract evolution equations. *Applied Analysis, 88*(7), 1079–1091.
8. Fan, Z., Liang, J., & Xiao, T. J. (2011). On Stepanov-like (pseudo) almost automorphic functions. *Nonlinear Analysis, 74*(8), 2853–2861.
9. Fan, X. L., & Zhao, D. (2001). On the spaces $L^{p(x)}(O)$ and $W^{m,p(x)}(O)$. *Journal of Mathematical Analysis and Applications, 263*, 424–446.
10. Han, Z., & Jin, Z. (2012). Stepanov-like pseudo almost periodic mild solutions to nonautonomous neutral partial evolution equations. *Nonlinear Analysis, 75*(1), 244–252.
11. Li, H. X., & Zhang, L. L. (2011). Stepanov-like pseudo-almost periodicity and semilinear differential equations with uniform continuity. *Results in Mathematics, 59*(1–2), 43–61.
12. Long, W., & Ding, H. S. (2011). Composition theorems of Stepanov almost periodic functions and Stepanov-like pseudo-almost periodic functions. *Advance in Difference Equations*, Article ID 654695, 12 p.
13. Nguyen, P. Q. H. (2011). *On variable Lebesgue spaces*. Thesis (Ph.D.) - Kansas State University. ProQuest LLC, Ann Arbor, MI, 63 pp.
14. Zhang, C. (2003). *Almost periodic type functions and ergocity*. Kluwer Academic Publishers. Dordrecht/ Boston/ London: Springer.

Group Circle Systems on Conics

Raymond R. Fletcher

Abstract A *group circle system* is a collection of points and circles in the Euclidean plane determined by the elements of an abelian group. Let G be an abelian group and g a fixed element of G. Let ϕ:G $\to \Pi$ be an injective mapping from G into the Euclidean plane Π such that no five points in ϕ(G) are cocyclic . If for each four element subset {a, b, c, d} of G such that a + b + c + d = g, the points {ϕ(a), ϕ(b), ϕ(c), ϕ(d)} are cocyclic, then we call the set of points ϕ(G) and the associated circles, a *(G, g) circle system*. The abelian group G is the *base* of the circle system. In this first paper on circle systems we will confine our attention to group circle systems all of whose points (or vertices) lie on a noncircular conic α. We will construct circle systems on α with base group Z, Z_n and Z $\times$ Z. These are subalgebras of an algebra with one ternary operation defined on α. A remarkable nine variable identity suggested by the circle systems is shown to hold in this algebra.

1 A Parallelism Property for Noncircular Conics

First we have a basic result concerning the intersection of a circle with a parabola. This will be useful when we discuss the intersection of a circle with a general noncircular conic. Basic properties of conics can be found in [1, 5].

Lemma 1 *Let p denote the parabola (1)* $y = a(x-h)^2 + k$, *and let c be a circle such that* $p \cap c$ *contains the four points* $A = (s, t)$; $B = (q, p)$; $C = (u, v)$ *and* $D = (m, n)$. *Then* $4h = s + q + u + m$. *Moreover, if* $A' = (-s, t)$ *and* $C' = (-u, v)$, *then BD is parallel to* $A'C'$.

R.R. Fletcher (✉)
Department of Mathematics and Computer Science, Virginia State University, 303 Drake Ave, Colonial Heights, VA 23834, USA
e-mail: rfletcher@vsu.edu

B. Toni (ed.), *New Frontiers of Multidisciplinary Research in STEAM-H (Science, Technology, Engineering, Agriculture, Mathematics, and Health)*, Springer Proceedings in Mathematics & Statistics 90, DOI 10.1007/978-3-319-07755-0_14,

Proof Let (2) $(x-d)^2+(y-e)^2=r^2$ be the equation of c and use (1) to substitute for y in (2) to obtain (3) $(x-d)^2+(a(x-h)^2+k-e)^2=r^2$. The coefficient of x^3 in (3) is $-4a^2h$. Since the points A, B, C, D satisfy (3), (3) must have the form: (4) $a^2(x-s)(x-q)(x-u)(x-m)$. The coefficient of x^3 in (4) is $a^2(-s-q-u-m)$, so we must have $-4a^2h=a^2(-s-q-u-m)$ which yields $4h=s+q+u+m$.

To prove the last statement of the Lemma we simply compute the slopes:

$$M_{BD}=\frac{p-n}{q-m}=\frac{\left(a(q-h)^2+k\right)-\left(a(m-h)^2+k\right)}{q-m}=a\,(q+m-2h)\,,$$

$$M_{A'C'}=\frac{t-v}{-s+u}=\frac{\left(a(s-h)^2+k\right)-\left(a(u-h)^2+k\right)}{-s+u}=a\,(-s-u+2h)\,.$$

Now by the first part of the Lemma: $M_{BD}=a(q+m-2h)=a(4h-s-u-2h)$ $=M_{A'C'}$. □

The parallelism property stipulated in Lemma 1 holds for any noncircular conic as we show next.

Theorem 1 *Let A, B, C, D be four points on a noncircular conic α with A, B, C distinct, and let A', C' be the reflections of A, C respectively across an axis of symmetry of α. Then A, B, C, D are cocyclic iff BD is parallel to $A'C'$.*

Proof Suppose A, B, C, D lie on a circle c with equation: (5) $(x-h)^2+(y-k)^2=r^2$. If α is a parabola, then BD is parallel to $A'C'$ by Lemma 1, so we may suppose that α is an ellipse with equation: (6) $x^2/a^2+y^2/b^2=1$, or a hyperbola with equation: (7) $x^2/a^2-y^2/b^2=1$. Suppose A', C' are reflections of A, C across the vertical axis of symmetry (y-axis) of α. If we solve (6) or (7) along with the equation of c by solving (5) for y^2 and substituting for y^2 in (6) or (7), we obtain the equation of a parabola (with vertical axis of symmetry) which contains the four points A, B, C, D. We may then invoke Lemma 1 to conclude that BD is parallel to $A'C'$. If A'', B'' are reflections of A, C across the horizontal axis of symmetry of α, then we obtain similarly that $A''C''$ is parallel to BD. Thus the lines BD, $A'C'$, $A''C''$ are mutually parallel.

Conversely, suppose BD is parallel to $A'C'$ and let E be the fourth point on α and on circle (A, B, C). By the first part of the Theorem, $A'C'$ is parallel to BE and thus BE and BD are parallel. We must then have that BE and BD are the same line. Any line through B meets the conic α in at most one point (besides B), so we must have $E=D$, and thus the points A, B, C, D are cocyclic. □

2 Construction of a (Z, 0) Circle System on a Noncircular Conic

In this Section we present the construction of a (Z, 0) circle system on a noncircular conic. If a, b are elements of the set Z of integers, we denote the line joining the corresponding points by [a, b].

Theorem 2 *Let α be a noncircular conic with axis of symmetry j. In the case of a hyperbola the transverse axis is selected for j so that in all cases j and α have a point in common. The following assignment of integer labels to points on α yields a (Z, 0) circle system: (i) Assign the integer 0 to a common point of j and α. (ii) Put vertex 1 on α but not on an axis of symmetry of α. (iii) Let vertex -1 be the reflection of vertex 1 across j, (iv) Let m denote the line $[0, -1]$ and draw a line through vertex 1 and parallel to m. Label the point where this line meets α with -2. (v) Let vertex 2 be the reflection of vertex -2 across j. (vi) In general, if elements $\{0, \pm 1, \pm 2, \ldots, \pm t\}$ have been assigned to points on α, assign $-(t+1)$ to the point on α and on the line through t and parallel to m, and let $t+1$ be the reflection of $-(t+1)$ across j. We assume that the generating vertex 1 is chosen on α so that no point on α is assigned more than one integer label.*

The construction described in Theorem 2 is illustrated in Fig. 1. Before proving Theorem 2 we need the following Lemma. We employ a variant of Pascal's Theorem which states that if hexagon (a, b, c, d, e, f) is inscribed in a conic and two pairs of opposite sides are parallel, then the third pair of opposite sides are also parallel. Pascal's Theorem for a hexagon inscribed in a circle can be found in [2], and the more general case in which the hexagon is inscribed in any conic can be found in [1].

Lemma 2 *In the proposed construction of a (Z, 0) circle system on a noncircular conic, if m, n, p, q are integers and $m+n=p+q$, then the lines [m, n] and [p, q] are parallel.*

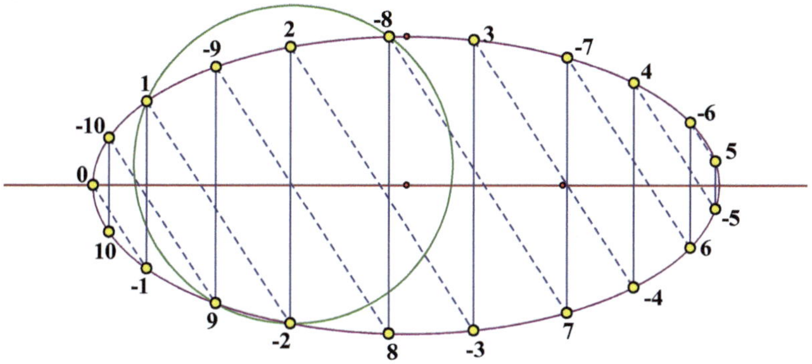

Fig. 1 (Z, 0) circle system on ellipse

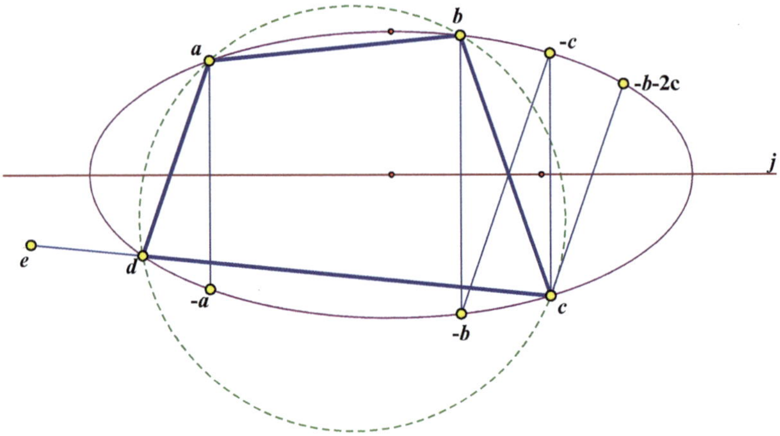

Fig. 2 Cyclic quadrilateral in (Z, 0) circle system

Proof Consider the inscribed hexagon (p, q, −q − 1, q + 1, p − 1, −p). Opposite sides [p, −p], [q + 1, −q − 1] are vertical hence parallel, and opposite sides [q, −q − 1], [p − 1, −p] are parallel by virtue of our construction. By Pascal's Theorem, the remaining pair of opposite sides [p, q], [p − 1, q + 1] are also parallel. Applying the same result with p − 1 in place of p and q + 1 in place of q, we obtain that the lines [p − 1, q + 1], [p − 2, q + 2] are parallel, and thus [p, q], [p − 2, q + 2] are parallel. Repeating this argument we obtain that all lines of the form {[p − t, q + t]: t∈Z} are mutually parallel. If we let t = p − m then we obtain that the lines [p, q], [m, n] are parallel. Note that the same proof works in case p = q, when line [p, q] is interpreted as the tangent to the conic at p. □

Now we may proceed with the proof of Theorem 2.

Proof (of Theorem 2) Let a, b, c, d be four distinct integers such that a + b + c + d = 0. We must show that the corresponding points on α are cocyclic. Consider: ∠(a, b, c) = ∠(−a, −b, −c) since −a, −b, −c are reflections of a, b, c across the axis of symmetry j. Since (−a) + (−b) = c + d, the lines [−a, −b], [c, d] are parallel by Lemma 2. Also, since (−b) + (−c) = c + (−b − 2c), the lines [−b, −c], [c, −b − 2c] are parallel by Lemma 2. It then follows that ∠(−a, −b, −c) = ∠(d, c, −b − 2c) as in Fig. 2. Since a + d = c + (−b − 2c), the lines [a, d], [c, −b − 2c] are parallel by Lemma 2. So if we extend segment [c, d] to point e as in Fig. 2, we have ∠(a, d, e) = ∠d, c, −b − 2c) = ∠(a, b, c). Consequently ∠(a, d, c) = 180° − ∠(a, d, e) = 180° − ∠(a, b, c), and it follows that quadrilateral (a, b, c, d) is cyclic. (The elementary result that a quadrilateral is cyclic iff its opposite angles are supplementary can be found in [3].) □

Let G be any abelian group and let 0 denote the identity in G. Next we show that any (G, 0) circle system on a noncircular conic must exhibit the properties inherent in our construction of a (Z, 0) circle system. The reader may find it useful to review some elementary properties of groups which can be found in [4]. The radical axes of three mutually nonconcentric circles concur at a point called the radical center of the three circles. This property of circles can be found in [2, 3].

Theorem 3 *Let Ω be a (G, 0) circle system whose vertices lie on a noncircular conic α and such that G contains distinct points* $\{g, -g, h, -h, k, -k\}$ *. Then one of the axes of symmetry of α has the property that g and −g are reflections with respect to it, for every g ∈ G. Moreover, if a, b, c, d ∈ G and* $a + b = c + d$*, then [a, b] is parallel to [c, d]. Conversely, if a set Ω of points on a noncircular conic is labeled with the elements of an abelian group G such that ∀ a, b, c, d ∈ G with* $a + b = c + d$*, the lines [a, b], [c, d] are parallel, then Ω is a (G, 0) circle system on the conic.*

Proof Let g, h, k be elements of G such that {g, −g, h, −h, k, −k} are distinct. Then $c_1 = (g, -g, h, -h)$, $c_2 = (h, -h, k, -k)$ and $c_3 = (k, -k, g, -g)$ are circles in Ω. The lines [g, −g], [h, −h], [k, −k] either meet at the radical center of c_1, c_2, c_3 or are mutually parallel. Let h′, −h′ denote the reflections of h, −h respectively across an axis j of symmetry of α. By Theorem 1, the lines [g, −g], [h′, −h′] are parallel. Considering circle c_2 we obtain similarly that the lines [k, −k], [h′, −h′]. It follows that all lines {[g, −g]: g ∈ G, g ≠ −g} are mutually parallel. Since [g, −g] is parallel to both [h, −h] and [h′, −h′], it follows that [h, −h] is parallel to [h′, −h′]. This can only happen if $h' = -h$ or [h, −h] is parallel to j. In the first case h and −h are reflections with respect to j, and in the second case h, −h are reflections with respect to a second axis of symmetry of α perpendicular to j. The circles $c_4 = [0, g, h, -(g + h)]$ and $c_5 = (0, -g, -h, g + h)$ belong to Ω and are reflections across an axis of symmetry of α. Since both circles contain vertex 0, we must have that 0 lies on this axis.

Now suppose g, −g are reflections across the axis j of symmetry of α for all g ∈ G, and suppose a, b, c, d ∈ G such that $a + b = c + d$. Then $a + b + (-c) + (-d) = 0$ and so (a, b, −c, −d) is a circle in Ω provided a, b, −c, −d are distinct elements of G. In this case we obtain [a, b] parallel to [c, d] by Theorem 1. If two of these points are identical, say $a = b$, then we interpret (a, a, −c, −d) as the circle (a, −c, −d) which is tangent to α at a. In this case Theorem 1 says that [a, a] = [a, b] is parallel to [c, d] where [a, a] denotes the tangent to α at a.

Conversely, suppose Ω is a set of points on a noncircular conic α, labeled with the elements of an abelian group G such that whenever a, b, c, d ∈ G and $a + b = c + d$, the lines [a, b], [c, d] are parallel. Let p, q, r, s ∈ G and suppose $p + q + r + s = 0$. Then $p + q = (-r) + (-s)$ and so the lines [p, q], [−r, −s] are parallel. But then, by Theorem 1, the points p, q, r, s are cocyclic and thus Ω is a (G, 0) circle system. □

3 A Ternary Operation Defined on a Noncircular Conic

Let α be a noncircular conic and let a, b, c be three points on α. Circle (a, b, c) must meet α in a fourth point which we denote by δ(a, b, c). This ternary operation is still defined in case two or even three of its arguments are identical; δ(a, a, b) is the remaining point on α and on the circle which is tangent to α at a and contains b, and δ(a, a, a) is the remaining point on α and on the unique circle which intersects α with multiplicity 3 at a. Regarding intersection multiplicity for algebraic curves, the reader may wish to consult [1]. These possibilities are illustrated in Figs. 3 and 4. A point x of α is *idempotent* if δ(x, x, x) = x. It can be shown that the idempotent elements of an ellipse consist precisely of the four points where the major and minor axes meet the ellipse. We have created a *ternary algebra* on α which we denote by (α, δ). The following identities hold in this algebra:

(i) δ(b, a, c) = δ(a, b, c) = δ(a, c, b)
(ii) δ[a, b, δ(a, b, c)] = c
(iii) δ(δ(a, b, c), δ(d, e, f), q) = δ(δ(a, b, d), δ(c, e, f), q)

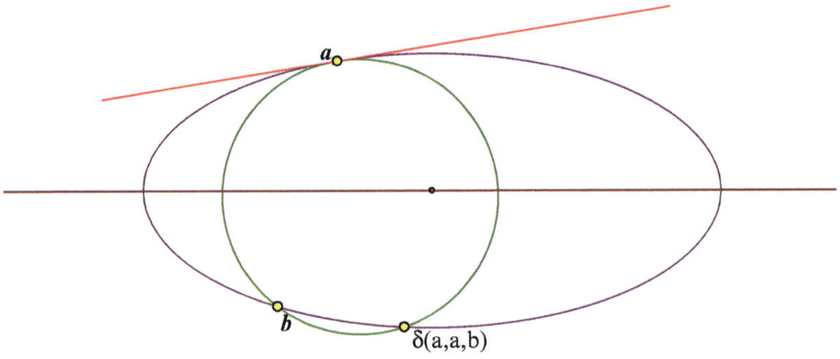

Fig. 3 δ operator with two identical arguments

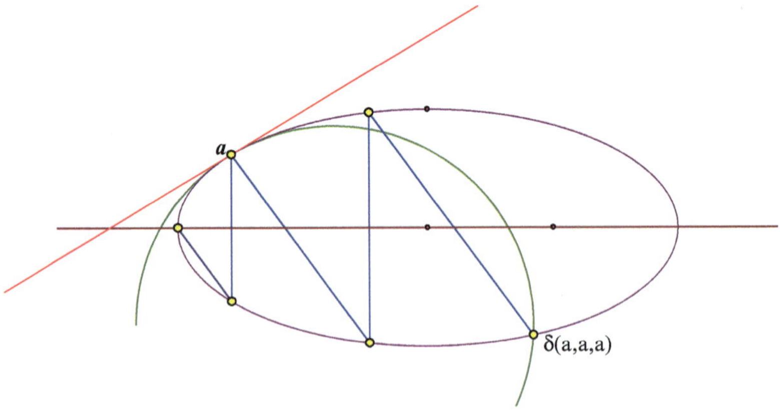

Fig. 4 δ operator with three identical arguments

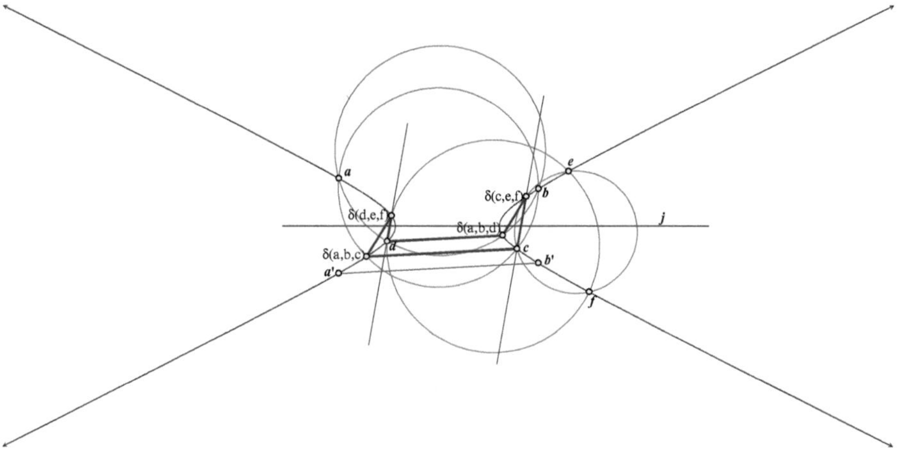

Fig. 5 Parallelism property for δ operator

Identities (i) and (ii) are obvious, and identity (iii) will be proved in this Section. A consequence of (i), (ii), (iii) is that the value of the expression δ[δ(a, b, c), δ(d, e, f), δ(g, h, i)] remains constant under any permutation of the nine arguments. We call (iii) the *tricubic identity*. We begin with yet another parallelism property.

Lemma 3 *Let a, b, c, d, e, f be six points on a noncircular conic* α. *Then the lines* $[\delta(a, b, c), \delta(d, e, f)]$ *and* $[\delta(a, b, d), \delta(c, e, f)]$ *are parallel.*

Proof Let j be an axis of symmetry of α and if x is any point on α, let x′ denote the reflection of x across j. Since the points {a, b, c, δ(a, b, c)} are cocyclic, the lines [c, δ(a, b, c)] and [a′, b′] are parallel by Theorem 1. Similarly, the lines [d, δ(a, b, d)] and [a′, b′] are parallel. Consequently the lines [c, δ(a, b, c)] and [d, δ(a, b, d)] are parallel as are the lines [d, δ(d, e, f)] and [c, δ(c, e, f)]. See Fig. 5. If we now consider the hexagon (δ(a, b, c), δ(d, e, f), d, δ(a, b, d), δ(c, e, f), c) inscribed on α, and apply Pascal's Theorem (variant), we obtain the desired result. □

Now we are prepared to show that the tricubic identity holds on any noncircular conic.

Theorem 4 *Let* α *be a noncircular conic and let a, b, c, d, e, f, q be any seven points on* α. *Then* $\delta(\delta(a, b, c), \delta(d, e, f), q) = \delta(\delta(a, b, d), \delta(c, e, f), q)$.

Proof Let x = δ(δ(a, b, c), δ(d, e, f), q) as in Fig. 6, and let j be an axis of symmetry of α. For each point w on α, let w′ denote the reflection of w across j. By Theorem 1 we have (8) [x, q] parallel to [δ(a, b, c)′, δ(d, e, f)′]. Let y = δ(δ(a, b, d), δ(c, e, f), q), then by Theorem 1 we have (9) [y, q] parallel to [δ(a, b, d)′, δ(c, e, f)′]. By Lemma 3 we have [δ(a, b, c), δ(d, e, f)] parallel to [δ(a, b, d), δ(c, e, f)], and by reflecting these lines across j we obtain (10) [δ(a, b, c)′, δ(d, e, f)′] parallel to [δ(a, b, d)′, δ(c, e, f)′]. From (8), (9), (10) we obtain that the lines [x, q], [y, q] are parallel. This implies that the points x, y, q are collinear. But any line through q meets the conic α in at most one other point, so we must have x = y. □

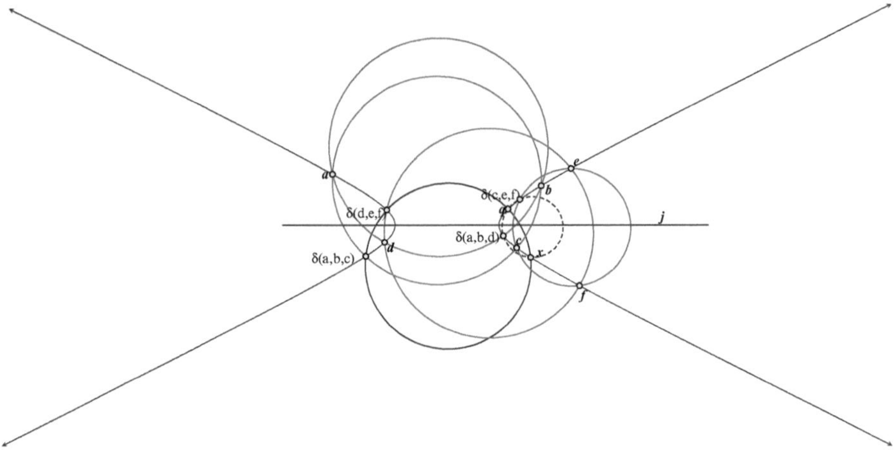

Fig. 6 The tricubic identity

4 Circle Systems on Z and Z_n

Let Ω_1 be a (G, g) circle system and Ω_2 be a (H, h) circle system on abelian groups G, H. If there exists a bijection ψ: $G \rightarrow H$ such that a, b, c, $d \in G$ and $a + b + c + d = g$, implies $\psi(a) + \psi(b) + \psi(c) + \psi(d) = h$, then we say that Ω_1 and Ω_2 are *equivalent* circle systems under the relabeling given by ψ. The relabeling ψ: $Z \rightarrow Z$ given by $\psi(x) = x + k$ adds the integer k to each vertex and thus adds 4k to each circle in a (Z, 0) circle system. Consequently every circle system in {(Z, 4k): $k \in Z$} is equivalent to a (Z, 0) circle system. Applying the same mapping, we obtain that every circle system on Z is equivalent to a (Z, 0), (Z, 1), (Z, 2) or (Z, 3) circle system. Besides adding a fixed integer to each vertex, we can also replace each vertex by its inverse. Thus a (Z, 1) circle system is equivalent to a (Z, −1) system which, by adding one to each vertex is, in turn, equivalent to a (Z, 3) system. We have proved the first part of our next result.

Theorem 5 *Every circle system on Z is equivalent to a (Z, 0), (Z, 1) or (Z, 2) circle system, and these are mutually nonequivalent.*

Proof Let Ω_0, Ω_1, Ω_2 be (Z, 0), (Z, 1), (Z, 2) circle systems respectively. We will show that Ω_0 is not equivalent to Ω_1. Suppose, to the contrary, that there exists a bijection ψ: $Z \rightarrow Z$ such that whenever $a + b + c + d = 0$, we have $\psi(a) + \psi(b) + \psi(c) + \psi(d) = 1$. Let: $\psi(0) = s$ and $\psi(1) = t$. Since $0 + 0 + 0 + 0 = 0$, we must have $4s = 1$ which cannot be satisfied by an integer. $4s = 2$ also cannot be satisfied by an integer, so a similar argument shows that Ω_0 and Ω_2 are not equivalent. The nonequivalence of Ω_1 and Ω_2 is more interesting. Suppose μ: $Z \rightarrow Z$ is mapping such that whenever a, b, c, $d \in Z$ and $a + b + c + d = 1$, then $\mu(a) + \mu(b) + \mu(c) + \mu(d) = 2$. We claim that such a mapping does exist and must have the form: (11) $\mu(k) = 2k - (4k - 1)s$ where $s = \mu(0)$

If $m(1) = t$, then since $0+0+0+1=1$, we must have $3s+t=2$ and thus $\mu(1)=2-3s$ in accordance with (11). Now assume inductively that (11) holds with $|k| \leq n$ and let $k = n+1$. Consider: $(n+1)+(-n)+0+0=1$ and so we must have $\mu(n+1)+\mu(-n)+2s=2$. By the inductive hypothesis $\mu(-n)=-2n+(4n+1)s$ so we obtain $\mu(n+1)=-4ns+2n-3s+2=2(n+1)-(4(n+1)-1)s$ in accordance with (11). Finally let $k=-n-1$ and consider: $(-n-1)+n+1+1=1$ and so we must have $\mu(-n-1)+\mu(n)+2t = 2$. Now replacing $\mu(n)$ by $2n-(4n-1)s$, we obtain $\mu(-n-1)=-2n+(4n-1)s-2t+2=4ns-2n-s-2t+2=4ns-2n-s-2(2-3s)+2=4ns-2n+5s-2=4ns+5s+2(-n-1)=2(-n-1)-(4(-n-1)-1)s$ in accordance with (11). However μ cannot be onto for any choice of s. The equation $2k-(4k-1)s=0$ has only the solution $k=s=0$ in the integers, i.e., the preimage of 0 under μ must be 0 , but then since $s=0$, $\mu(Z)=\{0, \pm 2, \pm 4, \pm 6, \ldots\}$, and so μ is not onto. □

It remains to show that (Z, 1) and (Z, 2) circle systems exist. If α is a noncircular conic, then a *subalgebra* of (α, δ) is a nonempty subset of α which is closed under the ternary operation δ. Clearly every circle system on α is a subalgebra of (α, δ). Let Ω_0 be a (Z, 0) circle system on α and let $S=\{m \in \Omega_0\colon m=4k+1$ for some integer $k\}$. If $a=4s+1$; $b=4t+1$ and $c=4q+1$ are elements in S, then $\delta(a, b, c)=-a-b-c=-(4s+1)-(4t+1)-(4q+1)=4(-s-t-q-1)+1 \in S$. So S is a subalgebra of Ω_0. Now define a mapping $\psi\colon S \Rightarrow Z$ by $\psi(4k+1)=k$ which is clearly bijective. We claim that $\phi(S)$ is a (Z, −1) circle system on α. For suppose $(a, b, c, -a-b-c)$ is a circle in S, then $[\psi(a), \psi(b), \psi(c), \psi(-a-b-c)]=(s, t, q, -s-t-q-1)$ is the corresponding circle in $\psi(S)$, whose vertices sum to −1. By multiplying each vertex in $\psi(S)$ by −1 we obtain a (Z, 1) circle system. If we let $T=\{m \in \Omega_0\colon m=4k+2$ for some integer $k\}$, then similarly T is a subalgebra of Ω_0, and the relabeling $\mu\colon T \to Z$ defined by $\mu(4k+2)=k$ yields a (Z, −2) circle system . The further operation of multiplying each vertex by −1 yields a (Z, 2) circle system on the points in $\mu(T)$. We have thus proved:

Theorem 6 *If α is a noncircular conic and Ω_0 is a (Z, 0) circle system on α, then every circle system on Z occurs as a subalgebra of Ω_0.* □

In the construction of a (Z, 0) circle system given in Theorem 2 suppose that the noncircular conic α is an ellipse with x and y axes as axes of symmetry. Let vertex 0 be chosen on the negative x-axis and let $k \geq 2$ be an integer. The placement of vertex 1 on the ellipse determines the positions of the remaining vertices and it is clear, from the construction given in Theorem 2, that vertex 1 can be positioned sufficiently close to vertex 0 so that vertices 1, … , k lie in the upper half of the ellipse and their reflections −1, … , −k lie on the lower half of the ellipse. By shifting vertex 1 a little to the right along the ellipse we can make vertices k and −k coincide at the other extremity of the major axis. This is illustrated for $k=5$ in Fig. 7. When this happens each pair of vertices which are congruent mod(2k) also coincide. The result is a $(Z_{2k}, 0)$ circle system on the ellipse. If instead we shift vertex 1 a little to the left we can make vertex k and vertex $-(k+1)$ coincide, resulting in a $(Z_{2k+1}, 0)$ circle system. Note that this construction is not possible on a parabola or hyperbola since these are not closed curves and our construction puts new vertices further and further away from vertex 0. We thus have the following result.

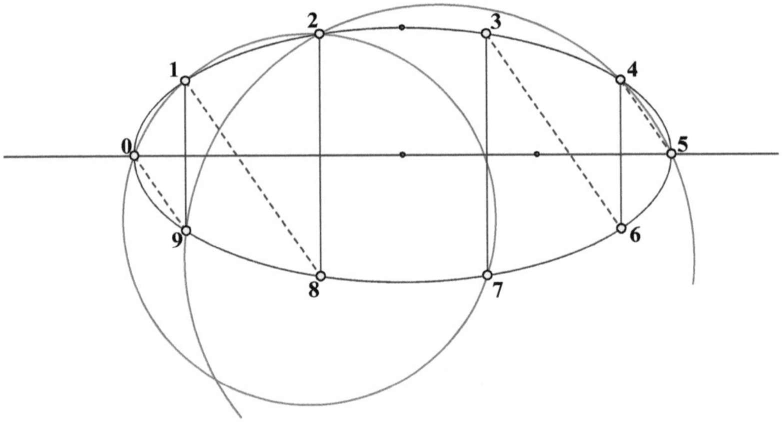

Fig. 7 (Z10, 0) circle system on ellipse

Theorem 7 *For any positive integer n, a* (Z_n, *0*) *circle system can be constructed on any ellipse.*

We have also determined the equivalence classes of circle systems on Z_n:

Theorem 8 (*i*) *If n is odd, then every* Z_n *circle system is equivalent to a* (Z_n, *0*) *circle system.* (*ii*) *If m is odd, then every* Z_{2m} *circle system is equivalent to a* (Z_{2m}, *0*) *or a* (Z_{2m}, *1*) *circle system.* (*iii*) *Every* Z_{4m} *circle system is equivalent to either a* (Z_{4m}, *0*), (Z_{4m}, *1*) *or* (Z_{4m}, 2) *circle system. The circle systems indicated in* (*ii*) *are nonequivalent and the three circle systems indicated in* (*iii*) *are mutually nonequivalent.*

Now suppose Ω is a (Z_{4m}, 0) circle system on an ellipse α and let $S = \{4k + 1 \in Z_{4m}: k = 0, 1, 2, \ldots, m - 1\}$. Then S is easily seen to be a subalgebra of (Ω, δ) and the mapping ϕ: $S \Rightarrow Z_m$ given by $\phi(4k + 1) = k$ is a relabeling which turns S into a (Z_m, -1) circle system on α. The further action of multiplying each vertex by -1 yields a (Z_m, 1) circle system. If we let $T = \{4k + 2 \in Z_{4m}: k = 0, 1, 2, \ldots, m - 1\}$, then T is a subalgebra of (Ω, δ) and the relabeling ψ: $S \Rightarrow Z_m$ given by $\psi(4k + 2) = k$ turns T into a (Z_m, -2) circle system. Then multiplying each vertex by -1 yields a (Z_m, 2) circle system on α. So we have:

Theorem 9 *Every circle system on* Z_n *can be constructed on any ellipse.*

We conjecture that the only finite abelian groups which support circle systems, (with vertices not necessarily on a conic), are the groups Z_n and $Z_2 \times Z_{2m}$ and $Z_2 \times Z_2 \times Z_2$ where m, n are any positive integers. The construction of circle systems on $Z_2 \times Z_{2m}$ and their properties will be included in a future work. Here we can show that no such circle system can exist on a noncircular conic. If Ω is any (G, g) circle system, let us refer to circles with a repeated vertex as *minor circles* of Ω. These are circles of the form (a, b, x, x) where $a + b + 2x = g$. If the points of Ω lie on a noncircular conic α, then this minor circle is tangent to α at x. Let Ω be a ($Z_2 \times Z_{2m}$) circle system, $m \geq 2$, and suppose Ω lies on a noncircular conic

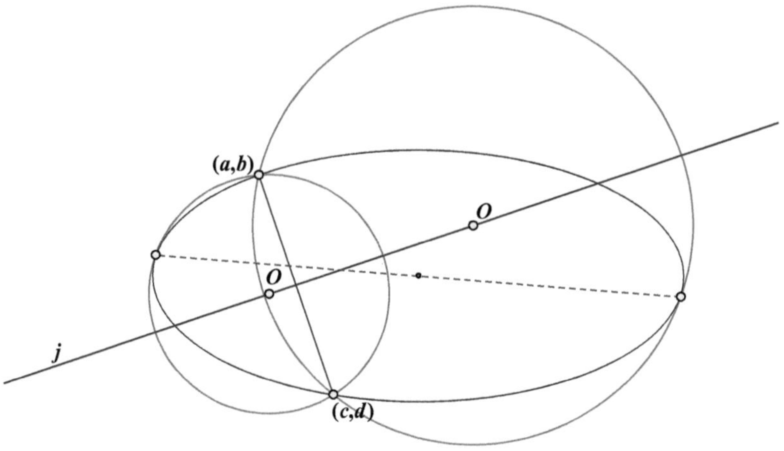

Fig. 8 Two circles through points (a, b), (c, d) which are tangent to ellipse

α. Since $Z_2 \times Z_{2m}$ contains a subgroup isomorphic to Z_{2m}, the conic α must be an ellipse. Now consider the minor circles: [(a, b), (c, d), (0, 0), (0, 0)], [(a, b), (c, d), (0, m), (0, m)], [(a, b), (c, d), (1, m), (1, m)], [(a, b), (c, d), (1, 0), (1, 0)], where (a, b) + (c, d) = g. These are four distinct circles each through the same two points (a, b), (c, d) and each tangent to the ellipse at a third point. But only two such circles exist: suppose β is a circle through (a, b) and (c, d), then its center O must lie on the perpendicular bisector j of the segment [(a, b), (c, d)]. As O moves along j from $+\infty$ there is a position for O where β is tangent to the ellipse. Similarly, as O moves along j from $-\infty$, there is a second position for O where β is tangent to the ellipse. These possibilities are illustrated in Fig. 8. We require four such circles but there are only two, so we conclude that no $Z_2 \times Z_{2m}$ circle system, $(m \geq 2)$ can exist on a noncircular conic.

5 Direct Product of Circle Systems

In this Section we show that a direct product construction can be used to create circle systems on a noncircular conic with base group $Z \times Z$. By extension, we can create circle systems with base group Z^n for any positive integer n.

Theorem 10 *Let Ω be a $(G, 0)$ circle system, and Ω' a $(H, 0')$ circle system on the same noncircular conic α, where $0, 0'$ are the identities of the abelian groups G, H respectively. We suppose, in accordance with Theorem 3, that the same point P on α and on an axis of symmetry j of α has been labeled with 0 and $0'$ and that $g, -g$ are reflections across j for every $g \in G$, and similarly $h, -h$ are reflections across j for every $h \in H$. Finally we suppose that Ω and Ω' have only the point P in common. Then a circle system $\Omega \times \Omega'$ with base group $G \times H$ and sum $(0, 0')$ can be defined* *on α as follows*: (*i*) *relabel each point* $g \in G$ *by* $(g, 0')$; (*ii*) *relabel each point* $h \in H$ *by* $(0, h)$. *Note that the point P will have label* $(0, 0')$; (*iii*) *let* $(g, h) = \delta[(-g, 0'), (0, -h), (0, 0')]$.

Proof Since the points $\{(g, h), (-g, 0'), (0, -h), (0, 0')\}$ are cocyclic, we have by Theorem 1: (12) the lines $[(g, h), (0, 0')]$ and $[(g, 0'), (0, h)]$ are parallel $\forall g \in G$, $\forall h \in H$. Now suppose $s, t \in H$ and $s + t = h$. Consider the inscribed hexagon $H_1 = [(g, 0'), (0, h), (0, 0'), (g, s), (0, t), (0, s)]$. We have that lines $[(0, h), (0.0')]$, $[(0, t), (0, s)]$ are parallel by Theorem 3. Also the lines $[(0, 0'), (g, s)]$, $[(0, s), (g, 0')]$ are parallel by (12). So we can apply Pascal's Theorem (variant) to H_1 to obtain (13) the lines $[(g, 0'), (0, h)]$, $[(g, s), (0, t)]$ are parallel. Then from (12), (13) we obtain (14) the lines $[(g, h), (0, 0')]$, $[(g, s), (0, t)]$ are parallel. Now let $H_2 = [(g, s), (g_0, t), (0, 0'), (g, h), (g_0, 0'), (0, t)]$ be a second hexagon inscribed on α where $g, g_0 \in G$. By (12) the lines $[(g_0, t), (0, 0')]$, $[(g_0, 0'), (0, t)]$ are parallel, and by (14) the lines $[(g, h), (0, 0')]$, $[(g, s), (0, t)]$ are parallel. Then applying Pascal's Theorem to H_2 we obtain that the lines $[(g, s), (g_0, t)]$, $[(g, h), (g_0, 0')]$ are parallel whenever $s + t = h$. Consequently we have (15) the set of lines $\{[(g, s), (g_0, t)]$: $s + t$ is a fixed element of $H\}$ are mutually parallel. Similarly (16) the set of lines $\{[(p, h), (q, h_0)]$: $h, h_0 \in H$, and $p + q$ is a fixed element of $G\}$ are mutually parallel. Now suppose (a, b), (c, d), (u, v), $(w, x) \in G|H$, and $(a, b) + (c, d) = (u, v) + (w, x)$ and thus $a + c = u + w$ and $b + d = v + x$. By (15) the lines $[(a, b), (c, d)]$, $[(a, v), (c, x)]$ are parallel, and by (16) the lines $[(a, v), (c, x)]$, $[(u, v), (w, x)]$ are parallel. We thus obtain that the lines $[(a, b), (c, d)]$, $[(u, v), (w, x)]$ are parallel. We have shown that our construction satisfies the parallelism property stipulated in Theorem 3 for a general 0-sum circle system. It then follows from Theorem 3, that $\Omega \times \Omega'$ is a $[G \times H, (0, 0')]$ circle system on α. □

Following the construction given in Theorem 2, we can construct a $(Z, 0)$ circle system on a noncircular conic α with zero vertex at a point P on α and on an axis of symmetry of α. In this construction vertex 1 is the generating vertex since its placement determines the positions of the remaining vertices. In principle we can construct a second $(Z, 0)$ circle system on α, putting its zero vertex at P and positioning its generating vertex so that the two circle systems have only the zero vertex in common at P. Then using the construction given in Theorem 10 we obtain a $[Z \times Z, (0, 0')]$ circle system on α. Constructing yet a third $(Z, 0)$ circle system on α and choosing the generating vertex so that this new circle system has only the zero vertex in common with the $[Z \times Z, (0, 0')]$ circle system, we obtain a zero sum circle system on α with base group $Z \times Z \times Z$. Repeating this process we can obtain a zero sum circle system on α with base group Z^n for any positive integer n. If α is an ellipse, then we can construct similarly zero sum circle systems on α with base group $Z_m \times Z^n$ for any positive integers m, n. Note that Theorem 10 cannot be used to construct a $[Z_2 \times Z_{2m}, (0, 0)]$ circle system on an ellipse, (which we have shown to be impossible), since in the construction given in Theorem 10, the $(Z_2, 0)$ circle system and the $(Z_{2m}, 0)$ circle system would have *two* points in common, namely the two points on the ellipse and on the axis of symmetry j.

6 Conclusion

In the previous Section we provide a construction for a (Z × Z, (0, 0)) circle system Ω_{00} on a noncircular conic. It is not difficult to show that all Z × Z circle systems are equivalent to a Z × Z circle system with sum g where g ∈ {(0, 0), (0, 1), (0, 2), (1, 1), (1, 2), (1, 3), (2, 2)}. We claim, similar to Theorem 6, that every circle system on Z × Z occurs as a subalgebra of (Ω_{00}, δ). We will only outline the proof of this claim for the creation of a (Z × Z, (1, 2)) circle system. Let S = {4s + 1: s ∈ Z} and let T = {4t + 2: t ∈ Z}. Then S × T is a subalgebra of (Ω_{00}, δ), and the mapping ϕ: S × T⇒Z × Z given by ϕ(4s + 1, 4t + 1) = (s, t) is a relabeling which turns S × T into a [Z × Z, (−1, −2)] circle system. Replacing each element in this circle system by its inverse yields a [Z × Z, (1, 2)] circle system. We conjecture that a similar result holds for circle systems on Z^n and $Z_m \times Z^n$. It remains an open question as to which infinite abelian groups besides these may support circle systems.

As an afterthought we observe that the construction of a Z_n circle system on an ellipse given in Theorem 7 is essentially unique once the vertex 0 is chosen on either the major or minor axis.

References

1. Bix, R. (2006). *Conics and cubics, a concrete introduction to algebraic curves*. New York: Springer.
2. Coxeter, H. S. M., & Greitzer, S. L. (1967). *Geometry revisited*. Washington: The Mathematical Association of America.
3. Isaacs, I. (2001). *Geometry for college students*. Pacific Grove: Brooks/Cole.
4. Papantonopoulo, A. (2002). *Algebra pure and applied*. Prentice Hall, Upper Saddle River, NJ
5. Sullivan, M. (2012). *Algebra and trigonometry* (9th ed.). Prentice Hall, Upper Saddle River, NJ

Remanufacturing Processes, Planning and Control

Jianzhi Li and Zhenhua Wu

Abstract This chapter provides a summary of critical issues in remanufacturing process and its planning and control. The chapter starts with an introduction of the special characteristics and the associated problems in remanufacturing. Typical remanufacturing processes such as cleaning, testing, and disassembly are then discussed in details. The chapter also provides a discussion of process sequencing for product disassembly to minimize cost and energy consumption. Due to stochastic nature in the material arrival process, production planning represents another main challenge for remanufacturers. Based on a case study of a business in Austin TX, a simulation model with a prioritized stochastic batch arrival mechanism, considering factors that affect the total profit, is also discussed. The chapter also presents a genetic algorithm (GA) algorithm to optimize the production planning and control policies for dedicated remanufacturing.

1 Introduction

Product remanufacturing develops rapidly in recent decades due to intensified environmental legislations and growing economic concerns. Through remanufacturing, products/components that would otherwise head to land-fill or incineration will instead go through a set of value and material recapturing processes, including collection and distribution, inspection,disassembly, cleaning, testing, repair,

J. Li (✉)
Manufacturing Engineering Department, The University of Texas–Pan American, Edinburg, TX 23834, USA
e-mail: jianzhili@utpa.edu

Z. Wu
Department of Engineering, Virginia State University, Petersburg, VA 23806, USA

B. Toni (ed.), *New Frontiers of Multidisciplinary Research in STEAM-H (Science, Technology, Engineering, Agriculture, Mathematics, and Health)*,
Springer Proceedings in Mathematics & Statistics 90, DOI 10.1007/978-3-319-07755-0_15,

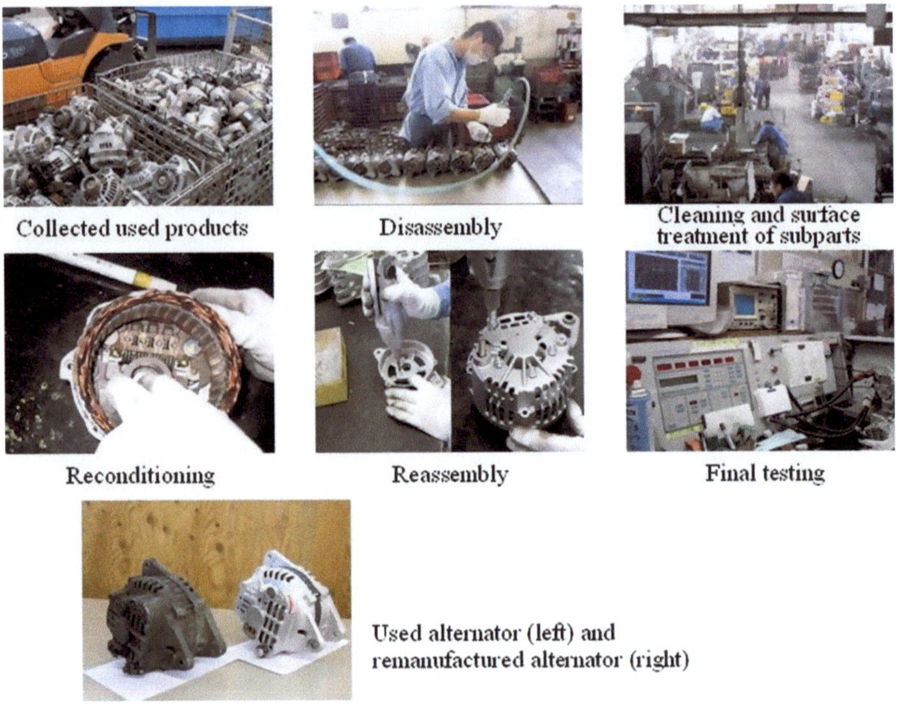

Fig. 1 Remanufacturing processes of alternators [8]

reassembly, redistribution, and remarketing or recycling. Remanufacturing allows for reusable components and recoverable materials reenter the supply chain for future reuse or new product fabrication.

Although there are numerous definitions of remanufacturing, Lund [5–7], describes remanufacturing as "... an industrial process in which worn-out products are restored to like-new condition. Through a series of industrial processes in a factory environment, a discarded product is completely disassembled. Usable parts are cleaned, refurbished, and put into inventory. Then the product is reassembled from the old parts (and where necessary, new parts) to produce a unit fully equivalent and sometimes superior in performance and expected lifetime to the original new product." Note that remanufacturing differs from simple repair or recovery in that a remanufactured product should meet the same customer expectation as new products in quality, warranties, life span, and functions.

Generally, the production processes of remanufacturing are comprised of following stages: product arrival, inspection, disassembly, cleaning, testing, repairing (reconditioning), reassembly, final testing, labeling, packaging, and shipping. The typical remanufacturing processes are demonstrated in Fig. 1.

Material arrival process for remanufacturing is a typical compound stochastic batch arrival process with varied product types and conditions. Thus, in receiving, incoming product to be remanufactured have to be classified according to its type and condition. Received products will then be briefly inspected to collect related information, and sent to inventory. For received products, cleaning processes are required to separate undesired dirt, coating, or other contaminants from the parts to be remanufactured. Following this, the testing process is carried out to investigate the condition of the product and assign appropriate remanufacturing processes, which include refurbishing, repair, reuse and material recycling. Finished goods will be labeled, packed and shipped for resale. Components that cannot be reused will be further disassembled and classified according to their material contents, and then shipped for material recycling.

2 Key Remanufacturing Processes

A research survey of remanufacturers [1] proved the importance of the cleaning process. As the graph illustrates, 29 % of the remanufacturers' highest expense is the cleaning process. The cost of the cleaning process is a significant factor for remanufacturers. Survey results are shown on the figure below (Fig. 2).

Part refurbishing process is generally different for different types of products, so our discussion of remanufacturing will focus on cleaning, testing and inspection, and disassembly/reassembly.

2.1 *Cleaning Processes*

Cleaning process of mechanical parts can be grouped into two categories: liquid based cleaning process and mechanical based processes. In liquid based processes, parts are cleaned by solutions through mechanisms such as wetting and other

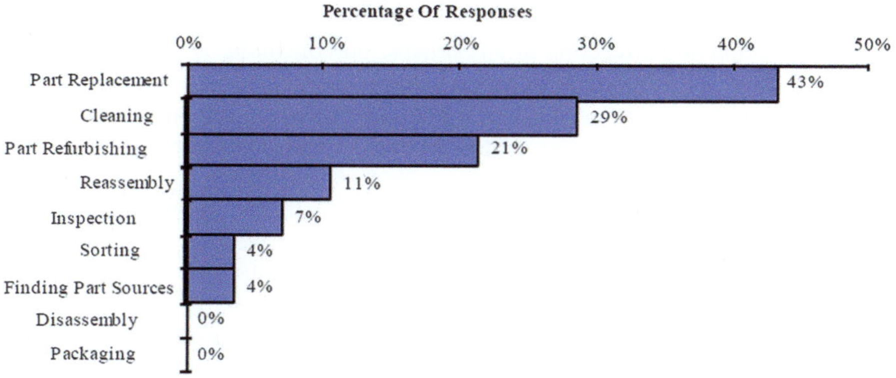

Fig. 2 Which is more costly in remanufacturing processes? [1]

chemical reactions. While in mechanical processes, external force is applied to the part being cleaned to separate undesired layers from the parts. It should be noted that liquid based cleaning can also employ mechanical energy to achieve a better cleaning effect.

2.2 *Liquid Based Cleaning Process*

2.2.1 Cleaning Mechanism of Liquid Based Cleaning Method

Most of the liquid based cleaning techniques rely on following mechanisms to achieve effective cleaning: wetting; emulsification; solubilization; saponification; deflocculation; and sequestration.

Wetting mechanism is essential to any liquid based cleaning. It delivers the cleaning chemistry to contaminants to be separated. Through wetting, substrate-soil bonds are broken, so that mechanical energy can be delivered to displace and remove the contaminants. Wetting can also reduce undesired surface and interfacial tensions, allowing cleaning agent to penetrate between the contaminant and the substrate.

Emulsification is the dispersion of oils to be removed in the solvent. The main factors of emulsification include types of oil and the surfactants selected. The pH level and temperature can also affect the level of emulsification. Mechanical energy, such as vibration, ultrasonic, and turbulence are generally employed to enhance the emulsion effect. Note that emulsification does not change the chemical characters of the contaminants, however it is essential for most cleaning process in effective separation of the contaminates from the substrate.

Solubilization is a process to enhance the solubility of the contaminants in a particular solution using surface-active agents. Solubilized contaminants are then dissolved into the solution. In a typical cleaning process, cleaning agents generally solubilize a certain amount of contamination while additional contaminant is held in suspension by emulsification.

Saponification is the reaction of any organic oil containing reactive fatty acids with free alkali to form soap. Alkaline cleaners containing saponifiers rely on this process to remove some oils, including vegetable and animal fats and their derivatives. The soaps that are generated are easily removed by subsequent rinsing with water.

Deflocculation causes the breakdown of contaminants into very small particles that are then dispersed in the liquid cleaning medium and swept away. This process is similar to emulsification except it happens on a larger scale.

Sequestration is a process where undesirable ions, such as Ca^{+2} or Mg^{+2}, and heavy metals are de-activated; preventing them from reacting with material that normally would form insoluble products. The classic example is the hard water scum formed when soaps are used. The scum formed is the reaction between the

Ca^{+2} or Mg^{+2} ions in hard water with soap. When the water is softened, the Ca^{+2} or Mg^{+2} ions become tied or sequestered and are unable to react.

For any cleaning processes, proper cleaning equipment is required to implement the cleaning mechanisms described previously. The cleaning equipment provides not only the site for accomplishing the cleaning process; it can also provide other desired functions such as separation and collection of removed coating or dirt. In addition, most of the cleaning equipment integrates heating or mechanical vibration to provide external agitation that enhances the cleaning effectiveness.

The goal of agitation of the cleaning solution is to apply external energy to the part surface so that the cycle time and effectiveness of the cleaning process can be enhanced greatly. Agitation can be achieved by simply stirring solution with rotary stirrers. Similar effect can be achieved by rotating parts inside the solution. Stirring agitation is gentle in general and does not significantly improve cleaning effectiveness unless the chemistry is very aggressive. Nonetheless, due to its simplicity and easy to implement, it can be applies in most processes. Ultrasonic agitation uses high-frequency sound waves to achieve mechanical agitation. Ultrasonic waves can also penetrate thin layers of metal and propagate around corners to clean work pieces inside and out. Ultrasonic cleaning is usually not appropriate for thick buildups of contaminant.

Based on the solution and external energy sources used, cleaning processes can be grouped as follows:

a) *Immersion cleaning*

Immersion cleaning refers to a group of the most applied cleaning methods for mechanical parts. It generally uses cleaners with high concentration. Convection current combined with external vibration, soils are removed from metal surface conveniently. This cleaning approach is particularly good for cleaning irregular shapes, box sections, tube and cylindrical configurations that cannot be penetrated using spray systems. The operation may vary from hand dipping a single part or agitating a basket containing several parts in an earthenware crock at room temperature to a highly automated installation operating at elevated temperature and using controlled agitation.

Several approaches of immersion cleaning are summarized below:

- Barrel cleaning: this approach is generally used for cleaning large quantities of small parts. Parts are placed and agitated inside a barrel that rotates in the cleaner solution.
- Moving conveyor cleaning: in this approach, parts are placed on a moving conveyor, which moves parts through solution flow.
- Mechanical contact: cleaner is applied with brushes or squeegees.
- Mechanical agitation: in this approach, parts are flooded with solution which is circulated using pumps, mechanical mixers, or ultrasonic waves.
- High pressure agitation: in this approach, a high pressure solution flow generated by pumps is applied to the parts to clean deep and blind holes as well as tubes with a small diameter.

b) *Ultrasonic Cleaning*
Ultrasonic cleaning employs high frequency ultrasonic waves (20–40 kHz) passing through liquid solutions to assist effective cleaning. Due to the gas bubbles created by ultrasonic waves inside the cleaners, ultrasonic cleaning can provide strong cleaning effects on the parts immersed in the solution. Ultrasonic cleaning is ideal for parts with complicated shapes, surfaces, and cavities that may not be easily cleaned by traditional immersion techniques.
The basic ultrasonic cleaning process generally is composed of following components: the cleaning tank, ultrasonic transducers, and the power supply.
Another similar cleaning technique is Megasonic cleaning. It uses a much higher frequency (700–1,000 kHz) acoustic energy to generate pressure waves in a liquid. Compared with ultrasonic, megasonic technique does not suffer from cavitations which is a typical drawback for ultrasonic. Less cavitations reduce the likelihood of surface damage.

2.3 Mechanical Cleaning

Another group of cleaning technology is based on employment of mechanical force to separate contaminants from the substrate. The mechanical force can be in the forms of air blowing or exhausting, vibration, abrasion using brushes or small hard particles blasted by air. Since no chemical reaction occurs during the cleaning process, one of the most attractive benefits of mechanical cleaning is less hazardous emissions. However, due to strong mechanical forces applied to parts to be cleaned, it is also possible to damage the substrates.

a) *Vibration cleaning*
Vibration cleaning utilizes high frequency rotary oscillation to create strong vibration that overcomes the adhesive force so that dirts are separated from the parts. The dirts separated can be exhausted to a special container and can be reused. Additional abrasive bush can be combined with the vibration movement to enhance the cleaning effectiveness and reduce cycling time
b) *Abrasive cleaning*
Abrasive cleaning use high speed propelling blade shot small hard particles on the part surface, thus cleaning contaminants by impact force. The particles used as abrasive media vary in types and sizes to meet specific cleaning scenarios. Abrasive cleaning is most commonly used method to remove heavy scale and paint on large easy to access parts. Major components of the Centrifugal blast machines include: blast wheel, work conveyor, abrasive recycling system, and a dust collection device.
c) *Dry-Blast cleaning*
Dry-blast cleaning is also called abrasive blasting cleaning. Dry blast cleaning is considered as the most efficient and environmentally effective method for abrasive cleaning. It generally employs a 685 kPa air supply system to propel

abrasive particles to separate contaminants from the parts. Different replaceable air-blast nozzles are developed with different shape and wear resistant materials. Although all metals can be cleaned abrasive blasting processes, one should carefully select suitable abrasive medium for soft and brittle metals such as aluminum, magnesium, copper, zinc, and beryllium, to avoid damage to the part itself.

With respect to the equipment available for dry blast cleaning, people developed several types based on different material handling approach:

d) Cabinet machines: A cabinet is used to contain the abrasive-propelling mechanism, holds the work in position, and confines flying abrasive materials and dust. Cabinet machines may be designed for manual, semiautomatic, or completely automated operation to provide single-piece, batch, or continuous-flow blast cleaning.

e) Continuous-flow machines: compared with cabinet machine, continuous flow machine uses proper conveying devices to continuously clean parts in the cabinets. These machines are used to clean coils and wires as well as castings and forgings at a high production rate. Combined with a abrasive particle recycling system, it can reuse the blast particles.

f) *CO_2 dry ice blasting*

   CO_2 dry ice blast is a special dry blasting method in that it uses frozen CO_2 particles or snow as abrasive media. Some parts may be sensitive to thermal changes from the pellets and should be tested first. While particles can be clean the surface at a faster rate, it can also damage the surface being cleaned. The advantage of the CO_2 dry ice blasting is that they sublimate on contact with the material to be cleaned.

2.4 *Testing and Inspection for Remanufacturing*

After parts are cleaned, inspection and testing procedures are followed to check if repair is required. Since parts to be remanufactured are always in different conditions, testing is generally unavoidable. Since the purpose of remanufacturing is to reuse the parts, most of the testing methods are not intended to create any damage to the part being tested. As such called, *nondestructive* testing is a commonly used technique to reveal flaws and defects in a material or device without damaging or destroying the test sample.

Since nondestructive testing (NDT) is a wide group of analysis techniques used in science and industry to evaluate the properties of a material, component or system without causing damage, currently commonly used NDT methods are summarized below.

2.4.1 Methods for Nondestructive Testing

NDT methods employ techniques such as microscope, electromagnetic radiation, sound, and combined with the inherent properties of materials to detect flaws in the parts to be remanufactured. Microscope method is generally used to examine external surfaces of the part being tested. To test the inside of the part, methods such as electromagnetic radiation and liquid penetrant testing are generally used to examine fatigue cracks. For liquid penetrant methods, a certain liquid is applies to penetrate and reveal the cracks. For non-magnetic material, fluid with fluorescent or non-fluorescing dyes is commonly used. For magnetic material, an externally applied magnetic field or electric current through the material is used. When parts have cracks, magnetic flux will leave at the area of the flaw, resulting in leakage of magnetic field at the flaw area. This leakage can be captured and used as an indication of flaws.

NDT can be further classified in to various methods and techniques. It is important to select the right method and technique for a specific part or material to ensure the performance of NDT.

Liquid Penetrant Inspection method usually takes following test procedure:

1. Pre-cleaning: cleaning methods discussed earlier are used to remove any dirt, oil, grease or any other contaminants to ensure that any defects are open to the surface, dry, and free of contamination.
2. Application of Penetrant: After parts are cleaned, penetrant is then applied to the surface. A certain period of time (5–30 min) is required to allow the penetrant to immerse into any flaws. The length of the penetration time depends on the penetrant being used, the type of material being testing, and the size of flaws being examined. Generally, smaller flaws require a longer penetration time. Excess penetrant has to be removed from the surface of the part being tested.
3. Application of developer: A developer is a chemical that draws penetrant from defects so that defects can be identified. From the stains that show up in the developer one can identify the positions and types of defects on the surface under inspection.
4. Inspection: In inspection, visible light is applied for visible dye penetrant. In contrary, for fluorescent penetrant, ultraviolet radiation is applied to the part surface being examined.
5. Post Cleaning: Cleaning is required to remove penetrant after inspection and recording of defects are finished.

As to magnetic penetrant testing, fine iron or magnetic particles, held in suspension in a suitable liquid, are used as penetrant. For better performance of the inspection, the particles are usually colored and coated with fluorescent dyes visible under ultraviolet light. To apply the penetrant, the liquid suspension is sprayed or painted on to the part, which is magnetized. Due to magnetic leakage at the defect area, the magnetic particles are attracted in the area of the defect. When UV light is applied, the location and size of the defect can be easily identified. Magnetic penetrant testing method is generally a low cost inspection method and is much faster than ultrasonic testing and radiographic testing.

Radiographic testing (RT) methods use short wavelength electromagnetic radiation to penetrate materials and reveal defect. Typical radiation source is an X-ray machine. Since the amount of radiation emerging from the opposite side of the material can be detected and measured, variations in the intensity of radiation are used to determine thickness or defect of material.

3 Disassembly Analysis and Disassembly Process Planning

One important step of remanufacturing is product disassembly [4]. A proper disassembly procedure can increase residual value recovery and reduce the environmental impact resulted in recycling processes. Disassembly analysis and planning in this regard, addresses three issues: (1) Optimal disassembly strategy that recovers maximum residual value, (2) Disassembly sequence planning, and (3) evaluation of disassembly time, cost, and disassembly difficulty rate, with component information provided.

The disassembly relationships among the components of a product to be remanufactured include component-fastener relationships and precedence relationships. Therefore, two types of graphs are needed in order to fully represent the relationships among the components of a product, namely, component-fastener relationship graph and precedence relationship graph.

Fasteners are used to attach one component to another for the purpose of assembly. Examples of fasteners include screws, rivets, inserts, etc. In a component-fastener graph $G_c = (V, E)$, The components are represented as the vertices $\mathbf{V} = \{v_1, v_2, \ldots, v_n\}$, where n is the number of components. Their relationships are represented as the edges $\mathbf{E} = \{e_1, e_2, \ldots, e_m\}$, where m is the number of edges. If two components vi and vj $(i \neq j)$ are joined by fasteners, then $(vi, vj) \in \mathbf{E}$; otherwise $(vi, vj) \notin \mathrm{E}$. The graph Gc is an undirected graph. Vertices and edges in graph Gc are modeled using object-oriented techniques. While the object vertex consists of component information including its name, weight, material type, etc., the object edge consists of fastener information including the number of fasteners, fastener type, etc. For example, Fig. 3a. shows component-fastener graph of a personal computer.

Precedence graph represents the precedence relationship among the components of a product, namely, a component cannot be disassembled before certain components. Figure 3b shows the precedence relationship graph.

Disassembly tree can then be constructed based on the component-fastener graph and precedence graph. The disassembly tree consists of vertices representing an assembly or a component and information such as its name, material type, weight/volume. A vertex is decomposed into child vertices representing its child sub-assemblies or components. An edge, linking a child vertex with its parent vertex, represents the disassembly relationship between two components and information about assembly method.

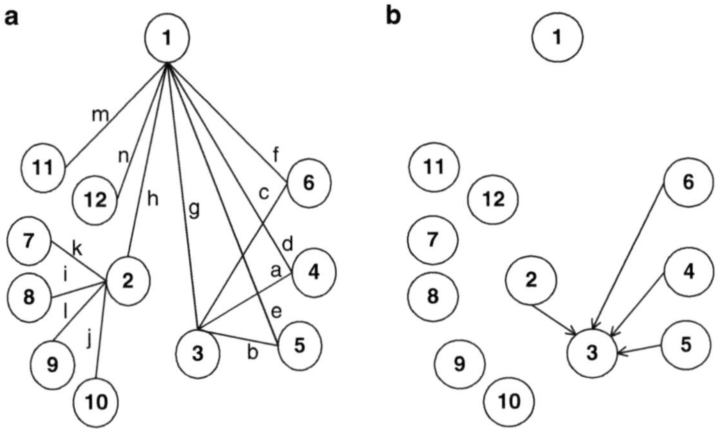

Fig. 3 (**a**): Component-fastener graph for the assembly, (**b**): Precedence relationship graph for the assembly

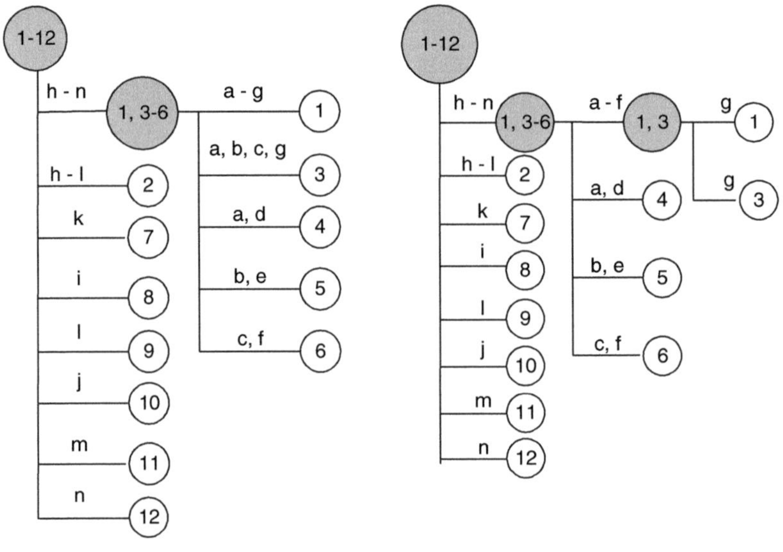

Fig. 4 (**a**): Pseudo-disassembly tree. (**b**): Disassembly tree for the assembly

The disassembly tree is constructed through searching of cut-vertices in the component-fastener graph. A cut-vertex is a vertex whose removal disconnects the graph. If a cut-vertex is found, the graph is split into two or more sub-graphs. The same procedure is repeated until no cut-vertices can be found. In this way, a pseudo-disassembly tree is generated which is showed in Fig. 4a.

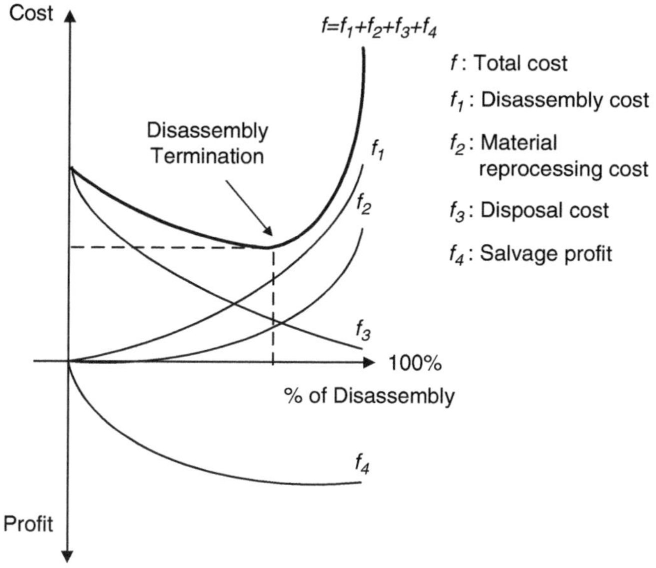

Fig. 5 Optimal disassembly termination analysis

The pseudo-disassembly tree is then modified by the precedence of the disassembly according to the precedence graph, and a disassembly tree can be obtained as illustrated in Fig. 4b.

In disassembly sequence planning, a popular assumption is that end-of-life products should be disassembled to the fullest extent possible. However, based on discussion with the recycling industry, such assumption is not practical in many cases due to the high cost of disassembly. It is very important to find the optimal level for disassembly where the benefit of reverse manufacturing is maximized and the cost is minimized. The disassembly sequence planning can be determined after such a termination point.

Optimal disassembly planning is determined based on the cost and profit. Three types of costs and one type of profit are addressed: (1) disassembly cost which includes labor and tooling cost, (2) material reprocessing cost, i.e. cost of recycling (3) disposal cost, which includes transportation fee and landfill cost, and (4) salvage profit, which is the profit gained by means of component reuse or recycling. The cost model for determining the termination of disassembly is illustrated in Fig. 5.

The total cost is calculated as the sum of disassembly cost, material reprocessing cost, disposal cost, and salvage profit. The lowest point of the curve (f) representing the total cost determines the termination of disassembly where the cost is minimized, in other words, the benefit of disassembly is optimized. Note that the obtained disassembly plan is optimized just from the viewpoint of economy and the plan is not always optimal from the environmental viewpoint.

4 Remanufacturing Production Planning and Optimization

The most significant characteristic of remanufacturing production system is its unstable and uncertain incoming flow [2]. The returned products generally have a high uncertainty in arrival pattern and high variation in product type with disparate residual value. Quantity, year of model, and quality of returned products are also subject to high uncertainty. For example, the product might come from a software company that updates its computers every 3 months or they might come from a family replacing its 10-year-old home computer. The consequence of high uncertainty and variation of the return flow is the difficulty associated in production planning and control of the remanufacturing, which leads to increased production cost and poor economic performance [3].

Another major challenge of remanufacturing comes from the distinct role of the receiving inventory. On one hand, it differs from traditional ones in that customers return their post consumer products to the inventory instead of taking the product away from the inventory. In this regard, inventory is used to meet the product return demand. A redistribution cost, which does not exist in a forward manufacturing system, is incurred when the remanufacturer finds no inventory space to handle the returns. On the other hand, receiving inventory can still act as a buffer to dampen the randomness of material arrival process, thus providing a relatively stable material flow for the reverse production. However, replenishment of stocks (post consumer products) is a stochastic process with high uncertainty, while remanufacturer has little control over it. This generally results in huge safety inventory for the remanufacturers.

As in forward manufacturing, operations and processes of remanufacturing should also be aligned and optimized to maximize the total profit. This leads to following three production planning problems that need to be addressed:

1) First of all, remanufacturing system has to handle substantial number of product types. Generally these different products share one production line. Therefore, a priority based switch rule has to be developed for production planning to determine how and when to switch between different production types. The priority mechanism is generally based on following concerns. The first concern is the depreciation rate of the products or components received. Products with the highest depreciation rate should be given first consideration. The second concern is the residual value of the product. Generally, products with a high residual value should be processed first. The third concern is the environmental impact. If the product has in-transition environmental impact, it should also be processed early. The fourth concern is the market demand. If the secondary-market demand for a certain remanufactured product or component is higher, these products should be handled first. In determining when to switch, production lot size for different products with different priorities have to be determined and optimized to reduce total holding cost, set up cost and redistribution cost.
2) The second issue for remanufacturing planning and control is determination of the optimal receiving inventory capacity and safety stock level. On one hand,

receiving inventory capacity set a constraint of safety stock level and the possible production run size (that is, the number of products to be produced in one run without changing the production configuration). It also has significant impact on both stability of production and redistribution cost. With more receiving inventory space, a higher level safety stock can be allocated to improve the stability of reverse production system. This could result in better efficiency. More receiving inventory space will also reduce the chance of redistribution and associated cost. Nonetheless, excessive inventory capacity also has shortcomings—large inventory capacity increases the space cost, while higher safety inventory results in higher inventory cost.

3) The third problem is to determine the optimal workforce level and production capacity. The unstable and uncertain incoming flow of the dedicated model requires workforce level and production capacity respond to the product return demand so that excessive capacity can be avoided. However, changing capacity of any production system will always incur costs.

Obviously, effective modeling and analysis of the production model of remanufacturing system is critical to attack the problems discussed. Approaches such as Queuing networks or other mathematical modeling techniques are possible options. However, due to the special stochastic characteristics of the arrival process and the priority based switching rules in production planning, the Queuing model has to consider both the compound bulk arrival and the priority Queuing. Analysis of priority queue with compound bulk arrival has shown to be very hard to solve. Optimization with simulation methods proved to be an effective approach and can be used in optimization of a system that possesses the characteristics described in a remanufacturing system [3].

4.1 General Simulation Model

The general simulation model developed for the remanufacturing is illustrated in Fig. 6. The remanufacturing system receives the products with stochastic, compound and batch arrival. It is assumed that a batch of n different products with random quantities (x_1, x_2, …, x_n) is transported via the same truck. It is also assumed that there is one production line that is capable of producing each of the different product types. The production line has N stations (or stages), with a queue in front of every station. Every station has one or more identical servers with a stochastic service time. There is a transit time between any two consecutive stations, which is assumed to be exponentially distributed. The production line does not have any coordination of job movement between stations. An available operator starts a job as soon as it is available and, upon completion, the job leaves the station provided there is room at the next station. This mode of operation may cause starvation and blocking of servers. A bulk of returned products will be accepted to the receiving inventory if enough receiving space is available to hold the entire load. Otherwise,

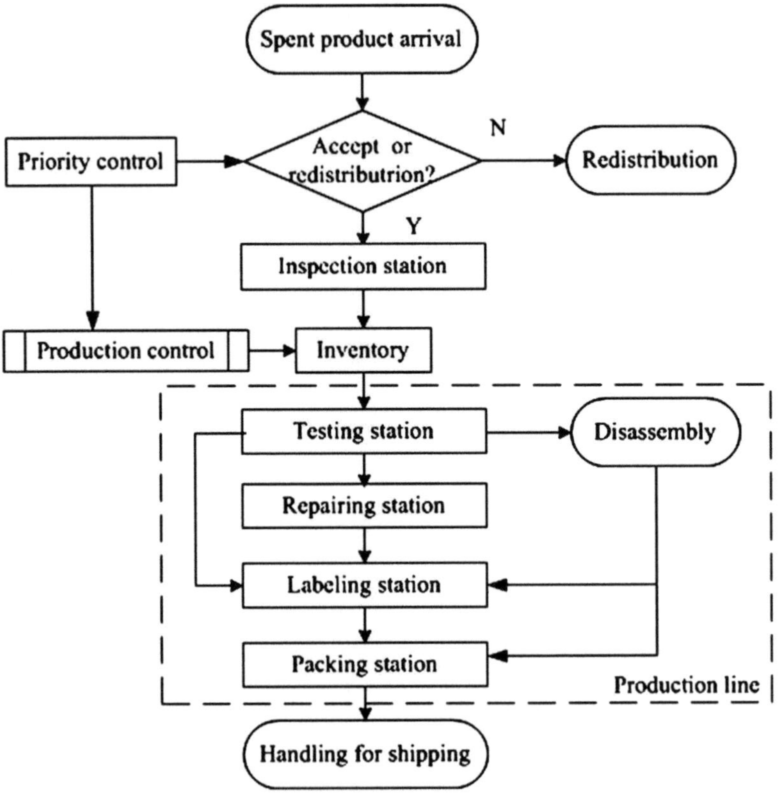

Fig. 6 General remanufacturing production flowchart

as many products as possible will be accepted and products with higher priority will be considered first. The rest of the load would be refused and a redistribution cost would be incurred.

4.2 Production Switch Rule

Since the production line is shared for processing different product types, a control mechanism is required to switch the line from one product type to another considering a specific run size and specific priorities. The switch rule can be either based on production batch or based on the product type. Production based switch is launched only when the inventory of products currently being processed is totally depleted. A variation of this rule is switching the production line after a certain period of time without referring to the current inventory level. For either case, all product types sharing the same production line implicitly have the same priority.

Another rule strictly follows the priority of each product type. Priority based rule could be impractical because it could switch the production line too frequently raising the setup costs. Consequently, it is rational to combine the product priority rule with the batch production rule. A pseudo code for the priority based batch switch control rule developed for the simulation model is given as follows:

Assume there are N types of products with distinct priority level, $N = 1, 2, 3, \ldots, n$, where the smaller number means higher priority. Let I_i denote the inventory level of product type i, *current* denote the type of product currently being processed, and q_i denote the run size of product i, where $i = 1, 2, 3, \ldots, n.$, then we have the following switch rule:

```
if (Production line is running and receiving inventory station is requested to send more
products to the production line) or (production line is idle and a new bulk arrives ) then

for i= 1 to n
     begin
     if i < current (i.e. product i has priority over current product) and I_i > q_i then
                   current = i
                   switch production line to product i
                   break
            else if i >= current and I_current > 0, then
                        continue sending current type of product to production line
                        break
            else
                   idle period begins, waiting till more spent products arrival
                   break
            end if
     end
end if
```

4.3 Optimization Problem Formulation

The control variables in the optimization model of the remanufacturing system are categorized into four types: inventory capacity, run size of each product type, number of workers in each manufacturing cell, and the buffer size of each manufacturing cells. The objective of the analysis is to find the optimal value of these decision variables that maximize the total net profit. The objective function of the remanufacturing system can be expressed as follows:

$$\max_{I \in D_I, W \in D_W, B \in D_B, Q \in D_Q,} f(I, W, B, Q) = TR - TC \tag{1}$$

Where

I = available inventory space for the receiving area.

W = $(w_1, w_2, \ldots, w_p)$ is the vector representing number of workers in manufacturing cell 1 to p,

$\boldsymbol{B} = (b_1, b_2, \ldots, b_p)$ is the vector representing buffer size of each manufacturing cell,
$\boldsymbol{Q} = (Q_1, Q_2, \ldots, Q_n)$ is the vector representing run size of product type 1 to n,
(D_I, D_W, D_B, D_Q) represents the feasible domain of (I, $\boldsymbol{W}$, $\boldsymbol{B}$, $\boldsymbol{Q}$).

Total expected profit can be derived as the difference between the expected total revenue (*TR*) and the expected total cost (*TC*) of a dedicated remanufacturing system. Total annual revenue is assumed to be

$$TR = \sum_{i=1}^{n} R_i \times V_i \tag{2}$$

where R_i is the residual value of product type i and V_i is the total volume (number) of product type i processed per year. The total cost is broken down into five major cost categories:

$$TC = C_c + C_L + C_M + C_H + C_F \tag{3}$$

Where

C_c: product collection cost. This includes the purchasing cost of used products from customers and the transportation cost during the collection process.
C_L: logistics cost. This is incurred during distribution and redistribution of the collected products. When the receiving inventory is full, redistribution cost is incurred in the re-transportation of returned products to other remanufacturing facilities.
C_M: remanufacturing processing cost. This includes labor cost, materials cost and utility cost, which are incurred in machine operating, line switch and setup, and line and operator idling.
C_H: inventory holding cost, which is incurred by holding received products in the inventory area and the production line.
C_F: fixed cost of running the factory regardless of the production level. This includes general utility, air-conditioning, insurance, and facility depreciation.

4.4 Hybrid GA Simulation Optimization Approach

Based on the objective function and the simulation model, a hybrid GA approach which combines the Fractional Factorial Design (FFD) with the GA method was developed.

As shown in Fig. 7, the optimizationprocedure starts with dividing the solution space into subspaces called cells. Each cell is considered as the local solution space of the FFD. The FFD is used to find the extrema of each cell. The results of the FFD provide the solution candidates for the GA. Based on the corresponding extrema of each cell produced by the FFD the GA will continue the search process until the termination condition is met. It is important that a well thought out fraction of the design be selected when the FFD is used to coordinate both efficiency and accuracy. A high fraction will increase the efficiency while losing accuracy as a trade-off.

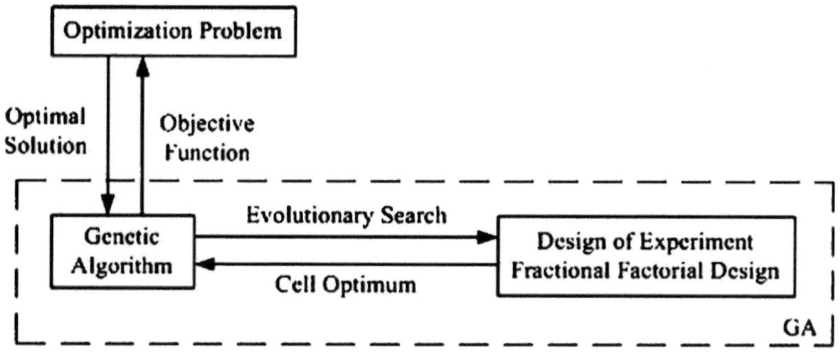

Fig. 7 The hybrid GA approach

Similarly, low fraction will increase the accuracy but will take more time to find the extrema. In general, the FFD will not only make sure that the output is the local optimum in the cell thereby improving the searching accuracy of the GA by considering all solutions in a subspace instead of a unique point, but also improve the searching efficiency due to its fractional runs. On the other hand, the GA can guarantee promising solutions due to its effective global searching performance. For a more thorough explanation of fractional factorial designs please refer to Montgomery [9].

4.5 *Case Study*

A case study is conducted based on a remanufacturing plant located in Austin Texas. The plant recovers, reuses, and recycles used laptops and desktops. The two types of products share the same reproduction line.

4.5.1 Model Parameters Assumptions

A number of parameters were obtained and assumptions were made in this case study based on conversations with the plant managers.

Truck arrivals are assumed to follow a Poisson process with the mean time between arrivals of 4 h per 8 h a day. Both laptop and desktop are contained in the same truck load. However, the proportion of desktop and laptop in a truck is not fixed and is a random number. The number of desktops and laptops in a single shipment satisfy the following equation: $0.5 \times$ (number of Laptops) $+ 1 \times$ (number of Desktops) $= 260$.

The number of laptops is a random integer variable with a uniform distribution between 0 and 520. The number of desktops is also a random number, which is complementary to the number of the laptops and equals $260 - 0.5 \times$ (number of laptops).

Based on the interview with the plant manager, we assume that 10 % of received computers will pass testing and be labeled directly; 10 % of them, which cannot be remanufactured, will be torn down for further recycling; the remaining 80 % of computers have to be fixed and labeled after repair.

The capacity of the receiving inventory was initially set to 500 sq. ft. If the returned products find no space available in the receiving area they will be redistributed. The redistribution fee is $450 each time regardless of how many computers are re-transported. It is assumed that all of the finished products will be immediately shipped and sold out after packaging. Hence, there is no inventory for finished goods.

There is only one production line that is shared by laptops and desktops. Based on the aforementioned production priority rule, the production line will switch with a setup time of 30 min. Letting q_L and q_D be the run size of laptop and desktop respectively, and assuming that the current production line is processing desktops. The priority based switch control can be stated as follows:

g) If I_L (number of laptops in the inventory) $> q_L$, the production line will be switched to process laptops.
h) Else if the I_D (number of desktops in the inventory) is greater than zero, the production line will keep processing desktops.
i) Otherwise, it will wait for the arrival of more computers, which causes an idle period.

The selling price of the remanufactured desktops is assumed to be $250 per unit and the selling price of the remanufactured laptops is assumed to be $400 per unit. Other parameters, assumptions and factory profile are summarized in Table 1 below.

4.5.2 Cost/Profit Evaluation

Based on the analysis outlined earlier, the costs for all related operations are summarized in Table 2.

The total inventory cost can be derived by summing all of the costs:

$$TIC = C_{space} + C_{utility} + C_{handling} + C_{equipment} \tag{4}$$

The fixed cost of the whole plant includes utility cost, as well as building and equipment expense that includes machine depreciation, building rental, taxes, insurance, fire protection, and general maintenance cost. The total building and equipment expense is $811,000 per year.

The total profit for the remanufacturer is the difference between gross revenue and total costs which is given by following equation: Profit = (Finished Desktops) × (Desktop Sell Price) + (Finished laptops) × (Laptop Sell Price) – Total Cost.

Table 1 Model assumptions

Labor	Working time	8 h per day	350 work days per year = 2,800 h per year
	Work efficiency	90 %	
Factory	Space	3,200 sq. ft.	
Truck	Arrival rate	Poisson (4 h)	
	Bulk capacity	260 sq. ft.	
	Returned products	Desktop	0.5 sq. ft. per unit, random # units per truck
		Laptop	1.0 sq. ft. per unit, random # units per truck
Inventory	For receiving	500 sq. ft.	
	For finished goods	0	
Production line	One production line shared by laptops and desktops		
	30 min setup time per switch		
	Run size	Laptop	q_L
		Desktop	q_D
Sale price	Laptop	$400 per unit	
	Desktop	$250 per unit	

4.6 Simulation Model

The simulation model for the remanufacturing operation is developed using Arena™ Software. The flowchart module is demonstrated in Fig. 8. The general purpose of the model is to analyze the effect of operational changes on the profit performance of this dedicated reverse manufacturing system. The ultimate goal of the simulation model is to find the optimal configuration of the production system resulting in maximum profit.

The distribution of the time for each operation modeled in the simulation model is assumed to be an exponential distribution, where the normal time listed in Table 3 represents the expected value. An exponential distribution is used for all service times in order to simulate the large range of possible values.

4.6.1 Optimization

In order to optimize the remanufacturing system, ten control variables are identified as important to the performance of the system under study. These parameters include: (1) receiving inventory capacity (I), (2) run size of laptops (q_L), (3) run size

Table 2 Cost analysis

Collection costs	Desktop	$150/unit	
	Laptop	$250/unit	
Logistic costs	Transportation cost from the supplier to the remanufacturer's site is included in the collection cost		
	Redistribution cost	$450 per redistribution	
Manufacturing costs	Labor costs	Operators	$9/worker/h
		Inspectors	$12/worker/h
		Supervisors	$15/worker/h
	Material costs	$1.6/day	
	Utility costs	Electricity	9 kWh/day
		Maintenance	$4/day
	Setup costs	30 min/switch	
Inventory holding cost	Space costs	$277.74/h/ 1,000 sq.ft.	
	Utility costs	Electricity	1 kWh/day
		Maintenance	$8/day
	Handling costs	$72/day	
	Inventory opportunity costs[a]	0.00882 %/h	
	Depreciation costs	Laptop	$4.8/h/unit
		Desktop	$3.2/h/unit
Fixed cost	Utility costs	21 kWh/day	
	Building	$277.74/h/ 1,000 sq.ft.	

[a]The presence of the goods in inventory means that they are not being sold, thus the opportunity of earning with the money invested in the inventory is lost

Table 3 Manufacturing cell analysis

		Receiving	Inspection	Inventory handling	Testing	Repairing	Labeling	Packing	Tear down	Shipping/handling
Cycle time (min)	Laptop	3.24	1.05	0.5425	6.5	15	5.66	9.1462	5.025	1.65
	Desktop	3.24	1.23	0.5425	7.32	20	5.66	9.1462	5.725	1.65

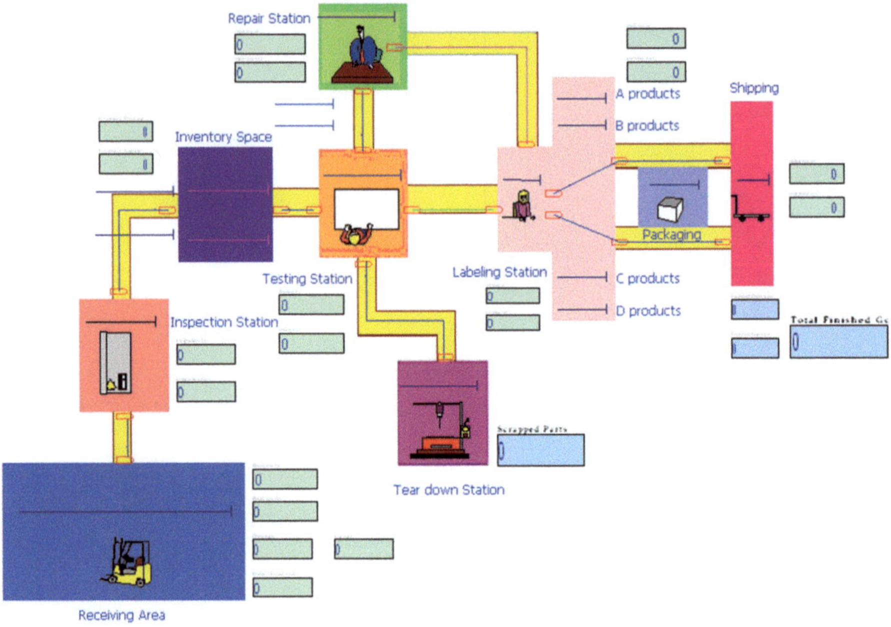

Fig. 8 Simulation model for reverse manufacturing

of desktops (q_D), (4) buffer size of repair stations (b_r), (5) buffer size of labeling area (b_l), (6) buffer size of packing station (b_p), (7) number of workers in the testing cell w_t, (8) number of workers in the repairing cell (w_r), (9) number of workers in the labeling cell (w_l), and (10) number of workers in the packing cell (w_p). Based on the consulting from the plan manager, the reasonable range for each control variable is also obtained:

$I = \{200, 300, 400, 500, 600, 700, 800\}$
$q_L = q_D = \{30, 40, 50, 60, 70, 80\}$
$b_r = b_l = \{2, 4, 6, 8, 10\}$
$b_p = \{10, 20, 30\}$
$w_t = \{4, 5, 6, 7, 8\}$
$w_r = \{10, 11, 12\ 13, 14, 15, 16, 17, 18, 19, 20, 21\}$
$w_l = \{3, 4, 5, 6\}$
$w_p = \{5, 6, 7, 8, 9, 10\}$

In order to use the GA approach presented earlier, these ranges were decomposed into smaller cells so that the two-level FFD algorithm can be implemented. Recall that each control variable has only one or two values. In doing this, the following new segments were obtained:

$I_1 = \{200, 300\}, I_2 = \{400, 500\}, I_3 = \{600, 700\}, I_4 = \{800\}$
$q_{L1} = \{30, 40\}, q_{L2} = \{50, 60\}, q_{L3} = \{70, 80\}$
$q_{D1} = \{30, 40\}, q_{D2} = \{50, 60\}, q_{D3} = \{70, 80\}$

Table 4 Initial population

Cell index	b_r	b_l	b_p	I	q_L	q_D	w_t	w_r	w_l	w_p
7,976	[2, 4]	[6, 8]	[10, 20]	[200, 300]	[30, 40]	[50, 60]	[8]	[16, 17]	[3, 4]	[7, 8]
22,517	[2, 4]	[10]	[30]	[800]	[30, 40]	[50, 60]	[6, 7]	[14, 15]	[5, 6]	[7, 8]
25,832	[6, 8]	[2, 4]	[10, 20]	[600, 700]	[50, 60]	[70, 80]	[6, 7]	[16, 17]	[3, 4]	[5, 6]
28,314	[6, 8]	[2, 4]	[30]	[400, 500]	[30, 40]	[50, 60]	[4, 5]	[14, 15]	[5, 6]	[9, 10]
43,652	[6, 8]	[10]	[30]	[200, 300]	[70, 80]	[70, 80]	[4, 5]	[16, 17]	[3, 4]	[7, 8]
18,526	[2, 4]	[10]	[10, 20]	[800]	[30, 40]	[30, 40]	[6, 7]	[16, 17]	[5, 6]	[5, 6]
1,166	[2, 4]	[2, 4]	[10, 20]	[400, 500]	[30, 40]	[50, 60]	[8]	[14, 15]	[3, 4]	[7, 8]
66,424	[10]	[10]	[30]	[200, 300]	[50, 60]	[30, 40]	[4, 5]	[10, 11]	[5, 6]	[5, 6]
16,444	[2, 4]	[10]	[10, 20]	[200, 300]	[70, 80]	[70, 80]	[4, 5]	[18, 19]	[5, 6]	[5, 6]
4,427	[2, 4]	[2, 4]	[30]	[200, 300]	[50, 60]	[50, 60]	[8]	[20, 21]	[5, 6]	[7, 8]

$b_{r1} = \{2, 4\}, b_{r2} = \{6, 8\}, b_{r3} = \{10\}$
$b_{l1} = \{2, 4\}, b_{l2} = \{6, 8\}, b_{l3} = \{10\}$
$b_{p1} = \{10, 20\}, b_{p2} = \{30\}$
$w_{t1} = \{4, 5\}, w_{t2} = \{6, 7\}, w_{t3} = \{8\}$
$w_{r1} = \{10, 11\}, w_{r2} = \{12, 13\}, w_{r3} = \{14, 15\}, w_{r4} = \{16, 17\}, w_{r5} = \{18, 19\}, w_{r6} = \{20, 21\}$
$w_{l1} = \{3, 4\}, w_{l2} = \{5, 6\}$
$w_{p1} = \{5, 6\}, w_{p2} = \{7, 8\}, w_{p3} = \{9, 10\}$.

Combining the segment for each parameter, we have a total of $4 \times 3 \times 3 \times 3 \times 3 \times 2 \times 3 \times 6 \times 2 \times 3 = 69{,}984$ cells, which compose the original domain. Index numbers are also assigned to each of the cells from 1 to 69,984.

The population size N of each generation is set to 10. Other important parameters for the GA approach are crossover rate, P_c, and mutation rate, P_m, which are set to 0.8 and 0.78 respectively. Therefore, in each generation, 8 ($=N \times P_c$) of ten individuals will be selected to crossover and generate eight new designs. Among these eight new designs, 6 ($=N \times P_c \times P_m$) will be chosen for mutation.

To initialize the start population, ten random integers are generated with a uniform distribution between 0 and 69,984; they are {7,976, 22,517, 25,832, 28,314, 43,652, 18,526, 1,166, 66,424, 16,444, 4,427}. The corresponding cells are listed in Table 4.

A $\frac{1}{16}$ Fractional Factorial Design is built for each of these cells with a run size of 64 ($2^{10-4} = 64$). The output of the simulation model is used in the FFD analysis to determine the optimum for each individual cell. The result provides the first generation listed in Table 5.

In crossover, eight of ten individuals, individuals 1, 2, 3, 4, 6, 7, 9, 10, in the first generation are randomly picked with probability P_c. Meanwhile, based on a uniform distribution U (0, 1), four random numbers, 0.46, 0.91, 0.33, and 0.78, are generated for λ. The result is shown in Table 6.

In mutation, six of the eight new designs, new designs 2, 3, 4, 5, 6, 7, 8, are randomly picked with probability P_m. Meanwhile, six random numbers, 0.86, 0.74, −0.01, −0.27, 0.42, and −0.67, are generated for ζ following normal distribution

Table 5 First generation

Individual	Control parameters b_r	b_l	b_p	I	q_L	q_D	w_t	w_r	w_l	w_p	Output ($)
I_1	2	6	20	300	40	50	8	17	4	7	18,970.26
I_2	4	10	30	800	30	50	7	14	5	8	28,161.37
I_3	8	4	20	600	50	70	5	16	4	7	19,443.51
I_4	8	4	30	500	40	60	7	15	6	10	29,153.52
I_5	8	10	30	300	70	70	5	16	4	7	18,896.89
I_6	2	10	20	800	30	40	6	16	6	6	19,505.26
I_7	4	2	10	500	30	60	5	14	4	7	18,346.21
I_8	10	10	30	300	60	30	5	11	5	6	17,914.16
I_9	4	10	20	300	80	70	5	19	5	5	15,543.76
I_{10}	2	2	30	300	50	60	8	21	5	8	24,729.08

N (0, 1). The resulting mutation is listed in Table 7. The last column contains the corresponding cell indexes.

The FFDs are built for these new designs. After running the simulation model, the output is analyzed using the FFDs to find the cell optima. After eight optima are obtained, the best ten was selected from them and the second generation as the population of the first generation, which are shown in Table 8.

Similar procedures of crossover, mutation and cell analysis are followed to generate the rest of the generations until the termination condition is met. In this case, the procedure stops if the optimum does not change for two generations or the differences among individuals in a generation is less than 5 %. Under this criterion, the GA approach stops after seven generations. Figure 9 shows the outputs of each generation. In each generation ten seeds are selected for mutation and crossover which lead the next generation. These ten selected ones are demonstrated in Fig. 9. As illustrated, the profit of each generation is increasing as the generation evolves. The final optimal solution of the remanufacturing system in this case study is 700 sq. ft. for receiving inventory, 40 for run size of laptops, 80 for run size of desktops, 18 workers in the repairing station withbuffer size of 8, 6 workers in the labeling station with buffer size of 6, 10 workers in the packing station with buffer size of 20 and 8 workers in the testing station.

5 Conclusion

This chapter summarizes the critical issues involved in remanufacturing. Typical remanufacturing processes including cleaning, testing and inspection, and disassembly are illustrated. Characteristics of remanufacturing production system and problems are also introduced. A GA optimization approach based on the simulation model is also developed to obtain the optimal production policy.

Table 6 Cross over of first generation

New design	Cross over function	Control parameters									
		b_r	b_l	b_p	I	q_L	q_D	w_t	w_r	w_l	w_p
N_1	$I_1 \times 0.46 + I_2 \times 0.54$	3.08	8.16	25.4	570	34.6	50	7.46	15.38	4.54	7.54
N_2	$I_1 \times 0.54 + I_2 \times 0.46$	2.92	7.84	24.6	530	35.4	50	7.54	15.62	4.46	7.46
N_3	$I_3 \times 0.91 + I_4 \times 0.09$	8	4	20.9	591	49.1	69.1	5.18	15.91	4.18	7.27
N_4	$I_3 \times 0.91 + I_4 \times 0.09$	8	4	29.1	509	40.9	60.9	6.82	15.09	5.82	9.73
N_5	$I_6 \times 0.33 + I_7 \times 0.67$	3.34	4.64	13.3	599	30	53.4	5.33	14.66	4.66	6.67
N_6	$I_6 \times 0.67 + I_7 \times 0.33$	2.66	7.36	16.7	701	30	46.6	5.67	15.34	5.34	6.33
N_7	$I_9 \times 0.78 + I_{10} \times 0.22$	3.56	8.24	22.2	300	73.4	67.8	5.66	19.44	5	5.66
N_8	$I_9 \times 0.22 + I_{10} \times 0.78$	2.44	3.76	27.8	300	56.6	62.2	7.34	20.56	5	7.34

Table 7 Mutation of first generation

New design	Mutation function	Control parameters										Corresponding cell index
		b_r	b_l	b_p	I	q_L	q_D	w_t	w_r	w_l	w_p	
N_1	–	3.08	8.16	25.4	570	34.6	50	7.46	15.38	4.54	7.54	13, 769
N_2	$N_2 + 0.86$	3.78	8.7	25.46	530.86	36.26	50.86	8.4	16.48	5.32	8.32	20, 615
N_3	$N_3 + 0.74$	8.74	4.74	21.64	591.74	49.84	69.84	5.92	16.65	4.92	8.01	56, 976
N_4	$N_4 - 0.01$	7.99	3.99	29.09	508.99	40.89	60.89	6.81	15.08	5.81	9.72	28, 350
N_5	$N_5 - 0.27$	3.07	4.37	13.03	598.73	29.73	53.13	5.06	14.39	4.39	6.4	2, 065
N_6	$N_6 + 0.42$	3.08	7.78	17.12	701.42	30.42	47.02	6.09	15.76	5.76	6.75	13, 775
N_7	$N_7 - 0.67$	2.89	7.57	21.53	299.33	72.73	67.13	4.99	18.77	4.33	4.99	12, 445
N_8	–	2.44	3.76	27.8	300	56.6	62.2	7.34	20.56	5	7.34	503

Table 8 Second generation

Individual	Control parameters										Output ($)
	b_r	b_l	b_p	I	q_L	q_D	w_t	w_r	w_l	w_p	
I_1	8	4	30	500	40	60	7	15	6	10	29,153.52
I_2	8	4	30	500	40	60	7	15	6	10	29,153.52
I_3	2	10	30	500	40	60	8	16	6	8	28,684.64
I_4	4	10	30	800	30	50	7	14	5	8	28,161.37
I_5	10	8	20	700	60	80	7	17	6	8	27,993.82
I_6	4	8	30	700	30	50	7	14	5	8	26,383.17
I_7	2	8	20	700	30	60	7	16	6	8	25,424.63
I_8	2	2	30	300	50	60	8	21	5	8	24,729.08
I_9	4	2	30	300	60	60	7	20	6	8	24,287.99
I_{10}	4	4	20	700	40	60	5	15	4	6	19,564.63

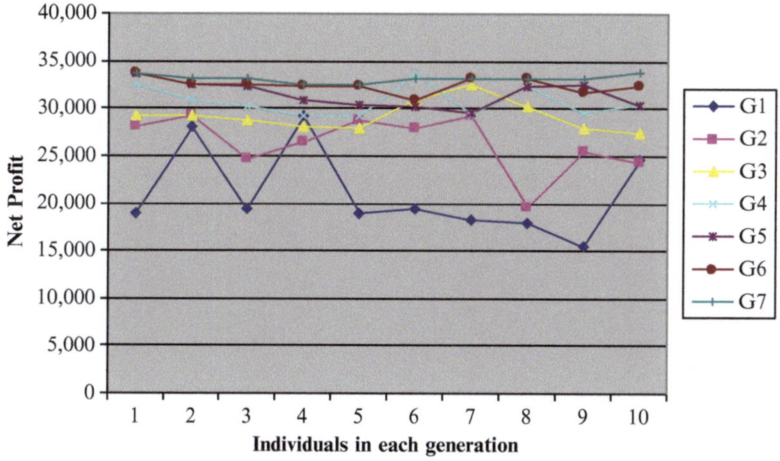

Fig. 9 Outputs of different generations

References

1. Hammond, R., Amezquita, T., & Bras, B. (1998). Issues in the automotive parts remanufacturing industry–A discussion of results from surveys performed among remanufacturers. *International Journal of Engineering Design and Automation – Special Issue on Environmentally Conscious Design and Manufacturing, 4*(1), 27–46.
2. Ilgin, M., & Gupta, S. (2010). Environmentally conscious manufacturing and product recovery (ECMPRO): A review of state of the art. *Journal of Environmental Management, 91*, 563–591.
3. Li, J., González, M., & Zhu, Y. (2009). A hybrid simulation optimization method for production planning of dedicated remanufacturing. *International Journal of Production Economics, 117*(2), 286–301.
4. Li, J., Puneet, S., & Zhang, H. C. (2004). A web-based system for reverse manufacturing and product environmental impact assessment considering end of life dispositions. *Annals of CIRP: Manufacturing Technology, 53*, 5–8.

5. Lund, R. (1983). *Remanufacturing: United States experience and implications for developing nations*. Washington: The World Bank.
6. Lund, R. T. (1984). Remanufacturing. *Technology Review, 87*(2), 19–29.
7. Lund, R. (1996). *The remanufacturing industry: Hidden giant*. Boston: Boston University.
8. Matsumoto, M., & Umeda, Y. (2011). An analysis of remanufacturing practices in Japan. *Journal of Remanufacturing, 1*(2), 1–11.
9. Montgomery, D. C. (2005). *Design and analysis of experiments* (6th ed., pp. 160–335). New York: Wiley.

Viscous Interfacial Motion: Analysis and Computation

Jin Wang

Abstract We consider the interfacial flows between two viscous incompressible fluids. After formulating the mathematical framework, we first present analytical solutions to the linearized problem, and discuss some results from linear asymptotic analysis. We then describe a numerical method for computing the nonlinear motion which ensures a high accuracy on and near the moving interface. Simulation results on viscous Stokes waves are presented to demonstrate the advantages of this method. In addition, as an example of nonlinear asymptotic study, we conduct a perturbation series analysis for Stokes waves with small viscosity, the results of which provide an analytical justification to the numerical observation.

1 Introduction

Interfacial motion between two viscous incompressible fluids is abundant in our world; common examples include water waves, bubbles, droplets, rain, cavitation, and oil spill, just to name a few. These phenomena span a wide range of scientific disciplines such as fluid dynamics, geophysics, oceanography, material science, mechanical engineering, and aerospace engineering, which underscores the importance of studying and understanding viscous interfacial flow problems. Mathematically, the motion in each fluid is governed by the incompressible Navier-Stokes equations, the solutions of which are connected through the interfacial conditions [3]. There are several difficulties associated with the study of such problems. First, the motion is strongly nonlinear and analytical solution is usually impossible to obtain. Second, the incompressibility condition has to be satisfied (somehow in an implicit manner) at all times [9]. In addition, the domain of interest

J. Wang (✉)
Department of Mathematics and Statistics, Old Dominion University, Norfolk, VA 23529, USA
e-mail: j3wang@odu.edu

B. Toni (ed.), *New Frontiers of Multidisciplinary Research in STEAM-H (Science, Technology, Engineering, Agriculture, Mathematics, and Health)*,
Springer Proceedings in Mathematics & Statistics 90, DOI 10.1007/978-3-319-07755-0_16,

contains an unknown interface which evolves in time and which must be determined as part of the solution. The interface plays a major role in defining the system and it is crucial to have an accurate representation of it.

Mathematical analysis to viscous flow problems is limited owing to their strong nonlinearity. In fact, the fundamental question on the proof of the existence and uniqueness of the solution to three-dimensional Navier-Stokes equations remains unresolved to date. Nevertheless, for some special types of such problems, mathematical analysis can provide deep insight into the fundamental mechanism involved. In particular, linearized motion (i.e., unsteady Stokes flow with a moving interface) can be solved in closed form and thoroughly analyzed, which, in turn, provides an important starting point for some nonlinear analysis as well as the development of computational methods. Meanwhile, asymptotic study based on perturbation series is a powerful analytical tool to investigate more complex problems.

For most of the interfacial flow problems, however, numerical methods have to be employed. Popular computational approaches for tracking or capturing interfacial motion include (but not limited to) the volume-of-fluid (VOF) [24], level set [26] and boundary integral [19]. In the VOF formulation, a volume fraction function C is defined and it satisfies an advection equation. At each time, the values of C are used to reconstruct an approximation to the interface and this approximate interface is then used to update the volume fractions at the next time. VOF methods provide a simple way to handle the topological changes of the interface and are relatively easy to extend from two-dimensional to three-dimensional domains. The level set approach was first proposed by Osher and Sethian [22] and has since been widely applied to many interfacial/free-surface problems [26]. In these methods, a level set function ϕ is introduced such that its initial value denotes the shortest distance between each point and the initial interface. The function ϕ then evolves in response to the propagation of the interface and, at anytime, the zero level set $\phi = 0$ gives exactly the location of the interface. The level set method does not require special procedures to treat topological changes of the interface and is relatively simple to generalize to three-dimensional problems. The boundary integral method was developed for computing inviscid potential flows, and notable work in this category was made by Longuet-Higgins and Cokelet [19], Vinje and Brevig [31], Baker et al. [2], etc. In the boundary integral formulation, Laplace's equation is solved by using Green's functions, leading to Fredholm integral equations of the second kind. The dynamic and kinematic surface conditions are integrated to update the interface at each time. A distinct advantage of this method is that the space dimension of the problem is reduced by one, thus they offer an efficient way for the computation of inviscid and irrotational flows. However, the method is not applicable to general viscous motion. In addition, there are several other well known computational methods including the marker-and-cell [10, 36], front tracking [8], phase field [28], and immersed interface [14, 16, 18]. A detailed review of these computational techniques for two-phase flows can be found in [4]. Although much success has been achieved by these numerical methods, all of them have their own strength and weakness. In particular, many of these methods encounter difficulty in resolving the fine-scale viscous boundary layers in interfacial flow computation, and are not suitable for an accurate investigation of surface details.

In this work, we discuss some of the analytical and computational aspects of viscous interfacial motion. We focus our attention on flows between two viscous incompressible fluids with distinct densities and viscosities, and each with an infinite depth. For ease of presentation, we restrict ourselves to two-dimensional (2D) settings, though most of the methods and results discussed here can be naturally extended to three-dimensional (3D) space. We start the presentation by considering the simplified, linearized problem which can be analytically solved and the analysis of which can be further augmented by a linear asymptotic study. We then describe a numerical method for the nonlinear problem that ensures both strong numerical stability and an accurate representation of the moving interface. By using this method, Stokes waves can be followed sufficiently in time to reveal the deep pattern of viscous effects on wave motion. In addition, as an example of nonlinear asymptotic study, we present a perturbation series analysis on Stokes waves with viscosity, which provides a theoretical verification of the numerical observation. Finally, conclusions are drawn and some discussion is made on related problems and research.

2 Basic Formulation

We first present the basic mathematical formulation for our moving interface problem in a two-dimensional setting. We denote the spatial coordinates by (x, z), the temporal coordinate by t, the velocity components by (u, w), and the pressure by p. The motion in each of the two fluids is described by the viscous incompressible Navier-Stokes equations [3, 15]

$$\rho u_t + \rho u u_x + \rho w u_z = -P_x + \mu(u_{xx} + u_{zz}) , \tag{1}$$

$$\rho w_t + \rho u w_x + \rho w w_z = -P_z + \mu(w_{xx} + w_{zz}) , \tag{2}$$

$$u_x + w_z = 0 , \tag{3}$$

where ρ is the density, μ is the dynamic viscosity, g is the gravitational acceleration, and where $P = p + \rho g z$ is referred to as the hydrodynamic pressure. The first two equations are referred to as the momentum equations, where the temporal derivatives describe the rate of change for the velocity. The nonlinear terms on the left-hand side represent the convection (or, advection) of the velocity field, whereas the second derivative terms on the right-hand side represent the diffusion. No time derivatives of the pressure appear here; instead, the pressure acts as a Lagrange multiplier in the Navier-Stokes equations [9]. In addition, Eq. (3) is the incompressibility condition, also referred to as the continuity equation.

Equations (1)–(3) hold in both the upper and lower fluids, and their solutions are connected through the interfacial conditions, to be provided in Eqs. (6)–(8). In addition, we will assume that solutions are periodic in the horizontal direction, and exponentially decay away from the interface in the vertical direction.

We represent the interface in the form

$$(x, z) = \big(x, h(x,t)\big). \tag{4}$$

The profile of h is determined by the kinematic condition

$$h_t + u^{(I)} h_x = w^{(I)}, \tag{5}$$

where $u^{(I)}$, $w^{(I)}$ are the interfacial velocity components. Essentially, this kinematic condition states that a fluid element on the interface must remain on and move with the interface at any time. Due to the presence of viscosity, we have the continuity of velocity at the interface for both the tangential and normal directions; in other words, both the horizontal and vertical velocities must be continuous across the interface:

$$u^{(1)} = u^{(2)} = u^{(I)}, \quad w^{(1)} = w^{(2)} = w^{(I)}, \tag{6}$$

where the superscripts (1) and (2) refer to the upper and lower domains, respectively. Moreover, the balance of stresses provides two more interfacial conditions [3]: one states that the tangential stress is continuous across the interface, whereas the other states that the normal stress is discontinuous at the interface and the jump in normal stress is balanced by the jump in pressure and the surface tension. In two-dimensional case, these two stress conditions yield

$$\begin{aligned}
&(h_x^2 - 1)\,[\mu^{(1)}(u_z^{(1)} + w_x^{(1)}) - \mu^{(2)}(u_z^{(2)} + w_x^{(2)})] \\
&\quad + 2h_x\,[\mu^{(1)}(u_x^{(1)} - w_z^{(1)}) - \mu^{(2)}(u_x^{(2)} - w_z^{(2)})] = 0, \qquad (7)\\
&(P^{(1)} - P^{(2)}) - gh(\rho^{(1)} - \rho^{(2)}) + h_x\,[\mu^{(1)}(u_z^{(1)} + w_x^{(1)}) - \mu^{(2)}(u_z^{(2)} + w_x^{(2)})] \\
&\quad - 2\,[\mu^{(1)} w_z^{(1)} - \mu^{(2)} w_z^{(2)}] - \gamma\kappa = 0, \qquad (8)
\end{aligned}$$

where γ is the surface tension and where κ is the mean curvature of the interface,

$$\kappa = \frac{h_{xx}}{(1 + h_x^2)^{3/2}}. \tag{9}$$

3 Linear Analysis

The strong nonlinearity of the Navier-Stokes equations and the presence of an unknown interface make the analytical solution impossible to find in general. Nevertheless, we may gain some insight by starting from the simpler, linearized problem.

The linearized Navier-Stokes equations, also referred to as the unsteady Stokes equations, take the form

$$u_t = -\frac{1}{\rho} P_x + \nu(u_{xx} + u_{zz}) \,, \tag{10}$$

$$w_t = -\frac{1}{\rho} P_z + \nu(w_{xx} + w_{zz}) \,, \tag{11}$$

$$u_x + w_z = 0 \,, \tag{12}$$

where $\nu = \dfrac{\mu}{\rho}$ is the kinematic viscosity. Meanwhile, by dropping all the nonlinear terms, the original interfacial conditions are reduced as

$$u^{(1)} = u^{(2)} \,, \tag{13}$$

$$h_t = w^{(1)} = w^{(2)} \,, \tag{14}$$

$$\rho^{(1)}\nu^{(1)}\big(u_z^{(1)} + w_x^{(1)}\big) = \rho^{(2)}\nu^{(2)}\big(u_z^{(2)} + w_x^{(2)}\big) \,, \tag{15}$$

$$(\rho^{(2)} - \rho^{(1)})gh + P^{(1)} - P^{(2)} - 2\big(\rho^{(1)}\nu^{(1)}w_z^{(1)} - \rho^{(2)}\nu^{(2)}w_z^{(2)}\big) = \gamma h_{xx} \,. \tag{16}$$

Solutions in the upper fluid domain have a positive vertical coordinate, $z > 0$, whereas those in the lower domain have a negative vertical coordinate, $z < 0$.

In order to simplify the calculations, we consider solutions in complex form. Real solutions can be generated by simply adding the complex conjugates. Let k be the wave number. For convenience of presentation, $k > 0$. Using the periodicity assumption on x and the normal mode analysis, we write solutions in the form

$$\begin{pmatrix} u \\ w \\ P \\ h \end{pmatrix} = e^{ikx}\, e^{\sigma t} \begin{pmatrix} \mathcal{U} \\ \mathcal{W} \\ \mathcal{P} \\ H \end{pmatrix} \tag{17}$$

where σ is referred to as the growth rate of the motion, H is a fixed number measuring the initial amplitude of the interface, and $\mathcal{U}$, $\mathcal{W}$, $\mathcal{P}$ all depend on the vertical coordinate z.

By substituting (17) into the Eqs. (10)–(16), we obtain

$$\sigma\mathcal{U} = -\frac{ik}{\rho}\mathcal{P} + \nu(-k^2\mathcal{U} + \mathcal{U}_{zz}) \,, \tag{18}$$

$$\sigma\mathcal{W} = -\frac{1}{\rho}\mathcal{P}_z + \nu(-k^2\mathcal{W} + \mathcal{W}_{zz}) \,, \tag{19}$$

$$ik\mathcal{U} + \mathcal{W}_z = 0 \,, \tag{20}$$

and

$$\mathcal{U}^{(1)} = \mathcal{U}^{(2)} \,, \tag{21}$$

$$H\sigma = \mathcal{W}^{(1)} = \mathcal{W}^{(2)} \,, \tag{22}$$

$$\rho^{(1)}\nu^{(1)}\big(\mathcal{U}_z^{(1)} + ik\mathcal{W}^{(1)}\big) = \rho^{(2)}\nu^{(2)}\big(\mathcal{U}_z^{(2)} + ik\mathcal{W}^{(2)}\big) \,, \tag{23}$$

$$(\rho^{(2)} - \rho^{(1)})gH + \mathcal{P}^{(1)} - \mathcal{P}^{(2)} - 2\big(\rho^{(1)}\nu^{(1)}\mathcal{W}_z^{(1)} - \rho^{(2)}\nu^{(2)}\mathcal{W}_z^{(2)}\big) = -k^2\gamma H \,. \tag{24}$$

Standard methods proceed by reducing the equations above into a single differential equation of 4th order. In contrast, our approach here is to reduce the order, while increasing the number, of the differential equations; that is, we convert the governing equations to a system of 4 first-order differential equations. To that end, we treat $\mathcal{U}_z$ as another unknown, and denote the unknown vector by $\mathbf{Y} = [\mathcal{U}, \mathcal{U}_z, \mathcal{W}, \mathcal{P}]^T$. Then Eqs. (18)–(20) can be rewritten as

$$\frac{d}{dz}\mathbf{Y} = \mathbf{B}\mathbf{Y} \triangleq \begin{bmatrix} 0 & 1 & 0 & 0 \\ \Omega^2/\nu & 0 & 0 & ik/\rho\nu \\ -ik & 0 & 0 & 0 \\ 0 & -\rho\nu ik & -\rho\Omega^2 & 0 \end{bmatrix} \mathbf{Y} \,, \tag{25}$$

where $\Omega = \sqrt{\sigma + \nu k^2}$. It is easy to find that the matrix $\mathbf{B}$ has four distinct eigenvalues:

$$\lambda_1 = k, \quad \lambda_2 = -k, \quad \lambda_3 = \Omega/\sqrt{\nu}, \quad \lambda_4 = -\Omega/\sqrt{\nu}. \tag{26}$$

Consequently, physically meaningful solutions (i.e., solutions that vertically decay away from the interface) to system (25) can be represented by

$$\mathbf{Y}^{(1)} = C_1 \begin{bmatrix} \Omega/\sqrt{\nu} \\ -\Omega^2/\nu \\ ik \\ 0 \end{bmatrix} \exp\Big[-\frac{\Omega z}{\sqrt{\nu}}\Big] + D_1 \begin{bmatrix} -ik \\ ik^2 \\ k \\ \rho\sigma \end{bmatrix} \exp[-kz] \tag{27}$$

in the upper domain ($z > 0$), where $\rho = \rho^{(1)}$ and $\nu = \nu^{(1)}$, and

$$\mathbf{Y}^{(2)} = C_2 \begin{bmatrix} -\Omega/\sqrt{\nu} \\ -\Omega^2/\nu \\ ik \\ 0 \end{bmatrix} \exp\Big[\frac{\Omega z}{\sqrt{\nu}}\Big] + D_2 \begin{bmatrix} -ik \\ -ik^2 \\ -k \\ \rho\sigma \end{bmatrix} \exp[kz] \tag{28}$$

in the lower domain ($z < 0$), where $\rho = \rho^{(2)}$ and $\nu = \nu^{(2)}$.

Those constants C_1, C_2 and D_1, D_2 can be expressed in terms of σ and H:

$$D_1 = \frac{HF\sigma}{k(F-E)}, \qquad C_1 = \frac{HE\sigma}{ik(E-F)},$$
$$D_2 = \frac{\sqrt{\nu^{(2)}}k + \Omega^{(2)}}{\sqrt{\nu^{(2)}}k - \Omega^{(2)}} D_1 + \frac{\sqrt{\nu^{(2)}}\Omega^{(1)} + \sqrt{\nu^{(1)}}\Omega^{(2)}}{\sqrt{\nu^{(2)}}k - \Omega^{(2)}} \frac{iC_1}{\sqrt{\nu^{(1)}}},$$
$$C_2 = C_1 - iD_1 - iD_2, \tag{29}$$

with

$$E = 2\rho^{(1)}\nu^{(1)}k^2 + 2\rho^{(2)}\sqrt{\nu^{(2)}}\,k\,\Omega^{(2)},$$
$$F = \rho^{(1)}\big(\sigma + 2\nu^{(1)}k^2\big) + \rho^{(2)}\Omega^{(1)}\sqrt{\frac{\nu^{(2)}}{\nu^{(1)}}}\big(\sqrt{\nu^{(2)}}k + \Omega^{(2)}\big)$$
$$- \rho^{(2)}\sqrt{\nu^{(2)}}k\,\big(\sqrt{\nu^{(2)}}k - \Omega^{(2)}\big). \tag{30}$$

Given an initial amplitude (H) of the interface, the value of σ has to satisfy certain condition, referred to as the dispersion relation, to ensure a nontrivial solution of our problem. Indeed, substitution of the interfacial conditions yields, after some algebra,

$$\Big[\rho^{(1)}\sqrt{\nu^{(1)}}\big(\Omega^{(1)} + \sqrt{\nu^{(1)}}k\big) + \rho^{(2)}\sqrt{\nu^{(2)}}\big(\Omega^{(2)} + \sqrt{\nu^{(2)}}k\big)\Big]$$
$$\times\big[(\rho^{(2)} - \rho^{(1)})gk + \gamma k^3 + (\rho^{(2)} + \rho^{(1)})\sigma^2\big]$$
$$+ 4\big(\rho^{(1)}\sqrt{\nu^{(1)}}\Omega^{(1)} + \rho^{(2)}\nu^{(2)}k\big)\big(\rho^{(2)}\sqrt{\nu^{(2)}}\Omega^{(2)} + \rho^{(1)}\nu^{(1)}k\big)\sigma k = 0. \tag{31}$$

Equivalent form of Eq. (31) can be found in [5], though derived by a different approach. This dispersion relation is nonlinear and cannot be solved analytically; instead, some approximation methods (e.g., numerical or asymptotic approaches) have to be used.

One possible way to conduct asymptotic study on the linear viscous interfacial motion is based on the method of multiple scales [11, 20, 33]; some details are presented below.

Let us introduce two dimensionless parameters

$$r = \frac{\rho^{(1)}}{\rho^{(2)}}, \qquad R = \sqrt{\frac{\nu^{(2)}}{\nu^{(1)}}}. \tag{32}$$

Since the boundary layers near the interface have thickness proportional to $\sqrt{\nu}$ [15, 20, 30], we introduce scaled vertical coordinates η_0, η_1 by

$$\eta_0 = \frac{z}{\sqrt{\nu}}, \qquad \eta_1 = \sqrt{\nu}\,\eta_0 = z. \tag{33}$$

Consequently,

$$\frac{\partial}{\partial z} = \frac{1}{\sqrt{\nu}} \frac{\partial}{\partial \eta_0} + \frac{\partial}{\partial \eta_1} ,$$
$$\frac{\partial^2}{\partial z^2} = \frac{1}{\nu} \frac{\partial^2}{\partial \eta_0^2} + \frac{2}{\sqrt{\nu}} \frac{\partial^2}{\partial \eta_0 \, \partial \eta_1} + \frac{\partial^2}{\partial \eta_1^2} . \tag{34}$$

Then we assume the following perturbation series expansions:

$$\mathcal{U} = u_0(\eta_0 , \eta_1) + \sqrt{\nu}\, u_1(\eta_0 , \eta_1) + \nu\, u_2(\eta_0 , \eta_1) + \cdots ,$$
$$\mathcal{W} = w_0(\eta_0 , \eta_1) + \sqrt{\nu}\, w_1(\eta_0 , \eta_1) + \nu\, w_2(\eta_0 , \eta_1) + \cdots ,$$
$$\mathcal{P} = P_0(\eta_0 , \eta_1) + \sqrt{\nu}\, P_1(\eta_0 , \eta_1) + \nu\, P_2(\eta_0 , \eta_1) + \cdots ,$$
$$\sigma = \sigma_0 + \sqrt{\nu}\, \sigma_1 + \nu\, \sigma_2 + \cdots . \tag{35}$$

Note that σ is the same in the upper and lower fluid domains but with different expansions. They are related by

$$\sigma_0^{(1)} = \sigma_0^{(2)} ; \qquad \sigma_m^{(1)} = R^m \, \sigma_m^{(2)} , \quad m = 1, 2, \cdots . \tag{36}$$

By substituting (35) into (18), we obtain

$$(\sigma_0 + \sqrt{\nu}\,\sigma_1 + \nu\,\sigma_2 + \cdots)\, u_0 + \sqrt{\nu}\,(\sigma_0 + \sqrt{\nu}\,\sigma_1 + \nu\,\sigma_2 + \cdots)\, u_1 +$$
$$\nu\,(\sigma_0 + \sqrt{\nu}\,\sigma_1 + \nu\,\sigma_2 + \cdots)\, u_2 + \cdots = -\frac{ik}{\rho} P_0 - \sqrt{\nu}\,\frac{ik}{\rho} P_1 - \nu\,\frac{ik}{\rho} P_2 + \cdots$$
$$+\nu \Big(- k^2 u_0 - \sqrt{\nu}\, k^2 u_1 - \nu\, k^2 u_2 - \cdots + \frac{1}{\nu} \frac{\partial^2 u_0}{\partial \eta_0^2} + \frac{2}{\sqrt{\nu}} \frac{\partial^2 u_0}{\partial \eta_0 \, \partial \eta_1} + \frac{\partial^2 u_0}{\partial \eta_1^2} +$$
$$\frac{1}{\sqrt{\nu}} \frac{\partial^2 u_1}{\partial \eta_0^2} + 2 \frac{\partial^2 u_1}{\partial \eta_0 \, \partial \eta_1} + \sqrt{\nu}\, \frac{\partial^2 u_1}{\partial \eta_1^2} + \frac{\partial^2 u_2}{\partial \eta_0^2} + 2\sqrt{\nu}\, \frac{\partial^2 u_2}{\partial \eta_0 \, \partial \eta_1} + \nu\, \frac{\partial^2 u_2}{\partial \eta_1^2} + \cdots \Big) . \tag{37}$$

Comparison of the coefficients of ν^n at each order, starting from the lowest, yields,

$$\text{order } \nu^0 : \quad \sigma_0\, u_0 = -\frac{ik}{\rho} P_0 + \frac{\partial^2 u_0}{\partial \eta_0^2} , \tag{38}$$

$$\text{order } \nu^{\frac{1}{2}} : \quad \sigma_1\, u_0 + \sigma_0\, u_1 = -\frac{ik}{\rho} P_1 + 2 \frac{\partial^2 u_0}{\partial \eta_0 \, \partial \eta_1} + \frac{\partial^2 u_1}{\partial \eta_0^2} , \tag{39}$$

$$\text{order } \nu^1 : \quad \sigma_2\, u_0 + \sigma_1\, u_1 + \sigma_0\, u_2 = -\frac{ik}{\rho} P_2 - k^2 u_0$$
$$+ \frac{\partial^2 u_0}{\partial \eta_1^2} + 2 \frac{\partial^2 u_1}{\partial \eta_0 \, \partial \eta_1} + \frac{\partial^2 u_2}{\partial \eta_0^2} . \tag{40}$$

If we substitute (35) into (19) and equate the coefficients of the terms with ν^n, we obtain,

$$\text{order } \nu^{-\frac{1}{2}} : \quad \frac{\partial P_0}{\partial \eta_0} = 0 , \tag{41}$$

$$\text{order } \nu^0 : \quad \sigma_0 \, w_0 = -\frac{1}{\rho}\frac{\partial P_0}{\partial \eta_1} - \frac{1}{\rho}\frac{\partial P_1}{\partial \eta_0} + \frac{\partial^2 w_0}{\partial \eta_0^2} , \tag{42}$$

$$\text{order } \nu^{\frac{1}{2}} : \quad \sigma_1 \, w_0 + \sigma_0 \, w_1 = -\frac{1}{\rho}\frac{\partial P_1}{\partial \eta_1} - \frac{1}{\rho}\frac{\partial P_2}{\partial \eta_0} + 2\frac{\partial^2 w_0}{\partial \eta_0 \, \partial \eta_1} + \frac{\partial^2 w_1}{\partial \eta_0^2} . \tag{43}$$

Similarly, the substitution of (35) into (20) yields,

$$\text{order } \nu^{-\frac{1}{2}} : \quad \frac{\partial w_0}{\partial \eta_0} = 0 , \tag{44}$$

$$\text{order } \nu^0 : \quad ik u_0 + \frac{\partial w_0}{\partial \eta_1} + \frac{\partial w_1}{\partial \eta_0} = 0 , \tag{45}$$

$$\text{order } \nu^{\frac{1}{2}} : \quad ik u_1 + \frac{\partial w_1}{\partial \eta_1} + \frac{\partial w_2}{\partial \eta_0} = 0 . \tag{46}$$

Meanwhile, we expand the interfacial conditions (21)–(24). Let $\nu = \nu^{(1)}$. Using (32), the substitution of (35) and (36) into (21) yields,

$$\text{order } \nu^0 : \quad u_0^{(1)} = u_0^{(2)} , \tag{47}$$

$$\text{order } \nu^{\frac{1}{2}} : \quad u_1^{(1)} = R \, u_1^{(2)} . \tag{48}$$

Substitution into (22) yields,

$$\text{order } \nu^0 : \quad H \, \sigma_0^{(1)} = w_0^{(1)} = w_0^{(2)} = H \, \sigma_0^{(2)} , \tag{49}$$

$$\text{order } \nu^{\frac{1}{2}} : \quad H \, \sigma_1^{(1)} = w_1^{(1)} = R \, w_1^{(2)} = HR \, \sigma_1^{(2)} . \tag{50}$$

Substitution into (23) yields,

$$\text{order } \nu^{-\frac{1}{2}} : \; r \, \frac{\partial u_0^{(1)}}{\partial \eta_0} = R \, \frac{\partial u_0^{(2)}}{\partial \eta_0} , \tag{51}$$

$$\text{order } \nu^0 : \quad r \left(\frac{\partial u_0}{\partial \eta_1} + \frac{\partial u_1}{\partial \eta_0} + ik w_0\right)^{(1)} = R^2 \left(\frac{\partial u_0}{\partial \eta_1} + \frac{\partial u_1}{\partial \eta_0} + ik w_0\right)^{(2)} . \tag{52}$$

Finally, substitution into (24) yields,

$$\text{order } \nu^0 : \quad (\rho^{(2)} - \rho^{(1)})gH + P_0^{(1)} - P_0^{(2)} = -k^2\gamma H \,, \tag{53}$$

$$\text{order } \nu^{\frac{1}{2}} : \quad P_1^{(1)} - R\,P_1^{(2)} - 2\Big(\rho^{(1)}\,\frac{\partial w_0^{(1)}}{\partial \eta_0} - R\rho^{(2)}\,\frac{\partial w_0^{(2)}}{\partial \eta_0}\Big) = 0\,. \tag{54}$$

Based on Eqs. (38)–(54), solutions can be determined order by order. The interfacial conditions at the lower orders are applied to determine the coefficients in the solutions, whereas secularity conditions in the higher order equations are used to determine the additional dependency of the solutions on the scaled variables η_0, η_1.

In particular, solution for σ at the lowest order, which represents the motion in inviscid fluids, is given by

$$\sigma_0^2 = -\Big(\frac{\rho^{(2)} - \rho^{(1)}}{\rho^{(2)} + \rho^{(1)}}\,gk + \frac{k^3\,\gamma}{\rho^{(2)} + \rho^{(1)}}\Big)\,, \tag{55}$$

where $\sigma_0 = \sigma_0^{(1)} = \sigma_0^{(2)}$. Equation (55) shows that when $r > 1$ (i.e., $\rho^{(1)} > \rho^{(2)}$), the motion is unstable for all wave numbers $0 < k < \sqrt{(\rho^{(1)} - \rho^{(2)})g/\gamma}$. This is the well-known Rayleigh-Taylor instability [5]. When $r < 1$, the motion is stable and σ_0 is purely imaginary; it determines the phase speed and does not change the interface amplitude, as can be naturally expected for the linear inviscid flow.

Solution for σ at the next order is given by

$$\sigma_1^{(1)} = R\,\sigma_1^{(2)} = -\frac{2kRr\,\sqrt{\sigma_0}}{(R + r)(1 + r)}\,. \tag{56}$$

Now the real part of σ_1 is nonzero and it determines the leading term of the decay rate for the wave amplitude due to viscous dissipation, while the imaginary part of σ_1 gives viscous correction to the inviscid phase speed. Furthermore,

$$\sigma_2^{(1)} = R^2\,\sigma_2^{(2)} = \frac{-2k^2}{(R + r)^2(1 + r)^2}\big[\,(1 + r)R^4 - 2r^2R^2 + r^3(1 + r)\,\big]\,. \tag{57}$$

This shows that σ_2 is real and only influences the wave decay rate, and has no contribution for the phase speed. When the viscosities are small, the first few terms in the series expansion, such as

$$\sigma_0 + \sqrt{\nu}\,\sigma_1 + \nu\,\sigma_2\,, \tag{58}$$

could provide a good approximation for σ.

In addition, we note that when r is very small (in a system of air and water, for example, $r \doteq 0.001$), Eq. (58) yields a simplified approximation

$$\sigma \doteq -2\nu^{(2)}k^2 \pm i\sqrt{gk + k^3\gamma/\rho^{(2)}}\,. \tag{59}$$

From (59), it is clear that the wave amplitude will decay exponentially as

$$H\exp[-2\nu^{(2)}k^2 t]\,, \tag{60}$$

so that the total energy dissipation rate per wavelength is given by

$$\frac{d}{dt}\left[\frac{1}{2}\rho^{(2)}k(He^{-2\nu^{(2)}k^2 t})^2c^2\right] = -2\rho^{(2)}kH^2c^2e^{-4\nu^{(2)}k^2 t}\,, \tag{61}$$

where c is the phase speed of the wave. This is consistent with the result in Sec. 348 of Lamb's classical textbook [15].

4 Numerical Calculation

We now turn to numerical study of the original nonlinear problem. In general, the moving interface $z = h(x,t)$ between the two fluids makes it a nontrivial task on the design of an accurate numerical method. To overcome this difficulty, we map the deformed geometry (due to the evolving interface) into a rectangular domain in new coordinates so as to facilitate accurate and efficient numerical discretization. The cost of doing this is that the details of the governing equations and the interfacial conditions are changed. Our numerical methods are then constructed on these mapped equations.

We introduce the new coordinates, (X, Z, τ), through the mapping [32]

$$x = X\,, \tag{62}$$

$$z = F(X, Z, \tau)\,, \tag{63}$$

$$t = \tau\,, \tag{64}$$

where

$$F(X,Z,\tau) \triangleq \begin{cases} Z + h(X,\tau)\exp(-\alpha Z), & Z \geq 0\,, \\ Z + h(X,\tau)\exp(\alpha Z), & Z \leq 0\,, \end{cases} \tag{65}$$

and where $\alpha > 0$ is a constant which can be used to adjust the grid spacing near the interface. Clearly, the coordinate line $Z = 0$ corresponds to the location of the interface $z = h(x,t)$. When far from the interface, Z is relaxing exponentially to the physical coordinate z so that the far-field boundary conditions can be easily handled.

We proceed to derive the mapped equations under the new coordinates. To that end we need to calculate the transformed derivatives and operators. If we define

$$G_0 = \frac{F_\tau}{F_Z}\,, \qquad G_1 = \frac{F_X}{F_Z}\,, \qquad G_3 = \frac{1}{F_Z}\,, \tag{66}$$

then the transformed first and second derivatives can be calculated by

$$\frac{\partial}{\partial t} = \frac{\partial}{\partial \tau} - G_0\frac{\partial}{\partial Z}\,, \tag{67}$$

$$\frac{\partial}{\partial x} = \frac{\partial}{\partial X} - G_1\frac{\partial}{\partial Z}\,, \tag{68}$$

$$\frac{\partial}{\partial z} = G_3\frac{\partial}{\partial Z}\,, \tag{69}$$

$$\frac{\partial^2}{\partial x^2} = \frac{\partial^2}{\partial X^2} + (G_1)^2\frac{\partial^2}{\partial Z^2} - 2G_1\frac{\partial^2}{\partial X\,\partial Z} + \left[G_1(G_1)_Z - (G_1)_X\right]\frac{\partial}{\partial Z}\,, \tag{70}$$

$$\frac{\partial^2}{\partial z^2} = (G_3)^2\frac{\partial^2}{\partial Z^2} + G_3(G_3)_Z\frac{\partial}{\partial Z}\,. \tag{71}$$

Let us further define

$$g_2 = (G_1)^2 + (G_3)^2\,, \quad g_3 = -2G_1\,, \quad g_4 = G_1\frac{\partial G_1}{\partial Z} + G_3\frac{\partial G_3}{\partial Z} - \frac{\partial G_1}{\partial X}\,. \tag{72}$$

Then we can write the Laplacian in the new coordinates as

$$\mathcal{L} \triangleq \frac{\partial^2}{\partial x^2} + \frac{\partial^2}{\partial z^2} = \frac{\partial^2}{\partial X^2} + g_2\frac{\partial^2}{\partial Z^2} + g_3\frac{\partial^2}{\partial X\,\partial Z} + g_4\frac{\partial}{\partial Z}\,. \tag{73}$$

We note that the coefficients G_i $(i = 0, 1, 3)$, g_i $(i = 2, 3, 4)$ are different in the upper and lower domains.

Now we substitute the transformation rules (67)–(73) into the basic equations (1)–(8) to obtain

$$u_\tau - G_0 u_Z + u(u_X - G_1 u_Z) + wG_3 u_Z = -\frac{1}{\rho}P_X + \frac{1}{\rho}G_1 P_Z + \nu\mathcal{L}\{u\}\,, \tag{74}$$

$$w_\tau - G_0 w_Z + u(w_X - G_1 w_Z) + wG_3 w_Z = -\frac{1}{\rho}G_3 P_Z + \nu\mathcal{L}\{w\}\,, \tag{75}$$

$$u_X - G_1 u_Z + G_3 w_Z = 0\,. \tag{76}$$

The kinematic condition becomes

$$h_\tau + u^{(I)}h_X = w^{(I)}\,. \tag{77}$$

Though the velocity interfacial conditions (6) stay the same, the two stress conditions are changed under the new coordinates:

$$\mu^{(1)}(G_3^{(1)}u_Z^{(1)} + w_X^{(1)}) - \mu^{(2)}(G_3^{(2)}u_Z^{(2)} + w_X^{(2)}) + (\frac{4h_X}{h_X^2 - 1} + \frac{G_1^{(1)}}{G_3^{(1)}})\mu^{(1)}(u_X^{(1)} - G_1^{(1)}u_Z^{(1)})$$

$$-(\frac{4h_X}{h_X^2 - 1} + \frac{G_1^{(2)}}{G_3^{(2)}})\mu^{(2)}(u_X^{(2)} - G_1^{(2)}u_Z^{(2)}) = 0\,, \tag{78}$$

$$(P^{(1)} - P^{(2)}) + (2 - \frac{4h_X^2}{h_X^2 - 1})[\mu^{(1)}(u_X^{(1)} - G_1^{(1)}u_Z^{(1)}) - \mu^{(2)}(u_X^{(2)} - G_1^{(2)}u_Z^{(2)})]$$

$$= gh(\rho^{(1)} - \rho^{(2)}) + \gamma\kappa\,. \tag{79}$$

We note that in order to obtain (78) and (79), we have eliminated $w_z^{(1)}$ and $w_z^{(2)}$ in Eqs. (7) and (8) by using the incompressibility condition (3).

We then write these mapped equations in the form of linear terms and nonlinear terms separately. Specifically, the linear terms are separated and put on the left-hand side of the equations, which recover the equations for the linear motion as presented in the previous section. Meanwhile, all the nonlinear terms, including the convection and the mapping associated terms, are put on the right-hand side of the equations, treated as perturbations to the corresponding linear equations. We note, however, that in most applications these nonlinear terms are strong and take a dominant role in the system, and one cannot simply apply a numerical linear solver to deal with such a strongly nonlinear problem. Nevertheless, the linear analysis presented before does provide insight into the development of nonlinear numerical discretization. The procedure is summarized below; for details, we refer to the work in [35].

The second-order backward difference formula (BDF) [1] is applied to update the motion in time. The method is fully implicit and so requires the solution of a nonlinear system of equations for the unknowns at the new time level. The linear terms on the left-hand side thus provide a simple iterative procedure. At each iterate, the Fourier transform is applied in the horizontal direction X, which possesses periodicity, to achieve spectral accuracy in X. Efficient implementation is achieved by using the Fast Fourier Transform (FFT), and a pseudo-spectral technique [21, 23] is employed simultaneously to handle those nonlinear terms. The temporal discretization and the Fourier transform in X result in a linear system of first-order differential equations with respect to the vertical coordinate, Z, at each time iteration. This first-order system is then computed by a second-order numerical integration technique (such as the trapezoid rule), together with the interfacial conditions and the far-field boundary conditions. The numerical integration is implemented by decoupling the growing and decaying modes (which correspond to the eigenvalues of opposite signs associated with the system) so as to catch the bounded (and physically meaningful) solutions. Once solved, the current iteration is complete and the procedure is repeated for the next cycle.

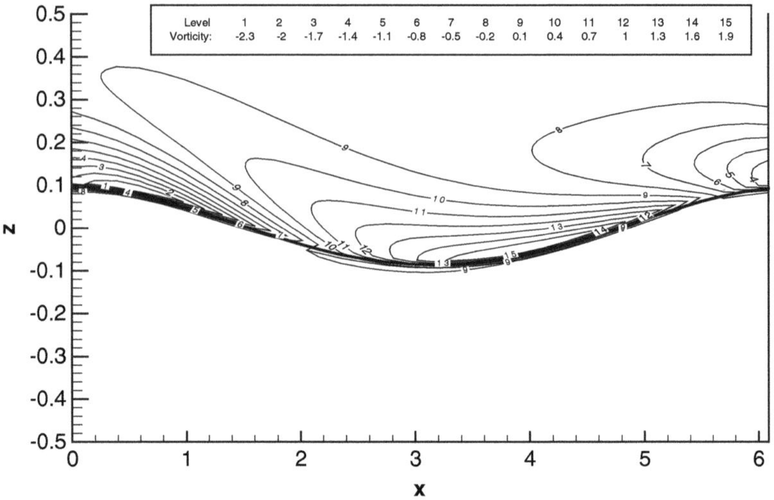

Fig. 1 Vorticity contours in air and water

Standard convergence tests have confirmed that this numerical method achieves spectral accuracy in the horizontal direction and second-order accuracy in the vertical direction and time marching, for both the velocity and pressure throughout the flow domain. The method is also capable of handling large density and viscosity jumps across the interface. A plot of the vorticity contours from a typical moving interface simulation involving air and water is presented in Fig. 1, where the horizontal domain is nondimensionalized to $[0,\ 2\pi]$. It shows that the viscous boundary layers are well resolved, and that the vorticity is much higher in value and more spread out in the boundary layer in air (the upper domain) than that in water (the lower domain).

One significant application of this numerical method is the simulation of Stokes waves in the presence of viscosity. Stokes waves are originally defined with inviscid fluids and refer to the motion of periodic, steady progressive free-surface or interfacial waves [6, 12, 17, 25, 27, 29]. Surface tension is neglected in this study. In a system with two inviscid fluids of infinite thickness, a Stokes wave can be expanded in a permanent form by a complex Fourier series

$$h(x,t) = \sum_{m=1}^{\infty} A_m(A)\, e^{m\alpha t}\, e^{imkx}\ , \tag{80}$$

where k is the wave number, A is a free parameter, and where $\alpha = ik\beta$ is referred to as the inviscid growth rate; it is purely imaginary and does not change the wave amplitude. The parameter β denotes the phase speed of the wave and is determined by

$$\beta^2 = \frac{g}{k}\frac{\rho^{(2)} - \rho^{(1)}}{\rho^{(2)} + \rho^{(1)}}\left(1 + \frac{(\rho^{(2)})^2 + (\rho^{(1)})^2}{(\rho^{(2)} + \rho^{(1)})^2} A^2 + \cdots\right), \tag{81}$$

where we assume $\rho^{(1)} < \rho^{(2)}$. The letter g denotes the gravitational acceleration. The first few coefficients in Eq. (80) are given in [29]; for example,

$$\begin{aligned} A_1 &= A\,, \\ A_2 &= \frac{\rho^{(2)} - \rho^{(1)}}{\rho^{(2)} + \rho^{(1)}} kA^2 + \cdots, \\ A_3 &= \frac{3(\rho^{(2)})^2 - 10\rho^{(2)}\rho^{(1)} + 3(\rho^{(1)})^2}{4(\rho^{(2)} + \rho^{(1)})^2} k^2 A^3 + \cdots, \\ &\vdots \end{aligned} \tag{82}$$

and in general, $A_m = O(A^m)$. It is certain that Stokes waves of such a permanent form can only exist in inviscid fluids. However, any fluid in nature has some viscosity. It is natural to ask what happens to a Stokes wave in the presence of viscosity, and, in particular, how the viscosity changes the Stokes' expansion given in Eq. (80).

Most of current computational methods for viscous interfacial motion introduce numerical smoothing which could mask the true effects of viscosity. In contrast, the numerical method presented here is capable of capturing a sharp interface with high accuracy, thus is suitable for a detailed study of viscous effects on Stokes waves.

For inviscid flow, the results in (82) suggest one way to view the family of Stokes waves is to consider the curves $|A_k|(|A_1|)$. Then the effects of viscosity can be studied by viewing the deviation of the numerical results from these curves. Thus, we draw the curves by using (82) for the modes $|A_2|$ versus $|A_1|$, $|A_3|$ versus $|A_1|$, $|A_4|$ versus $|A_1|$, $|A_5|$ versus $|A_1|$, etc., and refer to these curves as inviscid solutions. On the other hand, starting with a Stokes wave profile and running the simulation with viscosity by using the numerical methods described above, we obtain the numerical solutions which give the time evolution for the amplitude of each mode. We then plot these amplitudes in the same way as $|A_2|$ versus $|A_1|$, $|A_3|$ versus $|A_1|$, $|A_4|$ versus $|A_1|$, $|A_5|$ versus $|A_1|$, etc. In Fig. 2 we compare the numerical viscous solution to the analytic inviscid solution. The numerical solution is plotted from $\tau = 0$ and for every period, T, until $\tau = 20T$. Figure 2 gives the results in the air-water case for $A = 0.1$. Though not shown here, similar results are observed for different choices of the amplitude parameter A and the viscosities. These results suggest an interesting interpretation: viscous effects seem to reduce the magnitude of the Stokes wave while allowing it to remain a member of the family. Without viscosity, A is fixed. With viscosity it is reduced while maintaining the ratio of the amplitudes.

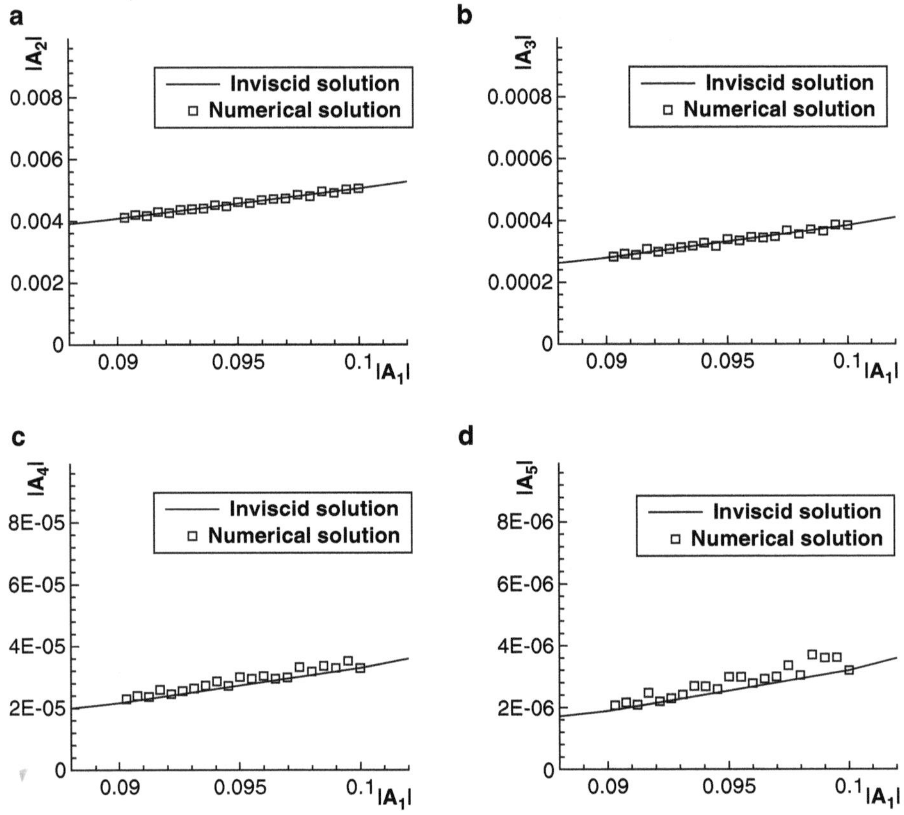

Fig. 2 Comparison between the inviscid solution and the numerical viscous solution of a Stokes wave in air and water. The numerical solution is displayed for 20 periods. (**a**) modes $|A_2|$ versus $|A_1|$; (**b**) modes $|A_3|$ versus $|A_1|$; (**c**) modes $|A_4|$ versus $|A_1|$; (**d**) modes $|A_5|$ versus $|A_1|$

5 Nonlinear Asymptotics

For some nonlinear viscous two-phase flow problems, asymptotic study can also provide a useful means to gain deeper understanding. Below we present an example for using perturbation series to analyze viscous effects on Stokes waves, the results of which can provide a verification of the numerical observation presented in the previous section.

We assume the interface h is expanded in terms of the amplitude parameter A as follows,

$$h = c_1 A e^{\sigma t} e^{ikx} + c_2 A^2 e^{2\sigma t} e^{2ikx} + \cdots + c_m A^m e^{m\sigma t} e^{mikx} + \cdots , \tag{83}$$

where $k > 0$ is the wave number and σ is the viscous growth rate. Each coefficient c_m is independent of A and, without loss of generality, we may set $c_1 = 1$. We expand the growth rate σ by

$$\sigma = \sigma_0 + A\,\sigma_1 + A^2\,\sigma_2 + \cdots + A^m\,\sigma_m + \cdots . \tag{84}$$

The velocities and pressure are also expanded in terms of A. For example,

$$u = u_1(z)Ae^{\sigma t}\,e^{ikx} + u_2(z)A^2e^{2\sigma t}\,e^{2ikx} + \cdots + u_m(z)\,A^m e^{m\sigma t}\,e^{mikx} + \cdots , \tag{85}$$

where u_m $(m = 1, 2, \cdots)$ are depending on the vertical coordinate z. Similar expansions hold for w and P.

For each order of A, we seek solutions in terms of small viscosity ν. For convenience of discussion, we again use the two dimensionless parameters r and R, first introduced in Eq. (32). We then expand each σ_m as follows, taking into account that the boundary layers have thickness proportional to $\sqrt{\nu}$ [20, 30],

$$\sigma_m = \sigma_{m,0} + \sqrt{\nu}\,\sigma_{m,1} + \nu\,\sigma_{m,2} + \cdots , \tag{86}$$

where we pick $\nu = \nu^{(1)}$, the viscosity of the upper fluid. Similarly, for each coefficient c_m in Eq. (83) we have

$$c_m = c_{m,0} + \sqrt{\nu}\,c_{m,1} + \nu\,c_{m,2} + \cdots . \tag{87}$$

For the velocities and pressure, we will need to consider the outer and inner solutions separately. In the regions outside the boundary layers, solutions are given by the regular perturbation series. For example,

$$u_m = u_{m,0}(z) + \sqrt{\nu}\,u_{m,1}(z) + \nu\,u_{m,2}(z) + \cdots . \tag{88}$$

Inside the boundary layers, we need singular perturbation series [20] to represent the solutions,

$$u_m = u_{m,0}(\eta) + \sqrt{\nu}\,u_{m,1}(\eta) + \nu\,u_{m,2}(\eta) + \cdots , \tag{89}$$

where

$$\eta = \frac{z - h(x,t)}{\sqrt{\nu}} \tag{90}$$

is referred to as the stretched coordinate. We make similar expansions for w_m and P_m.

Based on these asymptotic expansions, calculations can be performed from lower orders to higher ones. At each order, solutions are determined by solving the Navier-Stokes equations together with interfacial conditions, and by matching the outer and inner expansions through the well-known Van Dyke matching principle [20, 30, 34].

In particular, we obtain the growth rate σ at the inviscid level $\big($order $(\sqrt{\nu})^0\big)$ as follows,

$$\sigma_{0,0} = \pm i\sqrt{\frac{\rho^{(2)} - \rho^{(1)}}{\rho^{(2)} + \rho^{(1)}}\, gk}\,, \tag{91}$$

$$\sigma_{1,0} = 0\,, \tag{92}$$

$$\sigma_{2,0} = \pm i\sqrt{\frac{\rho^{(2)} - \rho^{(1)}}{\rho^{(2)} + \rho^{(1)}}\, gk}\;\frac{(\rho^{(2)})^2 + (\rho^{(1)})^2}{(\rho^{(2)} + \rho^{(1)})^2}\,, \tag{93}$$

$$\vdots$$

We observe again that the growth rate at the inviscid level is purely imaginary and has no effect on on the wave amplitude. It is easy to observe that the above results agree with the inviscid expansion of the phase speed in Eq. (81).

The viscous corrections to the growth rate are given by

$$\sigma_{0,1} = -\frac{2kRr\sqrt{\sigma_{0,0}}}{(R+r)(1+r)}\,, \tag{94}$$

$$\sigma_{0,2} = \frac{-2k^2}{(R+r)^2(1+r)^2}\big[\,(1+r)R^4 - 2r^2R^2 + r^3(1+r)\,\big]\,, \tag{95}$$

$$\sigma_{1,1} = 0\,, \tag{96}$$

$$\sigma_{1,2} = 0\,, \tag{97}$$

$$\vdots$$

where the square root of $\sigma_{0,0}$ is taken with positive real part. The real part of $\sigma_{0,1}$ is nonzero which contributes to the viscous dissipation of the wave amplitude, while the imaginary part of $\sigma_{0,1}$ gives viscous correction to the inviscid phase speed. The $\sigma_{0,2}$ is real and only influences the wave amplitude; it has no contribution to the phase speed. There is no viscous correction in the order A level; this pattern is consistent with the inviscid expansion (81).

The first few terms of the coefficient c_m in Eq. (83) are found as

$$c_1 = 1\,, \tag{98}$$

$$c_{2,0} = \frac{\rho^{(2)} - \rho^{(1)}}{\rho^{(2)} + \rho^{(1)}}\, k\,, \tag{99}$$

$$c_{2,1} = \frac{4kRr\sigma_{0,0}\sqrt{\sigma_{0,0}}}{(R+r)^2(1-r^2)g}\big[3(R-r^2) + (4\sqrt{2}-5)r(1-R)\big]\,, \tag{100}$$

$$\vdots$$

Equations (98) and (99) indicate that if we set $\nu = 0$ in the expansion form (83), we will recover the inviscid Stokes wave expansion (80) to the order A^2. The value of $c_{2,1}$ makes the major contribution of viscous correction to $c_{2,0}$. What we are most interested is perhaps the case $r \ll 1$, as in a system with air and water, for example. In such a case we have

$$c_{2,1} \doteq \frac{12kr}{g} \sigma_{0,0} \sqrt{\sigma_{0,0}} \,. \tag{101}$$

Using (101), we can obtain an estimate for the absolute value of $c_{2,1}$,

$$|\, c_{2,1} \,| \doteq \frac{12k^{7/4} r}{g^{1/4}} \,. \tag{102}$$

This shows that the correction term $\sqrt{\nu}\, c_{2,1}$ is very small, especially for long waves (where the wave number k is small and where we can reasonably neglect the surface tension). Indeed, some simple evaluation reveals that even with a wave number as large as 1,000, the viscous correction only counts 15 % of $c_{2,0}$. For small or moderate wave numbers, the viscous correction to $c_{2,0}$ is negligible (for instance, when $k = 10$, it is found $\frac{\sqrt{\nu}\, c_{2,1}}{c_{2,0}} \doteq 0.0047$). In such situations, we have

$$c_2 \doteq c_{2,0} \,, \tag{103}$$

which implies that viscous effects would almost be equivalent to replacing the inviscid growth rate α by the viscous growth rate σ in the inviscid Stokes wave expansion (80), up to the second order of A. It is as though small viscosity tends to keep the Stokes' expansion form. This result is consistent with the numerical observation presented in the previous section. It can be expected that the same pattern holds for higher-order expansions; this can be justified by carrying out the calculations to A^3 and higher levels.

6 Discussion

We have presented some mathematical analysis and numerical simulation to viscous interfacial motion associated with two-phase flows. Due to the strong nonlinearity and the presence of an unknown moving interface in such problems, analytical solutions are generally impossible to find. We started our discussion by considering the linearized problem where an analytical solution procedure can be formulated and can be further augmented by some linear asymptotic analysis. We then presented a numerical method to simulate the nonlinear interfacial flow problem. The method achieves fully second-order accuracy in the time marching and the vertical direction, and spectral accuracy in the horizontal direction, for both the velocity and pressure.

It allows us to treat viscosity jumps, large density ratios (about 1,000 to 1 in the water-air case), and reasonably high Reynolds numbers, without introducing unnecessary numerical smoothing. It is thus capable of capturing very thin boundary layers at an evolving interface in slightly viscous fluids. Our simulation results on viscous Stokes waves demonstrate these advantages of the method. Finally, as an illustration of nonlinear asymptotic analysis that can be possibly applied to some of the viscous interfacial flow problems, we performed an asymptotic study on Stokes waves with (small) viscosity, and the results provide a theoretical justification to the numerical observation of viscous effects on Stokes waves.

There are many related problems that involve viscous interfacial motion. For example, in fluid-structure interaction (FSI) problems [7], one or more solid structures interact with an internal or surrounding viscous fluid flow, and both fluid dynamics and structure mechanics are needed to understand the fundamental physics involved. A recent review of FSI computation can be found in [13]. One exciting area of current and future research is the simulation of coupled two-phase flow and FSI problems. Notable examples of applications include high-speed boats cruising on water, wind turbines floating in oceans, and energy buoys interacting with waves. A deeper understanding of the fluid and solid motion in these applications would enable more efficient and robust design of marine crafts and energy devices that can sustain strong wave impacts, and enhance the technological development in related industry. As mathematical analysis to such nonlinear, multi-physics problems are generally out of the question, and laboratory experiments are usually limited in scope, numerical simulation provides a very useful way to investigate such problems and to improve our understanding of the fundamental knowledge. Development of accurate and efficient computational methods for these problems is both important and challenging, and will benefit from interdisciplinary effort.

Acknowledgements The author is grateful to Drs. B. Toni and Z. Xie for an invited visit at Virginia State University. The author would also like to thank the anonymous reviewers for helpful comments. This work was partially supported by the National Science Foundation under Grant Numbers 1216936 and 1245769, and the Simons Foundation under Grant Number 208716.

References

1. Anderson, D. A., Tannehill, J. C., & Pletcher, R. H. (1984). *Computational fluid mechanics and heat transfer*. Washington, DC: Hemisphere Publishing Corporation.
2. Baker, G. R., Meiron, D. I., & Orszag, S. A. (1982). Generalized vortex methods for free surface flow problems. *Journal of Fluid Mechanics, 123*, 477–501,
3. Batchelor, G. K. (1967). *An introduction to fluid dynamics*. Cambridge: Cambridge University Press.
4. Caboussat, A. (2005). Numerical simulation of two-phase free surface flows. *Archives of Computational Methods in Engineering, 12*, 165–224.
5. Chandrasekhar, S. (1961). *Hydrodynamic and hydromagnetic stability*. Oxford: Clarendon Press.

6. De, S. C. (1955). Contribution to the theory of Stokes waves. *Proceedings of the Cambridge Philosophical Society, 51*, 713–736.
7. Dowell, E. H., & Hall, K. C. (2001). Modeling of fluid-structure interaction. *Annual Review of Fluid Mechanics, 33*, 445–490.
8. Glimm, J., McBryan, O., Menikoff, R. & Sharp, D. (1986). Front tracking applied to Rayleigh-Taylor instability. *SIAM Journal on Scientific and Statistical Computing, 7*, 230–251.
9. Gresho, P. M. (1991). Incompressible fluid dynamics: Some fundamental formulation issues. *Annual Review of Fluid Mechanics, 23*, 413–453.
10. Harlow, F. H., & Welch, J. E. (1965). Numerical calculation of time-dependent viscous incompressible flow of fluid with free surface. *Physics of Fluids, 8*, 2182–2189.
11. Hinch, E. J. (1991). *Perturbation methods*. Cambridge: Cambridge University Press.
12. Holyer, J. Y. (1979). Large amplitude progressive interfacial waves. *Journal of Fluid Mechanics, 93*, 433–448.
13. Hou, G., Wang, J., & Layton, A. (2012). Numerical methods for fluid- structure interaction - A review. *Communications in Computational Physics, 12*, 337–377.
14. Ito, K., Lai, M. C., & Li, Z. (2007). An augmented approach for Stokes equations with a discontinuous viscosity and singular forces. *Computers & Fluids, 36*, 622–635.
15. Lamb, H. (1945). *Hydrodynamics*. Dover Publications, New York.
16. LeVeque, R. J., & Li, Z. (1997). Immersed interface method for Stokes flow with elastic boundaries or surface tension. *SIAM Journal of Scientific Computing, 18*, 709–735.
17. Levi Civita, M. T. (1925). Determination rigoureuse des ondes permanentes d'ampleur finie. *Mathematische Annalen, 93*, 264–314.
18. Li, Z., & Ito, K. (2006). *The immersed interface method: numerical solutions of PDEs involving interfaces and irregular domains*. SIAM Frontiers in Applied Mathematics, *SIAM*, Philadelphia, vol. 33.
19. Longuet-Higgins, M. S., & Cokelet, E. D. (1976). The deformation of steep surface waves on water, I: A numerical method of computation. *Proceedings of the Royal Society Lodon A, 95*, 1–26.
20. Nayfeh, A. H. (1973). *Perturbation methods*. John Wiley & Sonsy, New York.
21. Orszag, S. A. (1971). Numerical simulation of incompressible flows within simple boundaries I: Galerkin (spectral) representations. *Studies in Applied Mathematics, 50*, 293–327.
22. Osher, S., & Sethian, J. A. (1988). Fronts propagating with curvature-dependent speed: Algorithms based on Hamilton-Jacobi formulations. *Journal of Computational Physics, 79*, 12–49.
23. Peyret, R. (2002). *Spectral methods for incompressible flow*. Springer, New York.
24. Scardovelli, R., & Zaleski, S. (1999). Direct numerical simulation of free surface and interfacial flow. *Annual Review of Fluid Mechanics, 31*, 567–603.
25. Schwartz, L. W. (1974). Computer extension and analytic continuation of Stokes' expansion for gravity waves. *Journal of Fluid Mechanics, 62*, 553–578.
26. Sethian, J. A. (2000). *Level set methods and fast marching methods*. Cambridge: Cambridge University Press.
27. Stokes, G. G. (1847). On the theory of oscillatory waves. *Transactions of the Cambridge Philosophical Society, 8*, 441–455.
28. Takada, N., Misawa, M., & Tomiyama A. (2006). A phase-field method for interface-tracking simulation of two-phase flows. *Mathematics and Computers in Simulation, 72*, 220–226.
29. Tsuji, Y., & Nagata, Y. (1973). Stokes' expansion of internal deep water waves to the fifth order. *Journal of the Oceanographical Society of Japan, 29*, 61–69.
30. Van Dyke, M. (1964). *Perturbation methods in fluid mechanics*. Academic Press, New York.
31. Vinje, T., & Brevig, P. (1981). Numerical simulation of breaking waves. *Advances in Water Resources, 4*, 77–82.
32. Wang, J. (2007). Computation of 2D Navier-Stokes equations with moving interfaces by using GMRES. *International Journal for Numerical Methods in Fluids, 54*, 333–352.
33. Wang, J. (2007). An asymptotic analysis of linear interfacial motion. *Methods and Applications of Analysis, 14*, 1–14.

34. Wang, J. (2008). An asymptotic expansion for Stokes waves with viscosity. *Fluid Dynamics Research, 40*, 155–161.
35. Wang, J., & Baker, G. (2009). A numerical algorithm for viscous incompressible interfacial flows. *Journal of Computational Physics, 228*, 5470–5489.
36. Welch, J. E., Harlow, F. H., Shannon, J. P., & Daly, B. J. (1966). The MAC method, Los Alamos Scientific Laboratory Report LA-3425, Los Alamos, New Mexico.

Index

B. Toni (ed.), *New Frontiers of Multidisciplinary Research in STEAM-H (Science, Technology, Engineering, Agriculture, Mathematics, and Health)*,
Springer Proceedings in Mathematics & Statistics 90, DOI 10.1007/978-3-319-07755-0,

MIX
Papier aus verantwortungsvollen Quellen
Paper from responsible sources
FSC® C105338

If you have any concerns about our products,
you can contact us on
ProductSafety@springernature.com

In case Publisher is established outside the EU,
the EU authorized representative is:
Springer Nature Customer Service Center GmbH
Europaplatz 3, 69115 Heidelberg, Germany

Printed by Libri Plureos GmbH
in Hamburg, Germany